State-of-the-Art Program on Compound Semiconductors 50 (SOTAPOCS 50) -and- Processes at the Semiconductor Solution Interface 3

Editors:

A. G. Baca
Sandia National Laboratories
Albuquerque, New Mexico, USA

J. Brown
RF Micro Devices
Greensboro, North Carolina, USA

P. Nam
Northrop Grumman
Los Angeles, California, USA

C. O'Dwyer
University of Limerick
Limerick, Ireland

D. N. Buckley
University of Limerick
Limerick, Ireland

A. Etcheberry
IREM Institut Lavoisier
Versailles, France

Sponsoring Division:

 Electronics and Photonics

Published by
The Electrochemical Society
65 South Main Street, Building D
Pennington, NJ 08534-2839, USA
tel   609 737 1902
fax 609 737 2743
www.electrochem.org

ecstransactions ™

Vol. 19 No. 3

Copyright 2009 by The Electrochemical Society.
All rights reserved.

This book has been registered with Copyright Clearance Center.
For further information, please contact the Copyright Clearance Center,
Salem, Massachusetts.

Published by:

The Electrochemical Society
65 South Main Street
Pennington, New Jersey 08534-2839, USA

Telephone 609.737.1902
Fax 609.737.2743
e-mail: ecs@electrochem.org
Web: www.electrochem.org

ISSN 1938-6737 (online)
ISSN 1938-5862 (print)

Printed in the United States of America.

Preface

The 50th State-of-the-Art Symposium on Compound Semiconductors was held at the 215th meeting of The Electrochemical Society on May 24-29, 2009 in San Francisco, California, USA. The symposium was sponsored by the Electronics and Photonics Division of The Electrochemical Society.

The symposium consisted of four half-day sessions and one poster session on topics relating to materials, processes, and compound semiconductor devices. This symposium has been held biannually since the fall of 1984. Over the years, this symposium has evolved from its inception as a unique forum for presenting processing-related material mostly for GaAs materials to its position today as a wide-ranging forum for research on a wide variety of compound semiconductors of which GaAs, InP, GaN, ZnO, and SiC are today the most prevalent. The special symposium includes invited talks and papers of a historical nature as well as contributed and invited papers on current research topics. This issue of *ECS Transactions* contains 17 of the papers presented, including invited papers by S. J. Pearton (University of Florida), C. Barratt (RF Microdevices), S. F. Yoon (Nanyang Technological University), and K. Smith (Raytheon).

The 3rd International Symposium on Processes at the Semiconductor-Solution Interface (PSSI 3) was held at the 215th meeting of The Electrochemical Society on May 24-29, 2009 in San Francisco, California, USA. The symposium was sponsored by the Electronics and Photonics Division of The Electrochemical Society.

The symposium covered topics at the forefront of semiconductor electrochemistry and solution-based processing including etching, patterning, passivation, porosity formation, electrochemical film growth, electrophoretic deposition, semiconductor surface functionalization, and other related processes. This issue of *ECST* contains 24 of the papers presented including invited papers by A. Etcheberry and A. Gonçalves (Institut Lavoisier, IREM), J. Stickney (Georgia Tech.), P. Allongue (LPMC - Ecole Polytechnique, CNRS), L. M. Peter (University of Bath), S. Ono (Kogakuin University), K. M. Ryan (University of Limerick), and H. Föll (Christian-Albrechts University Kiel).

This issue of ECST continues the PSSI tradition of being published before the meeting. Invited papers are denoted by an asterisk in the table of contents.

The editors gratefully acknowledge the authors for their efforts to submit the manuscripts on time, enabling this issue of *ECS Transactions* to be published before the meeting. We thank the organizers, the speakers, especially the invited speakers, and the session chairpersons for their contributions to the success of the symposia. Finally, we wish to express our appreciation to the staff of The Electrochemical Society for their efforts, which enabled the publication of this volume on a very tight schedule.

A. G. Baca, J. Brown, D. N. Buckley, P. Nam,
C. O'Dwyer, and A. Etcheberry

ECS Transactions, Volume 19, Issue 3
State-of-the-Art Program on Compound Semiconductors 50 (SOTAPOCS 50) -and-
Processes at the Semiconductor Solution Interface 3

Table of Contents

Preface   *iii*

State-of-the-Art Program on Compound Semiconductors 50 (SOTAPOCS 50)

Chapter 1
Plenary Session

Recent Progress in Dilute Nitride-antimonide Materials for Photonic and   5
Electronic Applications *
    S. F. Yoon, K. Tan, W. Loke, S. Wicaksono, K. Lew, T. Ng, Z. Xu, Y. Sim,
    A. Stöhr , S. Fedderwitz, M. Weiß, O. Ecin, A. Poloczek, A. Malcoci, D. Jäger,
    N. Saadsaoud, E. Dogheche, M. Zegaoui, J. Lampin, D. Decoster,
    C. Tripon-Canseliet, S. Faci, J. Chazelas, J. Gupta and S. McAlister

Detection of Cl- Ions with AlGaN/GaN High Electron Mobility Transistors   31
    S. Hung, B. Chu, C. Lo, B. Hicks, Y. Wang, C. Chang, S. J. Pearton,
    J. W. Johnson, P. Rajagopa, J. Roberts, E. Piner, K. Linthicum, G. Chi and
    F. Ren

Low Frequency Noise of Chlorine-treated GaN/AlGaN MSM-Photodetectors   39
    D. Lu, Y. Chiou, H. Lee and C. Lee

Chapter 2
III-V Compound Semiconductor Devices

The Evolution of Wide Bandgap Semiconductors at SOTAPOCS *   47
    S. J. Pearton and F. Ren

c-erbB-2 Sensing Using AlGaN/GaN High Electron Mobility Transistors For   57
Breast Cancer Detection
    K. Chen, B. Kang, H. Wang, T. Lele, F. Ren, Y. Wang, C. Chang,
    S. J. Pearton, D. Dennis, J. W. Johnson, P. Rajagopal, J. Roberts, E. Piner
    and K. Linthicum

Effect of the Flat and Pattern Surface Texturing on Light Extraction of GaN Flip-Chip Light-Emitting Diodes   65
R. Horng, Z. Liao, Y. Tsai, H. Hu, Y. Tsai, C. Hsu and M. Chu

Chapter 3
Compound Semiconductor Materials Devices and Applications

Electrical Performance of Gate-recessed AlGaN/GaN MOS-HEMTs Fabricated using Photoelectrochemical Method   73
Y. Chiou, N. Shiau, L. Huang, L. Lou and C. Lee

III-V Semiconductors, a History in RF Applications *   79
C. A. Barratt

AlGaN/GaN High Electron Mobility Transistors integrated into Wireless   85
Detection System for Glucose and pH in Exhaled Breath Condensate
B. Chu, B. Kang, C. Chang, F. Ren, A. Goh, A. Sciullo, W. Wu, J. Lin,
B. Gila, S. J. Pearton, J. W. Johnson, E. Piner and K. Linthicum

Electrodeposited Au-CdTe-Au Nanowires: Solution-based Control Over Cd/Te   99
Stoichiometry
L. O. Mair, K. Skinner, C. Donley and R. Superfine

Chapter 4
Compound Semiconductor Materials and Devices

GaN HEMT Reliability Through the Decade *   113
K. V. Smith, S. Brierley, R. McAnulty, C. Tilas, D. Zarkh, M. Benedek,
P. Phalon and A. Hooven

Real-time Detection of Botulinum Toxin with AlGaN/GaN High Electron   123
Mobility Transistors
Y. Wang, B. Chu, K. Chen, C. Chang, T. Lele, Y. Tseng, S. J. Pearton,
J. Ramage, D. Hooten, A. Dabiran, P. Chow and F. Ren

Annealing Effects on ZnO/NiSi Contact   129
R. Wei and H. Gong

Photoelectrochemical Splitting of Water to H2 and O2 at n-Fe2O3 Nanowires   137
and Nanocrystalline Carbon-Modified (CM)-n-Fe2O3 Thin Films
M. Frites and S. U. Khan

vi

Hydrothermally Grown ZnO Nanorods as Cell Adhesion Control Coating for   147
Implant Devices
   B. Chu, C. Chang, L. Leu, D. Norton, J. Lee, T. Lele, P. Jiang, Y. Tseng,
   S. J. Pearton, A. Gupte, B. Keselowsky and F. Ren

Chapter 5
Compound Semiconductor Materials and Devices Poster Session

Luminescent Properties of M-Al2O4:Eu2+ (M: Ca, Sr) Thin Films Prepared on   161
Sapphire Substrates
   J. Lee, C. Lee and Y. Kim

Observation of Self-limiting Regime in the Atomic Layer Deposition of ZnO   167
Films using Nitrous Oxide as the Oxygen Supply
   P. Chung, H. Lai, Y. Lin, K. Yen, C. Kung and J. Gong

vii

Processes at the Semiconductor Solution Interface 3

Chapter 6
Electrochemical Deposition of Semiconductors and Thin Films

Towards Sustainable Photovoltaic Solar Energy Conversion: Studies of New   179
Absorber Materials *
   P. J. Dale, L. M. Peter, A. Loken and J. Scragg

Electrochemical Growth of CuInSe2 Compounds on Polycrystalline Mo Films   189
Studied by Raman and Impedance Spectroscopy
   E. Chassaing, P. Grand, E. Saucedo, A. Etcheberry and D. Lincot

Electrochemical Growth Gold Buffer Layer on H-Si(111) Surfaces and Their   197
Applications *
   P. Allongue, F. Maroun, H. Jurca, R. Cortes, P. Prod'homme and
   N. Tournerie

Electrophoretic Deposition of Spherical and Rod-Shaped Nanocrystals into Close   209
Packed Superlattices *
   S. Ahmed, C. A. Barrett, C. O'Sullivan, A. Sanyal, H. Geaney, A. Singh,
   R. D. Gunning and K. M. Ryan

Spontaneous Deposition of Metallic Pt onto n-InP: An Electroless Process *   221
   A. Etcheberry, C. Mathieu, M. Bouttemy, J. Vigneron, P. Tran-Van and
   A. Goncalves

Characterisation of Self-Assembled Monolayers on Germanium Surfaces via   227
NEXAFS
   M. Lommel, F. Reinhardt, M. Kolbe, B. Beckhoff, M. Müller, P. Hönicke and
   B. Kolbesen

Electrochemical Stability of a Novel Inorganic Protective and Functionalizable   235
Monolayer onto InP
   O. El Ali, A. Goncalves, N. Mezailles, C. Mathieu, P. Le Floch and
   A. Etcheberry

Optimization of PbSe Nanofilms formation by Electrochemical Atomic Layer   245
Deposition (ALD) *
   D. Banga, Y. Kim, S. Cox, U. Happek and J. L. Stickney

viii

Morphology, Composition and Electrical properties of Thin Anodic Oxides on InP
  273
  N. Simon, C. Decorse Pascanut, L. Santinacci, A. Goncalves, A. Joudrier and A. Etcheberry

In Situ Infrared Kinetic Study of Multistep Chemical Modifications of Organic Monolayers at Silicon Surfaces
  283
  L. Touahir, A. Moraillon, S. Sam, A. Gouget-Laemmel, P. Allongue, J. Chazalviel, C. Henry de Villeneuve and F. Ozanam

Chapter 7
Porous Semiconductors

Deconvolution of the Potential and Time Dependence of Electrochemical Porous Semiconductor Formation
  295
  N. Quill, C. O'Dwyer, R. Lynch and D. N. Buckley

Effects of "P-N" Terminations on the Initial Stages of Pore Growth onto n-InP in HCl Aqueous Solution *
  305
  A. Goncalves, L. Santinacci, N. Simon, M. Bouttemy, C. Mathieu and A. Etcheberry

Unexpected Dissolution Process at Porous n-InP Electrodes
  313
  L. Santinacci, M. Bouttemy, I. Gerard and A. Etcheberry

Simulating Crystallographic Pore Growth on III-V Semiconductors
  321
  M. Leisner, J. Carstensen and H. Föll

Growth Mode Transition of Crysto and Curro Pores in III-V Semiconductors *
  329
  H. Föll, M. Leisner, J. Carstensen and P. Schauer

Production of High Aspect Ratio Single Holes in Semiconductors
  347
  M. Gerngroß, H. Föll, A. Cojocaru and J. Carstensen

Dynamics of Macropore Growth in n-type Silicon Investigated by FFT In-Situ Analysis
  355
  J. Carstensen, A. Cojocaru, M. Leisner and H. Föll

ix

Chapter 8
Electrochemical Characterization and Functionalization of Si Surfaces and Devices

Electrochemical Passivation of (100) Silicon in Alkyl Grignard Solutions   365
  S. Vegunta, J. Ngunjiri and J. C. Flake

Electronic Properties and pH Stability of Si(111)/Alkyl Monolayers   373
  D. Aureau, A. Moraillon, C. Henry de Villeneuve, F. Ozanam, P. Allongue,
  J. Chazalviel and J. Rappich

Phase-Relations between Photocurrent and In Situ Reflectance during   381
Photoelectrochemical Dissolution of Silicon
  M. Lublow and H. Lewerenz

Chapter 9
Properties and Patterning of Semiconductors and Related Compounds

Micro-Patterning of Semiconductors by Metal-Assisted Chemical Etching   393
through Self-Assembled Colloidal Spheres *
  S. Ono, F. Arai and H. Asoh

Surface Chemistry and Nanotopography of Step-Bunched Silicon Surfaces: 403
In-system SRPES and SPM Investigations
  T. Stempel, A. Munoz, K. Skorupska, M. Lublow, M. Kanis and H. Lewerenz

The Influence of Thermal Treatment on the Electronic Properties of a-Nb2O5   411
  F. La Mantia, M. Santamaria, H. Habazaki and F. Di Quarto

Novel Plasmaless Photoresist Removal Method in Gas Phase at Room   423
Temperature
  T. Miura, M. Kekura, H. Horibe, M. Yamamoto and H. Umemoto

Author Index   431

* *invited paper*

Facts about ECS

The Electrochemical Society (ECS) is an international, nonprofit, scientific, educational organization founded for the advancement of the theory and practice of electrochemistry, electrothermics, electronics, and allied subjects. The Society was founded in Philadelphia in 1902 and incorporated in 1930. There are currently over 7,000 scientists and engineers from more than 70 countries who hold individual membership; the Society is also supported by more than 100 corporations through Corporate Memberships.

The technical activities of the Society are carried on by Divisions. Sections of the Society have been organized in a number of cities and regions. Major international meetings of the Society are held in the spring and fall of each year. At these meetings, the Divisions and Groups hold general sessions and sponsor symposia on specialized subjects.

The Society has an active publications program that includes the following.

Journal of The Electrochemical Society — JES is the peer-reviewed leader in the field of electrochemical and solid-state science and technology. Articles are posted online as soon as they become available for publication. This archival journal is also available in a paper edition, published monthly following electronic publication.

Electrochemical and Solid-State Letters — ESL is the first and only rapid-publication electronic journal covering the same technical areas as JES. Articles are posted online as soon as they become available for publication. This peer-reviewed, archival journal is also available in a paper edition, published monthly following electronic publication. It is a joint publication of ECS and the IEEE Electron Devices Society.

Interface — *Interface* is ECS's quarterly news magazine. It provides a forum for the lively exchange of ideas and news among members of ECS and the international scientific community at large. Published online (with free access to all) and in paper, issues highlight special features on the state of electrochemical and solid-state science and technology. The paper edition is automatically sent to all ECS members.

Meeting Abstracts (formerly Extended Abstracts) — Abstracts of the technical papers presented at the spring and fall meetings of the Society are published on CD-ROM.

ECS Transactions — This online database provides access to full-text articles presented at ECS and ECS-sponsored meetings. Content is available through individual articles, or as collections of articles representing entire symposia.

Monograph Volumes — The Society sponsors the publication of hardbound monograph volumes, which provide authoritative accounts of specific topics in electrochemistry, solid-state science, and related disciplines.

For more information on these and other Society activities, visit the ECS website:

www.electrochem.org

SECTION 1

STATE-OF-THE-ART PROGRAM ON
COMPOUND SEMICONDUCTORS 50
(SOTAPOCS 50)

2

CHAPTER 1

PLENARY SESSION

4

Recent progress in dilute nitride-antimonide materials for photonic and electronic applications

S.F. Yoon, K.H. Tan, W.K. Loke, S. Wicaksono, K.L. Lew, T.K. Ng, Z. Xu, and Y.K. Sim

School of Electrical and Electronic Engineering, Nanyang Technological University, Nanyang Avenue, Singapore 639798, Republic of Singapore

A. Stöhr, S. Fedderwitz, M. Weiß, O. Ecin, A. Poloczek, A. Malcoci and D. Jäger,

ZHO/Optoelectronics, University Duisburg-Essen, 47048 Duisburg, Germany

N. Saadsaoud, E. Dogheche, M. Zegaoui, J. F. Lampin and D. Decoster

Institute of Electronics, Microelectronics and Nanotechnology (IEMN, UMR CNRS 8520, Universite des Sciences et Technologies de LilleBP 60069, 59652 Villeneuve d'Ascq Cedex, France

C. Tripon-Canseliet and S. Faci

Laboratoire d'Electronique et Electromagnétisme, Pierre and Marie Curie University, 3 rue Galilée, 94 200 Ivry sur Seine, France

J. Chazelas

Thales Airborne Systems, 2 Avenue Gay Lussac, 78852 Elancourt, France

J. A. Gupta and S. P. McAlister

Institute for Microstructural Sciences, National Research Council of Canada, 1200 Montreal Road, Ottawa, Ontario K1A 0R6, Canada

> This paper reviews the recent progress in GaNAsSb material for photonic and electronic applications. All the results and data presented in this review article are summarized from our previously published works in refs. 6-12. Photoresponsivity of 12A/W and cut-off frequency of 4.5GHz were achieved in the 1.3µm GaNAsSb based photodetector. A GaNAsSb/GaAs optical waveguide system was also demonstrated at 1.55µm. The GaNAsSb based photoconductive switch exhibits pulsed response with FWHM of 30ps and photoresponse of up to 1.6µm. The turn-on voltage of the device fabricated from GaNAsSb based HBT is ~330mV lower than that of a conventional AlGaAs/GaAs HBT.

1. Introduction

To operate at 1.3μm and 1.55μm, optoelectronic devices require an active layer material with bandgap of 0.9eV and 0.8eV, respectively. Previously, III-V material with bandgap less than 1eV was not possible to be realized on GaAs substrate platform and only available on InP susbtrate. This is due to the lack of small bandgap material, which is lattice-matched to the GaAs. However, with the introduction of GaNAs material at 1992,[1] incorporation of small amount of nitrogen atoms (0% to 4%) was found able to reduce the bandgap of GaAs. In 1996, Kondow *et al.*[2] reported the first material, which is lattice-matched to GaAs and have a small bandgap (1eV), using the GaInNAs material system.

GaNAsSb was firstly proposed by Ungaro *et al.*[3] in 1999 as a potential GaAs-based material for near infrared applications. In their report, a strain layer of GaNAsSb with bandgap ~0.8eV has been grown using the molecular beam epitaxy (MBE) system. In GaInNAs material system, Indium (In) atoms are used to counter-balance the tensile strain induced by the incorporation of nitrogen. However, in GaNAsSb material system, antimony (Sb) atoms are used instead of indium atoms. Thus, it is interesting to compare the bandgap shrinkage induced by indium and antimony under same amount of compressive strain. Harmand *et al.*[4] reported that Sb induces larger bandgap reduction compared to In at the same amount of compressive strain. Therefore, under the conditions of lattice-matched to GaAs and same amount of nitrogen atoms, Bandgap of GaNAsSb is smaller compared to that of GaInNAs. In other words, GaNAsSb system needs lesser nitrogen atoms to achieve 0.8eV (corresponding to wavelength of 1.55μm) compared to GaInNAs. Furthermore, the comparison between GaNAs, GaInNAs and GaNAsSb shows that the presence of Sb atoms promotes the incorporation of N atoms and decreases the nitrogen-related defects.[5] Due to such advantages, GaNAsSb is preferred by us in the photonic and electronic applications.

In this paper, we will review the recent progress in the GaNAsSb based photonic and electrical devices based on our previously published works in refs. 6-12. The review will be categorized into four sections, namely photodetector, optical waveguide, photoconductive switch and heterojunction bipolar transistor.

2. Photodetector

2.1 The Photoresponsivity

The GaNAsSb material system has attracted great interest for potential photodetector applications in the near infrared region (0.9-1.6μm).[13] Achieving high photoresponsivity is one of the key challenges in GaAs-based dilute nitride photodetector research. Until recently, reported results of GaAs-based dilute nitride photodetectors can be categorized into two groups: one based on quantum well (QW) absorption layers[14-16] and the other one based upon bulk absorption layers.[7,17-21] A thin dilute nitride layer (<10nm) is used for devices based on a quantum well absorption layer. While such QW devices enable the utilization of highly strained dilute nitride layers and thus offer a photo-response up to 1.6μm, their photoresponsivity is generally low[14,15] (typically less than 0.03A/W). Resonant cavity has been incorporated into the device structure to overcome this limitation.[16] On the other hand, photodetectors based on bulk dilute nitride absorption layers (>0.4μm thick) suffer from reduced photo-response at long wavelengths (the highest reported cut-off wavelength is ~1.4 μm[7]). Photodetectors based on bulk dilute nitride absorption layers exhibit a higher photoresponsivity (up to about 0.1A/W[17,18,22]) compared to QW-based devices. However, these photoresponsivity values are still much lower as compared to those of commercial InGaAs photodetectors, with a typical photoresponsivity of up to ~0.9A/W at 1.3μm.

In this section, we describe on a significant improvement in the photoresponsivity of GaNAsSb/GaAs photodetectors with a GaNAsSb bulk photoabsorption layer at 1.3μm wavelength. The device structure shown in Fig. 1 was grown using a molecular beam epitaxy (MBE) system in conjunction with a radio frequency (RF) N plasma-assisted source and a valved Sb cracker source. The i-GaNAsSb (bulk) photoabsorption layer was grown at 350°C, 400°C, 440°C and 480°C. Using the band anti-crossing (BAC) model,[23] the optical bandgap of the i-GaNAsSb layer was estimated to be ~0.9eV. Detail of growth conditions and experiment setup can be found elsewhere.[6]

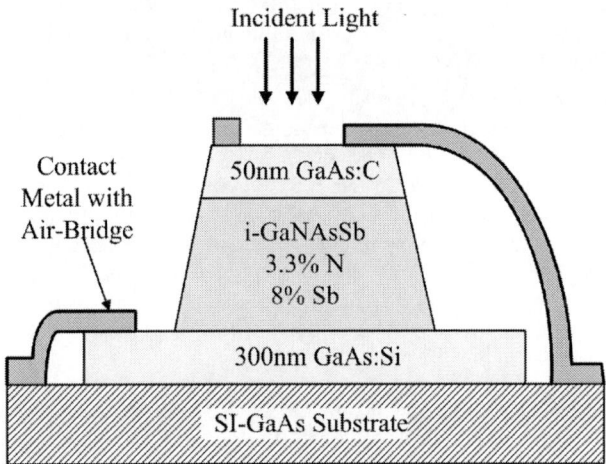

Figure 1. Schematic diagram of GaAs/GaNAsSb/GaAs p-i-n photodetector structure [ref. 6].

Figure 2(a) shows the plot of photoresponsivity at -3V *vs.* wavelength for the devices whose i-GaNAsSb layers were grown at 350°C to 480°C. The photodetectors show a photo-response up to wavelength of 1350nm. Figure 2(b) shows the photoresponsivity at different reverse biases measured at the wavelength of 1300nm. From Fig. 2(b), it is interesting to note that the photodetector with GaNAsSb layer grown at 350°C shows an extremely high photoresponsivity value of ~12A/W under -4.8V at 1300nm, ~ 2 orders higher than previously reported results. A photoresponsivity value of 12A/W implies a quantum efficiency value significantly larger than 1, possibly due to the presence of an avalanche carrier multiplication effect.

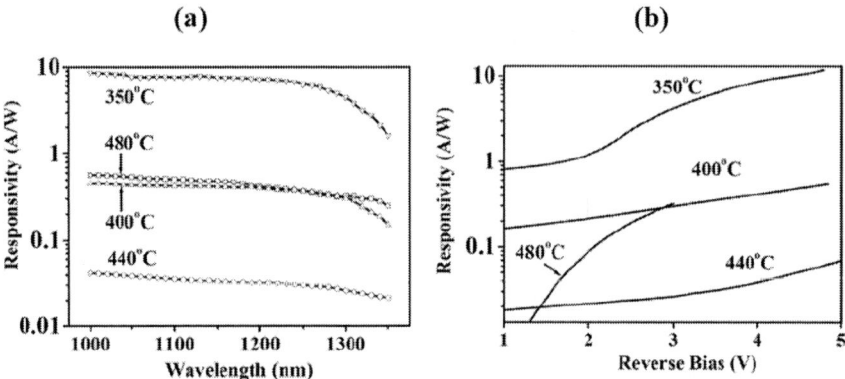

Figure 2. (a) Plot of photoresponsivity *vs.* wavelength measured at a reverse bias of 3V for devices with i-GaNAsSb layer grown at 350°C, 400°C, 440°C and 480°C. (b) Plot of photoresponsivity *vs.* reverse bias measured at 1300nm for devices with i-GaNAsSb layer grown at 350°C, 400°C, 440°C and 480°C. [ref. 6]

From capacitance-voltage (C-V) measurement, the unintentional doping concentrations in the i-GaNAsSb layer grown at 350°C, 400°C, 440°C and 480°C were experimentally determined to be $2 \times 10^{16} cm^{-3}$, $6 \times 10^{16} cm^{-3}$, $3 \times 10^{17} cm^{-3}$ and $1.5 \times 10^{18} cm^{-3}$, respectively. The depletion region width of the i-GaNAsSb layer at different reverse biases can be calculated using unintentional doping concentrations. The absorption coefficient, α of GaNAsSb has a value of $1.3 \times 10^4 cm^{-1}$ at the wavelength of 1300nm.[10] Using the measured photoresponsivity values, calculated depletion region widths and values of α, the photocurrent multiplication factor M for all devices at different electric field are calculated and shown in Fig. 3.

From Fig. 3, it can be seen that the photodetector with i-GaNAsSb layer grown at 350°C has a M value up to ~30 at electric field of <100KV/cm. This high value of M confirms our earlier suggestion of the presence of a photogenerated carrier multiplication due to the avalanche effect.

It is interesting to note that the photodetector with i-GaNAsSb layer grown at 350°C exhibits a high carrier multiplication factor at average electric field strengths of <100kV/cm. Even when considering a non-uniformly distributed electric field in the depletion region, the maximum electric field strength is ~100kV/cm and 180kV/cm at reverse bias of 1V and 5V, respectively. This electric field strength is unexpectedly low, considering the fact that GaAs or InGaAs based avalanche photodetectors only show carrier multiplication at electric field strength higher than ~200kV/cm.[24,25] These results suggest that the decrease in growth temperature of the i-GaNAsSb layer leads to a higher impact ionization coefficient in the material, resulting in initiation of the carrier avalanche process at low electric field.

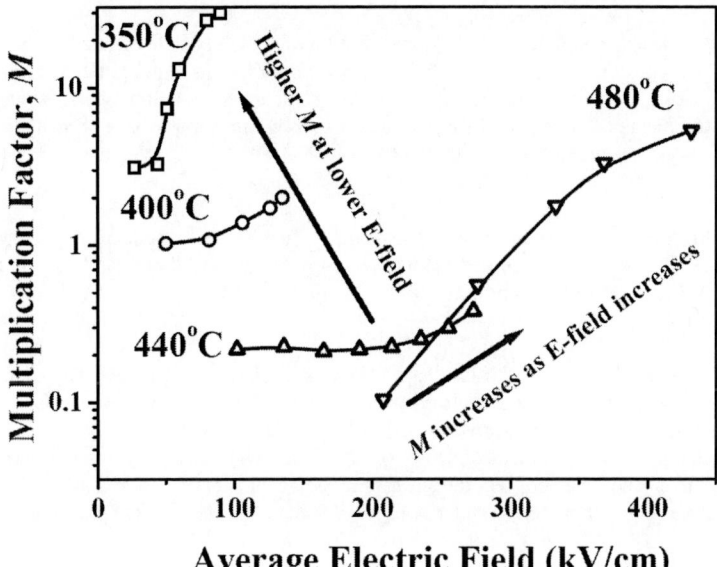

Figure 3. Plot of multiplication factor M as function of average electric field measured at 1300nm for devices with i-GaNAsSb layer grown at 350°C, 400°C, 440°C and 480°C. [ref. 6]

The high ionization coefficient in photodetector with low temperature grown i-GaNAsSb layer could be explained by the existence of mid-gap As antisite defects (As_{Ga}) in the material. It is known that dilute-nitride materials contain As_{Ga}[19,26] defects as they are grown at non-equilibrium low temperature (<500°C) growth conditions. We expected that the i–GaNAsSb grown at 350°C has the highest amount of As_{Ga} defects as content of these defects increases proportionally in response to the decrease in the growth temperature of dilute-nitride material.[26]

Generally, carriers in a p-n junction require energy of $\frac{3}{2}E_g$ to start an impact ionization and thus avalanche process.[27] E_g is the bandgap of the material. Mid-gap defects, such as As_{Ga} are reported[28] to enhance the impact ionization process by lowering the energy required in the impact ionization process. Instead of energy of $\frac{3}{2}E_g$, the impact ionization process through the mid-gap defects states requires only energy of $\frac{E_g}{2}$.[28] By lowering the required energy, the existence of mid-gap defects enables a more efficient impact ionization and carrier multiplication process at a lower electric field. This explains our observation that photodetectors, which have i-GaNAsSb layer with more As_{Ga} defects, have higher carrier multiplication and initialize the impact ionization process at a lower electric field.

2.2 The RF Performance

One of the key challenges in GaAs-based dilute nitride photodetectors is to obtain a high frequency response. Recent reports[16,29] on dilute nitride-based photodetectors have shown 800ps rise time in GaInNAs resonant-cavity-enhanced quantum well devices. Recent measurements of p-i-n GaNAsSb photodetectors by our group[21] have shown a RC-limited 3dB cut-off frequency f_T between 920MHz to 1.4GHz. So far, there has been no report on dilute nitride-based photodetectors with f_T exceeding these values.

In this section, a significant improvement in the high-frequency performance of GaNAsSb-based p-i-n photodetectors well into the multi-GHz region will be described. The growth conditions and fabrication of the photodetectors by MBE can be found elsewhere.[7]

The pulsed responses of the fabricated GaNAsSb photodetectors were measured using a femtosecond laser system. A frequency-doubled Nd:YVO$_4$ laser was used to pump a Ti:sapphire femtosecond Kerr-lens mode-locked laser generating 260fs pulses at 890nm wavelength with a repetition rate of 76MHz. The wavelength of 890nm ensures negligible photocarrier generation in the adjacent doped GaAs layers. The pulsed response of the photodetectors was observed using a 70GHz sampling oscilloscope.

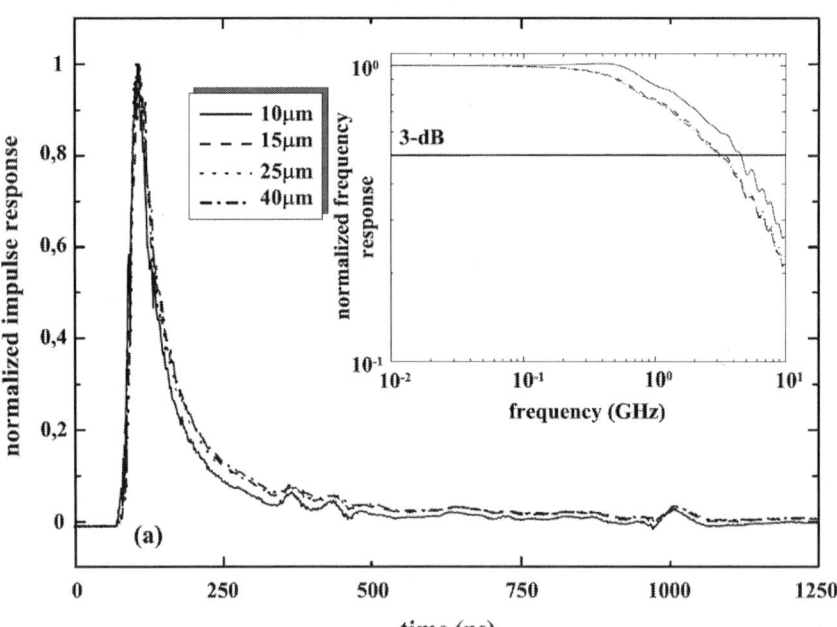

Figure 4. Normalized pulse response of GaAsNSb photodetectors with different diameters at -8V applied bias. Insets show the corresponding normalized frequency responses derived by FFT.[ref. 7]

In general, the photodetectors exhibit a very short rise time and a slightly longer fall time. Narrower total pulse widths were obtained for larger reverse biases and smaller diameters. The measured pulse responses of GaNAsSb photodetectors with different diameters of the optical window are shown in Fig. 4 for an applied bias of -8V. The pulse response of a 10μm diameter GaNAsSb photodetector is shown in Fig. 5 for different applied reverse voltages.

As can be seen from Fig. 4, the 10%-90% rise time accounts for about 17.4ps and the shortest pulse widths (FWHM) achieved with a 10μm diameter device at -8V applied bias is 40.5ps. This corresponds to a record bandwidth of about 4.5GHz which can also be seen from the normalized frequency response calculated by Fast-Fourier Transformation (inset in Fig. 4).

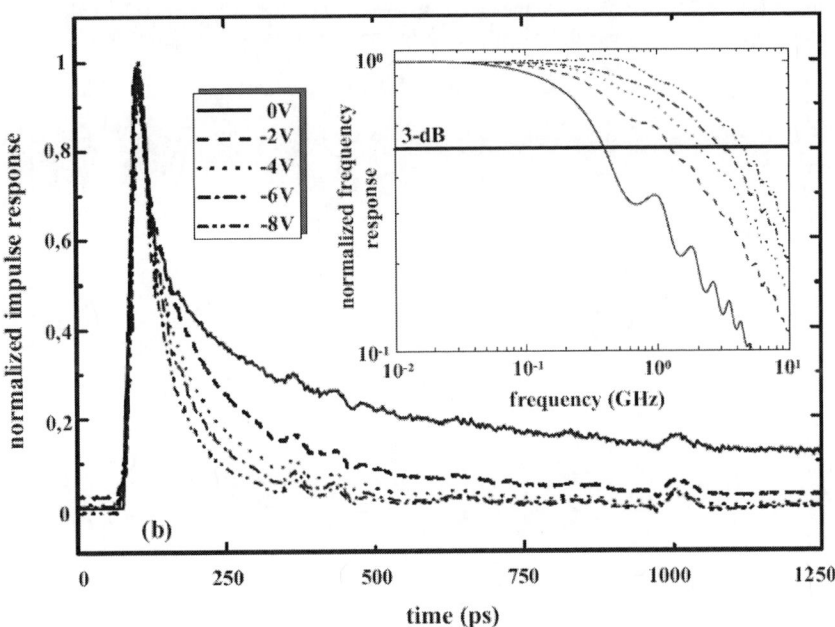

Figure 5. Normalized pulse response of a 10μm diameter GaAsNSb photodetector at different applied reverse biases Insets show the corresponding normalized frequency responses derived by FFT.[ref. 7]

From Figs. 4 and 5, it can be seen that the rise time is almost independent of the devices diameter and reverse bias, indicating a fast photocarrier generation process and electron drift in the photoabsorption layer. On the contrary, the fall time is significantly reduced by increasing the reverse bias to 8V (Fig. 5). Since we can neglect significant electron diffusion out of the 50nm thin and non-absorbing p-doped GaAs contact layer, the bias-dependent relaxation can be clearly attributed to carrier transit time effects i.e. a slow carrier transport in the n.i.d. absorption layer. According to drift-diffusion model simulation,[30] hole drift is the factor dominating the time constant here. The effective hole drift velocity is determined to be approximately 2×10^6cm/s, which is about 3-4 times slower than in GaAs. This low effective hole drift velocity can be explained by the fact that the GaNAsSb layer is not fully depleted. From capacitance-voltage measurement, we determined that GaNAsSb photoabsorption layer is of p-type with a doping concentration of ~3×10^{17}cm^{-3}. Thus, at -8V, the depletion width is only 0.24µm. Hence, the holes only drifted for a short distance of ~0.24µm, and have to diffuse across the remaining 1.8µm of the GaNAsSb layer at a slower velocity. This effectively increases the overall transit time of the holes across the GaNAsSb layer.

As expected, the RC time constant on the other hand does not significantly influence the device high-speed performance. This is due to the fact that the depletion layer thickness of 0.24µm results in an RC-limited cut-off frequency greater than 5GHz for all diameters. This fact is also confirmed from Fig. 4 which shows that the pulse widths are almost independent of the device diameter. Thus we can conclude that the high-speed performance of the fabricated devices is mainly dominated by carrier transit time.

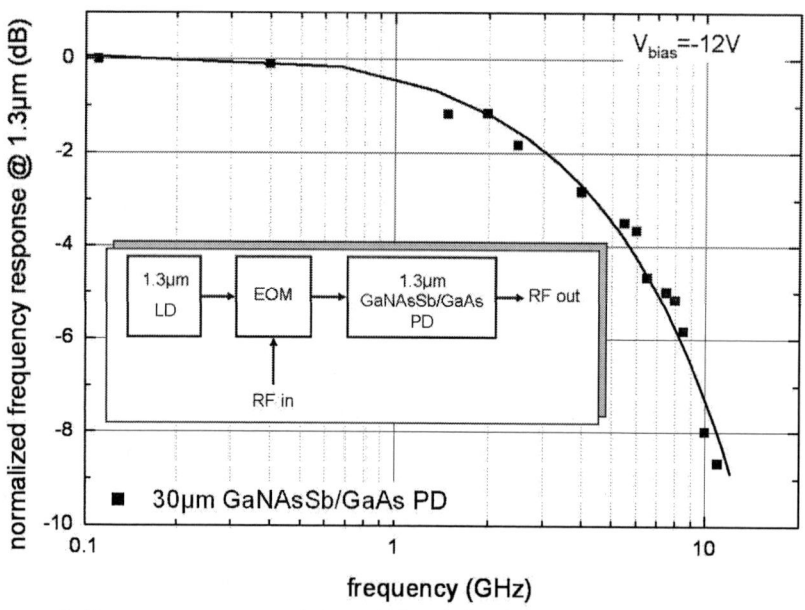

Figure 6. Frequency response characteristic of the GaNAsSb/GaAs photodetector at 1.3µm wavelength. The inset shows the setup of the frequency response measurement system. [ref. 8]

We have demonstrated bulk GaNAsSb photodetectors with ultra-fast pulsed response of 40.5ps at 890nm wavelength. To confirm there is no deterioration in the high-speed performance of these detectors at longer wavelength, especially at 1.3µm, we have performed frequency response measurements at 1.3µm wavelength using a CW laser and external Mach-Zehnder modulator (see inset in Fig. 6 for measurement set-up). The frequency response of the GaNAsSb/GaAs photodetector was measured up to 12GHz. As can be seen from Fig. 6, the 3dB cut-off frequency is 4.5GHz at -12V.

To carry out high-data rate transmission experiments at 1.3µm wavelength, a 12.5 Gb/s pseudo-random bit sequence (PRBS, non-return-to-zero (NRZ), 2^{31}-1) source was used in conjunction with a direct modulated 1.3µm laser. The RF output signal of the photodetector was measured using a 50GHz sampling oscilloscope. The measured eye diagram for data rate of 5 Gb/s at reverse bias of 12V and photocurrent of 0.1mA is shown in Fig. 7. The measured eye is clearly opened, verifying that the photodetectors are well suited for fiber-optic transmission up to 5 Gb/s.

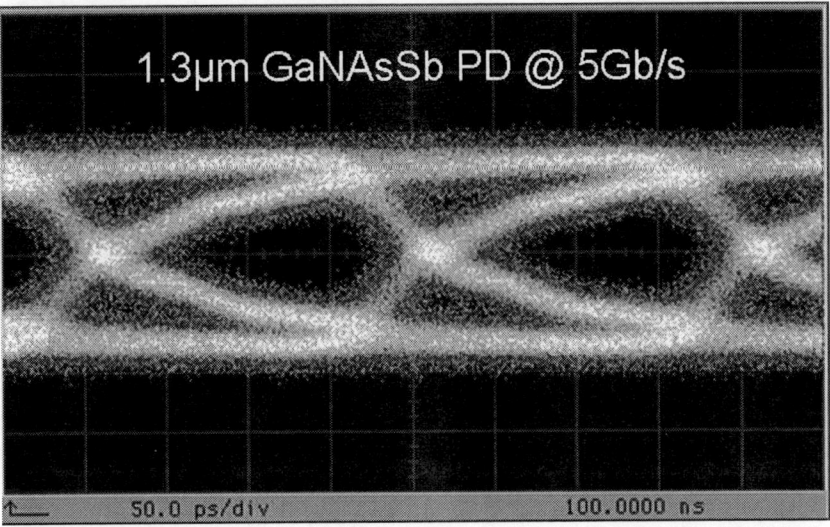

Figure 7. Eye diagram of the output signal from a 1.3µm GaNAsSb/GaAs photodetector with 30µm illumination window diameter at data rate of 5Gb/s.[ref. 8]

Next, bit-error-rate (BER) measurements were performed using a 12.5 Gb/s bit-error-rate tester (BERT). For comparison, two different photodetectors were used in this measurement; a commercial 12GHz InGaAs/InP photodetector with responsivity of 0.85A/W and the newly fabricated 30µm diameter GaNAsSb/GaAs photodetector. The measured BERs for the 30µm diameter GaNAsSb/GaAs photodetector at data rates of 2.5 Gb/s and 5 Gb/s, respectively are shown in Fig. 8. At 2.5 Gb/s, the measured receiver sensitivity at BER of 10^{-9} for the GaNAsSb/GaAs photodetector is 1.4dBm, whereas the measured receiver sensitivity also at BER of 10^{-9} and data rate of 2.5Gb/s for the InGaAs photodetector is -12.5dBm. The difference in receiver sensitivity between the InGaAs and GaNAsSb photodetectors can be attributed to the relatively higher responsivity of the InGaAs photodetector. As shown in Fig. 8, we observed some error-floor for the GaNAsSb/GaAs photodetector at data rate of 5 Gb/s. As the eye diagram at 5 Gb/s was fully opened, it is expected that the observed error-floor is mainly due to the performance limitation of the broadband amplifier used in the BER measurements.

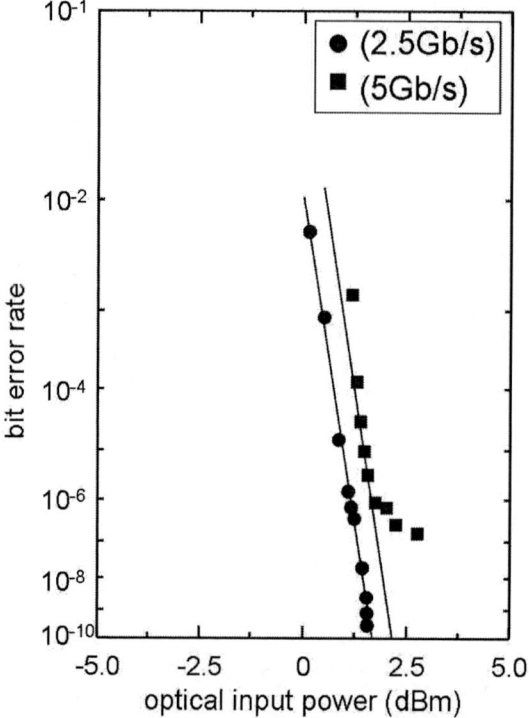

Figure 8. Dependence of the BER on the optical input power of the GaNAsSb/GaAs (30μm window diameter) photodetector at data rates of 2.5Gb/s and 5Gb/s, respectively. BER measurements were carried out at 1.3μm wavelength.[ref. 8]

2.3 Extending detectable wavelength to 1.6μm

It is potentially possible to make devices for 1.55μm wavelength detection if the nitrogen and antimony contents are increased to reduce the energy bandgap while keeping the crystal lattice-matched to GaAs. However, it was also found that by increasing the nitrogen (N) and antimony (Sb) content, more nitrogen-related defects will be incorporated, which degrades the crystal quality and device performance. An alternative approach is to use lattice mismatch system by increasing the Sb content only. However, the amount of light that can be absorbed (or quantum efficiency) will be constrained by its critical layer thickness. This constraint can be overcome by coupling light directly into the edge of the device based on a waveguide photodetector (WGPD) structure, where light absorption can be maximized for a thin GaNAsSb layer.

In this section, we describe the photoresponse of a GaNAsSb/GaAs p-i-n WGPD device. The inset of Fig. 9 shows the WGPD structure grown by molecular beam epitaxy in conjunction with radio frequency (RF) plasma-assisted nitrogen (N) source and a valved cracker source for antimony (Sb). The detail growth condition can be found elsewhere.[9] The GaNAsSb layer has N content of ~3.5% and Sb content of ~18% determined by x-ray diffraction (Fig.1), leading to bandgap of 0.77eV (or λ=1.61μm).

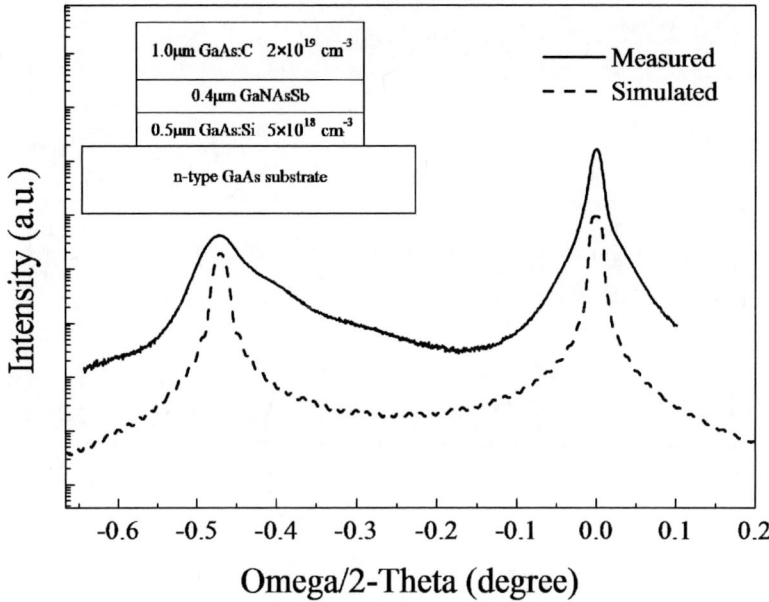

Figure 9. X-ray diffraction profile of GaNAsSb/GaAs p-i-n sample with inset showing its layer thickness.[ref. 9]

Figure 10 shows the photoresponse characteristic of the WGPD with 6.5μm ridge width and 500μm ridge length. The result shows the WGPD photoresponse has longer cutoff wavelength compared to that of the top-illuminated GaNAsSb/GaAs p-i-n device as reported in Ref. [7], with 2μm-thick $GaN_{0.033}As_{0.887}Sb_{0.08}$ as the absorption layer. This is because, with an increase in nitrogen and antimony content of the GaNAsSb layer in our WGPD device (i.e. $GaN_{0.035}As_{0.785}Sb_{0.18}$), the energy bandgap is reduced and hence resulted in longer cutoff wavelength.

To measure the absolute photoresponsivity, a 1.55μm laser source is coupled into the waveguide facet of the WGPD using a stripped and cleaved single mode fibre (10μm core diameter) as the input light. The photoresponsivity are measured at -1.5V under incident laser power of 100μW at 1.55μm. The results are shown in Fig.11 for two different ridge widths and several different ridge lengths.

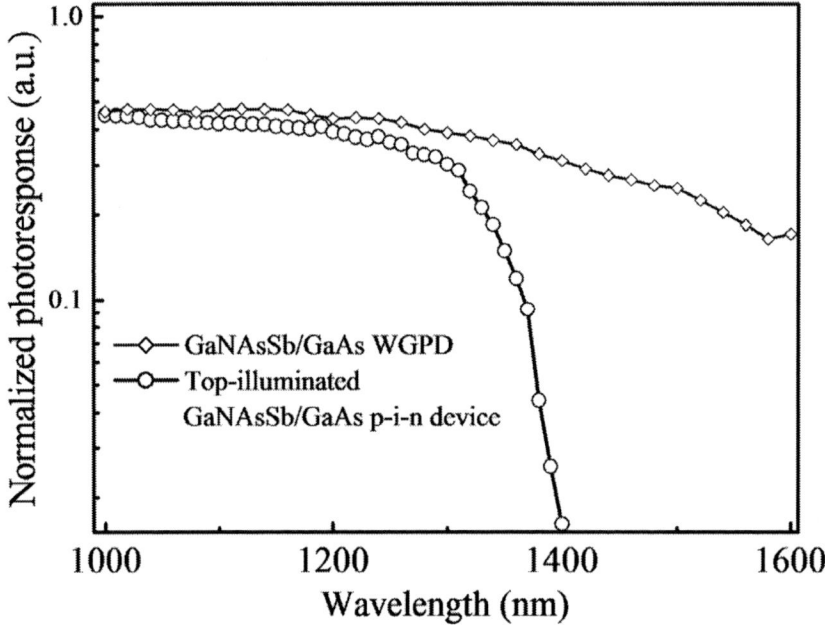

Figure 10. Normalized photoresponse characteristics of GaNAsSb/GaAs WGPD devices with 6.5μm ridge width and 500μm ridge length and a top-illuminated photodetector with the same material system from ref. 7. [ref. 9]

The photoresponsivity of the WGPD with 6.5μm ridge width saturates at 0.25A/W, while the device with larger ridge width of 10μm saturates at higher photoresponsivity of 0.29A/W. This is because larger ridge width reduces the coupling loss between the single mode fiber and waveguide facet. These photoresponsivity values are much higher than those from a top-illuminated photodetector using GaNAsSb/GaAs double quantum well structure as the absorption layer. Photoresponsivity values of 0.01-0.016A/W at $\lambda=1.55\mu m$ have been reported[15] for such GaNAsSb/GaAs double quantum well structures. The photoresponsivity values from our WGPD devices are also comparable to reported[16] photodetectors with GaInNAs absorption layer embedded in a resonant cavity enhanced (RCE) structure (0.4A/W). While the use of the RCE structure may enhance device photoresponsivity, it may not be suitable for high-speed applications.

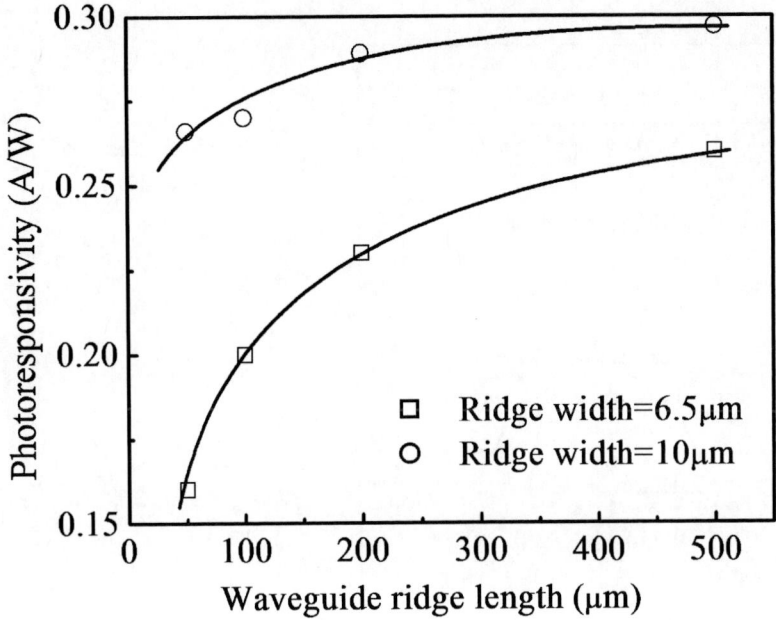

Figure 11. Photoresponsivity characteristics of GaNAsSb/GaAs WGPD devices with different ridge width and length measured at 1.55μm wavelength under reverse-biased voltage of 1.5V.[ref. 9]

3. Optical Waveguide

The use of optical waveguides based on III-V compound semiconductors is motivated by applications of monolithically integrated optoelectronic components. For optical waveguides on GaAs substrate, so far there have been numerous studies[31-38] on the AlGaAs/GaAs material system. The GaAs/GaNAsSb/GaAs system is an ideal 1.55μm optical waveguide system on GaAs substrate. Under the lattice- matched condition, the growth of a thick layer of $GaN_{0.035}As_{.0875}Sb_{0.09} >0.5μm$ can be achieved. Its bandgap of 0.9eV will ensure minimal absorption at transmission wavelength of 1.55μm. Furthermore, the 0.5eV bandgap difference between GaNAsSb and GaAs should give a high refractive index contrast. In the AlGaAs/GaAs/AlGaAs optical waveguide, a thick (>4μm) bottom AlGaAs cladding layer is needed to reduce the propagation loss.[34,37] However, in the GaAs/GaNAsSb/GaAs optical waveguide, the GaAs substrate acts as the bottom cladding layer, thus eliminating the need for another material to serve as the bottom cladding layer.

In this section, we describe a GaAs/GaNAsSb/GaAs optical waveguide fabricated on GaAs substrate at 1.55μm as an alternative to the AlGaAs/GaAs waveguide. The refractive index of the GaNAsSb layer is measured, and the propagation loss of the waveguide with different ridge widths is reported.

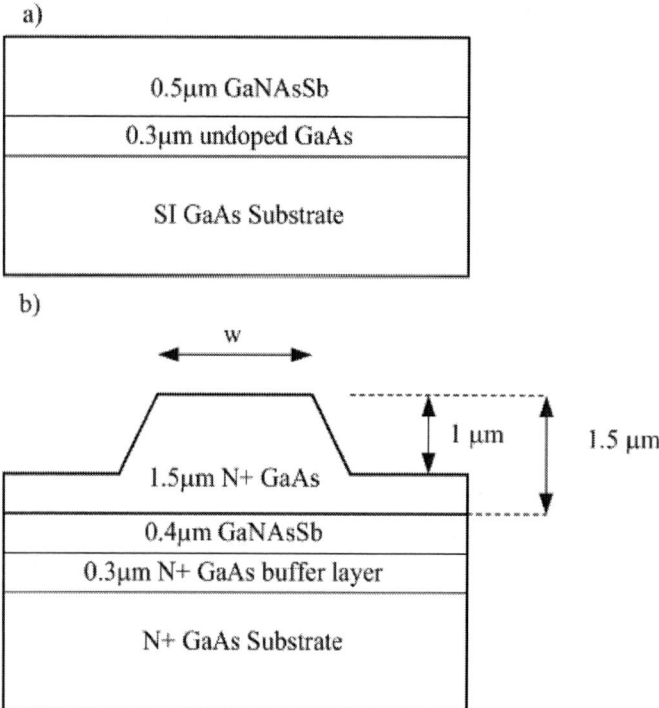

Figure 12. (a) Sample structure for refractive index measurement. (b) Cross section schematic of the GaAs/GaNAsSb/GaAs waveguide structure. The ridge widths of the waveguide are 4μm, 6.5μm and 8μm, respectively. The length of the waveguide is 2.1mm.[ref.10]

To perform the refractive index measurement, a sample with layer structure as shown in Fig. 12(a) was grown using molecular beam epitaxy (MBE) in conjunction with a radio frequency (RF) plasma-assisted nitrogen source and a valved antimony cracker source. The ridge optical waveguides structure as shown in Fig. 12(b) was also grown using MBE. The detail of growth conditions can be found elsewhere.[10]

The measurement results of refractive index (n) and extinction coefficient (k) of GaNAsSb are shown in Fig. 13. It can be seen that at 1.55μm, the real part of the refractive index (n) of GaNAsSb is 3.42, higher than that of GaAs (3.37)[39] as expected. Using the refractive index value obtained above, the optical confinement factor Γ can be calculated from the 3D beam propagation method[40] and has a value of 0.25.

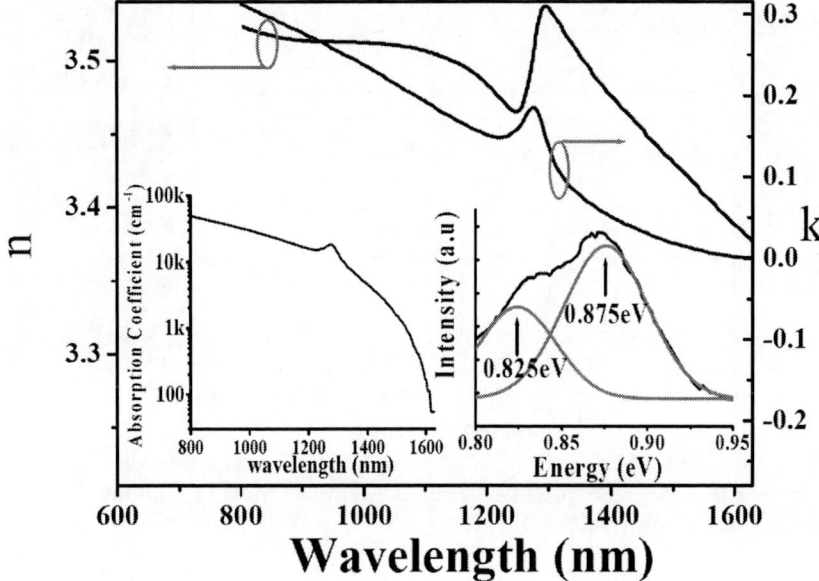

Figure 13. n-k plot of the GaNAsSb material from refractive index measurement. Inset on the bottom left is the absorption coefficient derived from the extinction coefficient (k) data set. Inset on the bottom right is PL spectrum (4K) of the GaNAsSb material. The grey lines are two Gaussian peaks used to fit spectrum. The Band-to-band transition gives rise to the main PL peak at 0.875eV. The shoulder peak at 0.825eV is attributed to transitions related to defects states.[ref.10]

The inset in the bottom left of Fig. 13 shows the corresponding absorption coefficient, α derived from the measured k in Fig. 13. It can be seen that there is absorption in the GaNAsSb layer up to 1.6μm (0.78eV). At 1.55μm, the value of α is ~700cm^{-1}. Apparently, this is inconsistent with the expected bandgap of GaNAsSb (0.88eV) mentioned previously. The PL measurement (4K) result shown in the inset in the bottom right of Fig. 13 provides an explanation for this discrepancy. The measured PL spectrum has been de-convoluted into two Gaussian peaks: a main peak centered at ~0.875eV and a broad shoulder centered at 0.825eV. The main peak position is close to the expected bandgap of the GaNAsSb layer, which is likely to be due to band-to-band transition, while the shoulder peak at 0.825eV is broad and extended below 0.8eV. The shoulder peak could be due to nitrogen-related defects, which is located at ~0.06eV above the valence band.[19] Due to the intra-bandgap transitions caused by the defect states, GaNAsSb exhibits absorption below its optical bandgap.

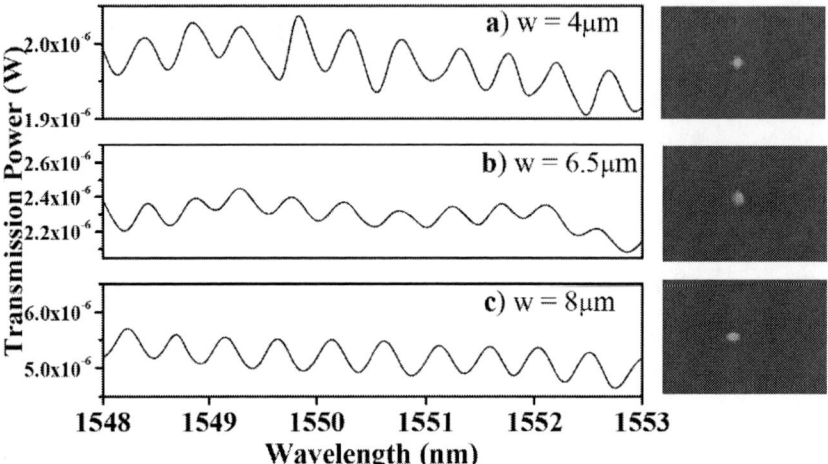

Figure 14. Fabry Perot resonance results and near-field optical intensity distribution pattern for GaAs/GaNAsSb/GaAs waveguides with ridge width of (a) 4μm, (b) 6.5μm and (c) 8μm, respectively.[ref.10]

Fig. 14 shows the Fabry Perot resonance results of the waveguides with 4μm, 6.5μm and 8μm ridge widths, respectively. Qualitatively, it can be observed in the near field optical intensity pattern that the transmitted light is properly confined in both vertical and lateral directions of the waveguides. From Fabry Perot resonance results, the propagation loss of these waveguides is found to be ~50 ± 10dB/cm and independent of the waveguide ridge width. This propagation loss is relatively high compared to AlGaAs/GaAs system. The nitrogen-related defects absorption is probably a major contributor to this high propagation loss.

4. Photoconductive switch in microwave switching application

Photoconductive switches are potential devices for THz optoelectronic applications due to its picosecond response to optical excitation. The photoconductive switch was first reported by Auston et al.[41] in 1984. Following that, in 1988, Smith et al.[42] demonstrated picosecond response in photoconductive switches based on low temperature grown gallium arsenide (LT-GaAs) as the active material. LT-GaAs is an ideal material for photoconductive switch applications due to its subpicosecond carrier lifetime and semi-insulating property. Besides LT-GaAs, ion implanted GaAs has also been demonstrated in photoconductive switch application, and has similar properties compared to LT-GaAs.[43,44]

However, the GaAs-based photoconductive switches have one major drawback, which is the optical bandgap of 1.43eV of GaAs. This prohibits the GaAs-based photoconductive switches to operate at common optical communication wavelengths of 1.3μm and 1.55μm. The emergence of dilute nitride materials for 1.3μm and 1.55μm applications has opened up another alternative for photoconductive switches. GaNAsSb can be tailored to lattice-match to GaAs and has bandgap from 1.43eV to 0.8eV, depending on the concentration of nitrogen and antimony atoms. With 3.5% of nitrogen and 9% of antimony, GaNAsSb is lattice-matched to GaAs and has bandgap of 0.88eV. Apart from its small bandgap, GaNAsSb is attractive for photoconductive switch application because of its short carrier lifetimes. The growth of dilute nitride materials, such as GaNAsSb and GaInNAs at low temperature, and the disparity between nitrogen atom and replaced group-V atoms lead to the formation of mid-gap arsenic antisite and nitrogen-related defects.[26,45-48] These defects contribute to such materials having short carrier lifetimes[49,50] and low photoluminescence efficiency.[51]

In this section, we describe the use of GaNAsSb as the active material for photoconductive switching. The pulsed response of the GaNAsSb photoconductive switch is measured at wavelength up to 1.6μm. We also report the performance of the GaNAsSb photoconductive switch in microwave switching application by measuring its ON/OFF ratio.

To fabricate the photoconductive switch, a sample with layer structure as shown in Fig. 12(a) was grown using MBE. The detail of growth conditions and fabrication process can be found elsewhere.[11] The dynamic properties of the GaNAsSb layer are investigated by characterizing the pulsed response and carrier lifetime of the material upon absorption of ultra short optical pulses.[52]

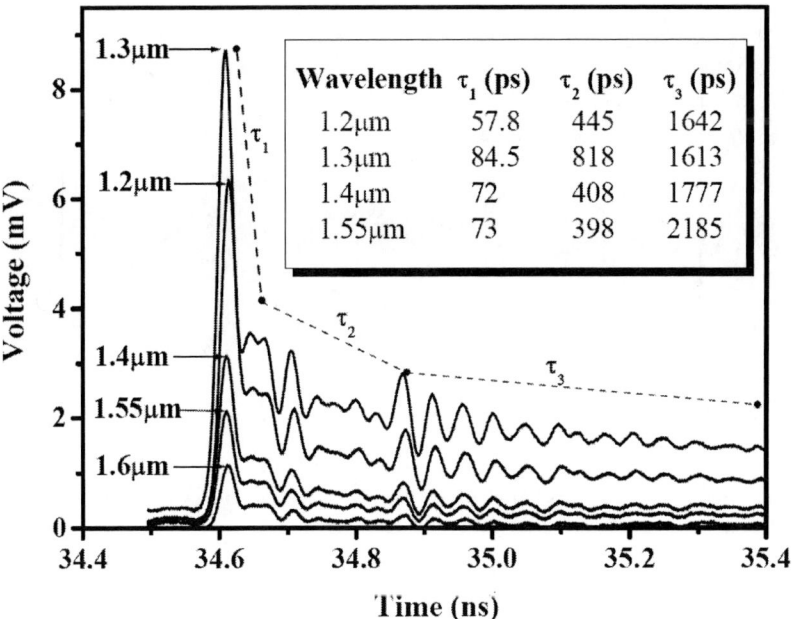

Figure 15. Pulsed response of GaNAsSb photoconductive switch at different laser excitation wavelengths. Inset shows carrier lifetimes of the GaNAsSb layer derived from pulsed measurement at different wavelengths.[ref. 11]

By applying a negative DC bias of 0.8V, the pulsed response of the GaNAsSb photoconductive switch was measured using a 50 GHz sampling oscilloscope shown in Fig. 15. The device shows photoresponse up to 1.6μm. The photoresponsivity of the device is highest at 1.3μm and gradually decreases at longer wavelength. So far, there is no other report on the photoresponse characteristic at >1.55μm for any lattice-matched dilute nitride material system on GaAs substrate. Due to the GaNAsSb bandgap of 0.88eV, the pulsed response at >1.55μm cannot be attributed to band-to-band transition and is likely due to defect-related absorptions. This is consistent with our observation in the refractive index and photoluminescence measurements mentioned earlier (see Fig. 13).

The measured FWHM of the pulses at different wavelengths is ~30ps. This value is close to the time response limit of the SMA alumina mounting. Thus, the real response time of the device could be lower than the measured response time. Furthermore, as shown in the inset of Fig. 15, there are three carrier recombination processes in the GaNAsSb layer at 1.2, 1.3, 1.4 and 1.55μm, giving rise to three values of carrier lifetimes, respectively. The two short carrier lifetimes, τ_1 and τ_2 are attributed to carrier recombination at defect states, while the long carrier lifetime, τ_3 is attributed to band-to-band recombination.

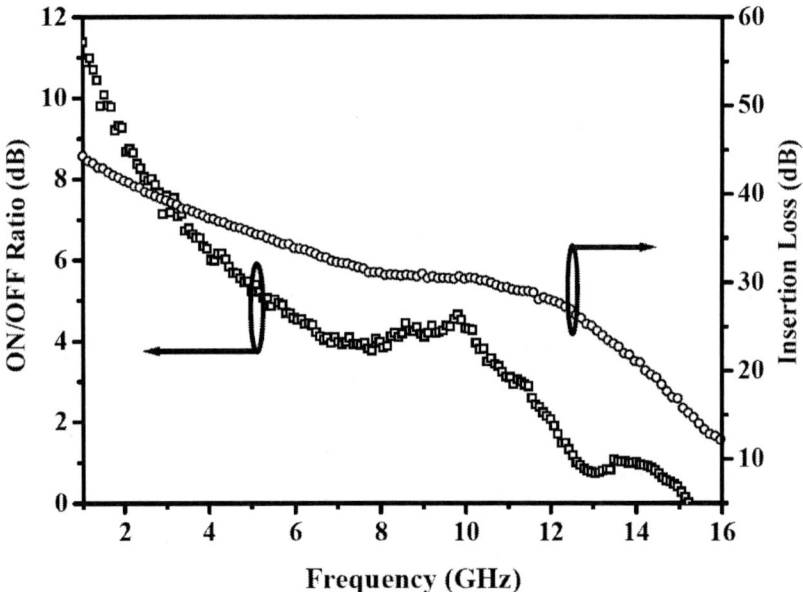

Figure 16.   Plot of ON/OFF ratio and RF insertion losses of the GaNAsSb photoconductive switch at different frequencies.[ref. 11]

The performance of the GaNAsSb photoconductive switch for microwave switching is characterized by its ON/OFF ratio measurement, where the S-parameters were extracted using a vector network analyzer with and without laser excitation. The laser has wavelength of 790nm and average power of 80mW. The ON/OFF ratio is derived from the difference in the S_{21} value with and without laser excitation. From Fig. 16, it can be seen that the ON/OFF ratio is ~11dB at 1GHz and maintained above 0dB up to 15GHz.

This value of ON/OFF ratio is significantly higher than that of the microwave photoconductive switch based on LT-GaAs.[53] Furthermore, the insertion loss of the GaNAsSb photoconductive switch is at least 16dB at 15GHz. Further improvement in microwave switching performance is expected following optimization of the device structure.

5. Heterojunction bipolar transistor

Modern-day portable electronics require high efficiency at low bias conditions to ensure a long battery life. Heterojunction bipolar transistor (HBT) technology using a small band gap base layer material to reduce the turn-on voltage is an attractive option to fulfill this requirement. While the InP HBTs having an InGaAs base work well at low voltage, because of their low base-emitter (B-E) turn-on voltages and large DC gain, they have disadvantages in term of higher substrate cost and mechanical fragility, when compared with their GaAs counterparts. The use of a dilute nitride (dilute N) material as the base layer in conventional GaAs HBTs provides a viable alternative to InP HBT technology.

Recently, the use of GaNAsSb as the base layer in a GaAs substrate HBT has been reported.[54,55] Subsequently, Lew et al.[56] reported the first AlGaAs/GaNAsSb HBT on a GaAs substrate with current gain of ~20, and established the benefit of rapid thermal annealing (RTA), albeit based on non-optimized condition, for improving the AlGaAs/GaAsNSb HBT performance. However, due to the low N concentration (N~0.5%, E_g ~1.2eV) used in the $GaN_{0.005}As_{0.915}Sb_{0.08}$ HBT base layer reported in ref. [56], there is significant room for improving the reduction in the turn-on voltage beyond the reported 80mV if higher N concentration coupled with optimized RTA conditions are used.

Layer	Material	Concentration	Thickness
Emitter cap	$In_{20\%}Ga_{80\%}As$ (Si)	$n=5\times10^{18}cm^{-3}$	200Å
Emitter cap	GaAs (Si)	$n=5\times10^{18}cm^{-3}$	2000Å
Emitter	$Al_{0.2}Ga_{0.8}As$ (Si)	$n=5\times10^{17}cm^{-3}$	1000Å
Spacer	GaAs	undoped	6 Å
Spacer layer	$GaN_{3.0\%}AsSb_{8\%}$	undoped	50Å
Base	$GaN_{3.0\%}AsSb_{8\%}$ (Be)	$p=1\times10^{19}cm^{-3}$	450Å
Spacer layer	$GaN_{3.0\%}AsSb_{8\%}$	undoped	100Å
Collector	GaAs (Si)	$n=3\times10^{18}cm^{-3}$	150Å
Collector	GaAs (Si)	$n=5\times10^{16}cm^{-3}$	4850Å
Sub-collector	GaAs (Si)	$n=5\times10^{18}cm^{-3}$	6000Å
GaAs Substrate			

Figure 17. Device structure of AlGaAs/GaNAsSb HBT.[ref. 12]

In this section, we describe the use of $GaN_{0.03}As_{0.89}Sb_{0.08}$ (N~3%, E_g~0.9eV) as the base layer of AlGaAs/GaNAsSb HBT fabricated on GaAs substrate. Under optimized annealing conditions, it will be shown that the device current gain increased by >300%, and a significant reduction in the turn-on voltage by >300mV compared to a standard AlGaAs/GaAs HBT was achieved.

The layer structure of the AlGaAs/GaNAsSb HBT is shown in Fig. 17. Two samples of AlGaAs/GaNAsSb HBT devices were fabricated. Detail of fabrication can be found elsewhere.[12] The first sample is an as-grown sample and the second sample is previously subjected to rapid thermal annealing.

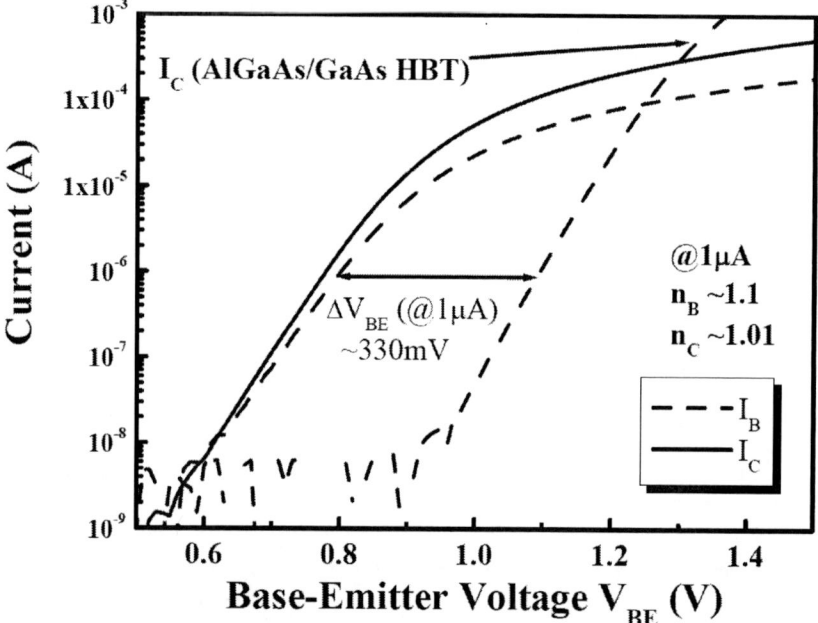

Figure 18. Gummel plot of the AlGaAs/GaNAsSb HBT. The dependence of the collector current (I_C) on the base-emitter voltage V_{BE} of a conventional $Al_{0.2}Ga_{0.8}As$/GaAs HBT is shown for comparison.[ref. 12]

Fig. 18 shows a Gummel plot for the AlGaAs/GaNAsSb HBT. The dependence of collector current (I_C) on the base-emitter voltage (V_{BE}) for a conventional $Al_{0.2}Ga_{0.8}As$/GaAs HBT is also shown in Fig. 18. The small band gap $GaN_{0.03}As_{0.89}Sb_{0.08}$ (N~3%) base layer in the HBT leads to a significant turn-on voltage reduction of up to 330mV compared to the conventional AlGaAs/GaAs HBT. The collector and base current ideality factors were 1.01 and 1.1, respectively. However, this device showed a DC gain of only ~4, which is low compared to that reported[56] for an HBT with $GaN_{0.005}As_{0.915}Sb_{0.08}$ (N~0.5%) base layer (DC gain ~8.5), and a device with nitrogen-free low band gap base layer[57] such as $GaAs_{0.92}Sb_{0.8}$ (DC gain ~20). Consistent with what has been reported for InGaAsN HBTs,[58] the DC gain degrades as the N concentration is increased. We believe that the gain reduction is attributed to an increase in N-related point defects, most likely in the form of N-As or N-N split interstitials.[59,60]

As reported by several groups,[23,61,62] rapid thermal annealing (RTA) plays an important role for improving the quality of GaNAsSb material. To determine the optimum RTA conditions for the GaNAsSb material, a test structure comprising a 500nm GaAs$_{0.89}$N$_{0.03}$Sb$_{0.08}$ layer capped with 50nm GaAs was grown. The wafer was cleaved into several pieces which were separately subjected to RTA at different temperatures under an N$_2$ ambient for 1min. Low temperature (4K) photoluminescence (PL) measurements show an improvement in the PL intensity when the RTA temperature is higher than 700°C, with a maximum PL intensity achieved at 750°C. Hence, RTA temperature of 750°C has been chosen for our AlGaAs/GaNAsSb HBT sample.

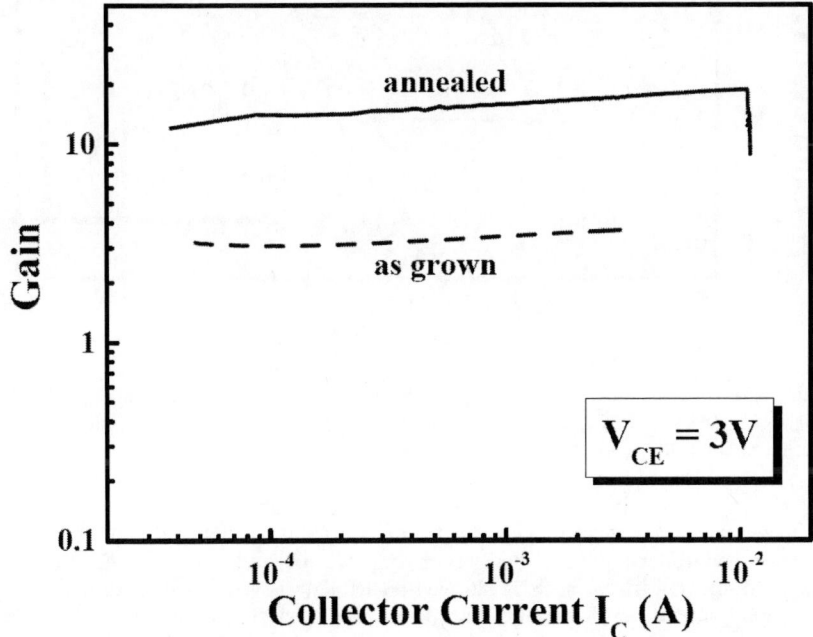

Figure 19. Plot of the current DC gain β vs. collector current (I$_C$) at room temperature for the annealed and as-grown AlGaAs/GaNAsSb HBT samples. [ref. 12]

The collector current dependence of the DC gain for the annealed and as-grown AlGaAs/GaNAsSb HBTs is shown in Fig. 19. The maximum current gain increased by more than 300% (from ~4 to ~18) in the device subjected to the RTA. Fig. 20 shows a comparison of the collector current vs. V$_{BE}$ characteristic for the AlGaAs/GaNAsSb HBT with and without annealing. The data for a conventional AlGaAs/GaAs HBT are included for comparison. The turn-on voltage of the annealed AlGaAs/GaNAsSb HBT is ~310mV lower than that of the conventional AlGaAs/GaAs HBT.

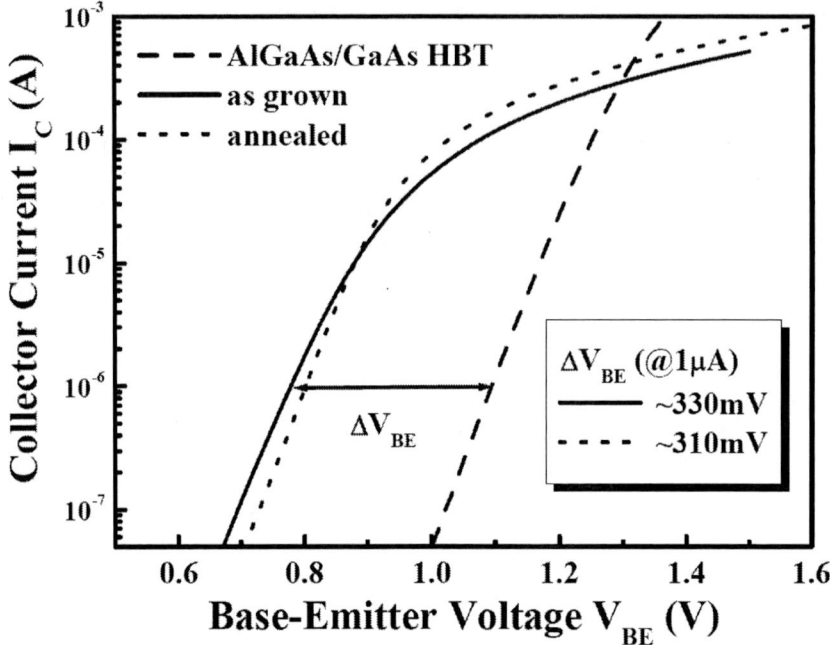

Figure 20. Dependence of collector current (I_C) on base-emitter voltage (V_{BE}) for AlGaAs/GaNAsSb HBTs with and without annealing. The characteristics for a conventional $Al_{0.2}Ga_{0.8}As$/GaAs HBTs are shown for comparison.[ref. 12]

Conclusion

We have reviewed the recent achievements of GaNAsSb material in the photonic and electronic applications based on our previously published works in refs. 6-12. It can be concluded that the defects, such as nitrogen related defects and arsenic antisite defects, play significant role in determining the device performance. In applications, such as photoresponsivity of photodetector and photoconductive switch, the defects are beneficial to the device performance. On the other hand, the defects deteriorate the performance of device in applications, such as waveguide, bandwidth of photodetector and HBT. Thus, precise engineering of defects level in GaNAsSb is needed to cater the requirement in different applications.

Acknowledgments

This work was supported by the European Commission within the European Network of Excellence ISIS, www.ist-isis.org under grant no. 26592. University Duisburg-Essen further acknowledges support by the European IPHOBAC project, www.ist-iphobac.org under grant no. 35317. The authors also acknowledge the partial financial support provided under the MERLION project no. 09.01.06 sponsored by the France Embassy in Singapore. Financial support by the THALES@NTU laboratory in Nanyang Technological University is also acknowledged.

Reference

1. M. Weyers, M. Sato, and H. Ando, Jpn. J. Appl. Phys. **31,** 853-5 (1992).
2. M. Kondow, K. Uomi, A. Niwa, T. Kitatani, S. Watahiki, and Y. Yazawa, Jpn. J. Appl. Phys. **35,** 1273-5 (1996).
3. G. Ungaro, G. Le Roux, R. Teissier, and J. C. Harmand, Electronics Letters **35,** 1246-8 (1999).
4. J. C. Harmand, A. Caliman, E. V. K. Rao, L. Largeau, J. Ramos, R. Teissier, L. Travers, G. Ungaro, B. Theys, and I. F. L. Dias, Semiconductor Science and Technology **17,** 778-84 (2002).
5. J. C. Harmand, G. Ungaro, L. Largeau, and G. Le Roux, Appl. Phys. Lett. **77,** 2482-4 (2000).
6. K. H. Tan, S. F. Yoon, W. K. Loke, S. Wicaksono, T. K. Ng, K. L. Lew, A. Stohr, S. Fedderwitz, M. Weiss, D. Jager, N. Saadsaoud, E. Dogheche, D. Decoster, and J. Chazelas, Optics Express **16,** 7720-7725 (2008).
7. K. H. Tan, S. F. Yoon, W. K. Loke, S. Wicaksono, K. L. Lew, A. Stohr, O. Ecin, A. Poloczek, A. Malcoci, and D. Jager, Appl. Phys. Lett. **90,** 183515-1 (2007).
8. S. Fedderwitz, A. Stohr, S. F. Yoon, K. H. Tan, M. Weiss, W. K. Loke, A. Poloczek, S. Wicaksono, and D. Jager, Appl. Phys. Lett. **93,** 033509-3 (2008).
9. W. K. Loke, S. F. Yoon, Z. Xu, K. H. Tan, T. K. Ng, Y. K. Sim, S. Wicaksono, N. Saadsaoud, D. Decoster, and J. Chazelas, Appl. Phys. Lett. **93,** 081102-3 (2008).
10. K. H. Tan, S. F. Yoon, W. K. Loke, S. Wicaksono, Z. Xu, T. K. Ng, K. L. Lew, N. Saadsaoud, M. Zegaoui, D. Decoster, and J. Chazelas, Appl. Phys. Lett. **92,** 113513 (2008).
11. K. H. Tan, S. F. Yoon, C. Tripon-Canseliet, W. K. Loke, S. Wicaksono, S. Faci, N. Saadsaoud, J. F. Lampin, D. Decoster, and J. Chazelas, Appl. Phys. Lett. **93,** 063509-3 (2008).
12. K. L. Lew, S. F. Yoon, S. Wicaksono, J. A. Gupta, and S. P. McAlister, submitted to *Electron Dev. Lett.* (2009).
13. S. Wicaksono, S. F. Yoon, K. H. Tan, and W. K. Loke, J. Vac. Sci. Technol. B **23,** 1054-9 (2005).
14. E. Luna, M. Hopkinson, J. M. Ulloa, A. Guzman, and E. Munoz, Appl. Phys. Lett. **83,** 3111-13 (2003).
15. H. Luo, J. A. Gupta, and H. C. Liu, Appl. Phys. Lett. **86,** 211121-1 (2005).
16. Q. Han, X. H. Yang, Z. C. Niu, H. Q. Ni, Y. Q. Xu, S. Y. Zhang, Y. Du, L. H. Peng, H. Zhao, C. Z. Tong, R. H. Wu, and Q. M. Wang, Appl. Phys. Lett. **87,** 111105-1 (2005).
17. W. K. Cheah, W. J. Fan, S. F. Yoon, D. H. Zhang, B. K. Ng, W. K. Loke, R. Liu, and A. T. S. Wee, IEEE Photon. Technol. Lett. **17,** 1932-4 (2005).
18. W. K. Loke, S. F. Yoon, S. Wicaksono, and B. K. Ng, Mater. Sci. Eng. B **131,** 40-44 (2006).
19. W. K. Loke, S. F. Yoon, K. H. Tan, S. Wicaksono, and W. J. Fan, J. Appl. Phys. **101,** 33122-1 (2007).
20. S. Wicaksono, S. F. Yoon, W. K. Loke, K. H. Tan, and B. K. Ng, J. Appl. Phys. **99,** 104502-1 (2006).
21. S. Wicaksono, S. F. Yoon, W. K. Loke, K. H. Tan, K. L. Lew, M. Zegaoui, J. P. Vilcot, D. Decoster, and J. Chazelas, J. Appl. Phys. **102,** 044505-7 (2007).
22. J. S. Ng, W. M. Soong, M. J. Steer, M. Hopkinson, J. P. R. David, J. Chamings, S. J. Sweeney, and A. R. Adams, J. Appl. Phys. **101,** 64506-1 (2007).

23. Y. X. Dang, W. J. Fan, S. T. Ng, S. Wicaksono, S. F. Yoon, and D. H. Zhang, J. Appl. Phys. **98**, 026102 (2005).
24. C. Hu, K. A. Anselm, B. G. Streetman, and J. C. Campbell, Appl. Phys. Lett. **69**, 3734-6 (1996).
25. J. S. Ng, J. P. R. David, G. J. Rees, and J. Allam, J. Appl. Phys. **91**, 5200-2 (2002).
26. W. M. Chen, I. A. Buyanova, C. W. Tu, and H. Yonezu, Physica B: Condensed Matter **376-377**, 545-551 (2006).
27. W. Shockley, Solid-State Electron. **2**, 35-67 (1961).
28. L. Partain, D. Day, and R. Powell, J. Appl. Phys. **74**, 335-40 (1993).
29. D. Jackrel, H. Yuen, S. Bank, M. Wistey, J. Fu, X. Yu, Z. Rao, and J. S. Harris, Semiconductor Photodetectors II, San Jose, CA, USA, 2005 (SPIE), p. 27-34.
30. A. Stohr, A. Malcoci, A. Sauerwald, I. C. Mayorga, R. Gusten, and D. S. Jager, J. Lightwave Technol. **21**, 3062-3070 (2003).
31. A. J. N. Houghton, D. A. Andrews, G. J. Davies, and S. Richie, Optics Communications **46**, 164-166 (1983).
32. H. Inoue, K. Hiruma, K. Ishida, T. Asai, and H. Matsumura, J. Lightwave Technol. **T-3**, 1270-6 (1985).
33. H. Hiruma, H. Inoue, K. Ishida, and H. Matsumura, Appl. Phys. Lett. **47**, 186-7 (1985).
34. E. Kapon and R. Bhat, Appl. Phys. Lett. **50**, 1628-30 (1987).
35. H. Takeuchi and K. Oe, J. Lightwave Technol. **7**, 1044-54 (1989).
36. M. Seto, A. Shahar, R. J. Deri, W. J. Tomlinson, and A. Yi-Yan, Appl. Phys. Lett. **56**, 990-2 (1990).
37. R. J. Deri and E. Kapon, IEEE J. Quantum Electron. **27**, 626-40 (1991).
38. B. Young Tae, P. Kyung Hyun, K. Sun Ho, C. Sang Sam, and L. Tong Kun, Appl. Optics **35**, 928-33 (1996).
39. D. T. F. Marple, J. Appl. Phys. **35**, 1241-1242 (1964).
40. J. Van Roey, J. van der Donk, and P. E. Lagasse, J. Optical Soc. of America **71**, 803-10 (1981).
41. D. H. Auston, K. P. Cheung, J. A. Valdmanis, and D. A. Kleinman, Physical Review Letters **53**, 1555-8 (1984).
42. F. W. Smith, H. Q. Le, V. Diadiuk, M. A. Hollis, A. R. Calawa, S. Gupta, M. Frankel, D. R. Dykaar, G. A. Mourou, and T. Y. Hsiang, Appl. Phys. Lett. **54**, 890-2 (1989).
43. A. Claverie, F. Namavar, and Z. Liliental-Weber, Appl. Phys. Lett. **62**, 1271-1273 (1993).
44. A. Krotkus, S. Marcinkevicius, J. Jasinski, M. Kaminska, H. H. Tan, and C. Jagadish, Appl. Phys. Lett. **66**, 3304-3306 (1995).
45. R. J. Kaplar, D. Kwon, S. A. Ringel, A. A. Allerman, S. R. Kurtz, E. D. Jones, and R. M. Sieg, Solar Energy Materials and Solar Cells **69**, 85-91 (2001).
46. W. K. Loke, S. F. Yoon, S. Wicaksono, K. H. Tan, and K. L. Lew, J. Appl. Phys. **102**, 054501-6 (2007).
47. S. Wicaksono, S. F. Yoon, W. K. Loke, K. H. Tan, K. L. Lew, M. Zegaoui, J. P. Vilcot, D. Decoster, and J. Chazelas, J. Appl. Phys. **102**, 044505-7 (2007).
48. W. K. Loke, S. F. Yoon, K. H. Tan, S. Wicaksono, and W. J. Fan, J. Appl. Phys. **101**, 033122-5 (2007).
49. A. J. Ptak, S. W. Johnston, S. Kurtz, D. J. Friedman, and W. K. Metzger, J. Cryst. Growth **251**, 392-8 (2003).

50. R. A. Mair, J. Y. Lin, H. X. Jiang, E. D. Jones, A. A. Allerman, and S. R. Kurtz, Appl. Phys. Lett. **76,** 188-90 (2000).
51. S. Wicaksono, S. F. Yoon, K. H. Tan, and W. K. Loke, Journal of Vacuum Science and Technology B: Microelectronics and Nanometer Structures **23,** 1054-1059 (2005).
52. D. Vignaud, D. A. Yarekha, J. F. Lampin, M. Zaknoune, S. Godey, and F. Mollot, Appl. Phys. Lett. **90,** 242104-3 (2007).
53. S. Faci, C. Tripon-Canseliet, N. Guldner, G. Alquie, S. Formont, and J. Chazelas, 2007 International Topical Meeting Microwave Photonics, Victoria, BC, Canada, 2007 (IEEE), p. 194-7.
54. K. L. Lew, S. F. Yoon, H. Wang, S. Wicaksono, J. A. Gupta, and S. P. McAlister, Journal of Vacuum Science & Technology B (Microelectronics and Nanometer Structures) **24,** 1308-10 (2006).
55. J. C. Harmand, R. M. L.Li, G. Ungaro, V. Sallet, L. Travers, G. Patriarche, L. Largeau, R. Kudrawiec, G. Sek , and J. Misiewicz, in *Dilute Nitride Semiconductor*, edited by M. Henini (Elsevier Ltd, 2005), p. 471-493.
56. K. L. Lew, S. F. Yoon, H. Wang, S. Wicaksono, J. A. Gupta, and S. P. McAlister, IEEE Electron Device Letters **28,** 1083-5 (2007).
57. T. Oka, T. Mishima, and M. Kudo, Appl. Phys. Lett. **78,** 483-5 (2001).
58. P. M. Asbeck, R. J. Welty, C. W. Tu, H. P. Xin, and R. E. Welser, Semiconductor Science and Technology **17,** 898-906 (2002).
59. W. J. Fan, S. F. Yoon, T. K. Ng, S. Z. Wang, W. K. Loke, R. Liu, and A. Wee, Appl. Phys. Lett. **80,** 4136-8 (2002).
60. T. Ahlgren, E. Vainonen-Ahlgren, J. Likonen, W. Li, and M. Pessa, Appl. Phys. Lett. **80,** 2314-16 (2002).
61. J. C. Harmand, L. Li, R. Mouillet, G. Ungaro, V. Sallet, L. Travers, G. Patriarche, L. Largeau, R. Kudrawiec, G. Sek, and J. Misiewicz, in *Dilute Nitride Semiconductor*, edited by M. Henini (Elsevier Ltd, 2005), p. 471-493.
62. S. Wicaksono, S. F. Yoon, W. J. Fan, and W. K. Loke, 2005 International Conference on Indium Phosphate and Related Materials, Glasgow, Scotland, UK, 2005 (IEEE), p. 421-3.

30

Detection of Cl⁻ Ions with AlGaN/GaN High Electron Mobility Transistors

S. C. Hung[1,4], B. H Chu[1], C. F. Lo[1], B. Hicks[1], Y. L. Wang[2], C. Y. Chang[2], S.J. Pearton[2], J. W. Johnson [3], P. Rajagopa[3], l, J. C. Roberts[3], E. L. Piner[3], K. J. Linthicum[3], G. C. Chi[4], and F. Ren[1]

1. University of Florida, Department of Chemical Engineering, Gainesville, FL 32611

2. University of Florida, Department of Materials Science and Engineering, Gainesville, FL 32611

3. Nitronex Corporation, Durham, NC 27703
4. Department of Physics, National Central University, Jhong-Li 320, Taiwan

> AlGaN/GaN high electron mobility transistors (HEMTs) with Ag/AgCl gate are found to exhibit significant changes in channel conductance upon exposing the gate region to various concentrations of chorine ion solutions. Ag/AgCl gate electrode, prepared by potentiostatic anodization, changes the electrical potential when encounters chorine ions. This gate potential changes lead to a change of surface charge in the gate region on the HEMT, induced a higher positive charge on the AlGaN surface, and finally increases the pizeo-induced charge density in the HEMT channel. These anions create an image positive charge on the Ag gate metal for the required neutrality, thus increase the drain current of the HEMT. The HEMT source-drain current showed a clear dependence on the chorine concentration. The limit of detection (LOD) achieved was 1×10^{-8} M using a 20μm × 50μm gate sensing area.

Introduction

Chlorine is widely used in the manufacture of many products and items directly or indirectly, i.e., in paper product production, antiseptic, dye stuffs, food, insecticides, paints, petroleum products, plastics, medicines, textiles, solvents, and many other consumer products. It is used to kill bacteria and other microbes in drinking water supplies and waste water treatment. Excess chlorine also reacts with organics and forms disinfection by-products such as carcinogenic chloroform, which is harmful to human health. Thus, to ensure the safety of public health, it is very important to accurately and effectively monitor chlorine residues, including those in the form of Cl− ions, during the treatment and transport of drinking water[1,2]. Although the treatment of water with chlorine for disinfection typically leads to the formation of hypochlorite ions, the chloride content in water may also be considerably increased.

The Cl− ion is an essential mineral for humans. The total body balance of chloride, particularly the plasma levels of this monovalent anion, is carefully regulated by the kidneys. Chloride is found in most, if not all, body fluids, including blood, plasma, serum, urine, saliva, and exhaled breath condensate. The normal blood concentration of chloride in adults is approximately 95– 105 mequivalent/l. Variations in the chloride concentration in blood may indicate the presence of a number of important diseases such as renal dysfunction, metabolic disturbances, adrenalism, and pneumonia. Therefore, the measurement of this parameter is clinically important[3]. Several analytical methods such as colorimetry[4], ion-selective electrode[5], activation analysis[6], x-ray fluorescence spectrometry[7], and ion chromatography[8,9] have been used for the analysis of chloride in various samples. However, these methods are not portable and require expensive instrumentation. An accurate and fast determination of the inorganic ion content of various aqueous samples at low detection limits has been of great interest for a long time.

Recently, AlGaN/GaN high electron mobility transistors (HEMTs) have shown promise for such applications[10-19]. This is due to their high electron sheet carrier concentration channel induced by both piezoelectric polarization and spontaneous polarization[20,21]. Unlike conventional semiconductor field effect transistors, there is no intentional dopant in the AlGaN/GaN HEMT structure. The electrons in the two dimensional electron gas (2DEG) channel are located at the interface between the AlGaN layer and GaN layer and there are positive counter charges at the HEMT surface layer induced by the 2DEG. Slight changes in the ambient can affect the surface charge of the HEMT, thus changing the 2DEG concentration in the channel.

Recently, AlGaN/GaN high electron mobility transistors (HEMTs) have shown promise for such applications [17-14]. This is due to a high electron sheet carrier concentration channel induced by both piezoelectric polarization and spontaneous polarization. Unlike conventional semiconductor field effect transistors, there is no intentional dopant in the AlGaN/GaN HEMT structure. The electrons in the two-dimensional electron gas (2DEG) channel are located at the interface between the AlGaN layer and GaN layer and there are positive counter charges at the HEMT surface layer induced by the 2DEG. Slight changes in the ambient can affect the surface charge of the HEMT, thus changing the 2DEG concentration in the channel.

In this work, we studied the effect of exposing the gate region of Ag/AgCl gated GaN/AlGaN HEMTs with different concentrations of Cl− ion solutions. The AgCl was fabricated with selective area potentiostatic anodization from a Ag thin film. We quantified the sensitivity, the temporal resolution and the limit of detection of the HEMT sensor for Cl− ion detection. The recyclability of the sensors with wash buffers between measurements was also examined.

Experimental

The HEMT structures consisted of a 2 μm thick undoped GaN buffer and 250 Å thick undoped $Al_{0.25}Ga_{0.75}N$ cap layer. The epilayers were grown by metal-organic chemical vapor deposition on 100 mm (111) Si substrate. Mesa isolation was performed with inductively coupled plasma (ICP) etching with Cl_2 /Ar based discharged at −90 V dc self-bias, ICP power of 300 W at 2 Hz and a process pressure of 5 mTorr. 50×50 $μm^2$ Ohmic contacts separated with gaps of 20 μm consisted of e-beam deposited Ti/Al/Pt/Au patterned by lift-off and annealed at 850 °C for 45 s under flowing N_2. Ti (10nm)/Ag (100nm) thin film was deposited as the gate metal. AZ resist was used to open a window on the Ti/Ag for AgCl potentiostatic anodization. Prior to anodization, samples were cleaned with acetone, isopropanol and DI water. The selective area Ag was anodized in 0.1 N HCl solution stirred continuously at 25°C with a constant bias voltage of 1V for 5 seconds and only a part of Ag thin film was anodized into AgCl and the thickness of the AgCl. The AgCl thickness was measured around 320 nm with scanning electron microscope (SEM). 500 nm thick polymethyl methacrylate was used to encapsulate the source/drain regions, with only the Ag/AgCl gate region opened using e-beam lithography. A plain view photomicrograph and a schematic device cross sectional view of the Ag/AgCl gated HEMT are shown in Fig. 1. The drain current characteristics of the HEMT sensor were measured at 25 °C using an Agilent 4156C parameter analyzer with the gate region exposed to water and different concentrations of NaCl solutions.

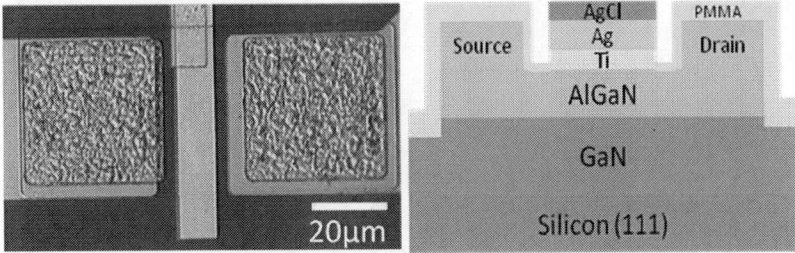

Figure 1. Plan view photomicrograph and a schematic device cross sectional view of the Ag/AgCl gated HEMT

Results and Discussion

The composition of the as-prepared Ag/AgCl thin film was characterized using energy dispersive x-ray spectroscopy (EDS) with a silicon substrate deposited with the same Ti (10nm)/Ag (100nm) thin film as on the gate area of the HEMT sample. There were Ag, Cl, Ti and Si signals detected with EDS, as shown in Figure 2.

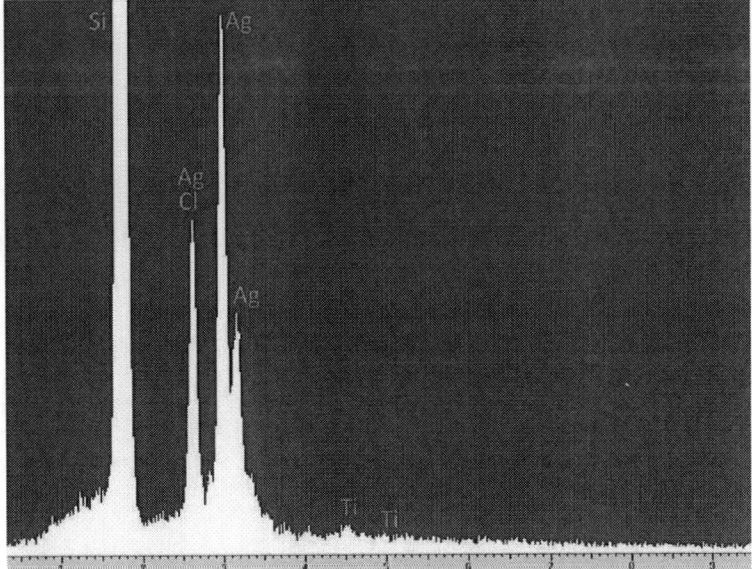

Figure 2. Energy dispersive spectroscopy image of AgCl thin film.

Atomic ratio of Cl to Ag was estimated to be 0.35 to 0.65, which validated the intentional partial anodization of the Ag thin film. The signals for Ti and Si element in the EDS spectrum were attributed to the Ti adhesion layer under the Ag thin film and the silicon substrate. Figure 3 shows the surface morphology of Ag thin film before anodization and AgCl layer after anodization. The grain size and the anodization rate of the AgCl mainly depended on the applied bias voltage during the anodization process.

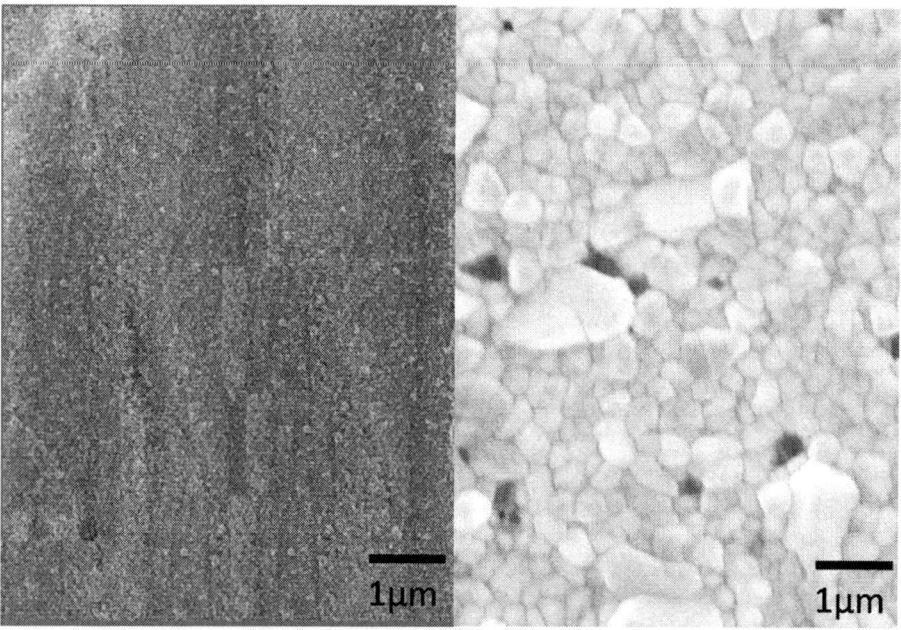

Figure 3. SEM pictures of Ag thin film before anodization (left) and AgCl layer after anodization (right).

Figure 4 shows the time dependence of Ag/AgCl HEMT drain current at a constant drain bias voltage of 500mV during exposure to solutions with different chlorine ion concentrations. The HEMT sensor was first exposed to DI water and no change of the drain current was detected with the addition of DI water at 100 second. This stability was important to exclude possible noise from the mechanical change of the NaCl solution. By sharp contrast, there was a rapid response of HEMT drain current observed in less than 30 seconds when target of 1×10^{-8} M NaCl solution was switched to the surface at 175 sec. The abrupt current change due to the exposure of chlorine in NaCl solution stabilized after the chlorine thoroughly diffused into water to reach a steady state. When Ag/AgCl gate metal encountered chorine ion, the electrical potential of the gate was changed, induced a higher positive charge on the AlGaN surface, and finally increases the pizeo-induced charge density in the HEMT channel. 1×10^{-7} M of NaCl solution was then applied at 382 second and it was accompanied with a larger signal corresponfing to the higher chlorine concentration. Further real time tests were carried out to explore the detection of higher Cl⁻ ion concentrations.

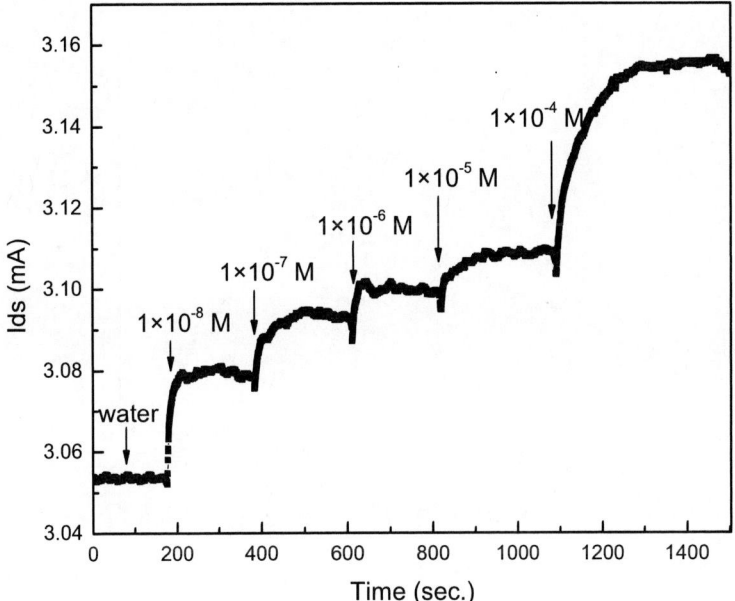

Figure 4. Time dependent current signal when exposing the HEMT to different concentration of NaCl solution.

The sensors was exposed to 1×10^{-8} M, 1×10^{-7} M, 1×10^{-6} M, 1×10^{-5} M and 1×10^{-4} M solutions continuously and repeated five times to obtain the standard deviation of source-drain current response for each concentration. The limit of detection of this device was 1×10^{-8} M chlorine in DI-water. Between each test, the device was rinsed with DI water. These results suggest that our HEMT sensors are recyclable with simple DI water rinse. Figure 5 shows the drain current change of Ag/AgCl gated HEMT with as a function of the Cl⁻ ion concentration. The presence of the Ag/AgCl gate leads to logarithmic dependence of current on the concentration for NaCl.

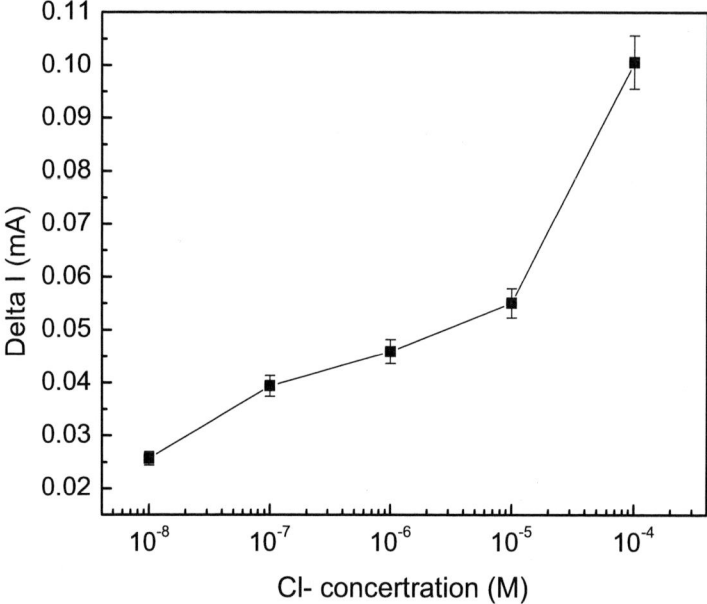

Figure 5. Current change in HEMT as a function of chorine concentration.

Conclusion

In conclusion, Ag/AgCl gated AlGaN/GaN HEMT, prepared by potentiostatic anodization in the solution of HCl, showed rapid changes in the drain current when encountered different concentrations of Cl⁻ ion solution. The results show the potential of Ag/AgCl gated AlGaN/GaN HEMT for a variety of chemical and biological sensing applications.

REFERENCES

1. J. P. Frejaville, and P. Kamoun (eds). Guide des examens de laboratoire, Paris, Flammarion (1981)
2. S. S. Raphael (ed). Lynch's medical laboratory technology, 4.ed., Philadelphia: Saunders (1983).
3. L. C. Oliveira, and C. B. Zamboni, Brazilian Journal of Physics, 35, 793 (2005)
4. J. Taylor and S. hong, Journal of Laboratory Medicine, 31-10: 563-568, (2000)
5. H. Shekhar, V. Chathapuram, S. H. Hyun, S. Hong, H. J. Cho, IEEE, 67 (2003)
6. R. Thadhani, M. Pascual, and J. V. Bonventre, N. Engl. J. Med. **334**, 1448 (1996).

7. G. M. Chertow, E. M. Levy, K. E. Hammermeister, F. Grover, J. Daley, Amer. J. Med. **104**, 343 (1998).
8. J. V. Bonventre, and J. M. Weinberg, J. Am. Soc. Nephrol. **14**, 2199 (2003).
9. T. Ichimura, J. V. Bonventre, V. Bailly, H. Wei, C. A. Hession, R. L. Cate, and M. Sanicola, J. Biol. Chem **273**, 4135 (1998).
10. V. S. Vaidya, V. Ramirez, T. Ichimura, N. A. Bobadilla, and J. V. Bonventre, Am. J. Physiol. Renal. Physiol. **290**, F517 (2006).
11. V. S. Vaidya, J. V. Bonventre, Expert Opin. Drug Metab. Toxicol. **2**, 697 (2006).
12. R. Lequin, Clin. Chem. **51** , 2415 (2005).
13. Dario A.A. Vignali, Journal of Immunological Methods **243**, 243 (2000).
14. R. J. Chen, S. Bangsaruntip, K. A. Drouvalakis, N. W. S. Kam, M. Shim, Y. Li, W. Kim, P. J. Utz, and H. Dai, Proc. Natl. Acad. Sci. USA **100**, 4984 (2003).
15. C. Li, M. Curreli, H. Lin, B. Lei, F. N. Ishikawa, R. Datar, R. J. Cote, M. E. Thompson, and C. Zhou, J. Am. Chcm. Soc. **127**, 12484 (2005).
16. G. Zheng, F. Patolsky, Y. Cui, W. U. Wang, and C. M. Lieber, Nature Biotechnology **23**, 1294 (2005).
17. F. Patolsky, G. Zheng, C. M. Lieber, Nanomedicine **1**, 51 (2006).
18. F. Patolsky, G. Zheng, C. M. Lieber, Nature Protocals **1**, 1711 (2006).
19. F. Patolsky, B. P. Timko, G. Zheng, C. M. Lieber, MRS Bulletin **32**, 142 (2007).
20. D. I. Han, D. S. Kim, J. E. Park, J. K. Shin, S. H. Kong, P. Choi, J. H. Lee, and G. Lim, Jpn. J. Appl. Phys. **44**, 5496 (2005).
21. G. Shekhawat, S. H. Tark, and V. P. Dravid, Science **311**, 1592 (2006)..
22. H.-T. Wang, B. S. Kang, F. Ren, R. C. Fitch, J. K. Gillespie, N. Moser, G. Jessen, T. Jenkins, R. Dettmer, D. Via, A. Crespo, B. P. Gila, C. R. Abernathy, and S. J. Pearton, Appl. Phys. Lett. **87**, 172105-1-3 (2005).
23. Y. Alifragis, A. Volosirakis, N. A. Chaniotakis, G. Konstantinidis, E. Iliopoulos, A. Georgakilas, Phy. Stat. Sol. (a) **204**, 2059 (2007).
24. M. Eickhoff, J. Schalwig, G. Steinhoff, O. Weidemann, L. Görgens, R. Neuberger, M. Hermann, B. Baur, G. Müller, O. Ambacher, and M. Stutzmann, Phys. Stat. Sol. (c) **0**, 1908 (2003).
25. B. S. Kang, S. Kim, F. Ren, J. W. Johnson, R. Therrien, P. Rajagopal, J. Roberts, E. Piner, K. J. Linthicum, S. N .G. Chu, K. Baik, B. P. Gila, C. R. Abernathy and S. J. Pearton, Appl. Phys.Lett. **85**, 2962 (2004)
26. B. S. Kang, F. Ren, L. Wang, C. Lofton, W. Tan, S. J. Pearton, A. Dabiran, A. Osinsky, and P. P. Chow, Appl. Phys. Lett. **87**, 023508 (2005).
27. B. S. Kang, H.T. Wang, F. Ren, B. P. Gila, C. R. Abernathy, S. J. Pearton, J. W. Johnson, P. Rajagopal, J. C. Roberts, E. L. Piner, and K. J. Linthicum, Appl. Phys. Lett. 91, 012110 (2007).
28. B. S. Kang, H.T. Wang, F. Ren, S. J. Pearton, T. E. Morey, D. M. Dennis, J. W. Johnson, P. Rajagopal, J. C. Roberts, E. L. Piner, and K. J. Linthicum, Appl. Phys. Lett. 91, 252103 (2007).
29. B. S. Kang, H.T. Wang, T. P. Lele, Y. Tseng, F. Ren, S. J. Pearton, J. W. Johnson, P. Rajagopal, J. C. Roberts, E. L. Piner, and K. J. Linthicum, Appl. Phys. Lett. 91, 112106 (2007).
30. H.T. Wang, B. S. Kang, F. Ren, S. J. Pearton, J. W. Johnson, P. Rajagopal, J. C. Roberts, E. L. Piner, and K. J. Linthicum, Appl. Phys. Lett. 91, 222101 (2007).

38

ECS Transactions, 19 (3) 39-44 (2009)
10.1149/1.3120684 ©The Electrochemical Society

Low frequency noise of chlorine-treated GaN/AlGaN MSM-photodetectors

De-En Lu[a], Ya-Lan Chiou[a], Hsin-Ying Lee[b], and Ching-Ting Lee[a]

[a] Institute of Microelectronics, Department of Electrical Engineering, National Cheng Kung University, Tainan, Taiwan, Republic of China
[b] Department of Electro-optical, National Cheng Kung University, Tainan, Taiwan Republic of China

> Prior to the deposition of the Ni/Au interdigital electrodes, the GaN surface of GaN/AlGaN metal-semiconductor-metal (MSM) ultraviolet photodetectors was chlorine-treated. The low frequency noise equivalent power of the chlorine-treated photodetectors, measured at a bias of 5 V, was 5.13×10^{-10} W, which was one order of magnitude lower than that of the photodetectors without chlorination surface treatment. The normalized detectivity of the chlorine-treated photodetectors was 6.16×10^8 cmHz$^{0.5}$W^{-1}, which was higher than that of untreated one. The dark current of chlorine-treated and untreated photodetectors, operated at 5 V, was 1.45×10^{-11} A and 3.68×10^{-11} A, respectively. The performance improvement was attributed to the passivation of GaN surface by chlorination treatment.

Introduction

Ultraviolet radiation detection has been attracted much attention for a wide range of traditional and emerging civil and military applications, such as chemical analysis, biological analysis, flame sensors, space-based optical communications, emitter calibration, and astronomical studies. Wide-band gap semiconductors of GaN-based compound semiconductors have become the promising materials due to their high thermal stability. In the recent years, various types of GaN-based photodetectors (PDs) have been reported, such as photoconductors[1-3], p-n junction diodes[4][5], p-i-n diodes[6][7], Schottky barrier photodetectors[8][9], metal-semiconductor-metal (MSM) photodetectors[10][11], phototransistor, photodetector array and avalanche photodetector. Among them, MSM photodetectors have an ultralow intrinsic capacitance and its fabrication process is compatible with field-effect-transistor (FET)-based electronics. Therefore, it can easily integrate GaN MSM-PDs with GaN FET-based electronics to realize GaN-based optoelectronic integrated circuits (OEICs). Because a lower dark current can efficiently reduce current noise, it is necessary to reduce dark current to achieve high-performances of MSM-PDs. In general, dark current originates from carrier leakage occurred at the metal/semiconductor interface. Recently, several methods have been developed to improve the Schottky characteristics of GaN, including KOH etching treatment[12] and chlorination surface treatment[14][15] et. al. The chlorination surface treatment is a useful technology to reduce the surface states by reducing Ga dangling bonds and passivating the N vacancies on the GaN surface. In this study, the responsivity, low-frequency noise and detectivity of chlorine-treated and untreated GaN/AlGaN MSM-PDs were measured and discussed.

Device structure and Fabrication process

Epitaxial layers were grown using a metalorganic chemical vapor deposition (MOCVD) system. First, a 700 nm-thick undoped GaN buffer layer and an 100 nm-thick undoped $Al_{0.2}Ga_{0.8}N$ buffer layer were grown on sapphire substrate followed by a 300 nm-thick undoped GaN absorption layer. After epitaxial growth, the UV-PDs were fabricated using a standard photolithography and lift-off technique. An 150 nm-thick SiO_2 isolation film was grown on the surface of the GaN absorption layer by electron-beam evaporator. The photosensitive region of the MSM-PDs was defined by opening a 100×100 μm^2 window in the SiO2 film, which was removed using buffer oxide etchant (BOE). Prior to deposition of the Ni/Au Schottky metals, the chlorination surface treatment was performed to remove the native oxide on the GaN surface. The aqueous solution of $1HCl+10H_2O$ was used as the electrolytic solution for the chlorination surface treatment. The GaN was placed under the Pt anodic electrode and a forward bias of 20 V was applied on it for 60 min. The produced chlorine near the Pt anodic electrode reacted with the GaN and the resulted GaO_x was formed on the GaN surface. The electron concentration of the GaN surface was reduced, when Ga vacancies were formed after chlorination surface treatment. After that, Ni/Au Schottky metals with an interdigitate finger geometry were formed on the photosensitive regions by standard photolithography and lift-off technique. The finger width and spacing were 2μm and 2μm, respectively. Ni/Au (100/100 nm) were deposited by electron-beam evaporator under a high vacuum environment (1×10^{-6} Torr).

Experimental results and Discussion

An HP-4156C semiconductor parameter analyzer was used to measure the current-voltage (I-V) characteristics of the MSM-PDs fabricated with and without chlorination surface treatment in the dark. Figures 1 shows the variation of the dark current with the bias voltage for both chlorine-treated and untreated MSM-PDs. The dark current of the chlorine-treated and untreated MSM-PDs operated at a bias of 5V was 1.45×10^{-11} A and 3.68×10^{-11} A, respectively. It can be seen that the dark current of chlorine-treated devices was obviously smaller than that of untreated ones.

A xenon (Xe) lamp through a calibrated monochromator was used as the optical pumping source. Figures 2 shows the spectra response of the MSM-PDs with and without chlorination surface treatment. The photoresponsivity of the MSM-PDs without chlorination surface treatment is larger than that of those with chlorination surface treatment. The photoresponsivity under a bias voltage of 5 V at a wavelength of 340 nm of the UV-PDs with and without chlorination surface treatment was 0.12 A/W and 0.52 A/W, respectively. The UV-visible rejection ratio of 10^3 for the chlorine-treated MSM-PDs is about one order in magnitude higher than those without chlorination surface treatment. The rejection ratio improvement of the chlorine-treated MSM-PDs is ascribed to the reduction of surface states.

Figures 3 shows the voltage-dependent product of quantum efficiency and internal gain at a wavelength of 340 nm of the chlorine-treated and untreated MSM-PDs. The product of quantum efficiency and internal gain of untreated-PDs is larger than 100%, indicating the existence of an internal gain. The internal gain is attributed to the reduction of the Schottky barrier height and the injection of additional electrons into the surface state traps[16]. The smaller internal gain in the chlorine-treated MSM-PDs is attributed to the effective reduction and passivation of surface states on the GaN surface.

Figures 4 shows the noise power spectra of the chlorine-treated and untreated MSM-PDs operated at 5 V. It was found that the level of the noise power density of the chloine-treated MSM-PDs was smaller than that of the untreated MSM-PDs. The total noise current power i_n can be estimated by integrating spectral density of the noise power $S_n(f)$ over the frequency range f:

$$\langle i_n \rangle^2 = \int S_n(f) df \tag{1}$$

Thus, the noise equivalent power (NEP) can be calculated by the following equation

$$NEP = \frac{\sqrt{\langle i_n \rangle^2}}{R} \tag{2}$$

where R is the responsivity of the MSM-PDs. The noise equivalent power for the chlorine-treated and untreated MSM-PDs, derived from the data of Fig.2, was 5.13×10^{-10} W and 4.88×10^{-9} W, respectively. The normalized detectivity (D*) can then be determined by the following equation

$$D^* = \frac{\sqrt{A}\sqrt{B}}{NEP} \tag{3}$$

where A is the area of the MSM-PDs and B is the bandwidth. For our case, the bandwidth was 1kHz and the detector area was 100×100 μm^2, the calculated D* value for the chlorine-treated and untreated MSM-PDs was 6.16×10^8 cmHz$^{0.5}$W^{-1} and 6.48×10^7 cmHz$^{0.5}$W^{-1}, respectively. The lower noise level for chlorine-treated MSM-PDs was attributed to the passivation of surface states by chlorination surface treatment.

Conclusions

Prior to the deposition of Ni/Au Schottky metals, GaN/AlGaN MSM photodetectors were fabricated with the GaN chlorination surface treatment. The resulted devices exhibited a lower noise level and a higher normalized detectivity. The NEP and D* of chlorine-treated photodetectors were 5.13×10^{-10} W and 6.16×10^8 cmHz$^{0.5}$W^{-1}, respectively, which is obviously better than those of the untreated MSM-PDs. In addition, the dark current of chlorine-treated MSM-PDs was also smaller than that of the untreated ones. The obtained performance improvement, including noise reduction as well as the decrease in dark current, can be attributed to the passivation of GaN surface by chlorine surface treatment.

Acknowledgments

This work was supported from the Advanced Optoelectronic Technology Center of National Cheng Kung University and the National Science Council of Taiwan, Republic of China under the contract number of NSC-96-2221-E-006-282-MY3.

References

1. J. C. Carrano, T. Li, P. A. Grudowski, C. J. Eiting, R. D. Dupuis and J. C. Campbell, *J. Appl. Phys.*, 83, 6148 (1998).
2. M. Misra, D. Korakakis, H. M. Ng and T. D. Moustakas, *Appl. Phys. Lett.*, 74, 2203 (1999).
3. T. N. Oder, J. Li, J. Y. Lin, and H. X. Jiang, *Appl. Phys. Lett.*, 77, 791 (2000).
4. Y. Z. Chiou, Y. K. Su, S. J. Chang, J. Gong, Y. C. Lin, S. H. Liu and C. S. Chang, *IEEE J. Quantum Electron.*, 39, 681 (2003).
5. E. Monroy, E. Munoz, F. J. Sáanchez, F. Calley, E. Calleja, B. Beaumont, P. Gibart, J. A. Munoz and F. Cusso, *Semicond. Sci. Technol.*, 13, 1042 (1998).
6. G. Y. Xu, A. Salvador, W. Kim, Z. Fan, C. Lu, H. Tang, H. Morkoc, G. Smith, M. Estes, B. Goldenberg, W. Yang and S. Krishnankutty, *Appl. Phys. Lett.*, 71, 2154 (1997).
7. N. Biyikli, I. Kimukin, O. Aytur and E. Ozbay, *IEEE Photon. Technol. Lett.*, 16, 1718 (2004).
8. S. L. Rumyantsev, N. Pala, M. S. Shur, R. Gaska, M. E. Levinshtein, V. Adivarahan, J. Yang, G. Simin and M. A. Khan, *Appl. Phys. Lett.*, 79, 866 (2001).
9. C. T. Lee, C. C. Lin, H. Y. Lee and P. S. Chen, *J. Appl. Phys.*, 103, 094504 (2008).
10. M. Mosca, J. L. Reverchon, F. Omnes and J. Y. Duboz, *Appl. Phys. Lett.*, 95, 4367 (2004).
11. Y. K. Su, P. C. Chang, C. H. Chen, S. J. Chang, C. L. Yu, C. T. Lee, H. Y. Lee, J. Gong, P. C. Chen and C. H. Wang, *Solid-State Electron.*, 49, 459 (2005).
12. J. Spradlin, S. Dogan, M. Mikkelson, D. Huang, L. He, D. Johnstone and H. Morkoc, *Appl. Phys. Lett.*, 82, 3556 (2003).
13. C. T. Lee, Y. J. Lin and D. S. Liu, *Appl. Phys. Lett.*, 79, 2573 (2001).
14. P. S. Chen and C. T. Lee, *J. Appl. Phys.*, 100, 044510 (2006)
15. P. S. Chen, T. H. Lee, L. W. Lai and C. T. Lee, *J. Appl. Phys.*, 101, 024507 (2007).
16. H. Jiang, N. Nakata, G. Y. Zhao, H. Ishikawa, C. L. Shao, T. Egawa, T. Jimbo and M. Umeno, *Jpn. J. Appl. Phys.*, 40, L505 (2001)

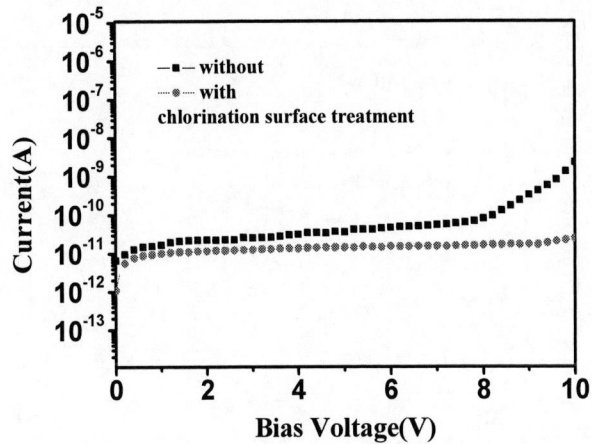

Figure 1. Dark current of GaN/AlGaN MSM photodetectors with and without chlorination surface treatment.

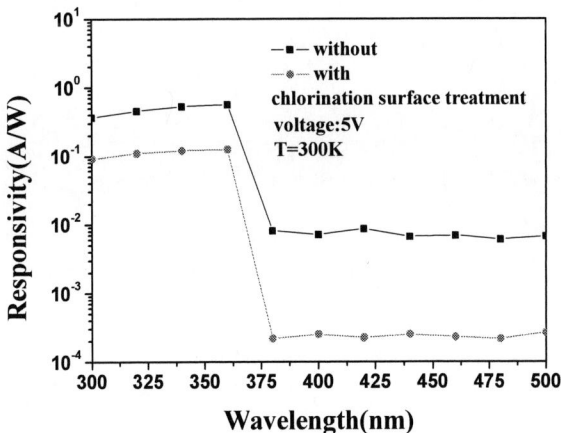

Figure 2. Spectra response of GaN/AlGaN MSM photodetectors with and without chlorination surface treatment.

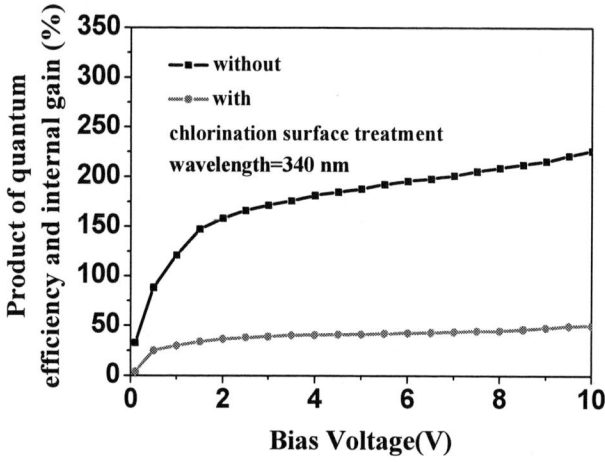

Figure 3. Voltage-dependent product of quantum efficiency and internal gain of GaN/AlGaN MSM photodetectors with and without chlorination surface treatment.

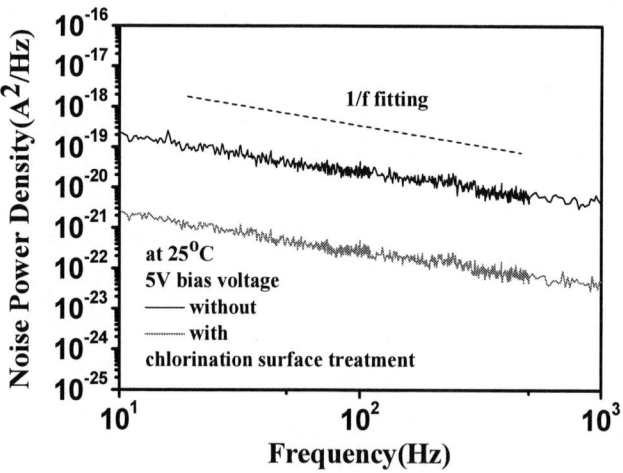

Figure 4. Noise power spectra of GaN/AlGaN MSM photodetectors with and without chlorination surface treatment.

CHAPTER 2

III-V COMPOUND SEMICONDUCTOR DEVICES

46

The Evolution of Wide Bandgap Semiconductors at SOTAPOCS

S.J. Pearton[a] and F. Ren[b]

[a] Department of Materials Science and Engineering, University of Florida, Gainesville FL 32611 USA
[b] Department of Chemical Engineering, University of Florida, Gainesville FL 32611 USA

At the 2009 San Francisco meeting, the State of the Art Program on Compound Semiconductors (SOTAPOCS) will be held for the 50th time. This manuscript provides some personal recollections of the place that wide bandgap materials have played after the early days of the symposium were focused mainly on GaAs and InP materials and devices and how this highlights the corresponding maturation of the compound semiconductor industry.

Introduction

SOTAPOCS was born during the frenzy to develop GaAs digital circuits in the mid 1980's. At the same time, there was a flourishing research effort on InP-based lasers and detectors for lightwave communication systems. Most of the early work focused on epi growth optimization studies, including many on the merits of MBE, MOCVD and later variants such as MOMBE and GSMBE and also process optimization of ion implantation, wet and dry etching, contact formation and surface passivation. Basic studies of many types of heterostructures and heterojunctions were a staple of early SOTAPOCS meetings, laying the groundwork for future advanced device concepts. The focus on basic GaAs MESFET technology eventually turned to HEMTs of all flavors and then HBTs as speed requirements were ratcheted up. This was an exciting period that saw the introduction of carbon as a more stable base dopant in HBTs, high density plasma etching methods, refractory metal contacts and many of the things that are now taken for granted in compound semiconductor device technology. The mid 1980's also saw a period where the growth of GaAs layers on Si substrates brought much debate and controversy on the relative merits of the approach and how one legitimately calculates defect densities. The shift from largely military and communication systems applications to consumer markets, especially cell phones, brought renewed focus on compound semiconductor materials and then the surge in high density data storage and solid state lighting with the breakthroughs in GaN growth and processing. This unique material has also provided new opportunities in UV detectors and high speed and high power electronics. Electronic oxides such as ZnO and related alloys have also become part of SOTAPOCS, with their potential application in multifunctional systems. A surprising recent development has been the achievement of high mobilities in amorphous transparent conducting oxides that can be used for making FETs on arbitrary substrates.

The maturation of compound semiconductor technology and the insertion of these devices in radar, cell phones, wireless communication systems, collision-avoidance (automobile) radar, high density DVDs, and traffic lights and other displays cable TV, fiber-optic nodes, cellular/ PCS / wireless local loop (WLL), wireless LAN, GPS, satellite cellular, electronic toll collection, point-to-point radio, very small aperture terminal (VSAT), satellite TV, broad-band satellite services, local multipoint distribution(LDMS), multipoint video distribution services (MVDS) and automotive radar-smart-cruise control has seen the market for compound satellite systems semiconductor devices rise to a total revenue for all compound semiconductor products in 2006 of ~$17.5B, about 10% that of Si. SOTAPOCS has played a major role in that process.

When the first SOTAPOCS symposium was held, ECS was already a leader in holding symposia related to compound semiconductors, with occasional series on dielectric films on compound semiconductors, analytical techniques for semiconductor materials and process characterization and aspects of crystal growth and plasma processing that contained the occasional presentation on GaAs. However, ECS was not that visible for the rapidly expanding compound semiconductor community and the main conferences were held under the aegis of IEEE (for devices), the conference series on Semi-Insulating III-V Materials that began in 1980 in Nottingham in the UK and the MRS. Most often, SOTAPOCS has been held in tandem with a specialized topic symposium that over the years has included such topics as "Superlattice Structures and Devices"(Montreal 1990), Metallization of III-V Compound Semiconductors" (Seattle 1990), Processes at the Compound Semiconductor Solution Interface(Quebec,2005), Wide Bandgap Semiconductors for Electronics, Photonics and Sensors (Chicago 1995, Washington 2001, Philadelphia 2002, Paris 2003, Orlando 2003, Honolulu 2005, Washington 2007, Cancun 2006, Honolulu 2008), High Speed III-V Electronics for Wireless Applications(San Antonio 1996, Phoenix 2000), Large Area Wafer Growth and Processing for Electronic and Photonic Devices (Washington 1991, San Francisco 1994, Compound Semiconductor Power Transistors (Boston 1998), Narrow Bandgap Optoelectronic Materials and Devices (Salt Lake City 2002), Light-Emitting Devices for Optoelectronic Applications (San Diego 1998) and ZnO, InZnO and InZnGaO Related Materials and Devices for Electronic and Photonic Applications (Phoenix 2008).Of course, this is not an exhaustive list, but serves to highlight that those companion topics deserved special emphasis in the particular time frame in which they were held, while SOTAPOCS contained presentations on a broader range of topics. The first meeting was held as part of the 1984 ECS meeting in Cincinnati, Ohio and the first I personally attended was in Las Vegas, NV in the Fall of 1985.The earliest SOTAPOCS proceedings volume available from ECS is the Twelfth that also contained papers from the symposium on Superlattice Structures and Devices, held in Montreal in 1990.That joint volume was edited by D.C. D'Avanzo from Hewlett-Packard, Ron Enstrom from the David Sarnoff Research Center in Princeton, NJ, Al Macrander, then of Bell Labs in Murray Hill, NJ, and D. DeCoster of the University of Lille in France. This nicely summarizes both the international flavor that has always been a feature of SOTAPOCS and also the disappearance of some of the major research labs that once provided much of the technical content and attendance at the meeting.

The electronic industry, driven by revolution in microelectronics has grown rapidly in past four decades. The first semiconductor transistor was invented by the scientists of Bell Labs in 1947. After that the concept of Integrated Circuit (IC) came up which was difficult to accept as it was reasoned that in order to achieve a working circuit all of the devices must work. Therefore, to have a 50% probability of functionality for a 20 transistor circuit, the probability of device functionality must be $(0.5)^{1/20} = 0.966$ or 96.6%. This was considered ridiculously optimistic at the time, yet today integrated circuits are built with billions of transistor. This is possible because each component or a device is many times reliable compared to a component in any other industry. Even though the very first semiconductor transistor was made from germanium (Ge), silicon (Si) became the semiconductor of choice as a result of the low melting point of Ge that limits high temperature processes and the lack of a natural occurring germanium oxide to prevent the surface from electrical leakage. Due to the maturity of its fabrication technology, silicon continues to dominate the present commercial market in discrete devices and integrated circuits for computing, power switching, data storage and communication. For high-speed and optoelectronic devices such as high-speed integrated circuits and laser diodes, gallium arsenide (GaAs) is the material of choice. It exhibits superior electron transport properties and special optical properties. GaAs has higher carrier mobility and higher effective carrier velocity than Si, which translate to faster devices. GaAs is a direct bandgap semiconductor, whereas Si is indirect, hence making GaAs better suited for optoelectronic devices. However, physical properties required for high power, high temperature electronics and UV/blue light emitter applications are beyond the limits of Si and GaAs. It is essential to investigate alternative materials and their growth and processing techniques in order to achieve these devices. The focus has shifted to semiconductors having wide bandgaps. They exhibit larger bandgaps, higher electron mobility and higher breakdown field strength making them suitable for high power, high temperature electronic devices and short wavelength optoelectronics.

Wide bandgap semiconductors offer the best technical promise for high power and high temperature transistors. Until recently, the most promising of these materials was silicon carbide (SiC). However, SiC have several technical shortfalls that have opened competition to the III-nitride materials. Thermal oxides in SiC power metal oxide semiconductor field effect transistors (MOSFETs) actually limit the temperature range of application since the gate contact degrades and becomes electrically leaky at high temperatures. The low electron mobility of only 400 cm^2/V.s yields lower PAE (<30%) for many transistors in the frequency range of 1 to 5 GHz. For silicon and SiC, amplifier efficiency decays rapidly with increase in frequency so that it drops below 25% for many devices operating above 2 GHz. GaN-base devices offer wider bandgap, greater chemical inertness and higher temperature stable operation than SiC. Table I shows a comparison of some of the properties of the most common semiconductor materials.

Table 1. Electrical properties of Si , GaAs and GaN.

Property		Si	GaAs	GaN
Bandgap energy (eV)		1.1(indirect)	1.4(direct)	3.4(direct)
Electron mobility (cm^2/Vs)		1400	8500	1000(bulk) 2000(2D-gas)
Hole mobility (cm^2/Vs)		600	400	30
Electron effective mass		0.98	0.067	0.19
Hole effective mass(light)		0.16	0.082	0.60

GaN and Related Materials

Interest in short-wavelength optoelectronic devices and high power electronic devices has triggered a myriad of research activities in GaN and related materials. GaN is a wide bandgap semiconductor which has numerous properties which makes it well suited for high temperature applications. It has a direct bandgap energy of 3.45eV (λ=359.37 nm) which is transparent to visible light and operates in ultra violet to blue wavelengths. Hall measurements at room temperatures show the Hall mobility of electron of 1000~1300 cm^2/V-s. It has saturation velocity little higher than GaAs. GaN like ZnO seems to be extremely stable at harsh environment of gamma radiations. It has little change in IV characteristic even after being irradiated by high energy proton radiation. This makes GaN very good candidate for outer space and nuclear application. Sapphire or SiC substrates are generally used for growing GaN. GaN also has different heterostuctures available with Al, In etc. (Al, Ga, In) N forms a continuous and direct band gap alloy from 1.92 eV (InN) to 6.2 eV (AlN) with potential for emission and detection in spectral range between visible and the ultraviolet wavelengths. In recent years, the intrinsic bandgap of InN is thought to be closer to 0.7-0.8 eV in samples with low background doping levels, well-controlled stoichiometry and low concentrations of oxygen.

GaN was a relatively backwater area of research until the early 1990's when Nakamura found an effective method for p-type doping(1), allowing for the realization of LEDs and eventually laser diodes. Papers in these areas started appearing in SOTAPOCS around 1994 and as mentioned earlier, there have been numerous joint nitride-SOTAPOCS symposia held at ECS over the years. Some of the advances covered include the use of MOCVD, use the now common sapphire as the substrate of choice, low temperature AlN and GaN buffers, which are now standard for obtaining good crystal quality of the GaN, high Al-AlGaN alloys for deep UV emitters and use of a double heterostructure approach to increase efficiency in blue LEDs. The high dislocation density had to be reduced for lasers and indirectly led to the use of Epitaxial Lateral Overgrowth and other approaches to reducing defect density, combined with various growth approaches involving multiple buffers and temperature cycling. In this method, a layer of GaN grown by MOCVD is covered with 100-200 nm of amorphous SiO_2 and Si_3N_4 with ex-situ techniques. Small circular or rectangular windows are then etched through to the underlying GaN. A GaN film is subsequently regrown under conditions such that growth occurs epitaxally only in the windows and not on the mask. If growth continues, lateral growth over the mask eventually occurs. Since most of the extended dislocations propagate in the growth direction through GaN, very few threading dislocations (TDs) are visible in the regrown GaN that extends laterally over the mask. Using LEO growth methods, TDs can be reduced to the 5 orders lower number of defects than in a normally grown GaN. A refined approach to a nearly dislocation free GaN substrate for devices can be employed by two successive LEO steps with the mask of the second step positioned over the opening defined by the mask of the first step. In addition to growing GaN films with low defect densities, another key requirement for fabricating devices is the ability to precisely control the desired electrical properties of the thin film. In general, wide bandgap semiconductors are difficult to dope due to native defects. When the enthalpy for defect formation is lower than the band gap energy, the probability of generating a defect increases with the bandgap, i.e. the energy released by donor-to-acceptor transition. Additional topics covered at SOTAPOCS involving GaN have included all the device process development such as Ohmic and Schottky contacts, dry and wet etching, gate dielectrics, surface passivation and a wide variety of devices including UV photodetectors, heterojunction bipolar transistors, high electron mobility transistors and sensors

ZnO and Related Materials

LEDs have numerous applications, including traffic signals, display, back-lighting in the electronic displays, automobile brake lights, indicators on electronic devices, UV Bio-detectors, and general lighting applications (Solid-State Lighting). White-light SSL sources will eventually replace traditional incandescent and fluorescent lamps in many general illumination applications due to its higher power efficiencies compared to traditional incandescent sources and generate substantial energy savings. LEDs also do not generate heat which makes them preferable in medical use for surgical lighting.

ZnO has attracted much attention for optoelectronic devices due to its direct wide bandgap of 3.37 eV at room temperature and high exciton binding energy of 60 meV(2- However, the difficulty of getting robust p-type due to asymmetric doping characteristics has hindered making pn junction device and thus p-type ZnO has been an issue for several years. [3] Recently, there have been some reports about an emission from ZnO based LED. [4-7] However, the bandedge EL emission was not sufficient for practical ZnO LED application and the reason for this problem is the absence of high quality and reliable p-type ZnO. There are several possible dopants such as group-I (Li, Na, and K) and group V (N, P, and As) elements for p-type doping. Until now it is unclear which dopant is the best for p-type even though theoretically nitrogen is the most promising candidate.[8] In this research, p-type ZnO film will be achieved by using phosphorus as a p-type dopant. This study will mainly focus on achieving reliable and robust p-type ZnO:P films and fabrication of p-n homojunction with the p-type ZnO:P layer.

Until now III-V system has dominated these LED applications. GaN is the most widely utilized wide bandgap semiconductor in industry. However, it is brittle and easily contaminated by environmental constituents. ZnO has a number of advantages over GaN, and thus it is considered as an alternative to traditional III-V based materials for next-generation short wavelength optoelectronic devices. ZnO has several important properties that make it a promising semiconductor material for optoelectronic devices and applications. The following features of ZnO make it a very good replacement for GaN

(a) A large exciton binding energy, 60 meV, compared with 26 meV for GaN and 19 meV for ZnSe. This large exciton binding energy provides more efficient exciton emission at room temperature and elevated temperature.

(b) Exceptional resistant to radiation damage by high energy radiation. High-energy radiation in semiconductors creates deep centers within the forbidden energy gap. These affect device sensitivity, response time, and read-out noise. Therefore, radiation hardness is very important as a device parameter for operation in harsh environments. In this point of view, ZnO is more suitable for space operation than other wide bandgap semiconductors.

(c) Large and high quality single-crystal wafers are commercially available. It is possible to grow homo-epitaxial ZnO-based devices that have low dislocation densities. Homo-epitaxial ZnO growth on ZnO substrates will alleviate many problems associated with hetero-epitaxial GaN growth on sapphire, such as stress and thermal expansion problems due to the lattice mismatch.

(d) Wet chemical etching is possible. It provides many advantages in device processing for UV detectors, LEDs, LDs, FETs, and other optoelectronic devices.

(e) A tunable bandgap can be realized by alloying ZnO with CdO, MgO, or BeO.

: The bandgap can be changed from 3 eV to 4.0 eV in $Zn_{1-x}Cd_xO$ and $Mg_xZn_{1-x}O$ alloy films with quite small lattice mismatch. This advantage makes it possible to realize strain-free and high quality multiple quantum well (MQW) device structures (see Figure 1).

(f) A very high breakdown electric field of $\sim 2 \times 10^6$ V/cm, which is about two times higher than the GaAs breakdown field. High operation voltage could be applied to ZnO-based devices for high power and gain.

(g) A large saturation velocity of 3.2×10^7 cm/sec at room temperature. This makes ZnO-based devices better for high frequency applications.

(h) A high melting temperature (~ 2000 ℃). It provides possibilities for high temperature treatments in post-growth processes such as annealing and baking during device fabrication

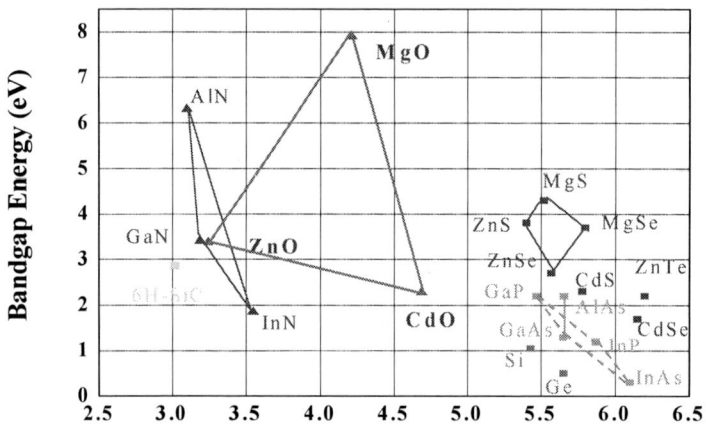

Lattice Constant (Å)

Figure 1. The bandgap versus lattice constant of various semiconductors.

In the case of ZnO, the ability to grow at low temperatures on cheap substrates is attractive for applications where GaN may still be too expensive. In that case, it would also be necessary to still employ a heterostructure of the types shown below, where the narrow gap layer would be ZnCdO. There is still much to be done on these alloys.

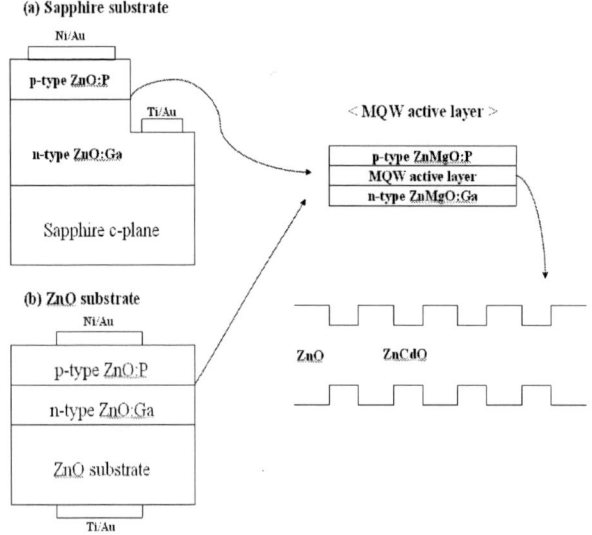

Figure 2. ZnO based pn junction LED on (a) sapphire substrate and (b) ZnO substrate

P-type Doping of ZnO and Related Alloys

One of the biggest recent areas of coverage of ZnO at SOTAPOCS has been the search for improved p-type doping. Known acceptors in ZnO include monovalent ions such as Li, Na, and K, and group-V elements such as N, P, and As. However, some groups theoretically predicted group-V elements usually form deep acceptors rather than shallow levels and thus it is difficult to contribute p-type conduction. In theory, if we consider shallowness of acceptor levels, group I elements could be better p-type dopants than groupV. However, group- I elements tend to occupy the interstitial sites due to their small atomic radii rather than substitutional sites and therefore act mainly as donors instead. Furthermore, significantly larger bond length for Na and K than ideal Zn-O bond length (1.93 Å) induces lattice strain forming native defects and compensates acceptors. Therefore, it has been believed that the most promising dopants for p-type ZnO are the group-V elements, although theory suggests some difficulty in achieving shallow acceptor level. Both P and As also have significantly larger bond lengths and therefore are more likely to form antisites to avoid the lattice strain and these antisite atoms play a role of donor compensating acceptors. Due to these reasons, attaining p-type doping in ZnO is still controversial issue even though several groups have reported the fabrication of ZnO p-n junctions by various means such as pulsed laser deposition, excimer-laser doping, and sputtering.

Wide Bandgap Semiconductor Nanowires

Semiconductor nanowire device structures are expected to have potential advantages in improved carrier confinement over their thin film counterparts. For GaN nanowires, there are possible applications in low power and high density field-effect transistors (FETs), solar cells, terahertz emitters and UV detectors. The high surface-to-volume ratio of nanowires means that if their surfaces are sensitive to external stimuli or can be functionalized to be sensitive to specific chemicals or biogens, then they are likely to be attractive for gas and chemical sensor arrays. ZnO is a piezoelectric, transparent wide bandgap semiconductor used in surface acoustic wave devices. The bandgap can be increased by Mg doping. ZnO has been effectively used as a gas sensor material based on the near-surface modification of charge distribution with certain surface-absorbed species. In addition, it is attractive for biosensors given that Zn and Mg are essential elements for neurotransmitter production and enzyme functioning. InN is currently receiving attention because of its recently reported considerably narrower direct bandgap (0.7 ~ 0.8 eV) and superior electron transport characteristics. This makes it a promising material for high efficiency IR emitters, detectors, and solar cells as well as high frequency electronic devices. Reports on the growth, properties, and applications of InN nanowires, however, are very limited.

To give a single example of the effectiveness of nanowire sensors, we consider single ZnO nanorods coated with Pt clusters by sputtering which are shown to selectively detect hydrogen at room temperature. The single nanorods operate at extremely low power levels of ~15-30 μW, which is approximately a factor of 25 lower than multiple ZnO nanorods operated under the same conditions. The addition of the Pt coatings increased the detection sensitivity of the nanorods for 500 ppm H_2 in N_2 by approximately an order of magnitude, which is about a factor of 2 improvement over the similar case for multiple nanorods. Pt-coated single ZnO nanorods showed relative responses of ~20 % and 50%, respectively, after 10 mins or 20 mins exposure, respectively, to 500 ppm H_2 in N_2. There was no response of either coated or uncoated nanorods to the presence of O_2 in the ambient at room temperature and indeed the I-V characteristics were independent of the measurement ambient for vacuum, air or pure N_2. By sharp contrast, the nanorods were sensitive to the presence of H_2 in the ambient, with the response being time-dependent. The nanorod resistance was still changing at least 15 mins after the introduction of the hydrogen. An Arrhenius plot of the rate of resistance change for the nanorods exposed to the 500 ppm H_2 in N_2 for 10 mins produced an activation energy of ~15KJ/mole. This is larger than that expected for typical diffusion processes and suggests that the rate-limiting step maybe chemisorption of hydrogen on the Pt surface. The reversible chemisorption of reactive gases at the surface of ZnO can produce a large reversible variation in the resistance. In addition, atomic hydrogen introduces a shallow donor state into ZnO and this may play a role in the increased conductance of the nanorods. The diffusion coefficient of the hydrogen is also much faster in ZnO than in any other wide bandgap semiconductor. Note the very low power consumption of the nanorod sensors, which is in the range 15-30 µW. This is approximately a factor of 25 lower than multiple ZnO nanorods operated under the same conditions and more than a factor of 50 lower than carbon nanotubes doped with Pd that were used for hydrogen detection. The low power consumptions is clearly of advantage in many types of remote sensing or long-term sensing applications. Figure 3 shows the as-grown nanowires (left) and completed sensor (right).

Figure 4 shows the time dependence of relative resistance change in both the uncoated and Pt-coated nanorods exposed to 500 ppm H_2 in N_2. The relative resistance responses were ~20 % and 50%, respectively, after 10 mins or 20 mins exposure. By comparison, the uncoated devices showed relative resistance changes of ~ 2% and 3%, respectively, after 10 min or 20 min exposure. The resistance change during the exposure to hydrogen was slower in the first few minutes, as is clear in Figure 3. This may be due to removal by the atomic hydrogen of native oxide on the Pt. As the effective surface area of the Pt would increase as the oxide was removed, the rate of change of resistance due to hydrogen adsorption should also increase.

Figure 3. SEM of ZnO multiple nanorods(top) and photograph of the nanorods contacted by Al/Pt/Au electrodes (bottom).The ZnO chip has edge length ~ 5mm in the bottom photo.

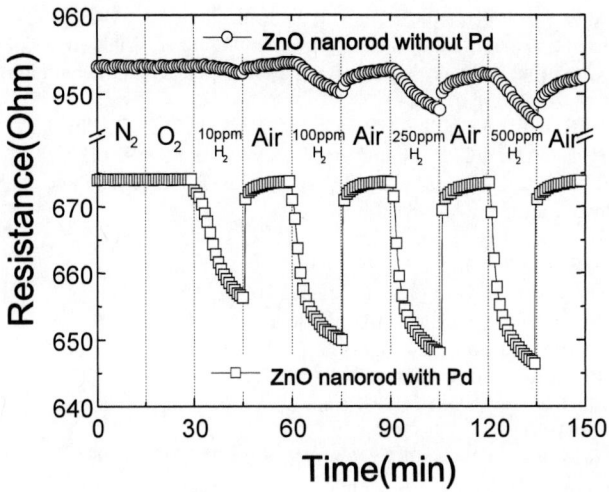

Figure 4.Time dependence of resistance of either Pd-coated or uncoated multiple ZnO nanorods as the gas ambient is switched from N_2 to various concentrations of H_2 in air (10-500 ppm) as time proceeds. There was no response to O_2.

It is likely that studies of nanowires and other nanostructures will play an increasingly important role in future SOTAPOCS symposia as their use in nano-electronics, sensors and photonics continues to increase. A key future need will be analysis methods that can examine very small areas and provide information on local stoichiometry and defect density.

Summary

SOTAPOCS has played a very important role in providing a forum for wide bandgap semiconductor materials and their physics and chemistry and their use in robust electronics and photonics. We look forward to the next 50 SOTAPOCS symposia with great interest.

Acknowledgments

The authors are very grateful to their many collaborators over the years who have participated in the SOTAPOCS series, including George Chu, Bert Schwartz, Swami Swaminathan, Al Macrander, Albert Baca, Randy Shul, Mark Overberg, Jennifer Wang, Pablo Chang, Jung Han, Bill Hobson, Cammy Abernathy, Rose Kopf, Jenshan Lin, David Norton, Kelly Ip, Jeff LaRoche, Hung-Ta Wang, B.S. Kang, L.C. Tien, H.S. Kim and many others. The work is supported by NSF (CTS-0301178 and DMR070416) and ARO (M. Gerthold and J.M. Zavada)

References

1. S. Nakamura, M. Senoh, N. Iwasa, S. Nagahama, Jpn. J. Appl. Phys., 34, L797(1995).
2. Z.K. Tang, G..K.L. Wong, P. Yu, Appl. Phys. Lett. 72, 3270 (1998)
3. S. B. Zhang, S. H. Wei, and A. Zunger, J. Appl. Phys. 83, 3192 (1998).
4. A. Tsukazaki, A. Ohtomo, T. Onuma, M. Ohtani, T. Makino, M. Sumiya, K. Ohtani, S. Chichibu, S. Fuke, Y. Segawa, H. Ohno, H. Koinuma, and M. Kawasaki, Nat. Mater. 4, 42 (2005)
5. W. Liu, S. S. Gu, J. D. e, S. M. Zhu, S. M. Liu, X. Zhou, R. Zhang, Y. Shi, Y. D. Zheng, Y. Hang, and C. L. Zhang, Appl. Phys. Lett. 88, 092101 (2006)
6. Y. Ryu, T. Lee, J. A. Lubguban, H. W. White, B. Kim, Y. Park, and C. Youn, Appl. Phys. Lett. 88, 241108 (2006)
7. J.-H.Lim, C.-K.Kang, K.-K. Kim,I.-K. Park, D.-K .Hwang and S.-J.Park, Adv. Mater 18, 2720 (2006)
8. T. Yamamoto and H. K. Yoshida, Physica B 302, 155 (2001)
9. E. M. Wong and P. C. Searson, Appl. Phys. Lett. 74, 2939 (1999)
10. A. Cimino, G. Mazzone, and P. Porta, J. Phys. Chem. 41, 154 (1964)
11. S. C. Abrahams and J. L. Beinstein, Acta Cryst. 25, 1233 (1969)
12. E. Kaldis, *Current Topic in Materials Science*, (North-Holland Publishing Co. New York, 1981)
13. D. C. Look and B. Clafin, Phys. Status Solidi B 241, 624 (2004)
14. S. B. Zhang, S.-H Wei, and A. Zunger, Phys. Rev. B 63, 075205 (2001)
15. C. H. Park, S. B. Zhang, and S. H. Wei, Phys. Rev. B 66, 073202 (2002)
16. D. C. Look, R. L. Jones, J. R. Sizelove, N. Y. Garces, N. C. Giles, and L. E. Halliburton, Phys. Status Solidi A 195, 171 (2004).
17. T.J. Anderson, H.T. Wang, B.S. Kang, F. Ren, S.J. Pearton, A. Osinsky, Amir Dabiran, P.P. Chow, Appl. Surf. Sci., 255, 2524, (2008).
18. W.Lim, J.S. Wright, B.P. Gila, S.J. Pearton, F.Ren, W. Lai, L.C. Chen, M.Hu and K.H. Chen, Appl. Phys. Lett. 93, 202109 (2008).
19. W.Lim, J.S.Wright, B.P.Gila, J.L.Johnson, A.Ural, T. Anderson, F.Ren and S.J. Pearton, Appl. Phys.Lett.93, 072110 (2008).
20. B. S. Kang, H. T. Wang, F. Ren, S. J. Pearton, T. E. Morey, D. M. Dennis, J. W. Johnson, P. Rajagopal, J. C. Roberts, E. L. Piner, and K. J. Linthicum, Appl. Phys. Lett. 91, 252103 (2007).
21. S.J. Pearton, B.S.Kang, B.P.Gila, D.P. Norton, O.Kryliouk, F.Ren, Y.W.Heo, C.Y.Chang, G.C.Chi, W.M.Wang and L.C.Chen, J.Nanosci.Nanotechnol.8, 99 (2008).
22. S.J.Pearton, B.Kang, L.Tien, D.P.Norton, Y.W.Heo and F.Ren, Nano 2, 201 (2007).
23. R A. Rosenberg, G. K. Shenoy, M. F. Chisholm, L.-C. Tien, D. Norton and S.J. Pearton, Nano Lett.7 1521(2007).
24. S. J. Pearton, D.P. Norton and F Ren, Small 3, 1144 (2007).
25. L.C. Tien, D.P. Norton, S.J. Pearton, Hung-Ta Wang and F. Ren, Appl. Surf. Sci._253, 4620 (2007).
26. L.C. Tien, D.P. Norton, B.P. Gila, S.J. Pearton, Hung-Ta Wang, B.S. Kang and F. Ren, Appl. Surf. Sci._253, 4748 (2007).

ECS Transactions, 19 (3) 57-63 (2009)
10.1149/1.3120686 ©The Electrochemical Society

c-erbB-2 Sensing Using AlGaN/GaN High Electron Mobility Transistors For Breast Cancer Detection

K.H. Chen, B. S. Kang, H. T. Wang, T. P. Lele, and F. Ren
Department of Chemical Engineering, University of Florida, Gainesville, FL 32611,
Y. L. Wang, C.Y. Chang, and S. J. Pearton
Department of Materials Science and Engineering, University of Florida, Gainesville, FL 32611
D. M. Dennis
Department of Anesthesiology, University of Florida, Gainesville, FL 32611
J.W. Johnson, P. Rajagopal, J.C. Roberts, E.L. Piner, and K.J. Linthicum
Nitronex Corporation, Raleigh, NC 27606

Antibody-functionalized, Au-gated AlGaN/GaN high electron mobility transistors (HEMTs) were used to detect c-erbB-2, a breast cancer marker. The antibody was anchored to the gate area through immobilized thioglycolic acid. The AlGaN/GaN HEMT drain-source current showed a rapid response of less than 5 seconds when target c-erbB-2 antigen in a buffer at clinically relevant concentrations was added to the antibody-immobilized surface. We could detect a range of concentrations from to 16.7 µg/ml to 0.25 µg/ml. These results clearly demonstrate the promise of portable electronic biological sensors based on AlGaN/GaN HEMTs for breast cancer screening.

Introduction

Currently, the overwhelming majority of patients are screened for breast cancer by mammography. This procedure involves a high cost to the patient and is invasive (radiation) which limits the frequency of screening. Work by Michaelson et al.[1] indicates a 96% survival rate if patients could be screened every three months. Thus, mortality in breast cancer patients could be reduced by increasing the frequency of screening. However this is not feasible presently due to the lack of cheap and reliable technologies that can screen breast cancer non-invasively.

There is recent evidence to suggest that salivary testing for makers of breast cancer may be used in conjunction with mammography[2-11]. Saliva-based diagnostics for the protein c-erbB-2, have tremendous prognostic potential[9,12]. Soluble fragments of the c-erbB-2 oncoprotein and the cancer antigen 15-3 were found to be significantly higher in the saliva of women who had breast cancer than in those patients with benign tumors[10]. Other studies have shown that epidermal growth factor (EGF) is a promising marker in saliva for breast cancer detection[12,13]. These initial studies indicate that the saliva test is both sensitive and reliable and can be potentially useful in initial detection and follow-up screening for breast cancer. However, to fully realize the potential of salivary biomarkers, technologies are needed that will enable facile, sensitive, specific detection of breast cancer.

AlGaN/GaN high electron mobility transistors (HEMTs) have shown promise for bio-sensing applications[14-17], since they include a high electron sheet carrier concentration channel induced by piezoelectric polarization of the strained AlGaN layer[14-25]. There are positive counter charges at the HEMT surface layer induced by the electrons located at the AlGaN/GaN interface. Any slight changes in the ambient can affect the surface charge of the HEMT, thus changing the electron concentration in the channel at AlGaN/GaN interface.

In this letter, we report the use of antibody-functionalized Au-gated AlGaN/GaN high electron mobility transistors (HEMTs) for detecting c-erbB-2 antigen. The c-erbB-2 antigen was specifically recognized through c-erbB antibody, anchored to the gate area. We investigated a range of clinically relevant concentrations from 16.7 μg/ml to 0.25 μg/ml.

Experimental Technique

The HEMT structures consisted of a 3 μm thick undoped GaN buffer, 30Å thick $Al_{0.3}Ga_{0.7}N$ spacer, 220Å thick Si-doped $Al_{0.3}Ga_{0.7}N$ cap layer. The epi-layers were grown by metal organic chemical vapor deposition (MOCVD) on thick GaN buffers produced on Si substrates Mesa isolation was performed with an Inductively Coupled Plasma (ICP) etching with Cl_2/Ar based discharges at –90 V dc self-bias, ICP power of 300 W at 2 MHz and a process pressure of 5 mTorr. 10×50 μm^2 Ohmic contacts separated with gaps of 5 μm consisted of e-beam deposited Ti/Al/Pt/Au patterned by lift-off and annealed at 850 °C, 45 sec under flowing N_2. 400-nm-thick 4% Polymethyl Methacrylate (PMMA) was used to encapsulate the source/drain regions, with only the gate region open to allow the liquid solutions to cross the surface. The source-drain current-voltage characteristics were measured at 25°C using an Agilent 4156C parameter analyzer with the gate region exposed.

A plan view photomicrograph of a completed device is shown in Figure 1 (top). The Au surface was functionalized with a specific bi-functional molecule, thioglycolic acid. We anchored a self-assembled monolayer of thioglycolic acid, $HSCH_2COOH$, an organic compound and containing both a thiol (mercaptan) and a carboxylic acid functional group, on the Au surface in the gate area through strong interaction between gold and the thiol-group of the thioglycolic acid. The devices were first placed in the ozone/UV chamber and then submerged in 1 mM aqueous solution of thioglycolic acid at room temperature. This resulted in binding of the thioglycolic acid to the Au surface in the gate area with the COOH groups available for further chemical linking of other functional groups. X-Ray Photoelectron Spectroscopy and electrical measurements confirming a high surface coverage and Au-S bonding formation on the GaN surface have been previously published[25]. The device was incubated in a phosphate buffered saline (PBS) solution of 500 μg/ml c-erbB-2 monoclonal antibody for 18 hours before real time measurement of c-erbB-2 antigen. Figure 1(bottom) shows a schematic device cross sectional view with thioglycolic acid followed by c-erbB-2 antibody coating.

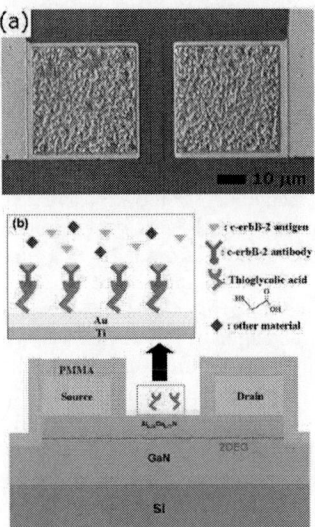

Figure 1. (top) Plan view photomicrograph of a completed device with a 5-nm Au film in the gate region. (bottom) Schematic of AlGaN/GaN HEMT. The Au-coated gate area was functionalized with c-erbB-2 antibody/antigen on thioglycolic acid.

After incubation with a PBS buffered solution containing c-erbB-2 antibody at a concentration of 1 µg/ml, the device surface was thoroughly rinsed off with deionized water and dried by a nitrogen blower. The source and drain current from the HEMT were measured before and after the sensor was exposed to 0.25 µg/ml of c-erbB-2 antigen at a constant drain bias voltage of 500 mV, as shown in Figure 2. Any slight changes in the ambient of the HEMT affect the surface charges on the AlGaN/GaN. These changes in the surface charge are transduced into a change in the concentration of the 2DEG in the AlGaN/GaN HEMTs, leading to the slight decrease in the conductance for the device after exposure to c-erbB-2 antigen.

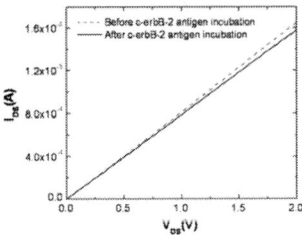

Figure 2. I-V characteristics of AlGaN/GaN HEMT sensor before and after exposure to 0.25 µg/ml c-erbB-2 antigen.

Results and Discussion

Figure 3 shows real time c-erbB-2 antigen detection in PBS buffer solution using the source and drain current change with constant bias of 500 mV. No current change can be seen with the addition of buffer solution around 50 seconds, showing the specificity and stability of the device. In clear contrast, the current change showed a rapid response in less than 5 seconds when target 0.25 μg/ml c-erbB-2 antigen was added to the surface. The abrupt current change due to the exposure of c-erbB-2 antigen in a buffer solution was stabilized after the c-erbB-2 antigen thoroughly diffused into the buffer solution. Three different concentrations (from 0.25 μg/ml to 16.7 μg/ml) of the exposed target c-erbB-2 antigen in a buffer solution were detected. The experiment at each concentration was repeated five times to calculate the standard deviation of source-drain current response. The limit of detection of this device was 0.25 μg/ml c-erbB-2 antigen in PBS buffer solution. The source-drain current change was nonlinearly proportional to c-erbB-2 antigen concentration, as shown in Figure 4. Between each test, the device was rinsed with a wash buffer of 10 μM, pH 6.0 phosphate buffer solution containing 10 μM KCl to strip the antibody from the antigen.

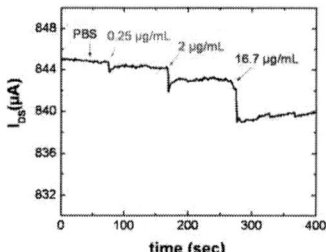

Figure 3. Drain current of an AlGaN/GaN HEMT over time for c-erbB-2 antigen from 0.25 μg/ml to 17 μg/ml.

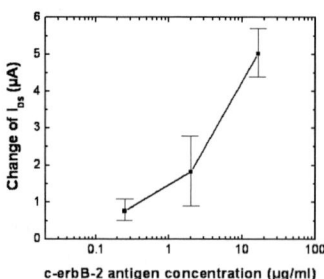

Figure 4. Change of drain current versus different concentrations from 0.25 μg/ml to 17 μg/ml of c-erbB-2 antigen

Clinically relevant concentrations of the c-erbB-2 antigen in the saliva and serum of normal patients are 4-6 µg/ml and 60-90 µg/ml respectively. For breast cancer patients, the c-erbB-2 antigen concentrations in the saliva and serum are 9-13 µg/ml and 140-210 µg/ml, respectively[9]. Our detection limit suggests that HEMTs can be easily used for detection of clinically relevant concentrations of biomarkers. Similar methods can be used for detecting other important disease biomarkers and a compact disease diagnosis array can be realized for multiplex disease analysis.

Conclusions

In summary, we have shown that through a chemical modification method, the Au-gated region of an AlGaN/GaN HEMT structure can be functionalized for the detection of c-erbB-2 with a limit of detection of ~ 0.25 µg/ml. This electronic detection of biomolecules is a significant step towards a compact sensor chip, which can be integrated with a commercial available hand-held wireless transmitter to realize a portable, fast response and high sensitivity breast cancer detector.

Acknowledgments

The work at UF was partially supported by ONR Grant N000140710982 monitored by Igor Vodyanoy, NASA Kennedy Space Center Grant NAG 3-2930 monitored by T. Smith, by NSF DMR 0400416 and the State of Florida, Center of Excellence in Nano-Bio Sensors.

REFERENCES

1. J. S. Michaelson, E. Halpern and D.B. Kopans, Radiology 212(2). 551 (1999).
2. T.Harrison, L.Bigler, M.Tucci, L.Pratt, F. Malamud, J.T. Thigpen, C. Streckfus and H. Younger, Spec. Care Dentist 18(3), 109 (1998) .
3. R. McIntyre, L. Bigler, T. Dellinger, M. Pfeifer, T. Mannery and C. Streckfus, Oral Surg Oral Med. Oral Pathol. Oral Radiol. Endod. 88(6), 687 (1999).
4. C. Streckfus, L. Bigelr, T. Dellinger, M. Pfeifer, A. Rose and J.T. Thigpen, Clin. Oral Investig. 3(3), 138 (1999)
5. C. Streckfus, L. Bigler, T. Dellinger, X.Dai, A. Kingman and J.T. Thigpen, Clin Cancer Res. 6(6), 2363 (2000).
6. C. Streckfus, L. Bigler, M. Tucci and J.T. Thigpen, Cancer Invest 18(2), 101 (2000).
7. C. Streckfus, L. Bigler, T. Dellinger, X. Dai, W.J. Cox, A. McArthur, A. Kingman and J.T. Thigpen, Oral Surg Oral Med Oral Pathol. Oral Radiol. Endod. 91(2), 174 (2001).
8. L. R. Bigler, C.F. Streckfus, L. Copeland, R. Burns, X. Dai, M. Kuhn, P. Martin and S. Bigler, J Oral Pathol. Med 31(7), 421 (2002).
9. C. Streckfus and L. Bigler, Adv. Dent. Res 18(1), 17 (2005).
10. C. F. Streckfus, L. R. Bigler and M. Zwick, J Oral Pathol. Med 35(5), 292(2006).
11. W.R. Chase, J Mich. Dent. Assoc. 82(2), 12(2000).
12. S.Z. Paige, C.F. Streckfus, Gen Dent 55(2), 156(2007).
13. M.A. Navarro, R. Mesia, O. Diez-Giber, A. Rueda, B. Ojeda and M.C. Alonso, Breast Cancer Res. Trea. 42(1), 83(1997).
14. A. P. Zhang, L. B. Rowland, E. B. Kaminsky, V. Tilak, J. C. Grande, J. Teetsov, A. Vertiatchikh and L. F. Eastman, J.Electron.Mater.32 388(2003).
15. O. Ambacher, M. Eickhoff, G. Steinhoff, M. Hermann, L. Gorgens, V. Werss, B. Baur, M. Stutzmann, R. Neuterger, J. Schalwig, G. Muller, V. Tilak, B. Green, B. Schafft, L. F. Eastman, F. Bernadini, and V. Fiorienbini, Proc. ECS 02-14, 27(2002).
16. R. Neuberger, G. Muller, O. Ambacher and M. Stutzmann, Phys. Stat. Soli. A185, 85(2001).
17. J. Schalwig, G. Muller, O. Ambacher and M. Stutzmann, Phys. Stat. Solidi A185, 39(2001).
18. G. Steinhoff, M. Hermann, W. J. Schaff, L. F. Eastman, M. Stutzmann and M. Eickhoff, Appl. Phys. Lett., 83, 177(2003).
19. M. Eickhoff, R. Neuberger, G. Steinhoff, O. Ambacher, G. Muller, and M. Stutzmann, Phys. Stat. Solidi B 228, 519(2001).
20. B. S. Kang, S. Kim, F. Ren, J. W. Johnson, R. Therrien, P. Rajagopal, J. Roberts, E. Piner, K. J. Linthicum, S. N .G. Chu, K. Baik, B. P. Gila, C. R. Abernathy and S. J. Pearton, Appl. Phys.Lett. 85, 2962(2004)
21. S. J. Pearton, B. S. Kang, S. Kim, F. Ren, B. P.Gila, C. R. Abernathy, J. Lin and S. N. G. Chu, J. Phys: Condensed Matter 16 R961(2004).
22. G. Steinoff, O. Purrucker, M. Tanaka, M. Stutzmann and M. Eickoff, Adv. Funct. Mater. 13,841(2003).
23. G. Steinhoff, B. Baur, G. Wrobel, S. Ingebrandt, A. Offenhauser, A. Dadgar, A. Krost, M.Stutzmann and M.Eickhoff, Appl. Phys. Lett.86, 033901(2005).
24. B. S. Kang, F. Ren, L. Wang, C. Lofton, Weihong Tan, S. J. Pearton, A. Dabiran, A. Osinsky, and P. P. Chow, Appl. Phys. Lett. 87, 023508(2005).

25. B. S. Kang, J.J. Chen, F. Ren, S. J. Pearton, J.W. Johnson, P. Rajagopal, J.C. Roberts, E.L. Piner, and K.J. Linthicum, Appl. Phys. Lett. 89, 122102 (2006).

Effect of the Flat and Pattern Surface Texturing on Light Extraction of GaN Flip-Chip Light-Emitting Diodes

R. H. Horng[a], Z. W. Liao[a], Y. L. Tsai[a], H. L. Hu[a], Y. J. Tsai[b], C. P. Hsu[b], M. T. Chu[a]

[a] Institute of Precision Engineering, National Chung Hsing University, Taichung 402, Taiwan, R.O.C.
[b] Electronics and Opto-Electronics Research Laboratories, Industrial Technology Research Institute, HsinChu 310, Taiwan, R.O.C.

Device performances of GaN-based flip-chip light-emitting diodes (FC LEDs) with planar and patterned sapphire substrates (PSS) were compared in this study. It was found that for the FC LED with planar sapphire, enhancement factor of luminous intensity can be raised to 107.5% after the processes of substrate removal and surface roughening. By contrast, for the FC LED with PSS, the intensity enhancement factor is already up to 169.5% without any post-processes as compared with the intensity of an as-fabricated conventional FC LED. Further intensity improvement to 205.1% can be achieved for the FC LED with PSS by employing subsequent processes such as substrate removal and surface roughening. These results indicate that the PSS approach is useful in improving light extraction of a nitride-based FC LED.

Introduction

GaN-based semiconductors have attracted much attention due to their variety applications such as flat panel displays and traffic light, and also the potential application in solid-state lighting. In order to meet the requirements for these applications, high extraction efficiency and excellent heat dissipation are very important for the GaN-based LEDs. In general, extraction efficiency is limited by total internal reflection originating from large refractive index contrast between semiconductor and air. In order to extract the light from active region, interfacial morphology is usually roughening-treated to minimize the total internal reflection loss. A simple and efficient approach for improving light extraction is by growing a nitride-based LED on a patterned sapphire (1-2). However, thermally induced performance deterioration at high current injection still exists in these LEDs due to poor thermal conductivity of sapphire. An approach that can simultaneously satisfy the requirements for light extraction and thermal management is the flip-chip technology (3-5). In FC LEDs, the elimination of wire bonds and the redistribution of the n-contact metallization result in significant improvement in light output power. Another advantage of the FC LEDs over the conventional LEDs is the lower refractive index contrast between sapphire and GaN or air. Recently, Shchekin et al. reported improved performance using a modified FC LED, called thin-film FC LED, where sapphire substrate was removed using an excimer laser and the exposed GaN was then photoelectrochemically roughened (6). To our knowledge, most of the currently proposed GaN-based FC LEDs were constructed on conventional planar sapphire. In this

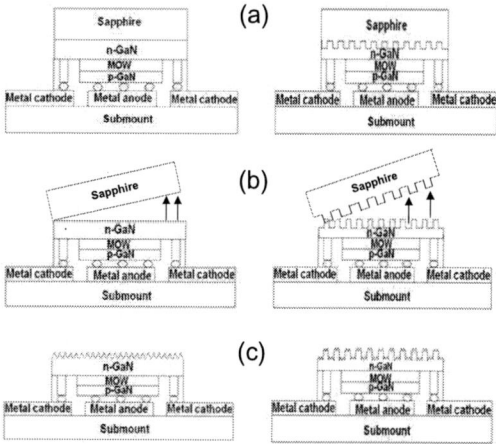

Figure 1. Schematic structural comparison of the C- and PSS-FC LEDs where 1(a) ~ 1(c) represent the processes of standard flip-chip, substrate removal, and surface roughening, respectively.

work, we report device performance of advanced GaN-based FC LEDs using the thin-film technique combined with the PSS approach.

Experiment

Growth of GaN-based LEDs was carried out on planar and patterned (0001) sapphire by metalorganic chemical vapor deposition. The LED structure consisted of a 20 nm-thick low temperature GaN buffer layer, a 1 μm-thick undoped GaN layer, a 2 μm-thick Si-doped GaN layer, an InGaN/GaN multiple-quantum-well stack (emission wavelength at 470 nm), and a 0.3 μm-thick Mg-doped GaN layer. The patterned sapphire used was formed with a protruded rod-like array by conventional photolithography combined with induced-coupled plasma-reactive-ion-etcher (ICP-RIE) dry etching process. The dimension of a protruded rod in the pattern is 3 μm in diameter. After the epitaxial growth, the wafers were then processed into flip-chip LEDs by conventional procedures. For simplicity, the FC LED grown on conventional planar sapphire is denoted as "C-FC LED", and that grown on PSS is called "PSS-FC LED". In order to enhance light extraction, subsequent processes such as substrate removal and surface roughening were performed on the C- and PSS-FC LEDs. Figs. 1 (a) ~ 1(c) compare schematic device structures of the C- and PSS-FC LEDs after the processes of standard flip-chip, substrate removal, and surface roughening, respectively. The removal of sapphire was carried out by laser lift-off (LLO) technique, and energy density of laser used here was in the range of 710 to 750 mJ/cm². The remaining gallium metal on the exposed surface was then removed with HCl etching solution. In the surface roughening process, the samples were dipped in NaOH solution (4M) at 80°C for 6 min.

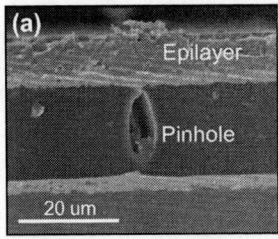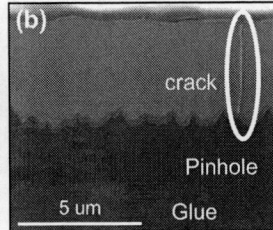

Figure 2. Cross-sectional SEM images showing the formation of (a) pinholes form underfill process and (b) pinhole-induced epilayer cracking.

It should be noted that pinholes which typically formed in the underfill process of a standard flip-chip procedure should be avoided. As shown in Fig. 2, the existence of the pinholes induces epilayer cracking during the subsequent laser lift-of process because the epilayer is free standing on these pinholes. These cracks typically provide a leakage path of injected current.

Results and discussion

As mentioned previously, light extraction efficiency of a FC LED strongly depends on the interfacial morphology between semiconductor and air, and thus it is interesting to compare the surface morphology of the FC LEDs grown on the planar and patterned sapphire. Typical surface images of a GaN-based C-FC LED after the processes of LLO and surface roughening are shown in Figs. 3 (a) and 3(b), respectively. Obviously, the C-FC LED after the LLO process presents relatively mirror-like surface. With the subsequent chemical etching by NaOH solution, the flat surface becomes roughened. As for the PSS-FC LED, the surface morphology of the device is quite different form the C-FC LED. As shown in Fig. 3 (c), the PSS-FC LED just after the process of substrate removal shows periodically roughened surface, and it is thus expected that the PSS-FC LED extracts light more efficient than the C-FC LED does. Further improvement in light extraction can be achieved for the PSS-FC LED when dipping its exposed GaN surface in an etching solution of NaOH, as shown in Fig. 3(d). One can see that surface roughening occurs not only on the top mesa area but also on the hole array. As will be shown later, such a roughened surface in Fig. 3 (d) exhibits the best light extraction efficiency.

Fig. 4 compares the on-axis luminous intensity of the C- and PSS-FC LEDs as a function of injected current. For the C-FC LED shown in Fig. 4 (a), the intensity enhancement factor can be raised to 107.5% after a series of processes of substrate removal and surface roughening. However, one can notice a slight decrement in luminous intensity after the LLO process. This deterioration may be originated from LLO-induced process damage which is considered as to be inevitable in practical fabrication (7). As for the PSS-FC LED, light output performance is remarkably superior to the C-FC LED. As shown in Fig. 4 (b), the intensity enhancement factor at injection current of 350 mA is already up to 169.5% for the *as-fabricated* PSS-FC LED without any post-process. With the subsequent process of surface roughening, the intensity of the PSS-FC LED can be further increased to 205.1% at the same driven current. The LLO process here seems to have negligible intensity deterioration. All the details about the comparison of intensity enhancement factor between the PSS-FC LED and the C-FC LED are shown in table I.

Since all of the FC LEDs were grown with the same epi-structures and processed into equal-dimensional chips using the same procedures, the superior intensity enhancement of the PSS-FC LED to the C-FC LED can be primarily attributed to the roughening-enhanced PSS approach. Physical origin for the intensity improvement is related to scattering of light. For light incident on an irregular interface, a roughened surface enables the emitted photons form active region to randomize effectively, and thus enhances the light extraction.

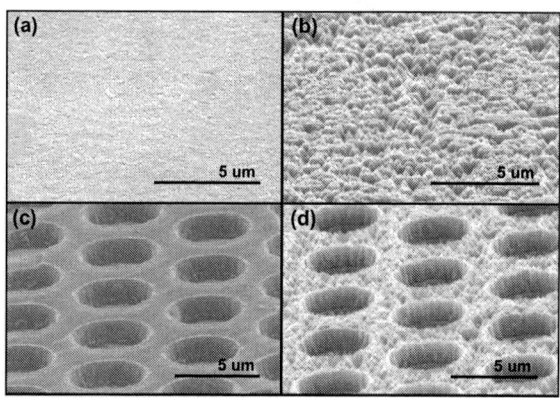

Figure 3. Surface morphology of a LED after the processes of substrate removal and surface roughening by NaOH solution for (a)-(b) the C-FC LED and (c)-(d) the PSS-FC LED.

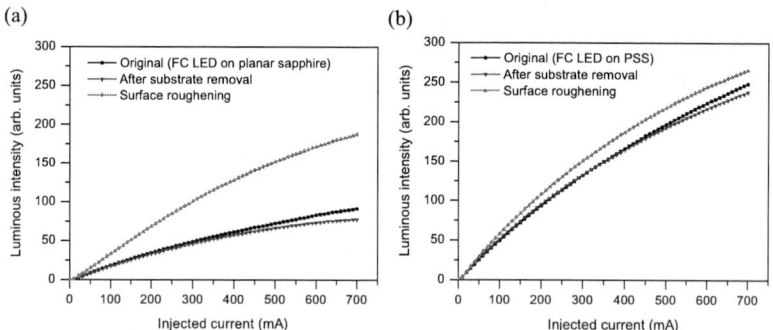

Figure 4. Luminous intensity of (a) the C-FC LED and (b) the PSS-FC LED as a function of injected current.

TABLE I. Enhancement factor (%) of luminous intensity for the FC LEDs grown on the planar sapphire and the PSS at injection current of 350 mA. The enhancement factor is calculated with respect to the reference luminous intensity of the as-fabricated FC LED grown on the planar sapphire.

	on Planar Sapphire	on PSS
As-fabricated	0%	+ 169.5%
After Substrate Removal	- 6.1%	+ 169.4%
Surface Roughening	+ 107.5%	+ 205.1%

Conclusions

Device performance of the advanced PSS-FC LED was compared with that of the conventional FC LED. It was demonstrated that the intensity enhancement factor at injection current of 350 mA for the *as-fabricated* PSS-FC LED is already up to 169.5% without any post-process, which is better than the 107.5% of the C-FC LED with the employment of subsequent surface roughening. The intensity of the PSS-FC LED can be further increased to 205.1% at the same driven current as processed with subsequent surface roughening. These results indicate that the PSS approach is useful in improving light extraction of a nitride-based FC LED.

Acknowledgments

This work was financially supported by the National Science Council and ministry of education of Taiwan under contract No. 97-ET-7-005-003-ET and ATU plane, respectively.

References

1. R C. H. Chiu, H. H. Yen, C. L. Chao, Z. Y. Li, P. Yu, H. C. Kuo, T. C. Lu, S. C. Wang, K. M. Lau, and S. J. Cheng, *Appl. Phys. Lett.*, **93**, 081108 (2008).
2. W. K. Wang, D. S. Wuu, W. C. Shih, J. S. Fang, C. E. Lee, W. Y. Lin, P. Han, R. H. Horng, T. C. Hsu, T. C. Huo, M. J. Jou, A. Lin, and Y. H. Yu, *Jpn. J. Appl. Phys.*, **44**, 2512 (2005).
3. J. J. Wierer, D. A. Steigerwald, M. R. Krames, J. J. O'Shea, M. J. Ludowise, G. Christenson, Y. C. Shen, C. Lowery, P. S. Martin, S. Subramanya, W. Goetz, N. F. Gardner, R. S. Kern, and S. A. Stockman, *Appl. Phys. Lett.*, **78**, 3379 (1993).
4. M. Koike, N. Shibata, H. Kato, and Y. Takahashi, *IEEE J. Sel. Top. Quantum Electron.*, **8**, 271 (2002).
5. J. O. Song, W. K. Hong, Y. Park, J. S. Kwak, and T. Y. Seong, *Appl. Phys. Lett.*, **86**, 133503 (2005).
6. O. B. Shchekin, J. E. Epler, T. A. Trottier, T. Margalith, D. A. Steigerwald, M. O. Holcomb, P. S. Martin, and M. R. Krames, *Appl. Phys. Lett.*, **89**, 071109 (2006).
7. S. J. Wang, K. M. Uang, S. L. Chen, Y. C. Yang, S. C. Chang, T. M. Chen, and C. H. Chen, *Appl. Phys. Lett.*, **87**, 011111 (2005).

70

CHAPTER 3

COMPOUND SEMICONDUCTOR MATERIALS DEVICES AND APPLICATIONS

72

Electrical performance of gate-recessed AlGaN/GaN MOS-HEMTs fabricated using photoelectrochemical method

Ya-Lan Chiou[a], Nan-Teng Shiau[a], Li-Hsien Huang[a], Li-Ren Lou[a], and Ching-Ting Lee[a]

[a] Institute of Microelectronics, Department of Electrical Engineering, National Cheng Kung University, Tainan, Taiwan, Republic of China

Gate-recessed AlGaN/GaN MOS-HEMTs were fabricated using a photoelectrochemical (PEC) method to form the recessed structure and to directly grow gate insulator on the recessed surface. The resulting devices exhibited better performances than conventional one without gate recess, including a saturation drain-source current of 642 mA/mm, and an off-state breakdown voltage larger than −100V. The normalized noise power spectra of both kinds of devices at the saturation region were well-fitted by the 1/f law and the Hooge's coefficient α in both devices was about 10^{-4}, demonstrating that PEC wet etching is an effective way to achieve a damage-free surface and to improve the DC characteristics of the MOS-HEMTs.

Introduction

GaN-based semiconductors are promising candidates for high power and high frequency applications because of their unique material properies, such as high-electron saturation velocity, wide energy bandgap, and better thermal and chemical stability. Furthermore, AlGaN/GaN metal-oxide-semiconductor high-electron mobility transistors (MOS-HEMTs) have a number of advantages, including small gate leakage current, high breakdown voltage and high drain-source current. In recent years, various gate insulators for MOS-HEMTs have been reported [1-5], and the AlGaN/GaN MOS-HEMTs with an oxidized layer grown directly by a photoelectrochemical (PEC) oxidation method have been successfully fabricated and demonstrated [6-8]. In addition, because the off-state breakdown voltage is a significant factor for high power applications, several methods, such as field plate [9], multiple field plate [10] and gate recess process [11], have been developed and reported. The dry etching methods of reactive ion etching (RIE) system or inductively coupled plasma (ICP) system have general been used for the gate recess process [12,13], because the GaN-based materials inherently possessed a better chemical stability and difficulty of wet etching property. However, the plasma etching process can lead to defects in the dry etched region, resulting in poorer performance.

In this study, the dry etching method was thus replaced by the PEC wet etching one which can etch GaN-based semiconductors successfully. The gate-recessed structure was developed with this method, and subsequently the gate insulators were grown directly on the region between the source and drain electrodes by the PEC oxidation method. The resulting oxide films were used as the gate oxide layer of the MOS-HEMTs. The PEC method was used to perform gate recess and grew oxide layer of AlGaN/GaN MOS-HEMTs. The DC characteristics and low frequency noise performance of the conventional and gate-recessed AlGaN/GaN MOS-HEMTs were measured and compared.

Experiments

The epitaxial structures of the AlGaN/GaN MOS-HEMTs were grown on sapphire substrates using a molecular-beam epitaxy (MBE) system, which consisted of a 20 nm-thick AlN nucleation layer, a 1.7 μm-thick carbon-doped GaN buffer layer, a 500 nm-thick undoped GaN layer, and a 50 nm-thick $Al_{0.2}Ga_{0.8}N$ layer. The structure has a sheet electron density of 8.89×10^{12} cm^{-2} and mobility of 1460 cm^2/Vs measured by Hall measurement.

The schematic configuration of the gate-recessed MOS-HEMTs is shown in Figure 1. A reactive ion etching (RIE) system was used to perform the mesa isolation of devices. The ohmic contact was formed by a rapid thermal annealing (RTA) system at 850°C in N$_2$ ambient for 2 mins after Ti/Al/Pt/Au (25/100/50/200 nm) ohmic metals were deposited. The gate-recessed structure was fabricated by a PEC wet etching method, in which a He–Cd laser with a wavelength of 325 nm and a H$_3$PO$_4$ chemical solution (pH value = 0.7) were used. The depth of the gate recess process was about 8 nm. The PEC oxidation process was then performed, using a H$_3$PO$_4$ solution (pH value = 3.5) to directly oxidize the recessed surface of the AlGaN layer to form gate insulator layer and to passivate its surface. The grown oxide films were then annealed at 700°C in O$_2$ ambient for 2 hours, and the thickness of the annealed oxide film was about 30 nm. After the annealing process, two-finger Ni/Au (20/100 nm) metals with 1-μm-length and 50-μm-width were deposited using an electron-beam evaporator.

Experimental Results and Discussion

Figure 2 shows the drain-source current (I$_{DS}$) as a function of the drain-source voltage (V$_{DS}$) for the conventional and gate-recessed MOS-HEMTs. It can be seen that the drain-source current of gate-recessed MOS-HEMTs was obviously enhanced. At V$_{GS}$ = 0V, the saturation drain-source current of the conventional and gate-recessed MOS-HEMTs was 509 mA/mm and 642 mA/mm, respectively. Furthermore, the threshold voltage of the conventional and gate-recessed MOS-HEMTs was −8V and −8.5V, respectively. The extrinsic transconductance (g$_m$) as a function of V$_{GS}$ at V$_{DS}$ = 10V is shown in Figure 3. It can be seen that the maximum trasconductance was improved from 78 mS/mm to 86 mS/mm, due to the decrease of the distance between the oxide/semiconductor interface and the two-dimentional electron gas (2DEG) channel. The gate-source leakage current of the conventional and gate-recessed AlGaN/GaN MOS-HEMTs operated at V$_{GS}$ = −100V was 9.32×10^{-6} A and 4.83×10^{-6} A, respectively. The gate leakage current level of both devices showed similar results.

The off-state breakdown behavior of the MOS-HEMTs is shown in Figure 4. The off-state breakdown voltage of gate-recessed MOS-HEMTs is greater than 100V, which is obviously better than that of the conventional one. This behavior is attributed to the improvement in electric field distribution at the gate edge [14]. Figure 5 and Figure 6 show the normalized noise power density (S$_{IDS}$/I$_{DS}^2$) of the conventional and gate-recessed MOS-HEMTs operated in saturation region, respectively, where S$_{IDS}$ is the noise power density and I$_{DS}$ is the drain-source current. Both of these were fitted well by the 1/f law in the low frequency region, and the Hooge's coefficient α was both around 10^{-4}, implying that the PEC treatment did not damage the AlGaN surface and induce interface-state.

Conclusion

The PEC method was used to perform the gate recess and directly grow the oxide layer of the gate-recessed AlGaN/GaN MOS-HEMTs. The saturation drain-source current was 642 mA/mm, the maximum trasconductance was 86 mS/mm, and the off-state breakdown voltage was larger than −100V. The gate-source leakage current operated at $V_{GS} = -100V$ was 4.83×10^{-6} A. Compared with conventional devices, the DC performance was enhanced for the gate-recessed MOS ones, while the low frequency noise behavior was similar. It can be deduced that the PEC method is a promising one for the fabrication of gate-recessed structure and for growing oxide layers for the MOS-HEMTs.

Acknowledgments

This work was supported by the National Science Council of Taiwan, Republic of China under the contract number NSC-096-2221-E-006-282-MY3. The epitaxial wafers were grown by Institute for Microstructural Sciences, National Research Council, Canada. The low noise measurement was supported from the National Nano Device Laboratories.

References

1. Y. Yue, Y. Hao, J. Zhang, J. Ni, W. Mao, Q. Feng, and L. Liu, *IEEE Electron Device Lett.*, **29**, 838 (2008).
2. V. Adivarahan, M. Gaevski, A. Koudymov, J. Yang, G. Simin, and M. A. Khan, *IEEE Electron Device Lett.*, **28**, 192 □(2007).
3. Y. C. Chang, W. H. Chang, H. C. Chiu, L. T. Tung, C. H. Lee, K. H. Shiu, M. Hong, J. Kwo, J. M. Hong, and C. C. Tsai, *Appl. Phys. Lett.*, **93,** 053504 (2008).
4. M. Hong, K. A. Anselm, J. Kwo, H. M. Ng, J. M. Baillargeon, A. R. Kortan, J. P. Mannaerts, A. Y. Cho, C. M. Lee, J. I. Chyi, and T. S. Lay, *J. Vac. Sci. Technol. B*, **18**, 1453 (2000).
5. C. J. Kao, M. C. Chen, C. J. Tun, G. C. Chi, J. K. Sheu, W. C. Lai, M. L. Lee, F. Ren, and S. J. Pearton, *J. Appl. Phys.*, **98**, 064506 (2005).
6. L. H. Huang, and C. T. Lee, *J. Electrochem. Soc.*, **154**, H862 (2007).
7. L. H. Huang, S. H. Yeh, C. T. Lee, H. Tang, J. Bardwell, and J. B. Webb, *IEEE Electron Device Lett.*, **29**, 284 (2008).
8. L. H. Huang, S. H. Yeh, and C. T. Lee, *Appl. Phys. Lett.*, **93**, 043511-1 (2008).
9. Y. Okamoto, Y. Ando, T. Nakayama, K. Hataya, H Miyamoto, T. Inoue, M. Senda, K. Hirata, M. Kosaki, N. Shibata, and M. Kuzuhara, *IEEE Trans. Electron Devices*, **51**, 2217 (2004).
10. Y. Dora, A. Chakraborty, L. McCarthy, S. Keller, S. P. Den Baars, and U. K. Mishra, *IEEE Electron Device Lett.*, **25**, 161 (2004).
11. T. Egawa, G. Y. Zhao, H. Ishikawa, M. Umeno, and T. Jimbo, *IEEE Trans. Electron Devices*, **48**, 603 (2001).
12. M. Miyoshi, A. Imanishi, T. Egawa, H. Ishikawa, K. Asai, T. Shibata, M. Tanaka, and O. Oda, *J. J. Appl. Phys.*, **44**, 6490 (2005).
13. C. H. Chen, S. Keller, E. D. Haberer, L. Zhang, S. P. DenBaars, E. L. Hu, U. K. Mishra, and Y. Wu, *J. Vac. Sci. Technol.*, **17**, 2755 (1999).
14. R. S. Qhalid Fareed, X. Hu, A. Tarakji, J. Deng, R. Gaska, M. Shur, and M. A. Khan, *Appl. Phys. Lett.*, **86**, 143512-1 (2005).

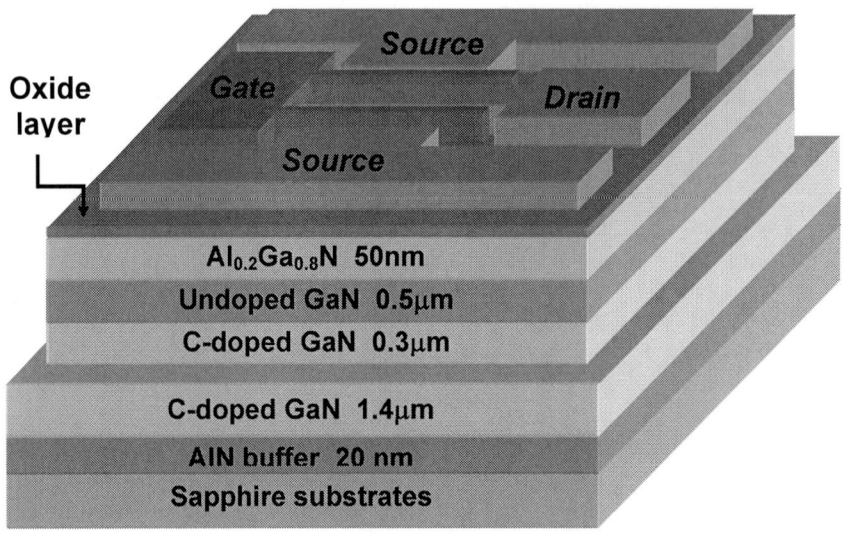

Figure 1. The schematic configuration of the gate-recessed MOS-HEMTs

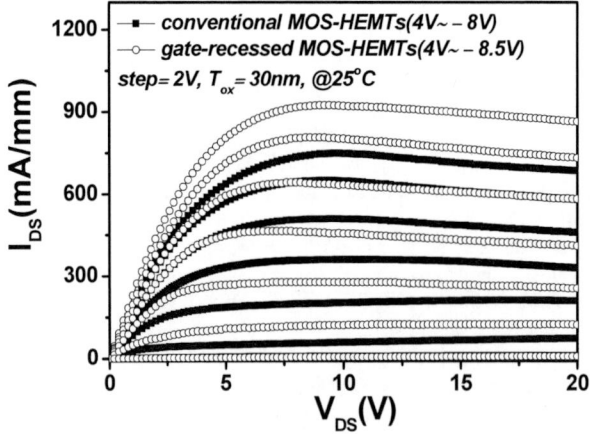

Figure 2. The I_{DS}-V_{DS} characteristics of conventional and gate-recessed MOS-HEMTs.

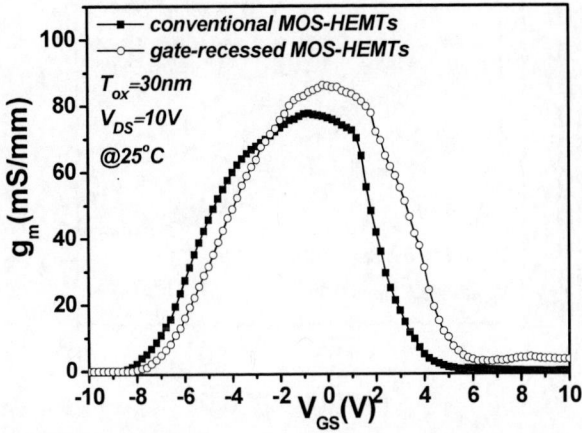

Figure 3. The g_m as a function of V_{GS} for conventional and gate-recessed MOS-HEMTs.

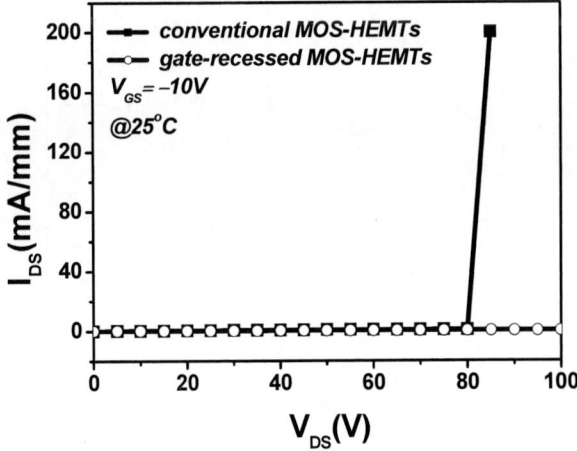

Figure 4. The off-state breakdown behavior of conventional and gate-recessed MOS-HEMTs.

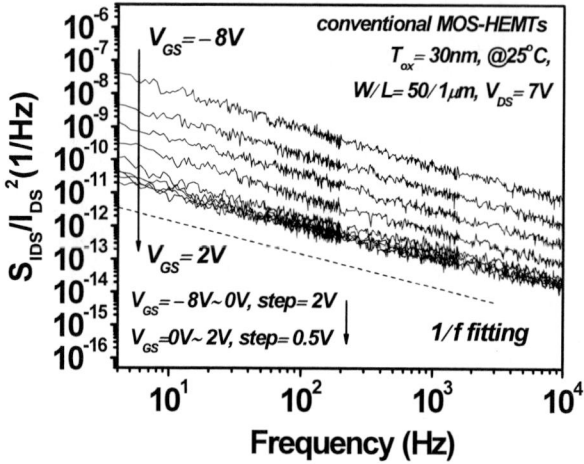

Figure 5. Normalize noise spectra of conventional MOS-HEMTs operated in saturation region.

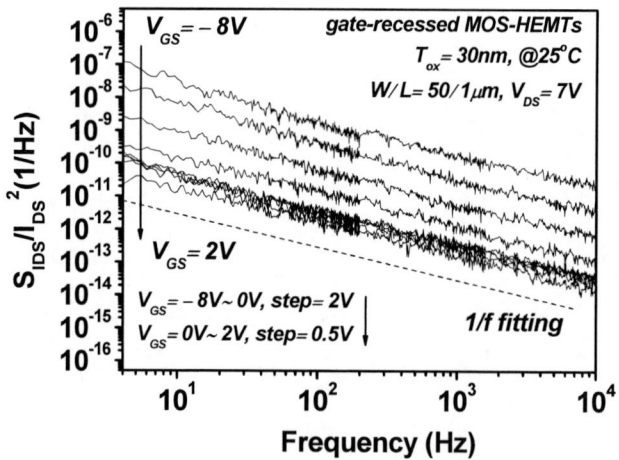

Figure 6. Normalize noise spectra of gate-recessed MOS-HEMTs operated in saturation region.

ECS Transactions, 19 (3) 79-84 (2009)
10.1149/1.3120689 ©The Electrochemical Society

III-V Semiconductors, a History in RF Applications

C.A. Barratt

RFMD, Greensboro, NC 27410, USA

The history of III-V semiconductors in RF applications is more than forty years long. It began in epitaxially prepared diodes based on unique physical properties for certain applications including the "Gunn" effect as well as the higher mobility of an electron when compared with silicon. This gave way to planar processing and ion implantation which permitted the realization of the monolithic microwave integrated circuits (MMICs). Digital wireless applications have become the most ubiquitous applications for III-V RF devices. This was facilitated by taking advantage of not simply the mobility advantages of III-V semiconductors but also the unique properties available through the use of different combinations of active layer alloy compositions. This has driven wide market acceptance for pHEMT and HBT based electronics used in every day life.

Historic Perspective

Two Terminal Niche Applications

III-V semiconductors were made particularly interesting for RF applications by the discovery of the transferred electron or "Gunn" effect by J.B. Gunn in 1962 while working at the Royal Signals and Radar Establishment in the UK[1]. This phenomenon allows that under specific circumstances, a two terminal device could produce the negative resistance necessary to create oscillation. These two terminal devices while called diodes, do not contain a semiconductor junction of any kind. They are designed to interact with a circuit, typically a wave guide fixture or cavity and can produce low efficiency RF energy directly from DC.

Figure 1. III-V Diodes in Ceramic Packages, a Horizontal Bridgeman Ingot Mounted for Sawing and Some Discreet Die

As these devices were commercialized they found applications in military radar and communications, microwave telecommunications and commercial applications such as motion detectors, police radar and police radar detectors. This unique physical property of GaAs and InP was not available from silicon based microwave diodes leading to investment in III-V diode technology for this purpose. These investments subsequently led to development and commercial exploitation of diodes made from similar materials that were available from silicon but with improved performance due to the superior mobility of III-V semiconductors. Diodes such as mixers, tuning varactors and impatts and even PIN diodes[2] were all developed because of synergies with the fabrication of Gunn Diodes. Figure 1 shows diode die and ceramic packaged diodes of the type used in these applications. The similarities were that these devices were all made using epitaxial deposition, typically vapor phase epitaxy (VPE), and mesa diode construction on conductive substrates. The substrates were prepared by Horizontal Bridgman. This technique involved moving a hot zone from a seed through a saturated melt mostly along the (111) plane and then (100) wafers were cut and polished. Figure 1 also shows a Horizontal Bridgman ingot mounted for sawing. This led to a "D" shaped wafer that were very irregular ingot to ingot and while these wafers made processing difficult, they made automation impossible.

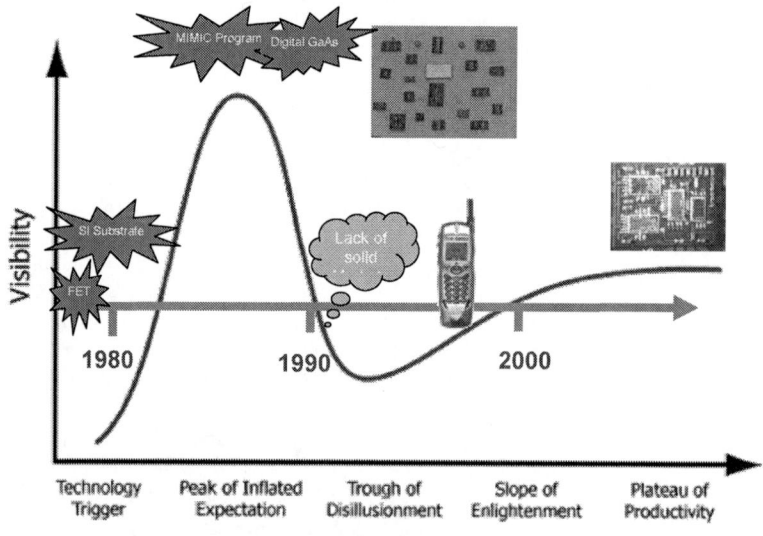

Figure 2. Gartner Hype Cycle applied to the III-V RF semiconductor industry over the last three decades.

There were minor construction variations introduced in some device configurations including beam lead devices for stripline applications and plated heat sink diodes where heat dissipation was a critical need. This industry model was directed at markets of limited size with unique requirements where the cost and performance for each application met market needs. The lack of automation and the absence of a path to multi-function integration meant there was no expectation that this industry model supported explosive growth or the economies of scale enjoyed by IC production.

III-V Integrated Circuits

In the decade of the 70s and into the 80s, significant interest was developing in gallium arsenide metal-semiconductor field effect transistors (GaAs MESFETs). While there was no native oxide on GaAs that would permit producing an equivalent to silicon MOSFETs, a schottky junction gate could be substituted and due to the material properties of GaAs, a high performance transistor could be produced. These devices were still fabricated on epitaxial active layers and through most of the 70s, the semi-insulating substrates were doped with trapping impurities principally chromium, and remained "D" shaped. This still led to an industry model that could not support the promise of massive scale and therefore low production costs but these devices found niche applications.

With the arrival of the high pressure Liquid Encapsulated Czochralsky (LEC) growth of GaAs substrate, SEMI standard round, undoped, semi-insulating wafers were now available. This gave promise to the expectation that planar processing of multi-functional devices integrated on the same die could be implemented using selective ion implantation. Combining that capability with the automation available with SEMI standard wafers led to a direction for the III-V RF semiconductor industry toward a silicon model and set off a "hype" cycle[3] (see figure 2). This hype resulted in enormous investments led by U.S. DoD and included the private sector. During the peak of the hype cycle there was a very strong technical direction imposed on the III-V RF semiconductor industry to emulate the silicon industry in order to compete and supplant that industry wherever the advantages of III-V materials might permit[4]. This led to a push for direct implant into LEC as the preferred active layer preparation technique.

Implant activation in III-V semiconductors was hindered by the fact that above the congruent sublimation temperature, GaAs loses arsenic from the surface preferentially over gallium. This added a requirement for either a dielectric film, usually silicon nitride be placed on the surface of the wafer prior to implant anneal or an over pressure of arsenic be present during annealing to prevent the decomposition of the surface as arsenic preferentially became volatile. Much activity and some commercial success was enjoyed on the basis of this approach but the lack of a true, passivating oxide for III-V material led to persistent problems with exposed semiconductor surfaces and the physical properties of P-type materials were a barrier to complementary operation for digital circuits. This placed III-V materials at a disadvantage in the highest volume applications. In addition, direct implant into semi-insulating GaAs eliminated the possibility to take advantage of the types of devices that can be formed using a combination of III-V alloys. This is a degree of freedom that led to the current industry model.

Contemporary Industry Status

The Cellular Handset

In the middle of the decade of the 90s cellular communication was emerging. Handsets were battery operated devices and the power amplifier was a major user of energy in the handsets. Further, battery systems moved from nickel metal hydride to lithium ion lowering battery voltage to below 4.0 volts. Suddenly there was a high growth application for which III-V semiconductors were uniquely suited and close in time to the winding down of the major DoD investments. This facilitated strong alignment between market need and the infrastructure necessary to serve it. Initially significant activity was directed at addressing these emerging handset applications with MESFET devices based on ion implantation. Power amplifiers and RF switches were provided based on a MESFET active device but the performance levels and reproducibility necessary for the high volumes these applications required were difficult to meet with this technology. These struggles compelled the industry to seek advantages available from device technologies that employed combinations of III-V alloys to improve raw performance and avoid reproducibility issues stemming from the III-V semiconductor surface.

The predominant device in use for handset power amplifiers is the hetro-junction bipolar transistor (HBT). This NPN emitter up device is constructed using a high band gap emitter of either AlGaAs or InGaP over a GaAs base and collector. Because current flow in an HBT is through the bulk semiconductor and not along the device surfaces, these surfaces are less critical to the device performance than more surface oriented devices like the MESFET. Surface recombination is an issue for HBT devices at the emitter-base junction but a depleted ledge is typical of HBT structures and reduces this effect.

The predominant device utilized for antenna switch components in handsets is the pseudomorphic high electron mobility transistor (pHEMT). This device involves a GaAs contact layer on top of a high bandgap layer, typically AlGaAs over a low bandgap layer, usually InGaAs. The word pseudomorphic comes from the fact that the InGaAs layer is not lattice matched to the substrate. Performance is improved over simple MESFETs because the current carrying charge for the device is created from dopant in the high bandgap material and confined in the low bandgap material where there is higher mobility. This allows for higher current gain and lower parasitic resistances in the device which provides for higher performance in switch applications.

Both types of devices are prepared by either molecular beam epitaxy (MBE) or metal-organic chemical vapor deposition (MOCVD). The movement back to epitaxial deposition as the technique of choice for III-V RFIC fabrication stems from the emphasis on using different alloys of III-V materials to create performance unavailable if any of these alloys were used separately. Alloys are also used to create etch stopping layers to aid in fabrication. Integration is possible in these systems because devices are easily isolated using ion implant bombardment and passive components are added using conventional IC interconnect processing. Integration paths are being extended by combining devices such as the BiFET and through creative alternative uses for existing epitaxial layers[5].

Plateau of Productivity

The market for III-V RF semiconductors is in excess of $3.6 billion as of August 2008[6]. RFMD has reached shipment rates of 2.0 million handset components per day containing both HBT and pHEMT devices.

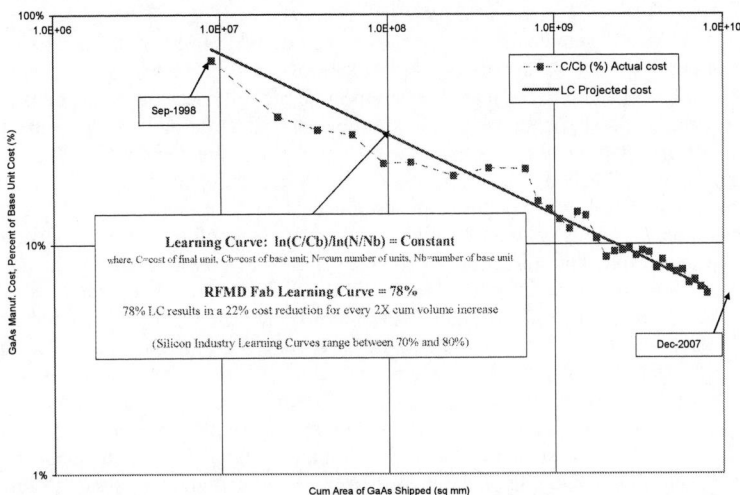

Figure 3. RFMD Learning Curve for III-V RF Semiconductors

The advantages of III-V RF semiconductors over alternative material systems include gain, power added efficiency, linearity; and all at operating voltages consistent with battery operated, handheld devices. There is also dominance in data communication applications commonly referred to as wireless local area networks (WLAN) covered by the IEEE 802.11 data communication standards. Technical performance enables these applications to be addressed by III-V RF semiconductors but only at the appropriate cost-performance balance. Figure 3 illustrates 10 years of relative production cost showing a reduction of 22% for every doubling of cumulative shipments. This agrees well with historic learning curve data for the silicon IC industry and speaks to the maturity and scale that now exists for these technologies.

Future Trends

The current and future needs of digital wireless communication systems are for increased bandwidth which can only be realized by packing more channels closer together in available spectrum and requiring more bits per frame which adds stringent distortion specifications to power amplifier and switch components. This trend will continue to fuel demand for III-V RF semiconductors. For high power applications III-V nitride based material systems have emerged because of the high voltages available from the higher band gaps of these materials. RFMD has produced AlGaN/GaN HEMT devices capable of operation at 48 volts with power densities between 7-10 watts/mm of device periphery[7] These devices again take advantage of the ability to team III-V alloys together to produce unique device performance unavailable from any single alloy.

Conclusion

III-V RF semiconductors began as an industry limited in size based on the limitations imposed by the specialty devices upon which the industry began. Unique epitaxial layers and devices that operated through the thickness of the die constrained the industry to niche applications and discreet devices. Electron mobility advantage was the property that distinguished the material system against others and allowed markets to be defined and established. As the capability to produce undoped semi-insulating substrates in SEMI standard form factor was introduced, expectations emerged that the III-V semiconductor industry would follow a similar technology path to silicon. This perception drove the direction of investment both private and DoD, toward ion implantation for active layer formation with the expectation that the mobility advantage would be an adequate differentiator. There was at the peak of the hype, a perception that III-V semiconductors might threaten silicon in some core applications. The lack of a suitable oxide that would thoroughly passivate the III-V semiconductor surface and an inability to support true complementary operation in digital applications, proved to be an intractable set of limitations for the path of emulating the silicon technology model. Digital wireless communication and low voltage battery systems in handheld devices have given high volume applications to III-V RF semiconductors although through the unique capability to use alloy composition as a degree of freedom. The capability to combine different III-V alloys in combination to create high performance devices has enabled the commercial success of the current industry. This technology trend has fueled a return to epitaxial crystal growth as the technology of choice for active layer preparation. This is largely driven by the need to produce complex, precise and in many cases, ultra-thin layers of differing alloys and impurity levels. If the native oxide passivation and complimentary capabilities of the silicon material system has enabled that industry, certainly the unique ability for a wide variety of alloys to coexist to produce differentiated devices performance has done so for III-V RF semiconductors.

Acknowledgments

I would like to acknowledge the members of the wafer fab team at RFMD for all that they do to make our industry so vibrant.

References

1. W. Liu, *Handbook of III-V Hetrojunction Bipolar Transistors* 1998 p. 58
2. J.L. Heaton, R.S. Posner, T.B. Ramachandran, R.E. Walline, *International Electronic Devices Meeting*, 1978
3. J. Fenn, *When to Leap on the Hype Cycle*, Gartner Group, 1995
4. C. Barratt, R. Boerstler, M.A. Shea, J. Heaton, *The Electrochemical Society SOTAPOCS*, 1985
5. B. Moser, W. Wohlmuth, M. Fresina, S. Nedeljkovic, W. Clausen, D. Halchin, R. Vass, *Digest of the International Symposium on Compound Semiconductor Manufacturing Technology*, 2008
6. A. Anwar, *GaAs Device Vendor Market Share 2007: North America*, Strategy Analytics, 2008
7. K. Krishnamurthy, J. Martin, B. Landberg, R. Vetury, M.J. Poulton, *IEEE MTT-S International Microwave Symposium Digest*, 2008

ECS Transactions, 19 (3) 85-97 (2009)
10.1149/1.3120690 ©The Electrochemical Society

AlGaN/GaN High Electron Mobility Transistors integrated into Wireless Detection System for Glucose and pH in Exhaled Breath Condensate

B.H. Chu[a], B.S. Kang[a], C. Y. Chang[a], F. Ren[a], Aik Goh[b], Andrew Sciullo[b], Wenhsing Wu[b], Jenshan Lin[b], B.P. Gila[c] and S.J. Pearton[c], J.W. Johnson[d], E.L. Piner[d] and K.J. Linthicum[d]

[a] Department of Chemical Engineering, University of Florida, Gainesville, FL 32611
[b] Department of Electrical and Computer Engineering, University of Florida, Gainesville, FL 32611
[c] Department of Materials Science and Engineering, University of Florida, Gainesville, FL 32611
[d] Nitronex Corporation, Raleigh, NC 27606

Peltier element cooling is demonstrated to be an effective method for collecting exhaled breath condensate (EBC) on AlGaN/GaN High Electron Mobility Transistors (HEMT). The HEMT sensors have functionalized gate areas for glucose and pH measurement. The current change measured in the HEMTs with EBC shows that the sensitivity of the glucose detection is lower than the glucose concentration in the EBC of healthy person and the pH measurement range includes 7- 8, typical of that for human blood. The sensors can be integrated into a wireless data transmission system that allows remote monitoring. Details of the transmitter and receiver design for the transmission system are given. Our work demonstrates the possibility of using AlGaN/GaN HEMTs for extended investigations of airway pathology without the need for clinical visits.

Introduction

There is significant interest in developing rapid diagnostic approaches and improved sensors for determining early signs of medical problems in humans. Exhaled breath is a unique bodily fluid that can be utilized in this regard [1-12]. While most applications will detect substances or diseases in the breath as a gas or aerosol, breath can also be analyzed in the liquid phase as exhaled breath condensate (EBC). Analytes contained in the breath originating from deep within the lungs (alveolar gas) equilibrates with the blood, and therefore the concentration of molecules present in the breath is closely correlated with those found in the blood at any given time [1-6]. EBC contains dozens of different biomarkers, such as adenosine, ammonia, hydrogen peroxide, isoprostanes, leukotrienes, peptide, cytokines and nitrogen oxide [8,9,12]. Analysis of molecules in EBC is noninvasive and can provide a window on the metabolic state of the human body, including certain signs of cancer, respiratory diseases, liver and kidney functions.

As a diagnostic, exhaled breath offers advantages since samples can be collected and tested with results delivered in real time at the point of testing. Another advantage is that the sample can be collected non-invasively by asking a patient to blow into the disposable portion of a handheld testing device. Therefore, the sample collection method is hygienic for both the patient and the laboratory personnel. Exhaled breath can be used to detect various drugs, medications, their metabolites and markers, and this can be valuable in measuring both medication adherence and in determining the blood levels of these drugs and medications. Some of today's blood and urine-based tests might be replaced with simple breath-based testing. In consumer healthcare, diabetics would be able to test their glucose level, replacing painful and inconvenient finger-prick devices. For roadside screening of driving impairment, a point-of-care (POC) device similar in function to a handheld breath alcohol analyzer will detect drugs of abuse such as marijuana and cocaine. In workplace drug testing, a similar desktop device might eliminate the cost, embarrassment and inconvenience of workplace urine screening. In the setting of chronic oral drug therapy (e.g., treatment of schizophrenia with atypical antipsychotic medications), mortality/morbidity and the cost of health care will be markedly reduced by developing breath-based systems that document that drugs were orally ingested and entered the blood stream.

The glucose oxidase enzyme (GOx) is commonly used in biosensors to detect levels of glucose for diabetics. By keeping track of the number of electrons passed through the enzyme, the concentration of glucose can be measured. Due to the importance and difficulty of glucose immobilization, numerous studies have been focused on the techniques of immobilization of glucose with carbon nanotubes, ZnO nano-materials, and gold particles [2-4]. ZnO based nano-materials are especially interesting due to its non toxic property, low cost fabrication and favorable electrostatic interaction between ZnO and GOx lever. However, the activity of GOx is highly dependent on the pH value of the solution. The pH value of a typical healthy person is between 7 and 8. It can vary significantly depending on the health condition of each individual, e.g. the pH value for patients with acute asthma was reported as low as 5.23 ± 0.21 (n=22) as compared to 7.65 ± 0.20 (n=19) for the control subjects. To achieve accurate glucose concentration measurement with immobilized GOx, it is necessary to determine the pH value and glucose concentration with an integrated pH and glucose sensor.

In this paper, we report on the use of AlGaN/GaN High Electron Mobility Transistors (HEMTs) for measurements of pH in EBC and glucose, integrate the pH and glucose sensor on a single chip, and demonstrate the integration of the sensors into a portable, wireless package for remote monitoring applications. HEMT devices are gaining increasing interest for use in detection of a range of bio-markers and medical conditions, including various forms of cancer [13-15] and kidney injury [16].

Experimental

The HEMT structures consisted of a 2 μm thick undoped GaN buffer and 250Å thick undoped $Al_{0.25}Ga_{0.75}N$ cap layer. The epi-layers were grown by Metal-Organic Chemical Vapor Deposition on 100 mm (111) Si substrates. The sheet carrier concentration was $\sim 10^{13}$ cm^{-2} and the mobility was 980 cm^2/V-s at 300K. For the pH sensing, 100Å Sc_2O_3 was deposited as a gate dielectric through a contact window of SiNx layer [18,19]. The Sc_2O_3 was deposited by rf plasma-activated Molecular Beam Epitaxy at 100 °C using elemental Sc evaporated from a standard effusion at 1130 °C and O_2 derived from an Oxford RF plasma source. Figure 1 shows an optical microscopy image of an integrated pH and glucose sensor chip and cross-sectional schematics of the completed pH and glucose device. The gate dimension of the pH sensor device and glucose sensors was $20 \times 50 \ \mu m^2$.

For the glucose detection, a highly dense 20-30 nm diameter and 2 μm tall ZnO nanorods were grown on the $20 \times 50 \ \mu m^2$ gate area, as shown in Figure 1. The lower right inset in Figure 1 shows closer view of ZnO nanorod arrays grown on the gate area. The total area of the ZnO was increased significantly with the ZnO nanorods. The ZnO nanorod matrix provides a microenvironment for immobilizing negatively charged GO_x while retaining its bioactivity, and passes charges produce during the GO_x and glucose interaction to the AlGaN/GaN HEMT. The GOx solution was prepared with concentration of 10 mg/mL in 10 mM phosphate buffer saline (pH value of 7.4, Sigma Aldrich). After fabricating the device, 5 μl GO_x (~100 U/mg, Sigma Aldrich) solution was precisely introduced to the surface of the HEMT using a pico-liter plotter. Then sensor chip was kept at 4 °C in the solution for 48 hours for GO_x immobilization on the ZnO nanorod arrays followed by an extensively washing to remove the un-immobilized GO_x.

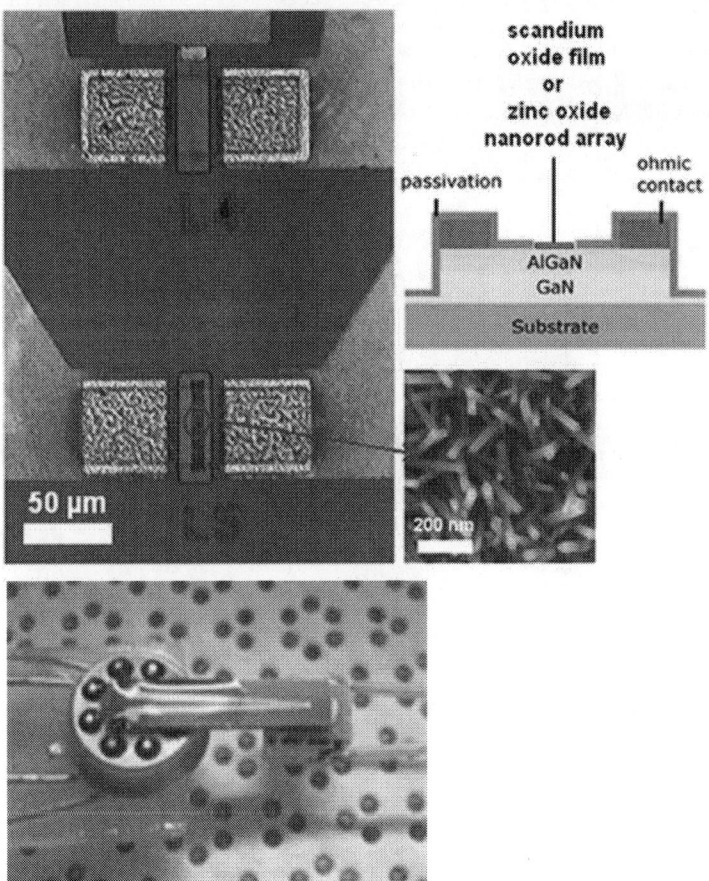

Figure 1. Optical microscope image and schematic of gateless HEMT (top), and optical microscope image of packaged sensor (bottom).

To take the advantage of quick response (less than 1 sec) of the HEMT sensor, a real-time EBC collector is needed[20-22]. The amount of the EBC required to cover the HEMT sensing area is very small. Each tidal breath contains around 3 µl of the EBC. The contact angle of EBC on Sc_2O_3 has been measured to be less than 45°, and it is reasonable to assume a perfect half sphere of EBC droplet formed to cover the sensing area 4×50 µm^2 gate area. The volume of a half sphere with a diameter of 50 µm is around 3×10^{-11} liter. Therefore, 100,000 of 50 µm diameter droplets of EBC can be formed from each tidal breath.

To condense entire 3 µl of water vapor, only ~ 7 J of energy need to be removed for each tidal breath, which can be easily achieved with a thermal electric module, a Peltier device, as shown in Figure 1(center and bottom). The AlGaN/GaN HEMT sensor is directly mounted on the top of the Peltier unit (TB-8-0.45-1.3 HT 232, Kryotherm), which can be cooled to precise temperatures by applying known voltages and currents to the unit. During our measurements, the hotter plate of the Peltier unit was kept at 21°C, and the colder plate was kept at 7 °C by applying bias of 0.7 V and 0.2 A. The sensor takes less than 2 sec to reach thermal equilibrium with the Peltier unit. This allows the exhaled breath to immediately condense on the gate region of the HEMT sensor.

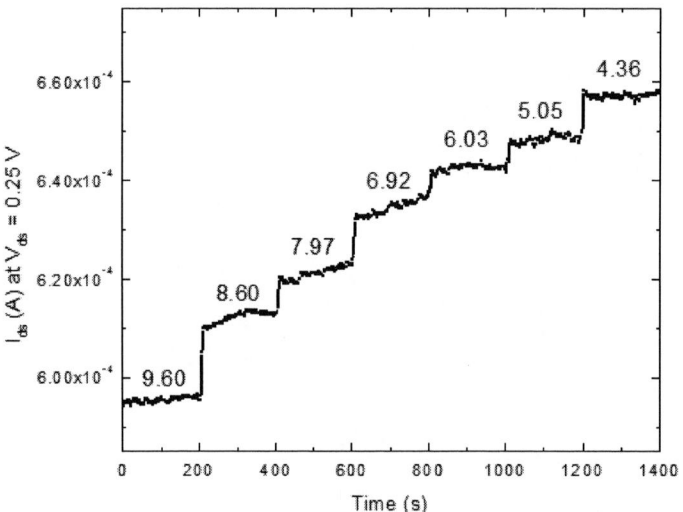

Figure 2. Change of drain current in gateless HEMT at fixed source-drain bias of 0.25 V with pH from 4-10.

Prior to pH measurements of the EBC, a Hewlett Packard soap film flow meter and a mass flow controller were used to calibrate the flow rate of exhaled breath. The HEMT sensors were also calibrated and exhibited a linear change in current between pH 3-10 of 37µA/pH, as shown in Figure 2. Figure 3 shows a photograph and schematic of the system or collecting the EBC. Due to the difficulty to collect the EBC with different glucose concentration, the samples for glucose concentration detection were prepare from glucose diluted in PBS or DI water.

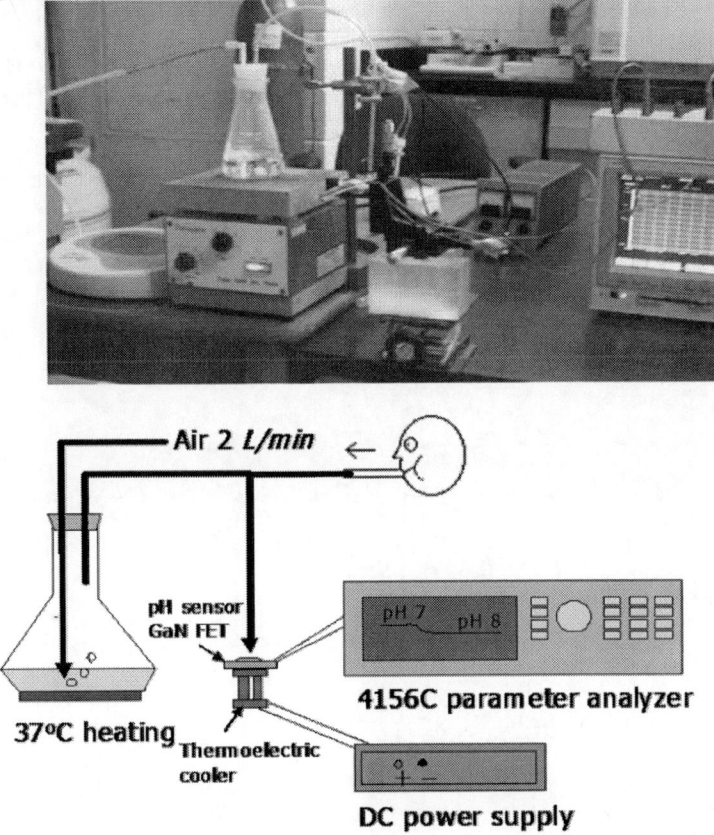

Figure 3. Photograph and schematic of the system for collecting EBC.

Results and Discussion

The HEMT sensors were not sensitive to switching of N_2 gas, but responded applications of exhaled breath pulse inputs from a human test subject, as shown at the top of Figure 4, which shows the current of a Sc_2O_3 capped HEMT sensor biased at 0.5V for exposure to different flow rates of exhaled breath (0.5-3.0 l/min). The flow rates are directly proportional to the intensity exhalation. Deep breath provides a higher flow rate. A similar study was conducted with pure N_2 to eliminate the flow rate effect on sensor sensitivity. The N_2 did not cause any change of drain current, but the increase of exhaled breath flow rate decreased the drain current proportionally from 0.5 L/min to a saturation value of 1 L/min. For every tidal breath, the beginning portion of the exhalation is from the physiologic dead space, and the gases in this space do not participate in CO_2 and O_2 exchange in the lungs. Therefore, the contents in the tidal breath are diluted by the gases from this dead space. For higher flow rate exhalation, this dilution effect is less effective.

Once the exhaled breath flow rate is above 1L/min, the sensor current change reaches a limit. As a result, the test subject experiences hyper ventilation and the dilution becomes insignificant. The bottom of Figure 4 shows the time response of the sensors to much longer exhaled breaths. The characteristic shape of the response curves is similar and is determined by the evaporation of the condensed EBC from the gate region of the HEMT sensor. The sensor is operated at 50 Hz and 10% duty cycle, which produces heat during operation. The energy required to vaporize the EBC on the HEMT sensing area is around 7×10^{-5} J, which is the same order of magnitude of the energy produced by the sensor. Therefore, it only takes a few seconds for the EBC to vaporize from the sensing area and causes the spike-like response. The principal component of the EBC is water vapor, which represents nearly all of the volume (>99%) of the fluid collected in the EBC[1]. The measured current change of the exhale breath condensate shows that the pH values are within the range between pH 7 and 8. This range is the typical pH range of human blood.

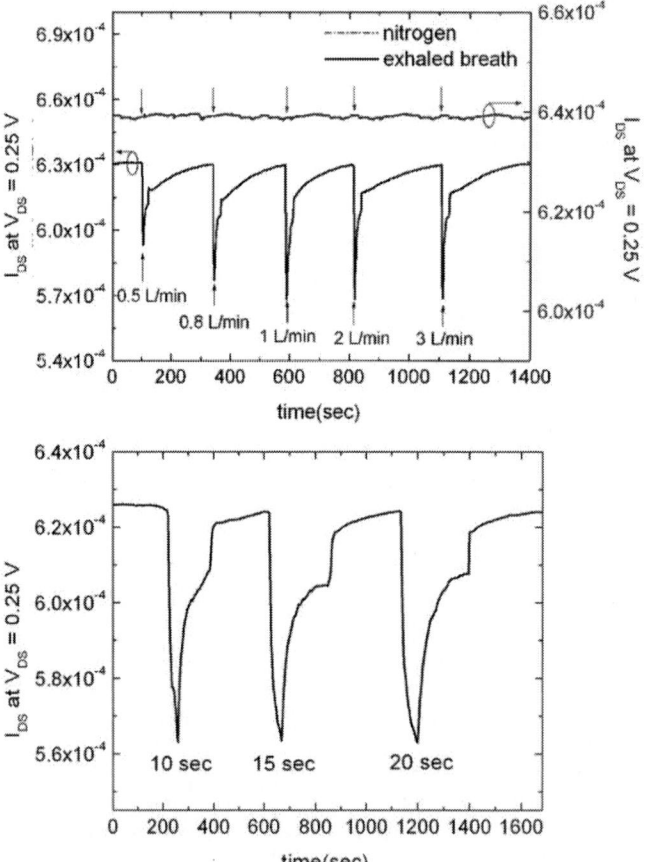

Figure 4. Changes of drain current for HEMT sensor at fixed drain-source bias of 0.5 V with different flow rates (top) or durations (bottom) of exhaled breath from tidal breath to hyperventilation. The duration of the breath is 5 secs in the top figure.

Figure 5 shows the real time glucose detection in PBS buffer solution using the drain current change with constant bias of 250 mV. No current change can be seen with the addition of buffer solution at around 200 sec, showing the specificity and stability of the device. By sharp contrast, the current change showed a rapid response of less than 5 seconds when target glucose was added to the surface. So far, the glucose detection using Au nano-particle, ZnO nanorod and nanocomb, or carbon nanotube material with GOx immobilization is based on electrochemical measurement[9, 10, 23, 24]. Since there is a reference electrode required in the solution, the volume of sample can not be minimized. The current density is measured when a fixed potential applied between nano-materials and the reference electrode. This is a first order detection and the range of detection limit of these sensors is 0.5-70 µM. Even though the AlGaN/GaN HEMT based sensor used the same GOx immobilization, the ZnO nanorods were used as the gate of the HEMT. The glucose sensing was measured through the drain current of HEMT with a change of the charges on the ZnO nano-rods and the detection signal was amplified through the HEMT. Although the response of the HEMT based sensor is similar to electrochemical based sensor, much lower detection limit of 0.5 nM was achieved for the HEMT based sensor due to this amplification effect. Since there is no reference electrode required for the HEMT based sensor, the amount of sample only depends on the area of gate dimension and can be minimized.

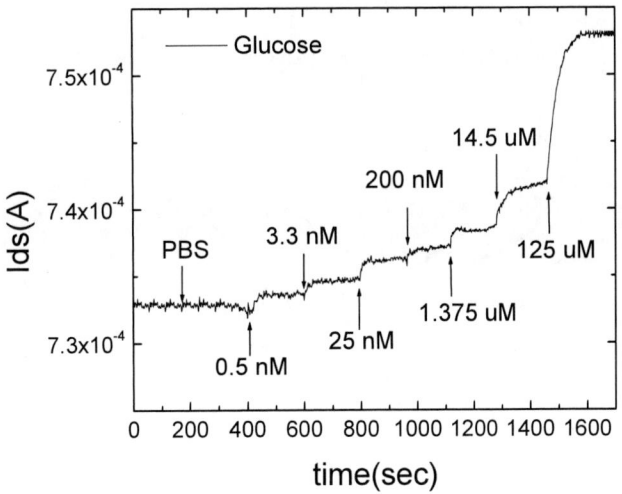

Figure 5   Plot of drain current versus time with successive exposure of glucose from 500 pM to 125 μ M in 10 mM phosphate buffer saline with a pH value of 7.4.

Although, measuring the glucose in the exhaled breath condensate (EBC) is a noninvasive and convenient method for the diabetic application, the activity of the immobilized GO_x is highly dependent on the pH value of the solution. The GOx activity can be reduced to 80% for pH = 5 to 6. If the pH value of the glucose solution is larger than 8, the activity drops off very quickly[25]. Figure 6 shows the time dependent source-drain current signals with constant drain bias of 500 mV for glucose detection in DI water and PBS buffer solution. 50 µl of PBS solution was introduced on the glucose sensor and no current change can be seen with the addition of buffer solution at 20 and 30 min. This stability is important to exclude possible noise from the mechanical change of the buffer solution. By sharp contrast, the current change showed a rapid response in less than 20 seconds when the sensor was dipped into the 100 ml of 10 mM glucose solution using DI water as the solvent. This sudden drain current increase indicated that GOx immediately reacted with glucose and oxygen was produced as a by-production of this reaction. However, the drain current did not maintain and gradually decreased. This was due to the oxygen produced in the GOx-glucose interaction reacting with water and changing the pH value adjacent the gate area. Since there was not agitation in the glucose solution, the solution around gate area became more basic and the activity of GOx decreased due to the high pH value environment from 60 to 85 min.

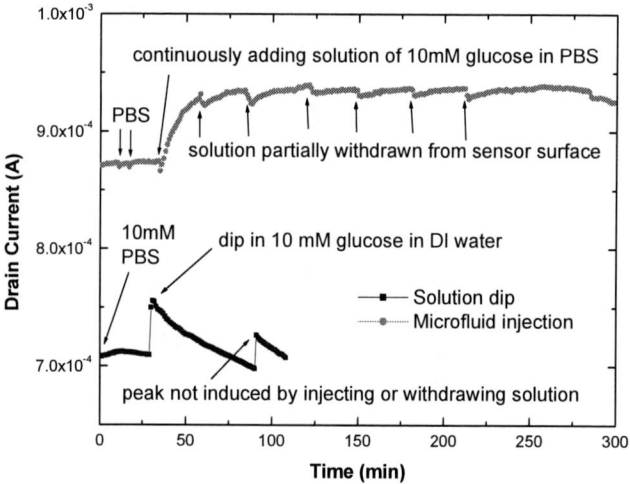

Figure 6. Plot of drain current versus time by dipping the glucose sensor in 10 mM of glucose dissolved in DI water (black line) and exposing the sensor to continuously flow of 10 mM of glucose dissolved in phosphate buffer saline with a pH value of 7.4.

Because the lower activity of GOx in the high pH value condition, the amount of oxygen produced from GOx and glucose decreased as well during the period of 60-85 min. Once the OH⁻ ions produce from reaction between oxygen and water diffused away the gate area, the pH value decreased. Thus Around 85 min, the pH value of the glucose solution around gate area decreasing low enough, the activity of GOx resumed and the drain current of the glucose sensor showed another sudden increase. Then, the same process happened again and drain current of the glucose current gradually decrease for a second time.

On the contrary, the glucose sensor used in a pH controlled environment, the drain current stayed fairly constant as shown in Figure 6. In this experiment, 50 µl of PBS solution was introduced on the glucose sensor to establish the base line of the sensor as previous experiment. Then, glucose of 10 nM concentration prepared in PBS solution was introduced to the gate area of the glucose sensor through a micro-injector. There was no glucose in the 50 µl PBS solution and the PBS solution added at 20 and 30 min. It took time for the glucose solution to diffuse to the gate area of the glucose sensor through the blank PBS and the drain current gradually increased corresponding the glucose diffusion process. Since the fresh glucose was continuously provide to the sensor surface and the pH value of the glucose was controlled. Once the concentration of the glucose reached equilibrium at the gate of the glucose sensor, the drain current of the glucose maintained except for the glucose solution, which was taken out from time to time using micro-pipette. There were small oscillation of the drain current observed, which could be eliminated by using microfluidic device for this experiment.

The human pH value can vary significantly depending on the health condition, eg. the pH value for patients with acute asthma was reported as low as 5.23 ± 0.21 (n=22) as compared to 7.65 ± 0.20 (n=19) for the control subjects [25]. Since we can not control the pH value of the EBC samples, we needed to measure the pH value while determine the glucose concentration in the EBC. With the fast response time and low volume of the EBC required for HEMT based sensor, a handheld and real-time glucose sensing technology can be realized.

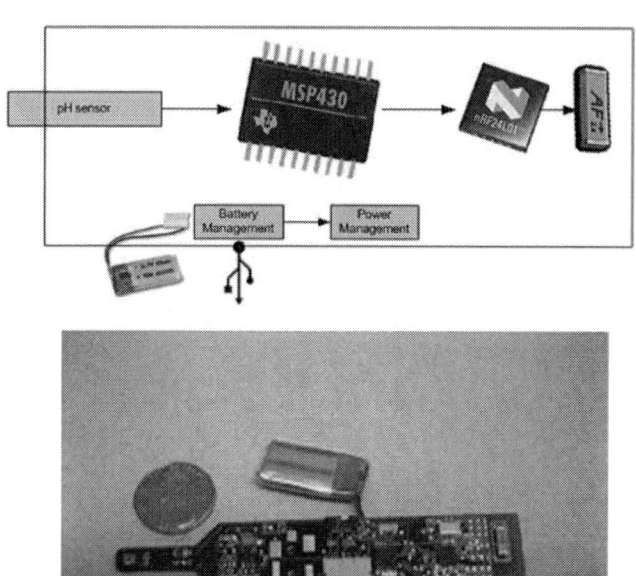

Figure 7. Schematic (top) and photograph (bottom) of transmitter.

As shown in the Figure 7, a pen-sized portable, re-configurable wireless transceiver integrated with pH sensor has been designed and fabricated. The wireless transmitter and receiver pair was designed to acquire EBC data and transmit it wirelessly. This system is able to interface multiple different sensors and consists of a transmitter and receiver pair. The transmitter is shown schematically in Figure 7 and was designed such that it is the size of a marker-pen so that it could be used as an ultra-portable lightweight handheld device. The transmitter is designed to be operated on an ultra-low-power mode. The transmitter is also equipped with an on-board rechargeing circuit, which can be powered by using a standard mini-USB cable. The transmitter consumes on average 80μA. The transmitter and receiver pair is designed to operate at 2.4GHz with range of up to 20ft line-of-sight. The receiver shown in Figure 8 has USB 2.0 connectivity, which relays EBC data from the transmitter to a PC while powering the receiver. The transmitter is designed to integrate with various different sensors through a connector. The transmitter can be reset for the required input signal range to trigger the alarm through the bi-directional wireless communication for a different sensing application. Thus this system is reconfigurable over-the-air. The wireless circuits only consume a power level around 1 μW. If the sensor consumes a similar power level, the battery installed on the transmitter package can last more than one month. This EBC sensing pair of devices can be mass-produced cost effectively well below $100 each pair. The sensor occupies the tip of the pen-shaped layout in Figure 7 and runs off a 75mA Li ion polymer rechargeable battery.

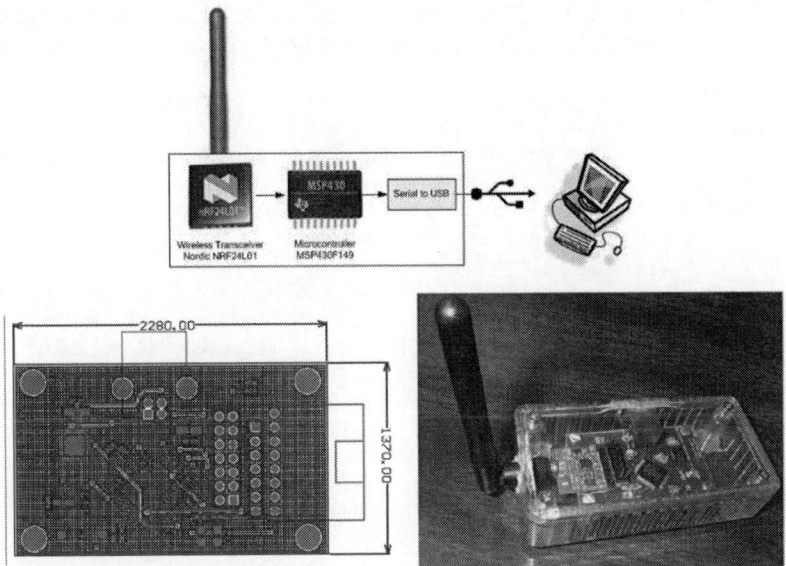

Figure 8. Schematic (top), circuit board layout (left) and photograph (right) of receiver.

Summary and Conclusions

In conclusion, ungated AlGaN/GaN HEMTs with Sc_2O_3 gate dielectric are sensitive to exposure to exhaled breath condensate. This electronic detection approach with rapid response and good repeatability shows potential for the investigation of airway pathology. There is still much work to be done in establishing the reliability of EBC analysis for a wide range of patients with different conditions. The devices can be integrated into a wireless data transmission system for remote monitoring applications.

Acknowledgments

The work at UF is supported by NSF (CTS-0301178 and DMR 0703340), ONR and State of Florida, Center of Excellence for Nano-Bio Applications.

References

1. T. Kullmann, I.Barta, B. Antus, M. Valyon and I. Horvath, "Environmental Temperature and Relative Humidity Influence Exhaled Breath Condensate pH", Eur. Respir. J.31 (2)474-475, Feb. 2008.
2. I. Horvath, J. Hunt and P. J. Barnes, "Exhaled breath condensate: methodological recommendations and unresolved questions", Eur. Respir. J. 26(9), pp.523-548, Sept. 2005.

3. K. Namjou, C. B. Roller and P. J. McCann, "The Breathmeter - A new laser device to analyze your health", IEEE Circuits & Devices Mag. September/October, Vol. 22(5), pp.22-28 (2006).
4. R. F. Machado, D. Laskowski, O. Deffenderfer, T. Burch, S. Zheng, P. J. Mazzone, T. Mekhail, C. Jennings, J. K. Stoller, J. Pyle, J. Duncan, R. A. Dweik and S. C. Erzurum, "Detection of Lung Cancer by Sensor Array Analyses of Exhaled Breath", Am. J. Respir. Crit. Care. Med. Vol.171(11), pp.1286-1291 June 2005.
5. T. Kullmann, I. Barta, Z. Lazar, B. Szili, E. Barat, M. Valyon, M. Kollai, I. Horvath, " Exhaled breath condensate pH standardised for CO_2 partial pressure", Eur. Respir.J. 29(3), pp.496-501 March 2007.
6. J. Vaughan, L. Ngamtrakulparit, T. N. Pajewski, R. Turner, T. A. Nguyen, A. Smith, P. Urban, S. Hom, B. Gaston, J. Hunt., " Exhaled breath condensate pH is a robust and reproducible assay of airway acidity", Eur. Respir. J. 22(12),pp.889-894, Dec 2003.
7. K. Kostikas, G. Papatheodorou, K. Ganas, K. Psathakis, P. Panagou, S. Loukides," pH in Expired Breath Condensate of Patients with Inflammatory Airway Diseases", Am. J. Respir. Crit. Care. Med. Vol.165(10), pp.1364-1370, May 2002.
8. G. E. Carpagnano, M. P. Foschino Barbaro, O. Resta, E. Gramiccioni, N. V. Valerio, P. Bracciale and G. Valerio, "Exhaled markers in the monitoring of airways inflammation and its response to steroid's treatment in mild persistent asthma", Eur. J. Pharmacol. Vol.519(1/2), pp.175-181, Sept. 2005.
9. C. Gessner, S. Hammerschmidt, H. Kuhn, H.-J. Seyfarth, U. Sack, L. Engelmann, J. Schauer and H.Wirtz, " Exhaled breath condensate acidification in acute lung injury", Respir. Med. Vol.97(11), pp.1188-1194, Nov.2003.
10. R. Accordino, A. Visentin, A. Bordin, S. Ferrazzoni, E. Marian, F. Rizzato, C.Canova, R. Venturini and P. Maestrelli "Long-term repeatability of exhaled breath condensate pH in asthma",Resp. Med., Vol. 102(3), March 2008, Pages 377-381.
11. K. Czebe, I. Barta, B. Antus, M. Valyon, I. Horváth and T. Kullmann, "Influence of condensing equipment and temperature on exhaled breath condensate pH, total protein and leukotriene concentrations", Resp. Med., Vol. 102(5), May 2008, Pages 720-725.
12. K. Bloemen, G. Lissens, K. Desager and G. Schoeters, "Determinants of variability of protein content, volume and pH of exhaled breath condensate", Resp. Med., Vol. 101(6), June 2007, pp. 1331-1337
13. B.S.Kang, H.T. Wang, F.Ren and S.J.Pearton, "Electrical Detection of Biomaterials Using AlGaN/GaN HEMTs", J.Appl.Phys. Vol.104(8), 031101 August 2008.
14. K.H. Shen, B.S.Kang, H.T.Wang, T.P.Lele, F.Ren, Y.L.Wang, C.Y.Chang, S.J.Pearton,   J.W. Johnson, P.Rajagopal, J.C.Roberts, E.L.Piner and K.J. Linthicum, "c-erB-2 Sensing Using AlGaN/GaN HEMTs For Breast Cancer Detection", Appl. Phys. Lett. 92,192103 (2008).
15. B. S. Kang, H. T. Wang, T. P. Lele, F. Ren, S. J. Pearton, J.W. Johnson, P. Rajagopal, J.C. Roberts, E.L. Piner and K.J. Linthicum, "Prostate Specific Antigen Detection using AlGaN/GaN High Electron Mobility Transistors", Appl. Phys. Lett. 91, 112106 (2007).
16. H.T. Wang, B.S. Kang, F. Ren, S.J. Pearton, J.W. Johnson, P. Rajagopal, J.C. Roberts, E.L. Piner and K.J.Linthicum, "Electrical Detection of Kidney Injury Molecule-1 With AlGaN/GaN High Electron Mobility Transistors", Appl. Phys.Lett.91, 222101(2007).
17. B. S. Kang, H. T. Wang, F. Ren, B. P. Gila, C. R.Abernathy, S. J. Pearton, D. M. Dennis, J. W. Johnson, P. Rajagopal, J. C. Roberts, E. L.Piner and K. J. Linthicum,

"Exhaled-Breath Detection Using AlGaN/GaN High Electron Mobility Transistors Integrated with a Peltier Element", Electrochem. Solid.State Lett.Vol.11, J19-21 (2007).

18. B.S. Kang, H.T. Wang, F. Ren, M. Hlad, B.P. Gila, C.R. Abernathy, S.J. Pearton, C. Li, Z.N. Low, J. Lin, J.W. Johnson, P. Rajagopal, J.Roberts, E.Piner and K.J. Linthicum, "Role of Gate Oxide in AlGaN/GaN HEMT pH Sensors", J. Electron. Mater.Vol. 37(5), pp.550-554 May 2008.

19. B. P. Gila, M. Hlad, A.H. Onstine, R. Frazier, G.T. Thaler, A. Herrero, E. Lambers, C.R. Abernathy, S.J. Pearton, T. Anderson, S. Jang, F. Ren, N. Moser, R.C. Fitch and M. Freund, "Improved Oxide Passivation of AlGaN/GaN HEMTs", Appl. Phys. Lett.Vol.87, 163503 (2005).

20. P. Montuschi and P. J. Barnes, "Analysis of exhaled breath condensate for monitoring airway inflammation", Trends Pharmacol. Sci. 23(5), pp.232- 237, May 2002.

21. Dam T. V. Anh, W. Olthuis and P. Bergveld, "A hydrogen peroxide sensor for exhaled breath measurement", Sensor Actuat. B, Vol.111/112(11), pp.494-499, Nov 2005.

22. G.M. Multu, "Collection and analysis of exhaled breath condensate in humans", Am. J. Res. Crit. Care Med. 164(11), pp. 731–737.Nov. 2001.

23. J. Hunt," Exhaled Breath Condensate: An Overview", Immunology and Allergy Clinics of North America, Vol. 27(4), Pages 597-606, Nov. 2007.

24. R. Accordino, A. Visentin, A. Bordin, S. Ferrazzoni, E. Marian, F. Rizzato, C. Canova, R. Venturini and P. Maestrelli, "Long-term repeatability of exhaled breath condensate pH in asthma", Respiratory Medicine, Vol.102(3), Pages 377-381,March 2008,

ECS Transactions, 19 (3) 99-109 (2009)
10.1149/1.3120691 ©The Electrochemical Society

Electrodeposited Au-CdTe-Au Nanowires: Solution-based Control over Cd/Te Stoichiometry

L. O. Mair[a*], K. Skinner[b,e], C. L. Donley[c], and R. Superfine[a,b,d]

[a] Curriculum in Applied Science and Engineering, [b] Department of Physics and Astronomy, [c] Chapel Hill Analytical and Nanofabrication Laboratory, [d] Department of Computer Science, University of North Carolina at Chapel Hill, Chapel Hill, North Carolina 27599
[e] Department of Electrical and Computer Engineering, Duke University, Durham, North Carolina 27708

*Corresponding author: lomair@email.unc.edu

Multisegmented, composition-adjusted Au-CdTe-Au nanowires were fabricated via templated electrodeposition. Cd/Te ratios were modified in these wires by varying the Te^{4+} concentration in the deposition electrolyte. These wires were characterized by energy-dispersive x-ray spectroscopy (EDS); subsequently, they were deposited onto Si/SiO_2 substrates and electrically contacted using Pt deposition in a focused ion beam tool. EDS and photoelectron x-ray spectroscopy (XPS) were used to characterize the composition and surface chemistry of CdTe films deposited from the various electrolytes.

Introduction

Cadmium telluride (CdTe) is a II-VI semiconductor which has potential for use in a wide array of applications. At room temperature CdTe has a direct band gap of 1.44 eV and a high absorption coefficient, making CdTe well-suited for photovoltaic applications[1]. Additionally, CdTe is chemically robust and has a simple phase diagram, making it a prime material for incorporation into industrial scale solar modules[2]. Cu, Zn, and Hg have each been implemented as alloying materials capable of engineering the CdTe band gap[3-5] and CdTe/CdS solar cells with greater than 10% efficiency have been fabricated on both rigid and flexible substrates[2,3,6]. Other cadmium chalcogenide-based nanomaterials such as CdS and CdSe have shown promise in electronic, optoelectronic, solar cell, and chemical sensing applications; consequently, CdTe has also received significant attention for use in these applications.

Bulk CdTe is typically grown using the Bridgman technique[7,8], however molecular beam epitaxy, vapor deposition, and electrodeposition have each been used to synthesize various CdTe materials. Among these techniques electrodeposition is viewed as a low cost method for producing large surface area compound semiconductors[1]. In addition to CdTe films, electrodeposition into the pores of anodized aluminum oxide (AAO) and porous polycarbonate membranes has resulted in CdTe nanowires with varying compositions and properties[5,9-15]. Recently, Kum et al. fabricated CdTe nanowires at elevated temperatures in the pores of polycarbonate membranes and measured electrical properties of individual wires by drop-casting nanowires onto pre-patterned gold electrodes[15]. Inherently, properties of quasi one-dimensional materials, particularly compound semiconductors, are dependent upon the stoichiometric compositions

99

incorporated into the nanomaterial. Successful engineering of these materials for specific applications, therefore, relies on controlling composition. Previously, Cd/Te stoichiometries in electrodeposited CdTe materials have been controlled using the deposition potential[11,12]. The effect of relative concentrations of Cd and Te in the deposition electrolyte has been less explored.

This transaction describes multisegmented Au-CdTe-Au wires grown via electrodeposition at room temperature. We show that both Cd- and Te-rich wires can be electrodeposited by varying Te^{4+} concentrations in the electrolytes. Additionally, we use chemically identical electrolytes to grow CdTe films on Au-coated glass slides. We collect energy dispersive x-ray spectroscopy (EDS) data on both wires and films while photoelectron x-ray spectroscopy (XPS) measurements were performed solely on films. Nanowires were also connected to photolithographically defined metallic pads for electrical characterization via electron beam induced Pt deposition using a focused ion beam microscope. Importantly, contact electrode geometries are well defined because our electrical connections were achieved by focused ion beam lithography. This well defined electrode geometry allows various parameters to be extracted from nanowire current-voltage data. Specifically, we obtain resistivity, majority carrier concentration, and majority carrier mobility values for multisegmented Au-CdTe-Au wires with three different Cd/Te ratios. Using various electrolytes to alter nanowire stoichiometry may allow future research to explore the electrical properties of continuously gradated compound semiconductor nanowires. By exploring the resulting Cd/Te ratios grown via uninterrupted deposition at a constant potential subsequent research may be able to grow nanowires with continuously and complexly varying compositions.

Stoichiometry-adjusted Au-CdTe-Au Nanowires and Films: Synthesis and Characterization

Similar fabrication techniques have been discussed previously[11,16,17]. Briefly, we used porous AAO membranes (Whatman Anodisc 13, Maidstone, UK) with nominal pore diameters of 200 nm. These templates have a pore density of $9x10^{12}$ pores/m^2. A 400 nm silver film was thermally evaporated onto one side of the templates, sealing the pores and providing a working electrode for deposition. The templates were placed into a custom designed electrodeposition chamber. Electrodeposition was carried out using a VoltaLab PST050 potentiostat operated in chrono coulometry mode. Ag was first plated as a sacrificial layer from a commercially available Ag electrolyte (1025 RTU, Technic Inc., Anaheim, California, USA). Subsequently, Au was deposited from a commercial solution (Orotemp 24 RTU, Technic, Inc.). CdTe was plated from one of several custom solutions, followed by a second Au deposition. The CdTe electrolytes contained 0.3M CdSO$_4$, 0.5M H$_2$SO$_4$, and either 0.25, 0.5, 2, 4, or 6mM Te(IV)O$_2$. All depositions were performed at 25°C. Semiconducting layers of CdTe were deposited at a constant -580 mV potential with respect to a standard Ag/AgCl reference electrode (Basi Inc., West Lafayette, Indiana, USA); a 0.5 mm thick platinum sheet was used as a counter electrode. Following deposition the template was removed from the chamber and all Ag segments were etched in dilute HNO$_3$; the multisegmented Au-CdTe-Au wires were released by etching the AAO template in 1M NaOH, then resuspended in ethanol. Wires were grown in an array of lengths, however wires with long Au segments (greater than 1.5 μm) experienced significantly higher rates of fracture at the metal-semiconductor junction than wires grown with Au segments shorter than 1.5 μm. Figure 1 shows multicomponent wires with majority CdTe segments (Figure 1a) and majority Au segments (Figure 1b). It should be noted that in some cases narrowing occurs in the CdTe segment at the Au-CdTe interface. This narrowing results when wires remain in contact with NaOH for excessive amounts of time, typically longer than 24 hours. This is likely the result of galvanic corrosion occurring at the Au-CdTe junction, however detailed analysis of this phenomenon is beyond the scope of this report.

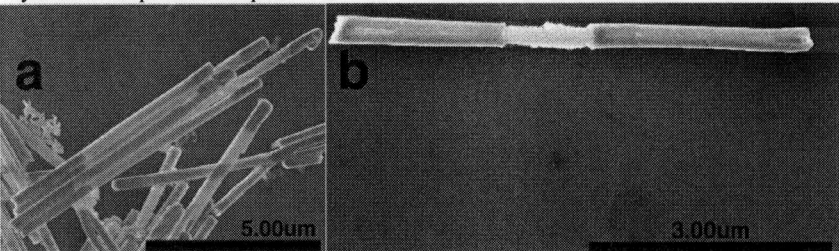

Figure 1: SEM images indicating Au-CdTe-Au wires deposited into the pores of Whatman Anodisc templates (nominal pore diameter 200 nm), released, and deposited onto a p^+-Si/SiO$_2$ substrate. (a) CdTe majority wires; (b) Au majority wire.

In order to compare template deposited CdTe wires and surface deposited films of varying stoichiometries we electrodeposited CdTe films onto Au-coated glass slides. Films and nanowires were electrodeposited using chemically identical electrolytes. CdTe film depositions were performed under the same conditions as wire depositions, namely, all depositions were performed at room temperature and at an applied potential of -580 mV with respect to a standard double barrel Ag/AgCl reference electrode. All deposition solutions were mixed using a magnetic stir bar.

Energy dispersive x-ray spectroscopy

In order to characterize the bulk composition of these wires, samples stored in ethanol were deposited onto 13 mm diameter graphite mounts (Ted Pella Inc., Redding, California, USA). A Hitachi S-4700 FESEM was used to image the wires and energy dispersive x-ray spectroscopy (EDS) data was collected using an Oxford Inca Energy EDS system. The XPP matrix correction scheme included in the Oxford Inca software package was used to analyze the data. This correction method is based on the common $\phi(\rho z)$ approach and corrects for atomic number effects, absorption, and fluorescence effects. Figure 2a shows a bundle of wires that have remained bound to one another due to non-specific aggregation. We performed Au, Te, and Cd atomic mapping on this bundle; Figures 2b-d indicate the spatial confinement of Au and CdTe segments within the wires.

The EDS data also indicates that varying the concentration of Te^{4+} in the electrolyte solution during deposition leads to easily adjustable nanowire Cd/Te stoichiometries. As shown in Figure 3, we fabricated both Cd- and Te-rich wires. The transition from Te-rich to Cd-rich nanowire deposition occurred between electrolytes with 0.5mM and 0.25mM Te^{4+}, respectively.

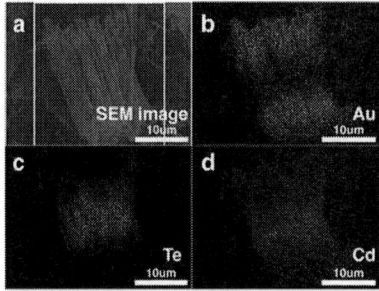

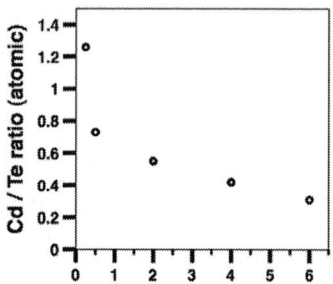

Te concentration in electrolyte (mM)

Figure 2: Energy-dispersive x-ray spectroscopy data indicating spatial distribution of elements Au, Te, and Cd. (a) SEM image of a Au-CdTe-Au wire bundle, (b) map of Au concentration, (c) map of Te concentration, (d) map of Cd concentration. These maps clearly indicate spatial confinement of Au and CdTe wire segments.

Figure 3: EDS data showing the concentration of tellurium incorporated into wires for each CdTe electrolyte. For an electrolyte with a Te^{4+} concentration of 0.25mM the resulting wires are cadmium rich (Cd/Te ratio of 1.26), however at 0.5mM Te^{4+} wires are slightly tellurium rich (Cd/Te ratio of 0.73).

Electrolyte designation	Electrolyte used (mM Te^{4+})	Cd/Te ratio in wires, EDS data (atomic ratio)	Cd/Te ratio in film, EDS data (atomic ratio)	Cd/Te ratio in film, XPS data (atomic ratio)
A	0.25	1.26	0.59	0.96
B	0.5	0.73	0.49	0.75
C	2	0.55	0.43	0.33
D	4	0.42	0.48	0.39
E	6	0.31	0.00	0.00

Table 1: EDS and XPS measurements taken on nanowires and films electrodeposited using identical electrolytes indicates clear differences in stoichiometry. CdTe deposition into AAO pores tends to produce wires that are Cd-rich relative to films depositions.

X-ray photoelectron spectroscopy (XPS) characterization of composition-adjusted CdTe films

In order to probe the atomic coordination of electrodeposited CdTe, we deposited films of CdTe onto glass slides coated with a 100 nm layer of thermally evaporated Au. (Films were analyzed instead of nanowires since it was much more difficult to locate the sparsely distributed nanowires within the large XPS analysis area.) Previous research has discussed the use of XPS as a technique capable of providing quantitative compositional surface analysis of various materials[18,19]. XPS studies were conducted with a Kratos Axis Ultra^{DLD} x-ray photoelectron spectrometer equipped with a monochromatic Al Kα source operated at 150 W (15 kV and 10 mA). Data was acquired at a pass energy of 20 eV and a step size of 0.1 eV for high resolution scans, and a pass energy of 80 eV and a step size of 1 eV for survey spectra. The system base pressure was ~$5x10^{-9}$ Torr, and the analyzed spot size was 300 x 700 μm. All experimental data was corrected to the C 1s line at a binding energy of 284.6 eV. XPS data was analyzed with Kratos Vision software and the quantitative analysis described below used the relative sensitivity factors contained within the Vision software.

The full spectrum data for samples A – D indicates the presence of tellurium and cadmium, along with oxygen and small amounts of adventitious carbon. The decreasing Cd 3d peak intensity and increasing Te 3p, Te 3d, Te 4d, and Te Auger peak intensities with increasing Te^{4+} concentration (in electrolyte) follows the expected trend. Cd 3p peaks also diminish with increasing Te^{4+} concentration in electrolyte and, as is expected given the EDS data taken on films, completely vanished for sample E.

In order to better resolve the Te 3d and Cd 3d peaks high resolution XPS data was collected in these binding energy ranges. The XPS data shown in Figure 5a indicates a splitting of the Te 3d peaks, indicating the presence of at least two forms of tellurium, presumably TeO_2 and CdTe at 572.5 eV and 576 eV, respectively[20-23].

The spectrum for sample E exhibited no Cd 3d peaks (Figure 5b) and the absence of cadmium in this sample was confirmed by EDS data for films (Table 1). We therefore attribute the peak at ~573 eV in this sample to Te^0, since previous reports indicate that it is difficult to resolve the Te 3d peaks in CdTe and Te^0 seeing as the individual contributions are only distant by 0.1 eV[24-27].

The XPS data and the EDS data on thin films show comparable trends (Table 1). The differences between EDS data on thin films and nanowires was unexpected, and indicates that differences in diffusion rates or electrical field differences between the two substrates plays an important role in the final film composition. Additionally, discrepancies in Cd/Te ratios exhibited in electrodeposited films have been previously noted and have been attributed to variation in stirring rate of the solution[28]. Interestingly, these discrepancies are not exhibited in the CdTe nanowire data.

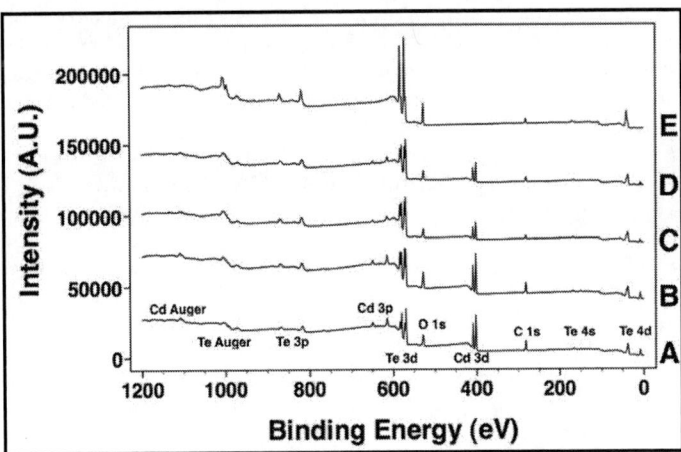

Figure 4: Full spectrum of XPS data. Relevant peaks are indicated on the spectrum for Sample A. The full spectrum clearly shows diminishing Cd 3d peaks through Sample E, at which point all Cd ceases being deposited. Labels A through E represent the electrolyte designations denoted in Table 1.

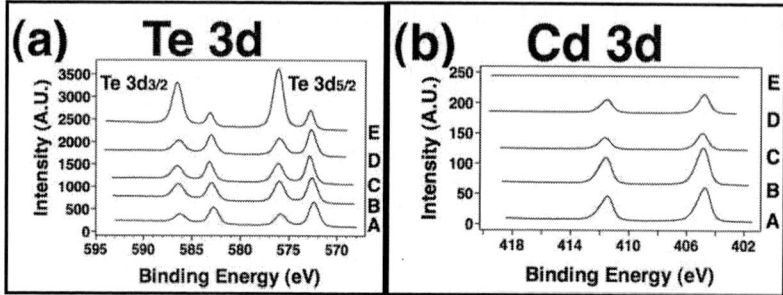

Figure 5: High resolution scans of Te 3d (Figure 5a) and Cd 3d (Figure 5b) XPS data. (a) The intensity shift occurring in Sample E is a result of increased Te^0 and TeO_2 concentrations and decreased TeCd concentrations. (b) Sample E shows complete disappearance of Cd 3d peaks.

The constant potential electrodeposition of CdTe produces the following two reactions:

$$HTeO_2^+ + 3H^+ + 4e^- \rightarrow Te + 2H_2O$$
$$Cd^{2+} + Te + 2e^- \rightarrow CdTe.$$

It is well documented that Te diffusion into the pores of a template is the rate-limiting process for growth of CdTe nanowires[13,15]. The first step, underpotential deposition of Te_s is diffusion controlled. The second step, deposition of Cd^{2+} to form CdTe, is kinetically controlled. Wire deposition occurs in the pores of a 50 μm thick AAO template and therefore is subject to what is essentially a 50 μm diffusion layer between pore opening and the working electrode (resistively evaporated Ag layer on one side of the template). This large diffusion layer explains differences in composition between nanowires and films. The film deposition process, although likewise Te-diffusion controlled, did not impose a 50 μm diffusion layer. This diffusion layer likely explains differences in Cd/Te ratios in electrodeposited nanowires and films.

Electrical Characterization of Au-CdTe-Au Nanowires

Focused ion beam (FIB) microscopy is currently under heavy investigation for nanofabrication purposes due to its ability to rapidly prototype nanoscale devices. A FIB microscope is similar to a scanning electron microscope (SEM) but operates by scanning an ion beam over a substrate instead of an electron beam. Typically, the ion source is generated from gallium ions due gallium's convenient physical properties, including having a low vapor pressure and being a liquid at near room temperatures[29]. Current FIB systems are often termed 'dual-beam systems' as they contain both electron and ion columns. These systems allow for patterning of a surface with either the ion or electron beam and even allow for the simultaneous patterning and imaging of a surface in real time.

We employed a Helios model 600 Nanolab dual beam FIB/SEM for patterning platinum electrodes onto individual nanowires in our investigation of elucidating the relationship between the elemental composition and the corresponding electrical properties of metal-semiconductor-metal Au-CdTe-Au nanowires. The use of a FIB for direct-write patterning permits rapid prototyping of nanostructures by directly structuring electrodes onto a given substrate and provides immediate feedback, eliminating the multistep and time consuming processing associated with conventional lithography. Patterning of metal is achieved through the injection of a precursor gas into the sample chamber through a needle near the surface of the substrate. The gas is decomposed to nanocrystalline metal contained in an amorphous carbon matrix on the substrate surface when gas molecules interact with the electron or ion beam[30]. A platinum precursor is used here, although tungsten and gold sources are also other alternatives which are typically available on these systems.

Direct-write patterning for electrical contacts via ion beam and electron beam deposition was performed on a dual beam FIB/SEM system based on a previously described method[30]. Briefly, a platinum metal-organic precursor is heated and injected into the path of the electron/ion source through an injection needle while the beam scans a surface oriented normal to the rasterization path of the beam. Electrical leads 10 µm in length were patterned with the electron beam while in close proximity to the nanowires of interest in order to avoid the unwanted implantation of Ga^+ ions into the semiconducting channel of the nanowire. At distances sufficiently far away from the semiconducting segment, additional platinum was deposited with the ion beam in order to form electrical connections between the electron beam deposited platinum connected to the nanowire and the photolithographically patterned macroscopic metallic pads. Electron beam platinum deposition was performed at 10 kV with a 1.4 nA aperture while ion beam platinum deposition was performed at 30 kV with a 48 pA aperture, with a dwell time of 1.0 µs and 200 ns for electron beam and ion beam platinum deposition, respectively. The internal pressure of the sample chamber was monitored between depositions and the electron beam was not re-enabled until the pressure was below 1.0×10^{-5} Torr to avoid any unwanted platinum being unintentionally deposited on the sample surface between depositions.

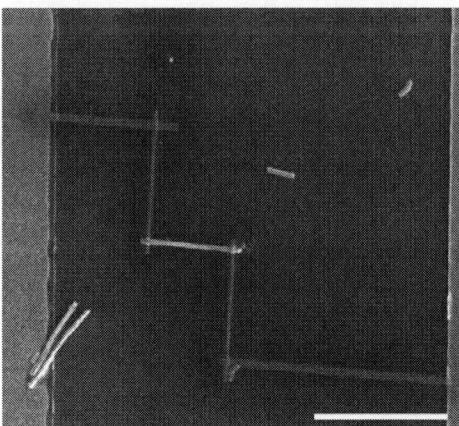

Figure 6: SEM image of an Au-CdTe-Au nanowire (6mM Te^{4+}) connected to larger photolithographically patterned pads via electron beam and ion-beam deposited platinum leads. Scale bar is 10 µm.

The electrical properties of Au-CdTe-Au nanowires were investigated by fabricating source-drain electrodes via a combination of platinum deposition in a combined FIB/SEM with both the electron and ion beams. Platinum deposition in FIB/SEM systems results in platinum ions embedded within a host matrix, where the matrix includes carbon from the precursor gas, gallium if the ion beam is used for deposition, and possibly small quantities of oxygen if the chamber pressure is not sufficiently low[31]. In order to avoid the unintentional doping of the semiconducting channel of the multisegmented nanowire, the electron beam was used to pattern initial contacts on the metallic ends of the nanowire. Ion beam induced platinum deposition was used in subsequent patterned electrodes that were farther than 10 μm away from the nanowire of interest, as the resistance between two parallel patterned electrodes has been shown to be larger than 10^{13} ohms[30].

Electrical characteristics of electrically contacted nanowires composed of varying stoichiometries of cadmium and tellurium were extracted using a quantitative parameter retrieval method[32]. Thermionic emission is the most widely used model when analyzing the forward-bias characteristics of quasi-one dimensional nanostructured devices containing Schottky contacts. This model also underestimates the reverse-bias characteristics which are non-negligible in nanostructured devices due to the lower measured device currents, although the model assumes the reverse bias currents are negligible in microstructured devices. Thus, the model used here and proposed by Zhang *et al.* is based on thermionic field emission theory developed by Padovani and Stratton[33], where the tunneling effects through the barrier layer are accounted for in a reverse-biased junction that gives rise to the reverse-biased current.

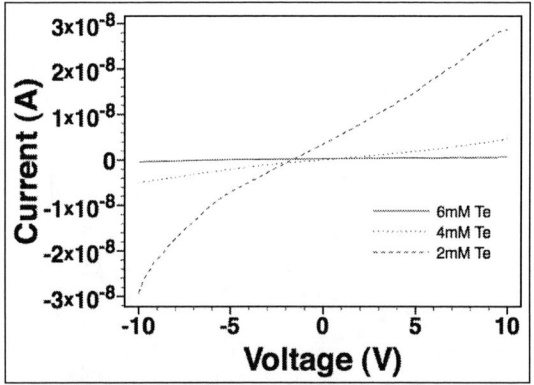

Figure 7: Current versus voltage (*I-V*) plots of Au-CdTe-Au nanowires on an Si/SiO$_2$ substrate contacted to photolithographically defined Au pads in a combined FIB/SEM microscope via electron and ion beam deposited platinum.

Figure 7 contains typical *I-V* curves for a subset of the 200 nm diameter nanowires composed of varying stoichiometries that are investigated in this work. The resistance can be obtained from the high bias portion of the *I-V* curve via the differential voltage method using the equation $R = a_{NW}(dV/dI) \approx dV/dI$, where a_{NW} is the voltage drop factor of the reverse biased Schottky barrier. At intermediate bias, where the reverse bias Schottky barrier dominates the transport characteristics, a logarithmic plot of *I* versus *V* gives a plot with an approximately straight line via the equation

$$\ln I = \ln(SJ) = \ln(S) + V\left(\frac{q}{kT} - \frac{1}{E_0}\right) + \ln J_s \ (1)$$

where S is the contact area of the Schottky barrier, J is the current density through this barrier, E_0 is a parameter that depends on the density of the majority carriers, and J_s is a slowly varying parameter based on the applied bias[32]. Since J_s is a slowly varying function, a line fitted to these plotted points has a slope of $q/kT - 1/E_0$, which subsequently yields the parameter E_0 and can be used to find the characteristic energy E_{00}[34], obtained from $E_0 = E_{00} \coth(qE_{00}/kT)$ with

$$E_{00} = \frac{qh}{4\pi}\left[\frac{N_d}{m^*\varepsilon_s\varepsilon_0}\right]^{1/2} \quad (2).$$

The donor concentration N_d can then be extracted from equation 2, which then permits the majority carrier mobility to be calculated using $\mu = 1/qn\rho$ where ρ is the resistivity of the nanowire. The calculated values extracted from various stoichimetry-adjusted CdTe nanowire $I\text{-}V$ curves in Figure 7 are summarized in Table 2, where the donor concentrations obtained are found to be in good agreement with values previously obtained for other CdTe materials[35]. The carrier mobilities are found to increase with decreasing starting concentrations of tellurium in the nanowire electrolyte, which results in a progressively increased concentration of cadmium in the final electrodeposited nanowire. Future work will examine the relationship between the majority carrier mobility and the stoichiometry of the nanowire as the stoichiometry approaches a 1.0:1.0 tellurium : cadmium ratio and ultimately crosses over into the cadmium-rich regime.

Table 2: Extracted parameters for semiconducting CdTe nanowires containing various Te electrolyte precursor concentrations.

Te concentration	2mM	4mM	6mM
Length (µm)	4.2	7.0	7.2
E_0 (meV)	26.2	26.1	26.0
Majority carrier concentration	7.5E18	5.3E18	3.2E18
Resistivity (Ω cm)	5.33E02	2.16E03	6.50E03
Mobility (cm^2/V s)	1.6E-03	5.4E-04	3.0E-04

Note: values for the effective electron mass ratio were taken from ref [35].

The low carrier mobilities observed here are comparable to those previously obtained in electrodeposited in-wire chalcogenide semiconductor nanowires, and it has been noted that much higher values may be achieved by improvements made to the metal-semiconductor interfaces as well as improvements made in selecting better metal contacts[36]. We have previously shown that improvements in the interfaces do result in markedly better carrier mobilities[37], and recent work also conducted in our laboratory demonstrates that careful selection of the electrodeposited metal also results in higher mobilities (to appear shortly in the literature).

Summary and Conclusions

We have presented a method for fabricating metal-semiconductor-metal nanowires with varying stoichiometries. We have characterized these wires using EDS and shown electrical property dependence upon nanowire composition. We calculated nanowire carrier concentration, resistivity, and mobility using a quantitative parameter retrieval method based on a commonly implemented thermionic emission model.

Acknowledgements

We gratefully acknowledge funding from the NIH and NSF for the work carried out in these experiments. Specifically, we acknowledge support from NIH grant number 5-P41-EB002025-21-25. Also, we would like to thank staff members of the Chapel Hill Analytical and Nanofabrication Laboratory.

References

1. Rakhshani, A. E. Electrodeposited CdTe - optical properties. *Journal of Applied Physics* 81, 7988 (1997).
2. Mathew, X., Enriquez, J. P., Romeo, A. & Tiwari, A. N. CdTe/CdS solar cells on flexible substrates. *Solar Energy* 77, 831-838 (2004).
3. Grecu, D., Compaan, A. D., Young, D., Jayamaha, U. & Rose, D. H. Photoluminescence of Cu-doped CdTe and related stability issues in CdS/CdTe solar cells. *Journal of Applied Physics* 88, 2490 (2000).
4. Zhou, S. M., Zhang, X. H., Meng, X. M., Wu, S. K. & Lee, S. T. Fabrication and characterization of Zn-doped CdTe nanowires. *Applied Physics A: Materials Science & Processing* 81, 1647-1650 (2005).
5. Gandhi, T., Raja, K. S. & Misra, M. Templated growth of cadmium zinc telluride (CZT) nanowires using pulsed-potentials in hot non-aqueous solution. *Electrochimica Acta* 51, 5932-5942 (2006).
6. Rakhshani, A. E. Heterojunction properties of electrodeposited CdTe/CdS solar cells. *Journal of Applied Physics* 90, 4265 (2001).
7. Casagrande, L. G., Di Marzio, D., Lee, M. B., Larson, D. J., *et al.* Vertical Bridgman growth and characterization of large-diameter single-crystal CdTe. *Journal of Crystal Growth* 128, (1993).
8. Rudolph, P., Rinas, U. & Jacobs, K. Systematic steps towards exactly stoichiometric and uncompensated CdTe Bridgman crystals. *Journal of Crystal Growth* 138, 249-254 (1994).
9. Zhao, A. W., Meng, G. W., Zhang, L. D., Gao, T., *et al.* Electrochemical synthesis of ordered CdTe nanowire arrays. *Applied Physics A: Materials Science & Processing* 76, 537-539 (2003).
10. Sima, M., Enculescu, I., Trautmann, C. & Neumann, R. Electrodeposition of CdTe nanorods in ion track membranes. *Journal of Optoelectronics and Advanced Materials* 6, 121-125 (2004).
11. Ohgai, T., Gravier, L., Hoffer, X. & Ansermet, J. P. CdTe semiconductor nanowires and NiFe ferro-magnetic metal nanowires electrodeposited into cylindrical nano-pores on the surface of anodized aluminum. *Journal of Applied Electrochemistry* 35, 479-485 (2005).
12. Enculescu, I., Sima, M., Enculescu, M., Enache, M., *et al.* Deposition and properties of CdTe nanowires prepared by template replication. *Physica Status Solidi Basic Research* 244, 1607 (2007).
13. Ion, Enculescu & Antohe Physical properties of CdTe nanowires electrodeposited by a template method, for photovoltaic applications. *Journal of Optoelectronics and Advanced Materials* 10, 3241-3246 (2008).
14. Wang, X. & Ozkan, C. S. Multisegment Nanowire Sensors for the Detection of DNA Molecules. *Nano Lett* 8, 398-404 (2008).
15. Kum, M. C., Yoo, B. Y., Rheem, Y. W., Bozhilov, K. N., *et al.* Synthesis and characterization of cadmium telluride nanowire. *Nanotechnology* 19, 325711 (2008).
16. Pena, D. J., Mbindyo, J. K. N., Carado, A. J., Mallouk, T. E., *et al.* Template Growth of Photoconductive Metal-CdSe-Metal Nanowires. *Journal of Physical Chemistry B* 106, 7458-7462 (2002).
17. Skinner, K., Dwyer, C. & Washburn, S. Selective Functionalization of Arbitrary Nanowires. *Nano Lett* 6, 2759 (2006).

18. Nebesny, K. W., Maschhoff, B. L. & Armstrong, N. R. Quantitation of Auger and x-ray photoelectron spectroscopies. *Analytical Chemistry* 61, (1989).
19. Tilinin, I. S., Jablonski, A. & Werner, W. S. M. Quantitative Surface Analysis by Auger and X-Ray Photoelectron Spectroscopy. *Progress in Surface Science* 52, 193-335 (1996).
20. Bahl, M. K., Watson, R. L. & Irgolic, K. J. X-ray photoemission studies of tellurium and some of its compounds. *The Journal of Chemical Physics* 66, 5526 (1977).
21. Sun, T. S., Buchner, S. P. & Byer, N. E. Oxide and interface properties of anodic films on HgCdTe. *Journal of Vacuum Science and Technology* 17, 1067 (1980).
22. Ricco, White & Wrighton X-ray photoelectron and Auger electron spectroscopic study of the CdTe surface resulting from various surface pretreatments: correlation of photoelectrochemical and capacitance-potential behavior with surface chemical composition. *J. Vac. Sci. Technol. A* 2, (1984).
23. El Azhari, M. Y., Azizan, M., Bennouna, A., Outzourhit, A., *et al.* Structural properties of oxygenated amorphous cadmium telluride thin films. *Thin Solid Films* 295, 131-136 (1997).
24. Shevchik, N. J., Cardona, M. & Tejeda, J. X-Ray and Far-uv Photoemission from Amorphous and Crystalline Films of Se and Te. *Physical Review B* 8, 2833-2841 (1973).
25. Polak X-ray photoelectron spectroscopic studies of $CdSe0.65Te0.35$. *Journal of Electron Spectroscopy and Related Phenomena* 28, 171-176 (1982).
26. Danaher, W. J., Lyons, L. E., Marychurch, M. & Morris, G. C. Chemical etching of crystal and thin film cadmium telluride. *Applied Surface Science* 27, 338-354 (1986).
27. Mandale & Badrinarayanan X-ray Photoelectron spectroscopic studies of the semimagnetic semiconductor system Pb1-xMnxTe. *Journal of Electron Spectroscopy and Related Phenomena* 53, 87-95 (1990).
28. Lepiller, C. & Lincot, D. New Facets of CdTe Electrodeposition in Acidic Solutions with Higher Tellurium Concentrations. *Journal of The Electrochemical Society* 151, C348 (2004).
29. Swart, J. W., Vaz, A. R., Silva, M. M. D., Leon, J. & Moshkalev, S. A. Platinum thin films deposited on silicon oxide by focused ion beam: characterization and application. *Journal of Materials Science* 3429-3434 (2008).
30. Gopal, V., Radmilovic, V. R., Daraio, C., Jin, S., *et al.* Rapid Prototyping of Site-Specific Nanocontacts by Electron and Ion Beam Assisted Direct-Write Nanolithography. *Nano Letters* 4, 2059-2064 (2004).
31. Tao, T., Ro, J., Melngailis, J., Xue, Z. & Kaesz, H. D. Focused ion beam induced deposition of platinum. *J Vac Sci Technol B* 8, 1826-1829 (1990).
32. Zhang, Z. Y., Yao, K., Liu, Y., Jin, C. H., *et al.* Quantitative analysis of current-voltage characteristics of semiconducting nanowires: Decoupling of contact effects. *Advanced Functional Materials* 17, 2478-2489 (2007).
33. Padovani, F. A. & Stratton, R. Field And Thermionic-Field Emission In Schottky Barriers. *Solid-State Electronics* 9, 695-695 (1966).
34. Horváth, J. Comment on "Analysis of I-V measurements on CrSI2-Si Schottky Structures in a wide temperature range". *Solid-State Electron.* 39, 176-178 (1996).
35. Marple, D. T. F. Effective Electron Mass in CdTe. *Physical Review* 129, 2466-2466 (1963).
36. Kovtyukhova, N. I., Kelley, B. K. & Mallouk, T. E. Coaxially gated in-wire thin-film transistors made by template assembly. *J Am Chem Soc* 126, 12738-12739 (2004).
37. Skinner, K., Dwyer, C. & Washburn, S. Quantitative analysis of individual metal-CdSe-metal nanowire field-effect transistors. *Applied Physics Letters* 92, 112105 (2008).

110

CHAPTER 4

COMPOUND SEMICONDUCTOR MATERIALS AND DEVICES

112

ECS Transactions, 19 (3) 113-121 (2009)
10.1149/1.3120692 ©The Electrochemical Society

GaN HEMT Reliability Through the Decade

K. V. Smith[a], S. Brierley[a], R. McAnulty[a], C. Tilas[a], D. Zarkh[a] M. Benedek[a], P. Phalon[a] and A. Hooven[a]

[a] Raytheon Company, Andover MA, 01810

The reliability of GaN-based transistors has improved over the last decade into a system insertion ready technology. Material quality and uniformity has played a strong role in this evolution. Identifying and understanding the physical evolution of the degradation mechanisms provides a basis for evaluating device lifetimes. Recent life testing has demonstrated sufficient lifetime for most applications.

Introduction

Since the early reports of GaN-based technologies in the 1990's, the wide band gap and high temperature stability has promised a disruptive technology. Published reports of high power densities [1] and high frequency operation [2-3] has pushed the expectations of these devices ever higher. Yet, concerns about the reliability continue to limit the inclusion of GaN devices into system applications.

The progress in GaN reliability has been very rapid. Fig. 1 illustrates the early reliability results from the beginning of this decade. In both Fig.1a) and b), the devices are degrading rapidly and reach failure criteria in 10's of hours. The voltage of these tests is significantly below the standard application voltages of 28V and 48V targeted for GaN power devices. In the few years since these reports, GaN reliability has reached the 10^6 hr reliability levels that are necessary for system insertion.

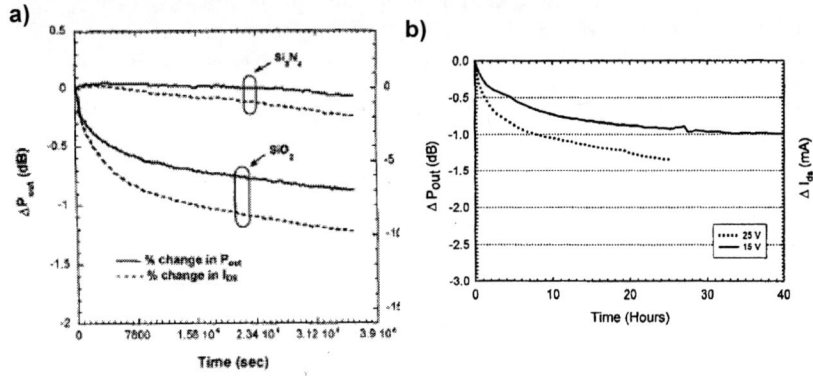

Figure 1: Early reliability reports: a) 20V RF stress on Si_3N_4 and SiO_2 reported by Kim, etal [4] (© [2001] IEEE) and b) RF stress at 15 and 20V reported by Lee, etal [5] (© [2002] IEEE)

113

Material Properties

One of the first challenges to developing a reliable GaN-based technology is the unique physics and physical properties of the III-nitrides. III-nitrides are very robust toward chemical processes and difficult to grow, resulting in a relatively simple device structure of an AlGaN layer on top of a GaN buffer layer. Figure 2 provides a simplified diagram of the device structure as well as some of the material variations that can and have affected device reliability. The common growth substrates of Al_2O_3, Si and SiC have large lattice mismatches to GaN that can result in high dislocation densities and other growth related non-uniformities. Initial growth efforts yielded densities of $>10^9$ per cm^2 and often much higher. Dislocation densities have been improved from $> 10^{10}$ cm^{-2} to 10^8 cm^{-2}, which is still significantly higher than acceptable levels in any other semiconductor system. The strong ionicity of the group III nitrogen bond results in strong polarization fields, which link material quality directly to device operation through the high sheet charge densities as described in [6-7]. Yet devices made from these materials are showing extremely good lifetimes and robustness as will be described later.

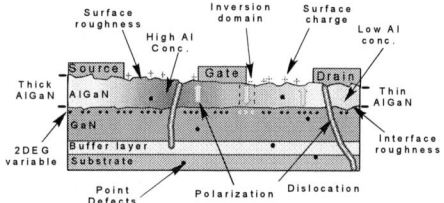

Figure 2: Simplified $Al_xGa_{1-x}N$/ GaN HEMT showing possible factors affecting performance and reliability.

In the early years dating back to 2000-2001, the non-uniformities of wafers and material defects were a significant challenge to reliability. An example of a 2" wafer grown by metal organic chemical vapor deposition in 2001 (Fig. 3a) had sheet resistivity variations of $> 15\%$, illustrating the non-uniformities of the material. On-wafer RF operational stress testing (Fig 3b) of < 100hrs of the wafer was performed in the shaded area of the wafer on multiple devices. Some of the devices reached the failure criteria of 1dB drop in output power with 10hrs, while others had projected lifetimes in excess of 100hrs and longer. The variation within a wafer and also between wafers was a significant problem in reliability testing and predictions

In the past 5 years, major strides have been taken both in improving the uniformity and quality of GaN-based HEMTs. The sheet resistivity map in Figure 3c of a 4" MBE-grown wafer from 2008 illustrates these improvements. The wafer has substantially better mobility and higher charge, but more importantly from a reliability perspective, the wafer has less than a 1% variation in uniformity across the wafer. Improvements such as these have made statistical testing and analysis possible in current sample populations.

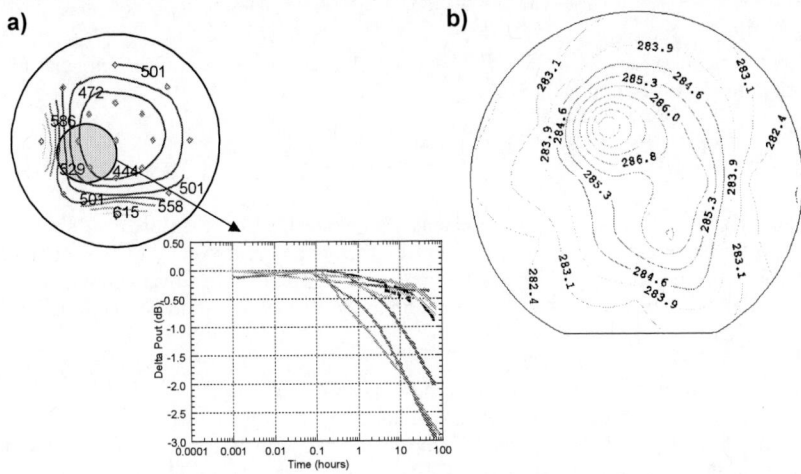

Figure 3: a) Sheet resistivity map of a 2" MOCVD-grown wafer from 2001 and b) RF stress of multiple devices from area indicated on map. c) Sheet resistivity map of 2008 MBE grown wafer showing significant improvement in uniformity.

Failure mechanisms

Throughout the development of III-nitrides, substantial effort has gone into identifying physical mechanisms responsible for device degradation Understanding the degradation mechanisms are an important part of any developing technology. Identifying a mechanism can be used to improve in material, process and/or device structures to extend device lifetime and improve robustness. Perhaps even more important, understanding the physical phenomena is essential to the reliability testing used to predict lifetimes. In an accelerated life test, the mean time to failure (MTTF) is related to the three major stress factors in electronics through the general reliability equation:

$$MTTF = C * AF(I) * AF(V) * AF(T)$$

where C is a constant, and AF() is an acceleration factor related to the 3 stress factors: current (I), voltage (V) and temperature (T). To accurately predict device lifetimes from accelerated testing, the functional form of the degradation mechanism needs to be understood. The most common form of acceleration (Arrhenius testing) is operation at elevated temperatures. Accurate predictions using the Arrhenius model requires several major assumptions. First, the acceleration factors related to current and voltage are constants. Common method for achieving this situation is to maintain the drain current constant throughout the test (a constant current test) by varying the gate voltage. The small changes in gate voltage are assumed to have no affect on the voltage acceleration factor, resulting in:

$$MTTF = C * C_i * C_v * AF(T)$$

where C_i and C_v are constants described above. The functional form of the temperature acceleration factor is assumed to be:

$$MTTF \propto e^{\frac{-Ea}{\kappa T}}$$

where Ea is the activation energy of the degradation mechanism, κ is Boltzman's constant and T is the channel temperature in the device. If any one of these conditions is not met, substantial discrepancies in predicted lifetimes will exist, most often over estimating device lifetime. These factors therefore are important to any discussion of degradation mechanisms

While degradation mechanisms can be specific to a process or material type, there have been several common themes throughout the development community. One of those mechanisms is the formation of trap states and trapping of charge in the drain region of the gate [4]. The large fields, high energy electrons and high device temperatures can result in the formation of microscopic defects in the passivation, active layers and in the buffer layer. Electrically active defect formation is difficult to identify as there are few techniques to physically "see" these trapped charges. The identification typically is inferred from changes in electrical behavior. A transmission electron microscope image (TEM) of a recent device exhibiting a trap-related degradation is shown in Fig.4a. There are no physical changes evident in the degraded device in a post stress test analysis, consistent with a trap creation phenomenon. Because trap related mechanisms have an activation energy related to the creation and filling of traps, an Arrhenius type behavior can be assumed for analysis of accelerated testing.

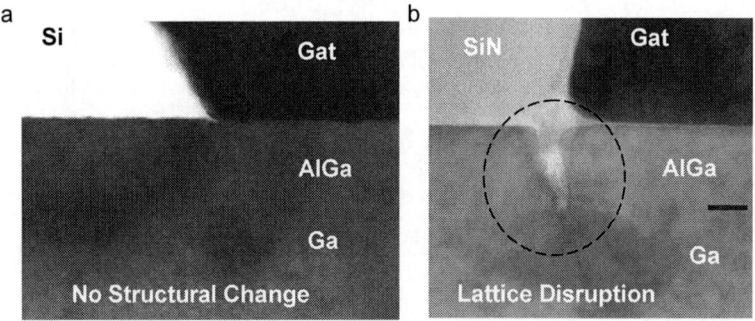

Figure 4: TEM images of degraded GaN devices showing a) a trapping mechanism and b) lattice disruptions

Another commonly reported mechanism is a structural change at the edge of the gate as seen in a device stressed at 28V high temperatures DC conditions (Fig. 4b). It should be noted that these devices demonstrated a lifetime in excess of 10^7 hrs at a channel temperature of 150°C (as shown in Fig. 7). First reported by Jimenez, etal [8] in

2006, these lattice disruptions (LD) have also been called pits and cracks [9]. While sometimes seen on the source side of the gate, the LD's are more prevalent and more severe at the drain edge of the gate.

Several factors appear to be necessary for this type of degradation mechanism. The only known reports of this mechanism have come from high temperature accelerated tests. We have analyzed devices tested under standard RF operating conditions for >17000hrs (Fig. 8) that did not have any indication of the LD's, although the devices were from the same wafer as the device shown in Fig.4b. Yet, devices from the same wafer were tested at high temperature heat storage testing (no bias) until failure and LDs were undetectable in those devices. This indicates that temperature is not sufficient for LD's and that electrical stimulus is important. Reports have linked the electrical stimulation to the LD's through a reverse piezo-induced stress [10]. The electric field interacts with the strong polarization fields in the AlGaN barrier layer increasing the stress in the strained AlGaN layer. This increase in strain is proposed to exceed the critical strain to induce cracking in the AlGaN layer.

The stress at the edge of the gate is a combination of several strains, shown in Figure 5. The biaxial strain of the epitaxial growth of AlGaN on GaN is important and can be calculated based on the Al composition in the active layer. As noted above, the inverse piezo-stress can be calculated using a field distribution (in our case taken from [11]) and the polarization constants from the literature. Detailed calculations (not shown here) indicate that the vertical field is dominant and that the lateral field (i.e. the applied drain voltage) will have no affect on the induced strain. The third stress is the edge strain created by depositing materials with different coefficients of thermal expansion (CTE) onto the epitaxial layer. A 2-dimensional strain field can be calculated using the methods of [12] and [13], which are similar to

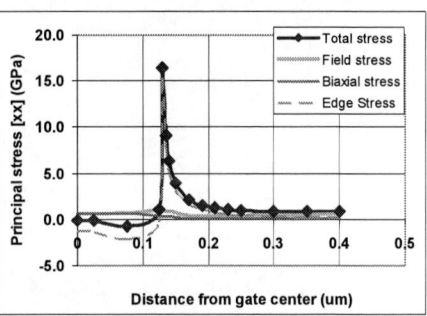

Figure 5: Calculated individual stresses and total stress at the AlGaN surface

finite element modeling of the process steps (including materials and temperatures of deposition),. Adding each of these components of the strain in Figure 5, the edge strain is clearly dominant. From a practical point of view, high temperature operation will reduce the edge strain (temperatures approach those in the depositions) and the increase due to the electric field would still be significantly less the strain of the device at room temperature with no bias. It is, therefore, our contention that the strain is not the dominant effect in LDs.

Detailed structural analysis of LDs indicate a etch pit type behavior. The lattice does not show the translational effects expected from a strain induced structural failure. In addition, the area within the defect has an amorphous material that forms a mound above the disruption. Micro-elemental analysis indicates that the area contains contaminants; most notably oxygen. The oxygen is isolated to the area of the disruption and is not present in the undisturbed lattice adjacent.

Combining all of these details, a complicated model for the formation of lattice disruptions is proposed in Figure 6. The defining characteristic of the process is an electro-chemical interaction similar to the anodization process reported by Miller, etal [14]. The total strain shifts the polarization charge and produces an electrical path for gate current through the material. The current interacts with the contaminant (believed to be oxygen) to "etch" the material along the strain gradients. The contaminant could be a part of the dielectric deposition and/or be diffusing from the atmosphere. Conway, etal [15] reported that stress testing in a nitrogen atmosphere (versus air) suppressed the formation of LDs, which has been supported by internal testing. The proposed model will be a diffusion limited process that will follow the Arrhenius model. It should be noted that a strain-induced cracking most likely would not have the necessary exponential dependence for an Arrhenius behavior, which would complicate lifetime predictions.

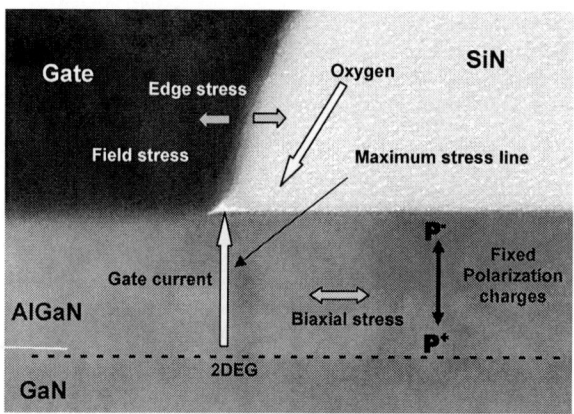

Figure 6: Theoretical representation of the formation of lattice disruption s

The final degradation mechanism for discussion has been reported NOT to be a concern. Metal migration near the gate (commonly referred to as gate sinking) is a significant reliability concern in GaAs technologies. There are no known reports of failure in GaN due to metal migration. Internal testing results (as has been mentioned in several presentations on GaN reliability studies) show that GaN is extremely robust and non-reactive to metals at standard operating temperatures up to and exceeding the common temperatures for accelerated testing.

Life test Results

The last 3 years have seen a significant increase in the number of reports of reliability results, both in quantity as well as in statistical significance of the tests themselves. Several companies have made such reports as well as providing parts for engineering samples and as off the shelf parts. While acknowledging these reports, the following will focus on results from Raytheon.

A common test for projecting lifetime in compound semiconductors is the 3 (or more) temperature DC test. Figure 7 illustrates a 28V DC test that we typically use for projecting lifetimes in III-V devices, where the MTTF is plotted as a function of the channel temperature. Each point represents 12 devices tested at constant current conditions (described above) to maintain 6W of dissipated power. This testing methodology results in a consistent stress condition and is more conservative compared to maintaining the gate and drain voltages constant over the course of the life test. For each point, $\geq$ 70% of the devices reached the failure criteria of 20% reduction in drain current measured at a gate voltage of 0V (Idss) measured at room temperature. This change in Idss has been correlated to $\sim$ a 1 dB reduction in RF output power.

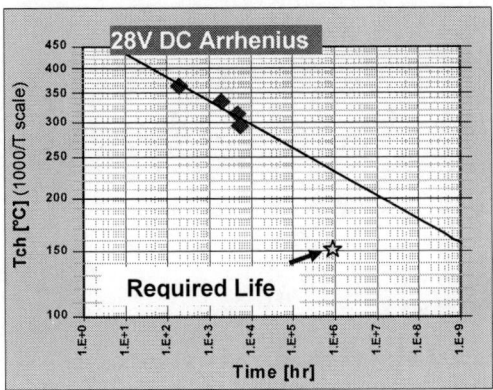

Figure 7: Results of high temperature DC testing utilizing the Arrhenius model.

Estimated lifetime is determined by fitting a line to the MTTF points using as the Arrhenius model. In Figure 7, the required lifetime of 10^6 hours is denoted with a star. The extrapolated lifetime of these parts is far in excess of the required lifetime at 150°C (required operational channel temperature). The activation energy is $\sim$1.7eV, which is consistent with reported values that range from 1 to >2 eV. The variation of the MTTF (sigma) ranges from between 0.5 and 1.3. Typical values for production processes are <0.7. The parts in this test were processed in 2005; further improvements in process and material has reduced the sigmas and improved overall lifetime. Due to limitations in number of test positions and length of time necessary to reach failure of the improved process, a full 3 temperature test is currently underway but has not been completed.

The second major test to determine reliability of a GaN HEMT is a RF operational life test (RFOLT). The parts are run at the intended use conditions: in this case at 28V, 3dB compressed and a channel temperature of 150°C. The test is designed to ensure no new failure mechanisms become dominate under RF operation. The test is typically 1000 hrs in duration, but can be longer during development of technology. Figure 8a demonstrates a RFOLT on the same wafer, whose HT DC results were reported above. The devices are showing stable operation in excess of 2 years, which is the longest reported RFOLT in GaN to our knowledge. The channel temperature of these devices was $\sim$ 170°C to 200°C, due to limitations in test station and device fixturing. Figure 8b illustrates the consistency of our process through RFOLT results on parts from 3 separate lots, showing stable operation for $\sim$ 1000hrs up to durations of > 4000hrs.

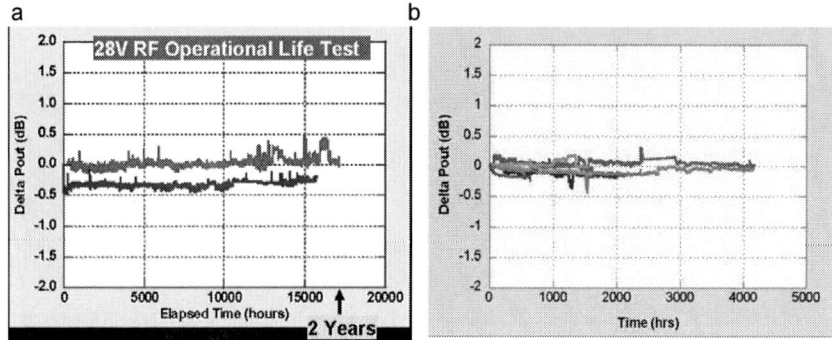

Figure 8: RFOLT results from a) same wafer as DC results in Fig 7 and b) samples from 3 lots

Summary

GaN-based technology has transitioned from an immature technology in the early 2000's to a demonstrated reliable technology with estimated lifetimes at 28V of $> 10^7$ hrs, providing a viable option for high power applications in less than a decade. The improvements in materials have and will play a key role in the continuing improvement in device lifetime as well as performance. Much of the progress can be attributed to understanding the degradation mechanisms and how they relate to the accelerated testing

Acknowledgements: work has been partially funded by DARPA under the Wide Band Gap Semiconductors (WBGS) program (*Contract # FA8650-04-C-7146*). The authors would like to thank the PM Mark Rosker and COTR Chris Bozada for their support

References

1) Y.F. Wu, A. Saxler, M. Moore, R.P. Smith, S. Sheppard, P.M. Chavarkar, T. Wisleder, U.K. Mishra and P. Parikh, *Elect. Dev. Lett.*, **25**(3) 117(2004).
2) T. Palacios, Y. Dora, A. Chakraborty, C. Sanabria, S. Keller S.P. DenBaars and U.K. Mishra, *Phys. Stat. Sol. (A)*, **203**(7)1845(2006).
3) M. Micovic, A. Kurdoghlian, H.P. Moyer, P. Hashimoto, M. Hu, M. Antcliffe, P.J. Willadsen, W.S. Wong, R. Bowen, I. Milosavljevic, Y Yoon, A. Schmitz, M. Wetzel, C. McGuire, B. Hughes and D.H. Chow, *Compound Semiconductor Integrated Circuits Symposium*, 2008.
4) H. Kim, V. Tilak, B.M. Green, H, Cha, J.A. Smart, J.R. Shealy and L.F Eastman, *Reliability Physics Symposium, 2001. Proceedings. 39th Annual. 2001 IEEE International,* 30 April-3 May 2001 Page(s):214 – 218.
5) C. Lee, L. Witkowski, M. Muir, H.Q. Tserng, P. Saunier, H. Wang, J. Yang and M.A. Khan, *High Performance Devices, 2002. Proceedings. IEEE Lester Eastman Conference on,* 6-8 Aug. 2002 Page(s):436 – 442.

6) E.T Yu, X.Z Dang, P.M Asbeck .and S.S Lau, *J. Vac. Sci. Tech. B*, **17**(4), 1742 (1999).
7) O. Ambacher, J. Majewski, C. Miskys, Al Link, M. Jermann, M. Eickhoff, M. Stutzmann, F. Berardini, V. Fiorentini, V. Tilack, B. Schaff and L.F. Eastman, *J. Phys. Condens. Matter*, **14**, 3399(2002).
8) J. Jimenez, U. Chowdhury, C. Lee, P Saunier, and T. Balistreri, *Proceeding of Rel. Of Compound Semiconductors*, Nov. 12, 2006.
9) U. Chowdhury, J.L. Jimenez, C. Lee, E. Beam, P Saunier, T. Balistreri, S.Y. Park, T. Lee, J. Wang, M.J. Lim, J. Joh, and J.A. del Alamo, *Electron Dev. Lett.*, **29**(10), 1098(2008).
10) J. Joh and J.A. del Alamo, *IEDM Tech. Dig.*, Dec. 2006, p. 415.
11) S. Karmalkar, M.S. Shur, G. Simin and M.A. Khan, *Trans. On Elect. Dev.*, **52**(12) 2534 (2005).
12) P.A. Kirkby, P.R. Selway and L.D. Westbrook, *J. Appl. Phys.*, **50**(7), 4567 (1979).
13) P.M. Asbeck, Chien-Ping Lee; M.-C.F Chang,. *Electron Devices, IEEE Transactions on*, **31**(10), 1377 - 1380(1984).
14) E. J. Miller, D. M. Schaadt, E. T. Yu, P. Waltereit, C. Poblenz, and J. S. Speck. *Appl. Phys. Lett*, **82**(8), 1293 (2003).
15) A.M. Conway, M. Chen, P. Hashimoto, P.J. Willadsen, and M. Micovic, *CS MANTECH Conference*, May 14-17, 2007

122

Real-time Detection of Botulinum Toxin with AlGaN/GaN High Electron Mobility Transistors

Yu-Lin Wang[a], B. H. Chu[b], K. H. Chen[b], C.Y. Chang[b], T. P. Lele[b], Y. Tseng[b], S. J. Pearton[a], J. Ramage[c], D. Hooten[c], A. Dabiran[d], P. P. Chow[d], and F. Ren[b]

[a] Department of Materials Science and Engineering, University of Florida, Gainesville, FL 32611, USA
[b] Department of, Chemical Engineering ,University of Florida, Gainesville, FL 32611, USA
[c] Constellation Technology Corp, Largo, FL 33777, USA
[d] SVT Associates, Eden Prairie, MN 55344, USA

AlGaN/GaN high electron mobility transistors (HEMTs) were used to detect botulinum toxin. The gate area was deposited with Au and thioglycolic acid. The antibody was anchored on the gate area. The AlGaN/GaN HEMT drain-source current showed a rapid response of less than 5 seconds when the target toxin in a buffer was added to the antibody-immobilized surface. Different concentrations (from 0.1 ng/ml to 100 ng/ml) of the exposed target botulinum toxin in a buffer solution were detected. The sensor showed low detection limit of less than 1ng/ml and saturates above 10ng/ml of the toxin. The sensor was recycled with the phosphate buffered saline (PBS) after the toxin detection and still showed the same sensitivity.

Introduction

Biological weapons are particularly attractive tools for terror because biological agents are available and easy to manufacture, small amounts are required to cause large-scale effects, and attacks can easily overwhelm existing medical resources. Reliable detection of biological agents in the field and in real time has proved to be challenging. Clostridium botulinum neurotoxins are among the more deadly toxins and are listed as a NIAID— Category A agent for bioterrorism potential. The lethal dose in unvaccinated humans is estimated at 1ng/kg[1,2]. Conventional methods of detection involve the use of HPLC, mass spectrometry and colorimetric ELISAs; but these are impractical because such tests can only be carried out at centralized locations, and are too slow to be of practical value in the field[3-15]. Another test for botulinum toxin detection is the 'mouse assay', which relies on the death of mice as an indicator of toxin presence[16]. Clearly, such methods are slow and impractical in the field. AlGaN/GaN high electron mobility transistors (HEMTs) have shown promise for bio-sensing applications[17-22], since they include a high electron sheet carrier concentration channel induced by piezoelectric and spontaneous polarization of the strained AlGaN layer [17-30]. There are positive counter charges at the HEMT surface layer induced by the electrons located at the AlGaN/GaN interface. Any slight changes in the ambient can affect the surface charge of the HEMT, thus changing the electron concentration in the channel at AlGaN/GaN interface

In this study, we report the use of antibody-functionalized Au-gated AlGaN/GaN high electron mobility transistors (HEMTs) for detecting botulinum toxin. The botulinum toxin was specifically recognized through botulinum antibody, anchored to the gate area. We investigated a range of concentrations from 0.1 ng/ml to 100 ng/ml.

Experimental Technique

The HEMT structures consisted of a 3 μm thick undoped GaN buffer, 30Å thick $Al_{0.3}Ga_{0.7}N$ spacer, 220Å thick Si-doped $Al_{0.3}Ga_{0.7}N$ cap layer. The epi-layers were grown by a molecular beam epitaxy system (MBE) on sapphire substrates. Mesa isolation was performed with an Inductively Coupled Plasma (ICP) etching with Cl_2/Ar based discharges at −90 V dc self-bias, ICP power of 300 W at 2 MHz and a process pressure of 5 mTorr. 10×50 μm^2 Ohmic contacts separated with gaps of 5 μm consisted of e-beam deposited Ti/Al/Pt/Au patterned by lift-off and annealed at 850 °C, 45 sec under flowing N_2. 400-nm-thick 4% Polymethyl Methacrylate (PMMA) was used to encapsulate the source/drain regions, with only the gate region open to allow the liquid solutions to cross the surface. The source-drain current-voltage characteristics were measured at 25°C using an Agilent 4156C parameter analyzer with the gate region exposed.

Figure 1 shows a schematic device cross sectional view with the immobilized thioglycolic acid, followed by botulinum antibody coating. The Au surface was functionalized with the specific bi-functional molecule, thioglycolic acid. We anchored a self-assembled monolayer of thioglycolic acid, $HSCH_2COOH$, an organic compound and containing both a thiol (mercaptan) and a carboxylic acid functional group, on the Au surface in the gate area through strong interaction between gold and the thiol-group of the thioglycolic acid. The devices were first placed in the oxygen plasma chamber and then submerged in 1 mM aqueous solution of thioglycolic acid at 4 °C. This resulted in binding of the thioglycolic acid to the Au surface in the gate area with the COOH groups available for further chemical linking of other functional groups. X-Ray Photoelectron Spectroscopy and electrical measurements confirming a high surface coverage and Au-S bonding formation on the GaN surface have been previously published [28]. The device was incubated in a phosphate buffered saline (PBS) solution of 200 μ g/ml botulinum polyclonal rabbit antibody for 18 hours before real time measurement of botulinum toxin subtype A acquired from Metabiologics Inc.

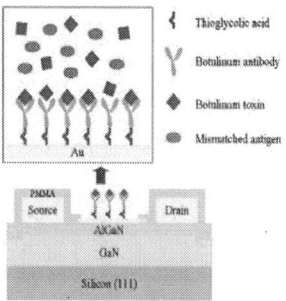

Figure 1. Schematic of AlGaN/GaN HEMT sensor. The Au-coated gate area was functionalized with botulinum antibody/antigen on thioglycolic acid.

Results and Discussion

After incubation with a PBS buffered solution containing botulinum antibody at a concentration of 200 μ g/ml, the device surface was thoroughly rinsed off with PBS and dried by a nitrogen blower. The source and drain current from the HEMT were measured before and after the sensor was exposed to 100 ng/ml of botulinum toxin at a constant drain bias voltage of 500 mV, as shown in Figure 2 (Left). Any slight changes in the ambient of the HEMT affect the surface charges on the AlGaN/GaN. These changes in the surface charge are transduced into a change in the concentration of the 2DEG in the AlGaN/GaN HEMTs, leading to the decrease in the conductance for the device after exposure to botulinum toxin.

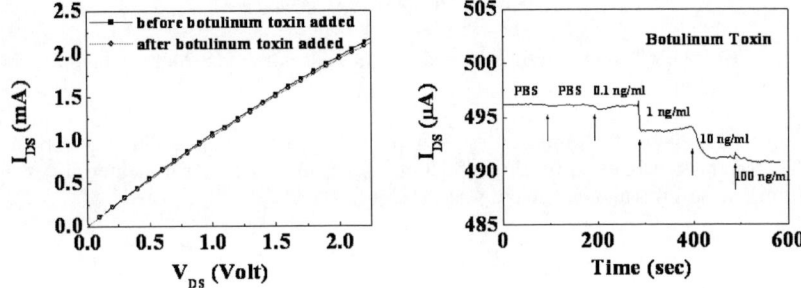

Figure 2. (Left) I-V characteristics of AlGaN/GaN HEMT sensor before and after exposure to 100ng/ml botulinum toxin. (Right) Drain current of an AlGaN/GaN HEMT versus time for botulinum toxin from 0.1 ng/ml to 100 ng/ml.

Figure 2 (Right) shows a real time botulinum toxin detection in PBS buffer solution using the source and drain current change with constant bias of 500 mV. No current change can be seen with the addition of buffer solution around 100 seconds, showing the specificity and stability of the device. In clear contrast, the current change showed a rapid response in less than 5 seconds when target 1 ng/ml botulinum toxin was added to the surface. The abrupt current change due to the exposure of botulinum toxin in a buffer solution was stabilized after the botulinum toxin thoroughly diffused into the buffer solution. Different concentrations (from 0.1 ng/ml to 100 ng/ml) of the exposed target botulinum toxin in a buffer solution were detected. The sensor saturates above 10ng/ml of the toxin. The experiment at each concentration was repeated four times to calculate the standard deviation of source-drain current response. The limit of detection of this device was below 1 ng/ml of botulinum toxin in PBS buffer solution. The source-drain current change was nonlinearly proportional to botulinum toxin concentration, as shown in Figure 3.

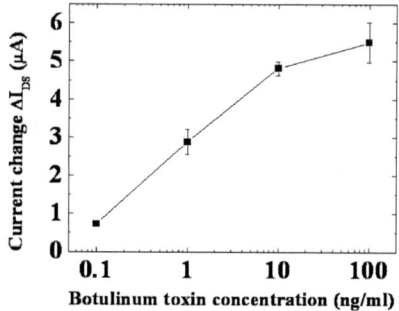

Figure 3. Change of drain current versus different concentrations from 0.1 ng/ml to 100 ng/ml of botulinum toxin.

Figure 4 shows a real time test of botulinum toxin at different toxin concentrations with intervening washes to break antibody-antigen bonds. This result demonstrates the real-time capabilities and recyclability of the chip.

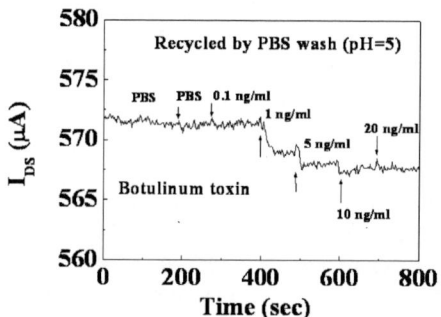

Figure 4. The real-time test from a used sensor which was washed with PBS in pH 5 to refresh the sensor.

Conclusions

In summary, we have shown that through a chemical modification method, the Au-gated region of an AlGaN/GaN HEMT structure can be functionalized for the detection of botulinum toxin with a limit of detection less than 1 ng/ml. This electronic detection of biomolecules is a significant step towards a field-deployed sensor chip, which can be integrated with a commercial available wireless transmitter to realize a real-time, fast response and high sensitivity botulinum toxin detector.

Acknowledgments

The work at UF was partially supported by ONR Grant N000140710982 monitored by Igor Vodyanoy, NASA Kennedy Space Center Grant NAG 3-2930 monitored by T.

Smith, by NSF DMR 0400416 and the State of Florida, Center of Excellence in Nano-Bio Sensors.

References

1. S.S. Arnon , R. Schechter and T.V.Inglesby T. V. JAMA, 285, No. 8, 256 (2001).
2. R.A. Greenfield , B.R. Brown , J.B. Hutchins J.J. Iandolo , R. Jackson, L.N. LSlater, and M.S. Bronze, The American J. Of The Medical Sciences, 323, 326 (2002).
3. J. S. Michaelson, E. Halpern and D.B. Kopans, Radiology 212(2). 551 (1999).
4. T. Harrison, L.Bigler, M.Tucci, L.Pratt, F. Malamud, J.T. Thigpen, C. Streckfus and H. Younger, Spec. Care Dentist 18(3), 109 (1998) .
5. R. Mclntyre, L. Bigler, T. Dellinger, M. Pfeifer, T. Mannery and C. Streckfus, Oral Surg Oral Med. Oral Pathol. Oral Radiol. Endod. 88(6), 687 (1999).
6. C. Streckfus, L. Bigelr, T. Dellinger, M. Pfeifer, A. Rose and J.T. Thigpen, Clin. Oral Investig. 3(3), 138 (1999).
7. C. Streckfus, L. Bigler, T. Dellinger, X.Dai, A. Kingman and J.T. Thigpen, Clin Cancer Res. 6(6), 2363 (2000).
8. C. Streckfus, L. Bigler, M. Tucci and J.T. Thigpen, Cancer Invest 18(2), 101 (2000).
9. C. Streckfus, L. Bigler, T. Dellinger, X. Dai, W.J. Cox, A. McArthur, A. Kingman and J.T. Thigpen, Oral Surg Oral Med Oral Pathol. Oral Radiol. Endod. 91(2), 174 (2001).
10. L. R. Bigler, C.F. Streckfus, L. Copeland, R. Burns, X. Dai, M. Kuhn, P. Martin and S. Bigler, J Oral Pathol. Med 31(7), 421 (2002).
11. C. Streckfus and L. Bigler, Adv. Dent. Res 18(1), 17 (2005).
12. C. F. Streckfus, L. R. Bigler and M. Zwick, J Oral Pathol. Med 35(5), 292(2006).
13. W.R. Chase, J Mich. Dent. Assoc. 82(2), 12(2000).
14. S.Z. Paige and C.F. Streckfus, Gen Dent 55(2), 156(2007).
15. M.A. Navarro, R. Mesia, O. Diez-Giber, A. Rueda, B. Ojeda and M.C. Alonso, Breast Cancer Res. Trea. 42(1), 83(1997).
16. K.Bagramyan, J.R. Barash, S.S. Arnon and M. Kalkum M, Matrices PLoS ONE, 3, 2041(2008).
17. A. P. Zhang, L. B. Rowland, E. B. Kaminsky, V. Tilak, J. C. Grande, J. Teetsov, A. Vertiatchikh and L. F. Eastman, J.Electron.Mater.32 388(2003).
18. O. Ambacher, M. Eickhoff, G. Steinhoff, M. Hermann, L. Gorgens, V. Werss, B. Baur, M. Stutzmann, R. Neuterger, J. Schalwig, G. Muller, V. Tilak, B. Green, B. Schafft, L. F. Eastman, F. Bernadini, and V. Fiorienbini, Proc. ECS 02-14, 27(2002).
19. R. Neuberger, G. Muller, O. Ambacher and M. Stutzmann, Phys. Stat. Soli. A185, 85(2001).
20. J. Schalwig, G. Muller, O. Ambacher and M. Stutzmann, Phys. Stat. Solidi A185, 39(2001).
21. G. Steinhoff, M. Hermann, W. J. Schaff, L. F. Eastman, M. Stutzmann and M. Eickhoff, Appl. Phys. Lett., 83, 177(2003).
22. M. Eickhoff, R. Neuberger, G. Steinhoff, O. Ambacher, G. Muller, and M. Stutzmann, Phys. Stat. Solidi B 228, 519(2001).
23. B.S. Kang, H.T. Wang, F. Ren and S.J.Pearton, J.Apl.Phys.104, 031101(2008).
24. B. S. Kang, S. Kim, F. Ren, J. W. Johnson, R. Therrien, P. Rajagopal, J. Roberts,

E. Piner, K. J. Linthicum, S. N .G. Chu, K. Baik, B. P. Gila, C. R. Abernathy and S. J. Pearton, Appl. Phys.Lett. 85, 2962(2004)

25. S. J. Pearton, B. S. Kang, S. Kim, F. Ren, B. P.Gila, C. R. Abernathy, J. Lin and S. N. G. Chu, J. Phys: Condensed Matter 16 R961(2004).

26. G. Steinoff, O. Purrucker, M. Tanaka, M. Stutzmann and M. Eickoff, Adv. Funct. Mater. 13,841(2003).

27. G. Steinoff, B. Baur, G. Wrobel, S. Ingebrandt, A. Offenhauser, A. Dadgar, A. Krost, M.Stutzmann and M.Eickhoff, Appl. Phys. Lett.86, 033901(2005).

28. B. S. Kang, H. T. Wang, F. Ren, S. J. Pearton, T. E. Morey, D. M. Dennis, J. W. Johnson, P. Rajagopal, J. C. Roberts, E. L. Piner, and K. J. Linthicum, Appl. Phys. Lett., 91, 252103 (2007).

29. B. S. Kang, F. Ren, L. Wang, C. Lofton, Weihong Tan, S. J. Pearton, A. Dabiran, A. Osinsky, and P. P. Chow, Appl. Phys. Lett. 87, 023508(2005).

30. B. S. Kang, J.J. Chen, F. Ren, S. J. Pearton, J.W. Johnson, P. Rajagopal, J.C. Roberts, E.L. Piner, and K.J. Linthicum, Appl. Phys. Lett. 89, 122102 (2006).

ECS Transactions, 19 (3) 129-135 (2009)
10.1149/1.3120694 ©The Electrochemical Society

Annealing Effects on ZnO/NiSi Contact

Ren Wei[1]*, Hao Gong[2]

[1]College of Sci., Univ. of Shanghai for Science and Technology, Shanghai 200093, China
[2]Mater. Sci. & Eng. Dep., National University of Singapore, Singapore, 119620

Abstract: NiSi as the electrode of ZnO films was annealed for different temperatures. It was found that with the increase of the temperatures, although the structure quality of ZnO films does not change greatly, the optical properties of the films do indicate the introduction of the oxygen defects into ZnO films by annealing process.

Keywords: ZnO; NiSi; anneal; photoluminescence; Raman spectroscopy

Introduction

Zinc oxide (ZnO) has been extensively investigated as a potential II-VI semiconductor material for UV light emitting diodes and detectors due to its interesting optoelectronic properties such as a wide band gap (3.37 eV) [1] at room temperature and a large exciton binding energy of 60 meV, when compared to those values of GaN (21 meV) or ZnSe (20 meV) [2-3]. Although ZnO films have been grown on various materials such as sapphire, GaAs, GaN, and glass, Si as substrates of ZnO has unique advantages, such as low background noise of its optical properties and low cost [4]. We compared the structure and optical properties of the ZnO/Si with the ZnO/NiSi/Si samples in previous work [4]. The aim of this work is to further explore the annealing effects on these two types of samples.

Experimental

A p-type, 6 – 10 $\Omega \cdot cm$ Si (001) wafer was cut into small pieces (1 × 1 cm^2) and dipped into diluted hydrofluoric acid solution (1 HF: 5 H$_2$O) for 2 minutes to remove native oxide. And then they were blown-dried by N$_2$ gas. Following that, a layer of Ni film (30-nm-thick) was deposited on some of these pieces by dc sputtering from a pure Ni target (99.99%). The samples with Ni films were annealed to form NiSi films by rapid thermal annealing process at 500 °C for 1 minute. Immediately after that, the samples with and without NiSi films were simultaneously loaded into a radio frequency (r.f.) sputtering system for growth of ZnO films. Finally, annealing these samples at 200, 300 and 400 °C was done in air for 3 minutes. These samples were characterized by X-ray diffraction (XRD), atomic force microscopy (AFM), secondary ion mass spectroscopy (SIMS), room temperature micro-Raman spectroscopy (under a backscattering geometry), and photoluminescence (PL) spectroscopy.

129

Results and discussion

Fig. 1(a) and (b) show XRD patterns (under the Bragg-Brentano scanning mode) of the ZnO/Si and ZnO/NiSi/Si samples annealed at different temperatures. In (a), the strong (002) peaks of ZnO are clearly observed at 2θ angles of 34.4° through all the temperature curves, indicating favorable growth of the ZnO grains along [002] orientation [5]. In (b), ZnO (002) peaks are similarly observed in spite of much lower intensity. The full width half maximum (FWHM) values of these peaks in (a) are 0.257° for no anneal, 0.256° for 200 °C, 0.258° for 300 °C and 0.254° for 400 °C curves; in (b) are 0.251°, 0.242°, 0.249° and 0.240°, respectively. It can be seen that the general FWHM values in (b) are less than in (a), which may suggest that the averaged ZnO grain size of the ZnO/NiSi/Si samples are bigger compared with the ZnO/Si counterparts. Another point is that the FWHM values were supposed to decrease with the increasing annealing temperature because the crystalline quality should be improved with the increase of temperatures. However, according to above data, the temperature of 300 °C is special, where the (002) peaks' FWHM values increase instead. The reason for this special increase is not clear. In addition, there are additional peaks: the 2θ angles of 31.57°, 36.12°, 45.8°, 47.27°, 51.53°, and 56.3° in (b) correspond to NiSi (011), (111), (112), (211), (103) and (013), respectively. Multiple XRD peaks of NiSi reveal that the NiSi film is polycrystalline. Also, the intensity of NiSi peaks is much less than that of ZnO ones, which is explained as the ZnO layer's absorption and the reflection of the X-ray beam before the beam enters the NiSi layer.

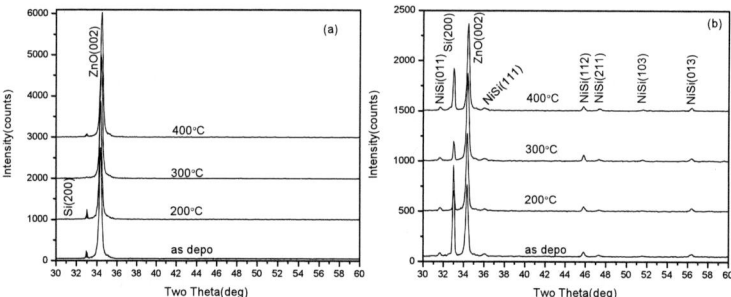

Figure 1. XRD patterns of: (a) ZnO/Si and (b) ZnO/NiSi/Si samples annealed in air at different temperatures for 3 minutes

Fig. 2 (a) and (b) illustrate the AFM images of the ZnO/Si and ZnO/NiSi/Si samples annealed at 400 °C (the AFM images of the other temperatures look similar and not shown). The annealing does not impact the surface morphology of the ZnO films. The difference of these two types of ZnO films lies in: the root-mean-square (RMS) values of the surface roughness in (a) vary between 2.5 and 3.0 nm; and the averaged planar size of ZnO nano-columns is about 50 – 80 nm while the corresponding values in (b) are in the range of 5.5 – 6.0 nm, and 75 – 80 nm respectively. This difference indicates that the ZnO growth on NiSi is more favorable than on Si.

Figure 2. AFM images of (a) ZnO/Si and (b) ZnO/NiSi/Si samples annealed in air at 400 °C for 3 minutes.

Fig. 3 (a) and (b) show the SIMS depth profiles of the ZnO/Si and ZnO/NiSi/Si samples annealed at 400 °C (other profiles of different temperature annealed samples are similar and not shown). According to the depth profiles in (a) and (b), the curves of Zn and oxygen ions, as well as Ni and Si ions are parallel, which indicate that ZnO and NiSi films are formed already and their contents vertical to Si surface are uniform. Particularly, at the interfacial regions of the ZnO/Si and ZnO/NiSi interfaces, no inter-diffusion is observed. As for the NiSi/Si interface of (b), the profile curve is normal and indicates that some Ni diffuses into Si, which was reported in [6]. As for the thickness of the layers, the ZnO films in both types of samples are around 165 nm and similarly they do not change with annealing temperatures. This value is consistent with our previous cross-sectional SEM results [4]. However, several points should be noted that (1) the thickness of the NiSi interlayer in (b) is around 149 nm, which is much more than our previous result of ~ 120 nm. This is possibly because that the sputtering rate for NiSi and ZnO layers are different, if under the same sputtering conditions during SIMS measurement. (2) It is very interesting to see that the intensity of zinc ions at the ZnO/NiSi interface is higher than that at the ZnO/Si ones, while the intensity is same for oxygen ions in both samples. Although the interfacial effect of SIMS profiling might be responsible for the difference of Zn ions at interfacial regions since the properties of ZnO/Si and ZnO/NiSi are different, it is more possible that more Zn ion do accumulate at the ZnO/NiSi interface, which may lead to the larger planar columnar size of ZnO film grew on NiSi than on Si. Since the annealing temperatures are relatively low (only 400 °C), these Zn ions or atoms do not diffuse with the temperatures.

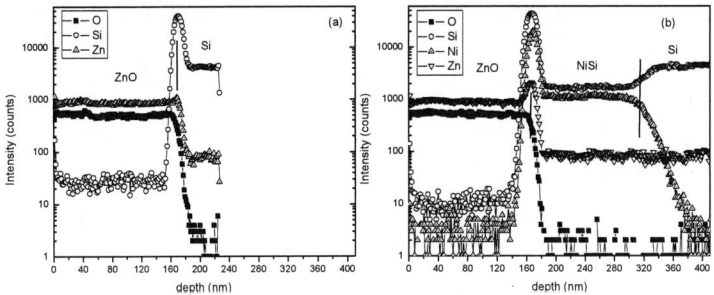

Figure 3. SIMS depth profiles of (a) ZnO/Si and (b) ZnO/NiSi/Si samples annealed in air at 400 °C for 3 minutes.

Fig. 4 (a) and (b) indicate the Raman spectra of ZnO/Si and ZnO/NiSi/Si samples annealed at different temperatures. In (a), ZnO peaks of A_1 (LO) band (575 cm^{-1}) and E_2 bands (low at 98 and high at 436 cm^{-1}) are observed. It is noticed that the two E_2 peaks have 3cm^{-1}-red-shifted, compared with their bulk values (101 and 439 cm^{-1}). Because low frequency E_2 mode represents the vibration of heavy Zn sublattice, while the high frequency E_2 mode reflects the vibration of oxygen atoms, the Raman redshift may be explained as the extrinsic stress, induced by the thermal expansion coefficient mismatch and lattice mismatch between ZnO films and substrate [4]. In addition, the peak widths of E_2 shrink with the increase of annealing temperature, which indicates the improvement of ZnO films quality [7]. In (b), although the ZnO peaks of A_1 (LO) and E_2 (low and high) are still observable, more complicated peaks appear and are ascribed to ZnO and NiSi, respectively. Among them, the 550 cm^{-1} peak is assigned to the interaction of photons with free carriers on the ZnO/NiSi interface [8]; 336 cm^{-1} ascribed to the ZnO second order mode [9-12]; while, 276 and 401 cm^{-1} still need further explanations. Also, NiSi peaks are observed at 197, 213, 256, 293, and 363 cm^{-1}. Although annealing seems not to impact most of these Raman peaks, it does change the A_1 and free carrier peaks: the spectral peaks of A_1 mode shift to higher frequency with the temperature increase (575.7, 576.3, 576.8 and 578.9 cm^{-1}); the peak intensities of free carriers decrease greatly and disappear at 400 °C. According to previous SIMS results, there are more Zn at the ZnO/NiSi interface, which might not combine with oxygen and act as the interstitial atoms due to more vacancy and defects located at the interfacial region. These Zn interstitial atoms may be responsible for the 550 cm^{-1} peaks as well as the peak shift of A_1 mode. When samples were annealed, the environmental oxygen diffuses to the ZnO/NiSi interface and bonds with Zn interstitial atom. Correspondingly, the carrier density and resistivity of ZnO films decrease, until these Zn interstitial atoms were bonded completely.

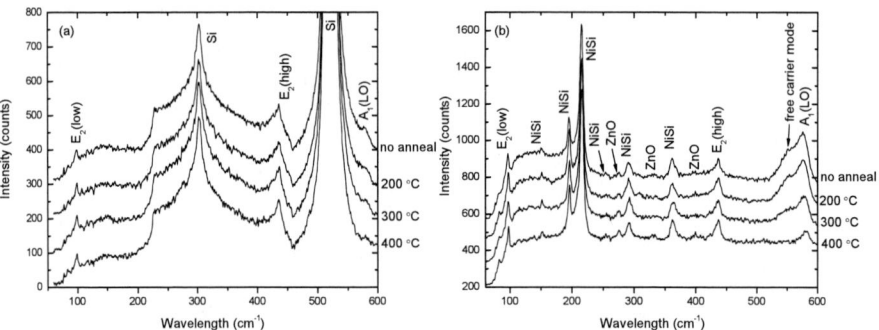

Figure 4. Room temperature micro-Raman spectra: (a) ZnO/Si and (b) ZnO/NiSi/Si samples annealed in air at no-anneal, 200, 300 and 400 °C for 3 minutes.

Fig. 5 (a) and (b) illustrate room temperature PL spectra of the ZnO/Si and ZnO/NiSi/Si samples annealed at different temperatures. Two band-edge spectral peaks are apparently observed at ~3.276 eV (or 378.8 nm) with a FWHM value of ~160 meV, and the ~3.347 eV (or 371.2 nm) with only 9 meV peak width. While the narrow peak-width of the 3.347 eV peak was explained as the emission from some specific-structured ZnO nanocolumns or the Fabry-Perot lasing cavity [4], the 3.276 eV peak represents the phonon-assited free exciton emission [13], and its intensity improvement should indicate the improvement of film crystalline quality. However, checking both peak intensities followed with temperatures, the results seem to be more complicated. Firstly, in both ZnO/Si and ZnO/NiSi/Si samples, the intensity of 3.276 eV peak decreases slightly when the temperature increases to 200 °C, and then starts to increase slightly with the temperature increasing to 400 °C. Their peak widths shrink with the temperature. Further, for the 3.347 eV peak of the ZnO/Si samples, its intensity tends to increase as the temperature increases to 200 °C, and then keeps stable after that; in the case of ZnO/NiSi/Si, this peak's intensity firstly rocks up to the highest value at 200 °C, and then falls down slowly till 400 °C. Since the peak width of 3.347 eV peak is only 9 meV, it is difficult to determine the trend with the annealing temperature. If comparing these two band edge peaks in both ZnO/Si and ZnO/NiSi/Si samples, it is apparent that the peak intensity of the ZnO/NiSi/Si sample is much higher. It seems that the band edge emission represented by 3.276 eV peak will be reduced at the very beginning of the annealing process by some defects thermally introduced into ZnO films, which can be proved by the intensity decrease of the 3.276 eV peak at the 200 °C curves in both samples. After that, with the increase of the annealing temperature, the band edge emission is improved by grain recrystallization of the ZnO films although more defects, probably only formed as interstitial atoms, might be introduced into ZnO films at higher temperatures. However, it is still not clear about the intensity evolution of the 3.347 eV peak with temperatures.

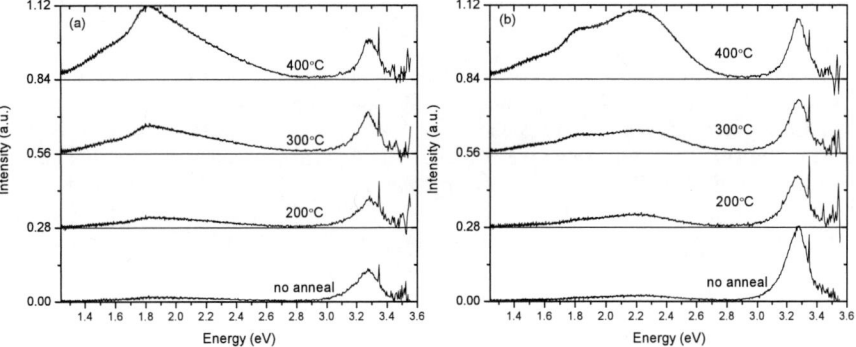

Fig 5. Room temperature PL spectra: (a) ZnO/Si and (b) ZnO/NiSi/Si samples annealed in air at no-anneal, 200, 300 and 400 °C for 3 minutes.

As mentioned above, the annealing process also brings some defect levels into the films, which induce the emissions of the red and green luminescence bands. In Fig. 5 (a), a broad peak at around 1.8 eV (red peak) with a range of 1.3 to 2.9 eV can be clearly seen. The peak intensity as well as the peak width increases with the temperature rising. In (b), except for 1.8 eV peak, another peak at around 2.2 eV (green peak) with higher intensities than 1.8 eV one is noticeable. Its intensity increases with the rising temperature as the red peak does. In addition, comparing the red and green peaks of the ZnO/NiSi/Si samples, the green peak's intensities are generally higher than those of red peak, especially at 400 °C, which indicates that the green luminescence emission process takes more important role than red counterparts, and even overrules the near band edge emission at 400 °C. We explain this special emission as following: in the ZnO/Si case, because the oxygen partial pressure in the air is not enough high, the annealing process might lead to the evaporation of oxygen atoms of ZnO surface into air and generate some surface oxygen vacancy. According to [14], the oxygen vacancy in ZnO film has an emission energy level of 1.62 eV. If considering the surface condition, the energy level of 1.62 eV may increase to 1.8 eV, which is the origin of the red luminescence peak. With the increase of the annealing temperature, more oxygen vacancies were generated and the intensity of red luminescence peaks rises correspondingly. In the case of ZnO/NiSi/Si case, it is surprisingly noticed that a new green luminescence peak was observed. Apparently, its origin should be different with the 1.8 eV peaks. Considering that NiSi/Si samples were exposed into the air for some time, before they were loaded into r.f. sputtering chamber for ZnO film growth, it is possible that the NiSi surface had absorbed some water or oxygen molecules. Since we did not heat these samples during ZnO growth, the water/oxgen molecules stayed at the ZnO/NiSi interface and were activated as oxygen interstitial atoms when samples were annealed. These oxygen interstitial atoms become the origin of the green luminescence peaks. And their energy level of 2.2 eV matches well with the calculation of reference [14].

Conclusion

ZnO films were deposited on Si as well as on NiSi/Si by RF sputtering, and annealed at different temperatures. Probably because the temperatures are relative low, the annealing process does not change the structural properties of ZnO films greatly. However, it still introduces some defects into ZnO films and changed their optical properties, especially inducing the red PL emission from the ZnO/Si samples as well as the red and green emission from the ZnO/NiSi/Si samples. The origins of these defects are probably due to the oxygen atoms evaporating into the air for 1.8 eV peak, and the interfacial oxygen of ZnO/NiSi, activated by the annealing process.

References

1. C. Klingshirn, *Phys. Status Solidi B*, **71**, 547 (1975).
2. D.M. Bagnall, Y. F. Chen, Z. Zhu, T. Yao, S. Koyama, M. Y. Shen, et al, *Appl Phys. Lett.*, **70**, 2230 (1997).
3. H. Cao, H. Cao and Y. G. Zhao, S. T. Ho, E. W. Seelig, Q. H. Wang, et al, *Phys. Rev. Lett.*, **82**, 2278 (1999).
4. W. Ren, H. Gong, D.Z. Chi, *Electrochem. Solid-State Lett.*, **12(4)**, H98 (2009).

5. N. Fujimura, T. Nishihara, S. Goto, J. Xu and T. Ito, *J. Crystal. Growth.*, **130**, 269 (1993).
6. W. Ren, D.Z. Chi, *Electrochem. Solid-State Lett.,* **10(9),** H260 (2007).
7. Y. Zhang, G. Du, X. Yang, B. Zhao, Y. Ma, T. Yang, et al, *Semicond. Sci. Technol.* **19**, 755 (2004).
8. J. Xu, G. Cheng, W. Yang and Y. Du, *J. Phys. B: At. Mol. Opt. Phys.*, **29**, 6227 (1996).
9. J. Zuo, C. Xu, L. Zhang, B. Xu and R. Wu, *J. Raman Spectro.*, **32**, 979 (2001).
10. J.M. Calleja and M. Cardona, *Phys. Rev. B,* **16**, 3753 (1977).
11. M. Rajalakshmi, A.K. Arora, B.S. Bendre and S. Mahamuni, *J. Appl. Phys.*, **87**, 2445 (2000).
12. X. Wang, S. Yang, J. Wang, M. Li, X. Jiang, G. Du, et al, *J. Crys. Grow.*, **226**, 123 (2001).
13. D.M. Bagnall, Y.F. Chen, Z. Zhu, T. Yao, M. Y. Shen and T. Goto, *Appl. Phys. Lett.*, **73**, 1038 (1998).
14. B. Lin, Z. Fu, Y. Jia, *Appl. Phys. Lett.*, **79**, 943 (2001).

136

ECS Transactions, 19 (3) 137-145 (2009)
10.1149/1.3120695 ©The Electrochemical Society

Photoelectrochemical Splitting of Water to H_2 and O_2 at n-Fe_2O_3 Nanowire Films and Nanocrystalline Carbon-Modified (CM)-n-Fe_2O_3 Thin Films

Mourad Frites and Shahed U. M. Khan*
Department of Chemistry and Biochemistry
Duquesne University, Pittsburgh, PA 15282
* email: khan@duq.edu

Abstract

Iron oxide n-Fe_2O_3 nanowire thin films were synthesized by thermal oxidation of Fe metal sheet (Alfa Co. 0.25 mm thick) in an electric oven then tested for their photoactivity. The photoresponse of the n-Fe_2O_3 nanowires was evaluated by measuring the rate of water splitting reaction to hydrogen and oxygen, which was found to be proportional to photocurrent density, j_p. The optimized electric oven-made n-Fe_2O_3 photoelectrodes showed photocurrent densities of 1.32 mA cm^{-2} at measured potential of 0.0 V/SCE with photoconversion efficiency of 1.69 % at applied potential of 0.70 V vs E_{aoc} (electrode potential at open circuit conditions) under illumination intensity of 100 mW cm^{-2} from a Solar simulator with a global AM 1.5 filter. The photoactivity was improved upon incorporation of carbon into the lattice of n-Fe_2O_3 by flame oxidation at 850°C. The carbon modified (CM)-n-Fe_2O_3 showed enhanced photocurrent response to 3.14 mA cm^{-2} at a measured potential of 0.0 V/SCE with an efficiency of 2.23% at applied potential of 0.52 V vs E_{aoc}. The nanocrystalline CM-n- Fe_2O_3 and n-Fe_2O_3 nanowires thin films were characterized using photocurrent density measurements under monochromatic light illumination, UV-Vis spectra, X-ray diffraction (XRD) and scanning electron microscopy (SEM).

1. Introduction

Iron (III) oxide (n-Fe_2O_3) is a low cost semiconductor, environmentally friendly-non-toxic, abundant, and corrosion resistant. Moreover, iron oxide semiconductor has a band gap of 2 - 2.2 eV[1], therefore, it can absorb a considerable part of the solar light up to 620 nm which consists nearly 40% of the sunlight photons at global AM 1.5[1-3]. These properties are attractive to investigate the possible use of iron oxide semiconductor for photoelectrochemical water splitting to produce hydrogen gas; the fuel of the future.

Iron (III) oxide photoelectrodes were prepared by different routes e.g pressing of polycrystalline Fe_2O_3 in palette and followed by sintering it [8, 9], spray pyrolysis of a solution of $Fe(NO_3)_3$ or $FeCl_3$ on a hot conductive glass[10, 11], and by chemical vapor deposition (CVD)[12]. However, this semiconductor exhibits a limited photocurrent when used for water splitting reaction under artificial or solar simulated light. McGregor et al[8]

137

investigated the flat band potential of the sintered polycrystalline n-Fe_2O_3 pressed into pellet and found it to be insufficiently negative to be able to produce hydrogen from water, therefore, an external voltage was needed to induce the photoactivity or the semiconductor must be coupled with another p-type semiconductor[11]. Depending on preparation method of Iron (III) oxide or its crystallinity the band gap of the semiconductor is reported to be 1.9-2.2 eV with a suitable valence band edge for oxygen evolution reaction (OER)[2]. But unfortunately the conduction band doesn't match for hydrogen evolution reaction (HER) [8].

The photoactivity n-Fe_2O is hampered by its low charge carrier mobility, high resistivity and slow charge transfer across the interface that leads to a high probability of recombination between the photogenerated electron-hole pairs[4]. The challenge is to minimize these limitations and improve the photoactivity of this semiconductor. It was reported that the anionic or cationic dopant to either n-type[5, 6] or p-type[7] increases the light-to-hydrogen conversion efficiency. Other alternatives would be the synthesis of nanowires to minimize the effect of high resistivity of iron oxide by reducing the transport distance of photogenerated carriers prior to their reactions with species in solution (e.g., OH⁻ ions) and also modify its optical and electrical behavior by incorporation of carbon in its lattice. Hence, in this present study we report the photoresponse of n-Fe_2O_3 nanowires and nanocrystalline carbon modified (CM)-n-Fe_2O_3 thin films during water splitting reactions. These samples were characterized in terms of their photocurrent density under white light and monochromatic light illuminations, x-ray diffraction (XRD) study, UV-Vis absorption spectra and scanning electron microscopy (SEM).

2. Experimental

2.1. Synthesis of n-Fe_2O_3 nanowires and carbon modified n-Fe_2O_3 (CM-n-Fe_2O_3) thin films

Iron metal sheets of 0.25 mm thick (Alfa Co.) were cut to an area of ~ 1.0 cm^2 samples. These metal samples were cleaned in a sonicator for three-fifteen minute intervals with: (i) Acetone; (ii) acetone: double de-ionized water (1:1); (iii) double de-ionized water. After drying in air, Iron (III) oxide n-Fe_2O_3 photoelectrodes were prepared by oxidation of Fe metal sheets in an electric oven at different temperatures (550-825°C) for 15 minute oxidation time. The synthesis of nanocrystalline carbon modified (CM)-n-Fe_2O_3 photoelectrodes was carried out by flame oxidation of Fe metal sheet at different temperatures (700 to 850°C) and oxidation time (3 to 15 min), using a custom designed large area flame (Knight, model RN. 3.5 xa wc) under controlled oxygen and natural gas flow.

2.2. Preparation of n-Fe_2O_3 nanowire and CM-n-Fe_2O_3 photoelectrodes:

The oxide layers in the top portion on both back and front sides of the Fe sheet were removed using a file to make an electrical contact. The back side and part of the front side were covered with a nonconductive epoxy adhesive. The uncovered area of the photoelectrode (~ 0.2 cm^2) was calculated using a computer program, "Image J". The Electrical connection was made using a copper clip to exposed surface of Fe metal on the top.

2.3. Measurements of photocurrent density

The photoresponse of both thin film photoelectrodes was evaluated by measuring the rate of water splitting reaction to hydrogen and oxygen in terms of photocurrent density, j_p. Photocurrents were measured using a scanning potentiostat (EG & G Princeton Applied Research model 362) and recorded by an X-Y recorder (EG & G Houston Model RE 0092). The measurements of photocurrent density, j_p, as a function of measured potential, E_{meas} were performed in a three-electrode configuration single-compartment cell, with $n-Fe_2O_3$ or $CM-n-Fe_2O_3$ as working electrode and platinum wire as a counter electrode. The potential of working electrode was measured using saturated calomel electrode (SCE) as a reference electrode. A solution of 5 M KOH was used as the electrolyte. The photoelectrodes were illuminated by light intensity of 100 mW cm^{-2} from a Solar simulator with a global AM 1.5 filter. The intensity of light was measured with a Silicon detector (UDT Sensors Inc., Model 10DP/SB).

2.4. UV-Vis spectroscopic measurements

UV-Vis absorption spectra of $n-Fe_2O_3$ (CM- $n-Fe_2O_3$) samples were measured using a Varian Carry 1E UV-Visible Spectrophotometer with a Varian GRID 36is-25 microprocessor and a Dell Optiplex PC. Labsphere (Model DRA-Ca-30I) with a reflectance standard (I.D.USRS-99-010) was used.

2.5. Scanning electron microscopic (SEM)measurements

Scanning electron micrograms (SEM) for CM- $n-Fe_2O_3$ was performed using a ZEISS EVO-SEM in which conductive carbon tape was used to make a contact with the conductive Fe surface with the aluminum sample deck, for both oven-made $n-Fe_2O_3$ nanowires and flame-made nanocrystalline CM- $n-Fe_2O_3$ thin films.

3. Results and Discussion

3.1. Dependence of photocurrent density of $n-Fe_2O_3$ nanowire and $CM-n-Fe_2O_3$ film on measured electrode potential

The photoresponse of the $n-Fe_2O_3$ nanowires was evaluated by measuring the rate of water splitting reaction to hydrogen and oxygen, which is proportional to photocurrent density j_p. The optimum conditions to fabricate $n-Fe_2O_3$ nanowires electrodes were found to be 700°C and 15 min oxidation time in an electrical oven. We designate this electrical oven made $n-Fe_2O_3$ nanowire film synthesized under these optimum conditions as sample 1.   For flame-made-film the optimum conditions were found to be flame temperature of 850°C and oxidation time of 3 min.   We refer this flame made CM-n-Fe_2O_3 film prepared under these optimum conditions as sample 2. The optimized electric oven-made iron oxide nanowire photoelectrode (sample 1) showed a photocurrent density of 1.32 mA cm^{-2} at a measured potential of 0.0 V/SCE (see figure 1).   The nanowire form of the oven made iron oxide (see Fig. 7a) reduced the transport distance of the photogenerated minority carriers (holes, h$^+$) from the bulk to the interface of $n-Fe_2O_3$ nanowires-electrolyte solution and hence generated higher photocurrent compared to earlier results on Si-doped $n-Fe_2O_3$ thin film [12].

The photocurrent density improved to 3.14 mA cm^{-2} at CM-n-Fe$_2$O$_3$ photoelectrode (sample 2) at a measured potential of 0.0 V/SCE upon incorporation of carbon in iron(III) oxide during flame oxidation (see Figure 1). Carbon incorporation was found to be responsible to much higher photoresponse for water splitting reaction at lower applied potential compared to undoped n-Fe$_2$O$_3$. This can be attributed mainly to increase in the conductivity and as well as lowering the band gap by carbon incorporation in CM-n-Fe$_2$O$_3$ film.

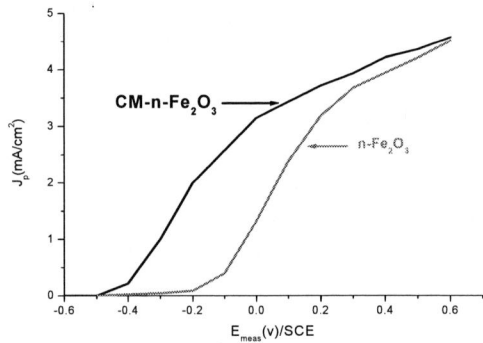

Figure 1. Photocurrent density, j$_p$, as a function of measured potential, E$_{meas}$ (V vs. SCE) for CM-n-Fe$_2$O$_3$ and n-Fe$_2$O$_3$ synthesized by thermal oxidation of Fe metal sheet at the optimum conditions under white light intensity of 100 mW/cm^2 from a Solar simulator with a global AM 1.5 filter.

Photoconversion efficiency:

The photoconversion efficiency was calculated using the expression given earlier as,[15-17]

$$(\varepsilon\%) = [\, j_p\,(E_{rev} - |V_{app}|)\, /\, I_o] \times 100 \qquad (1)$$

where j$_p$ is the photocurrent density (mA cm^{-2}), I$_0$ is the total incident light intensity (mW cm^{-2}), E$_{rev}$ is the standard reversible potential (which is 1.23 V for water splitting), E$_{app}$ is the applied potential; which can be obtained as E$_{app}$ = E$_{meas}$ - E$_{aoc}$ where E$_{meas}$ (vs. SCE) is the potential at which the photocurrent density was measured, E$_{aoc}$ (vs. SCE) is the electrode potential at open circuit condition in the same electrolyte solution under the same illumination intensity at which the photocurrent density was measured.

Using Equation 1 the photoconversion efficiencies as a function of applied potential for sample 1 and 2 were determined (see Figure 2). The maximum photoconversion efficiency of 1.69 % at an applied potential of 0.69 V vs E$_{aoc}$ and 2.23% at an applied potential of 0.52V vs E$_{aoc}$ were observed for sample 1 and 2 respectively.

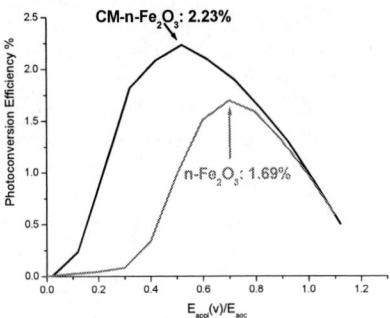

Figure 2. Photoconversion efficiency as a function of applied potential E_{appl} vs E_{aoc} for n-Fe_2O_3 (sample 1) and CM-n-Fe_2O_3 (sample 2). The E_{aoc} values were found to be – 0. 35 V/SCE and – 0.50 V/SCE for sample 1 and 2 respectively.

3.2. Monochromatic photocurrent – wavelength dependence:

Figure 3 shows the monochromatic photocurrent density, $j_p(\lambda)$, Vs. wavelength, λ at measured potential of 0.1 V/SCE under monochromatic light from a global AM 1.5 solar simulator (model 91192), for both sample 1 and 2. A spectra-physics monochromator, model 7725 and Keithley multimeter were used for monochromatic photocurrent measurements. The plot $J_p(\lambda)$ Vs. wavelength illustrates that doping iron oxide (Fe_2O_3) using carbon can improve the photoactivity of the semiconductor particularly in the UV region between 225 to 450 nm. It is to note that the photocurrent density values of 2.1 mA/cm^2 and 3.2 mA/cm^2 at the same measured potential of 0.1 V/SCE under white light illumination agrees reasonably well with the integrated values of monochromatic photocurrent densities of 2.064 mA/cm^2 and 2.27 mA/cm^2 for sample 1 and 2 respectively.

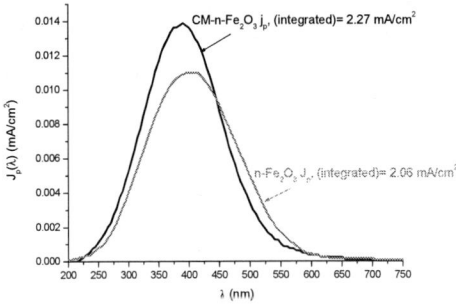

Figure 3. Monochromatic Photocurrent density, $j_p(\lambda)$, Vs. wavelength at measured potential of 0.1 V/SCE and under monochromatic light illumination from a global AM 1.5 solar simulator, for both sample 1 (n-Fe_2O_3) and sample 2 (CM-n-Fe_2O_3).

3.3. Band gap energy:

Figure 4 shows the Tauc plot of $(\eta\ hv)^{1/2}$ versus (hv) for oven made n-Fe_2O_3 (sample 1) and for flame made CM-n-Fe_2O_3 (sample 2). The quantum efficiency, η of the sample was calculated using $\eta(\lambda) = j_p(\lambda)/[eI_0(\lambda)]$, where $J_p(\lambda)$ is wavelength dependent photocurrent density, e is the electronic charge, $I_0(\lambda)$ is wavelength dependent intensity of incident light, and hv represents frequency dependent photon energy. The $(\eta\ hv)^{1/2}$ versus hv plot shows the indirect band gap of the iron oxide between 2.2 eV for n-Fe_2O_3 nanowire (sample 1) and 1.95 eV for CM-n-Fe_2O_3 (sample 2). Thus carbon incorporation reduces the band gap energy of CM-n-Fe_2O_3 by 0.25 eV.

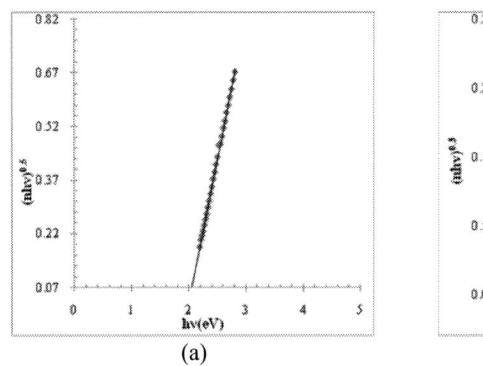 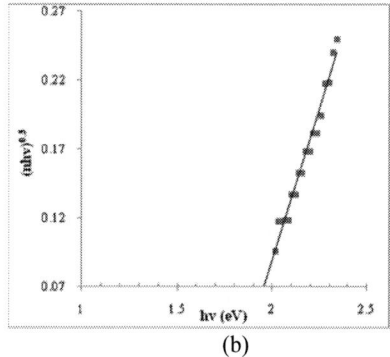

(a)   (b)

Figure 4. Plots of $[\eta(\lambda)\ hv\ ]^{1/2}$ vs. hv to determine the band gap of (a) n-Fe_2O_3 nanowire film (sample 1) and (b) CM-n-Fe_2O_3 sample (sample 2).

XRD pattern

Figure 5 illustrates the XRD patterns of n-Fe_2O_3 (sample 1) and CM- n-Fe_2O_3 (sample 2). Nanowire thin film sample shows the signature of Hematite (α-Fe_2O_3) and also the presence of impurities of magnetite. The XRD patters of CM-n-Fe_2O_3 also showed hematite structure. Presence of magnetite and Iron carbide impurities were also observed in its XRD patterns.

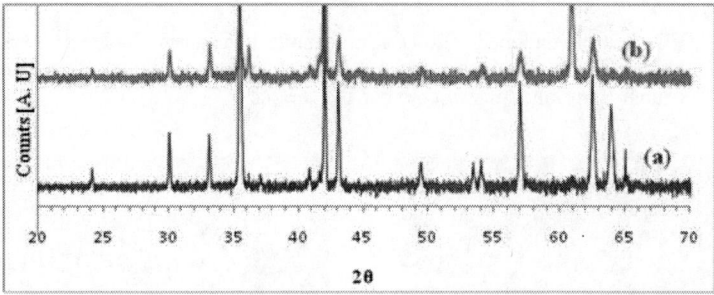

Figure 5. X-ray diffraction patterns for: (a) undoped n-Fe$_2$O$_3$ nanowires film (sample 1) and (b) Carbon modified iron oxide CM-n-Fe$_2$O$_3$ (sample 2).

UV-Vis spectra:

UV-vis spectra for n-Fe$_2$O$_3$ (sample 1) and CM-n-Fe$_2$O$_3$ (sample 2) are presented in figure 6. The samples have a clear absorption edge at around 600 nm, which conforms to the band gap of hematite (iron (III) oxide). The CM-n- Fe$_2$O$_3$ films (sample 2) showed an enhanced absorption in both UV and visible regions extending up to 700 nm. This extended absorption in the visible region up to 700 nm explains its higher photocurrent response and photoconversion efficiency of sample 2 compared to undoped n-Fe$_2$O$_3$. Nanowires film (sample 1).

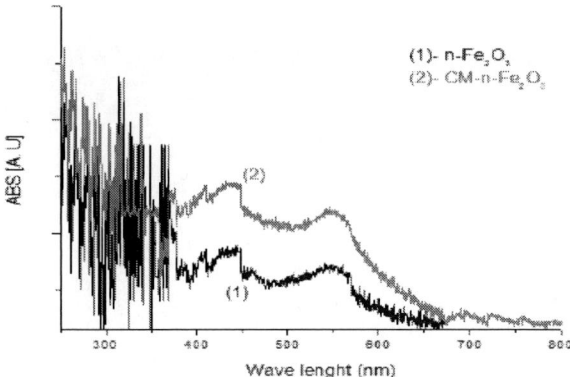

Figure 6. UV-Vis absorption spectra for (1) n-Fe$_2$O$_3$ nanowire film (sample 1) and (2) nanocrystalline CM-n-Fe$_2$O$_3$ thin film (sample 2).

Scanning electron micrograms (SEM) results:

Figure 7 shows that the undoped n-Fe$_2$O$_3$ film (sample 1) consists of nanowires oriented perpendicularly to the surface of the substrate (iron). The CM-n-Fe$_2$O$_3$ film (sample 2) consists of nanocrystals randomly oriented on the surface.

(a) (b)

Figure 7. SEM images for (a) electric oven-made n-Fe$_2$O$_3$ nanowire film (sample 1)   and (b) flame-made nanocrystalline CM- n-Fe$_2$O$_3$ film (sample 2)

Conclusions

The n-Fe$_2$O$_3$ nanowire films can be synthesized by oxidizing iron metal sheet in an electric oven and such films showed enhanced photoresponse for water splitting reaction compared to those made by spray pyrolysis or by CVD. Carbon modified iron (III) oxide (CM-n-Fe$_2$O$_3$) synthesized by flame oxidation of iron metal sheet showed lower band gap by 0.25 eV compared to n-Fe$_2$O$_3$ nanowire films and generated much higher photoresponse compared to undoped n-Fe$_2$O$_3$ and also those made by spray pyrolysis and by silicon doping.

References

1. A. Goetzberger V.U. Hoffmann, Photovoltaic Solar Energy Generation. Springer, **2005**.
2. A. Kleiman-Shwarsctein, Y. Hu, A. J. Forman, G. D. Stucky, and E. W. McFarland, J. Phys. Chem. 112(**2008**)15900-15907
3. Y. Hu, A. Kleiman-Shwarsctein, A. J. Forman, D. Hazen, J. Park, and E. W. McFarland, Chem. Mater. 20(**2008**) 3803–3805
4.   C. Sanchez, K. D. Sieber, G. A. Somorjai, J. electroanal. Chem. 252(**1988**) 269-290
5. I. Cesar, A. Kay, J. A. Gonzalez, and M. Gratzel, J. Am. Chem. Soc. 128(**2006**) 4582-4583
6. S. U. M. Khan, and Akikusa, J. phys. Chem. B, 103(**1999**) 7184-7189
7. W. B. Ingler Jr, S. U. M. Khan, Int J. Hydrogen Energy, 30(**2005**) 821-827
8. K. G. McGregor, M. Calvin, J. W. Otvos, J. Appl. Phys. 50 (**1979**) 369-373
9. M. M. Khader, N. N. Lichtin, G. H. Vurens, M. Salmeron, G. A. Somorjai, Langmuir 3 (**1987**) 303-304.

10. V. R. Satsangi, S. Kumari, A. P. Singh, R. Shrivastav, S. Dass, Int. J. Hydrogen Energy, 33 (**2008**) 312– 318
11. W. B. Ingler, S. U. M. Khan, J. Electrochem. and Solid-State Letters, 9(**2006**) G144-G146
12. A. Kay, I. Cesar, and M. Gratzel, J. Am. Chem. Soc. 128 (**2006**)15714-15721
13. H. Wang, T. Deutsch, and J. A. Turner, J. Electrochem. Soc. 155(**2008**)91-96
14. L. Vayssieres, C. Sathe, S. M. Butorin, D. K. Shuh, J. Nordgren, and J. Guo, Adv. Mater. 17(**2005**) 2320-2323
15. S.U.M. Khan, M. Al-Shahry, W.B. Ingler Jr., Science 297 (**2002**) 2243-2245
16. Y. A. Shaban, S. U. M. Khan, Chem. Phys., 339 (**2007**) 73-85.
17. Y. A. Shaban, S. U. M. Khan, Int. J. Hydrogen Energy, 33 (**2008**) 1118-1126.

146

ECS Transactions, 19 (3) 147-157 (2009)
10.1149/1.3120696 ©The Electrochemical Society

Hydrothermally Grown ZnO Nanorods as Cell Adhesion Control Coating for Implant Devices

B.H. Chu[a], C.Y. Chang[a], L.C. Leu[b], D. Norton[b] , J. Lee[a], T. Lele[a], P. Jiang[a], Y. Tseng[a], S. J. Pearton[b], A. Gupte[c], B. Keselowsky[d], and F. Ren[a]

[a] Department of Chemical Engineering, University of Florida, Gainesville, FL 32611
[b] Department of Material Science, University of Florida, Gainesville, FL 32611
[c] Depaertment of Gastroenterology, Hepatology and Nutrition, University of Florida, Gainesville, FL 32611
[d] Department of Biomedical Engineering, University of Florida, Gainesville, Florida 32611

> Recent studies show that hydrothermally grown ZnO nanorods can modulate cell adhesion and survival on various substrates. Hydrothermal growth of ZnO nanorods on biliary stents and nitinol wires used in metal stents is demonstrated. The nanorods can be conformably coated with silicon dioxide using low temperature plasma enhanced chemical vapor deposition (PECVD) system to reduce immunogenic response. In addition, the control of nanorod density is demonstrated by using a surfactant, Triton X-100. These results show promise of using nanorod array to control cell adhesion and viability on implantable devices.

Introduction

Zinc oxide (ZnO) is a II-VI semiconductor with wide bandgap energy of 3.37 eV and exciton binding energy of 60 meV at room temperature, which has attracted great interest for possible applications in optoelectronic devices such as room temperature lasers, light emitting diodes, transparent field effect transistors, ultra-violet (UV) detectors and solar cells [1-4]. One-dimensional ZnO in the form of nanorods and nanowires has also gained much interest due to its numerous applications in areas such as optoelectronics and bio-sensors [5-20]. Growth of ZnO nanostructures by the hydrothermal method [21-29] is very attractive for biomaterials applications compared to methods such as vapor-liquid-solid epitaxy (VLSE), metal organic chemical vapor deposition (MOCVD), pulsed laser deposition (PLD) because it does not involve high growth temperature, high operating cost or sophisticated equipment.

Recently, hydrothermally grown ZnO nanorods have been used to modulate cell adhesion and survival on various substrates [30]. A variety of umbilical vein and capillary endothelial cells were found to adhere far less to nanorods than to flat ZnO substrates. The nanorods appear to be useful as adhesion-resistant biomaterials capable of reducing viability in anchorage-dependent cells[30]. The ability to minimize cell adhesion is desirable for the success of implanted devices such as stents.

More cancer patients than ever are using stents as part of their treatment. A key challenge with the use of current stents in palliation of advanced esophageal, gastric, hepatobiliary, colon, pancreatic and pulmonary cancers is occlusion caused by tumor cell adhesion, ingrowth and overgrowth on the stent. Biofilm formation and cellular debris

147

can also cause occlusion of these stents. An occluded stent, based on its area of insertion, will subsequently cause difficulty in swallowing and breathing as well as intestinal obstruction, jaundice with severe infection and possible death. This necessitates the need for frequent stent exchanges and increases patient discomfort and hospital visits. Nanorod coating on the stents may reduce these problems significantly. However, stents have complicated three-dimensional geometries, which present challenges for the growth of well defined monolayer of nanorods.

While ZnO is a promising material for bio-sensing application, however it may potentially create immune responses if installed in the body [28-31]. To alleviate this problem, it becomes necessary to coat ZnO nanorods with less immunogenic and more benign materials. SiO_2 was chosen because glass cover-slips are commonly used for cell culture [31].

In this paper, we report the hydrothermal growth of ZnO nanorods on Cotton-Leung biliary stents from Wilson-Cook Medical Inc. and nitinol wires, a material that is widely used in metal stents. In addition, we investigated the density control of nanorods with a surfactant and the coating of SiO_2 on ZnO nanorods with a plasma enhanced chemical vapor deposition (PECVD) system.

Experimental

The hydrothermal growth of ZnO nanorods on stents and nitinol wires was achieved in two steps: the first involved coating the substrate with a ZnO nanocrystal solution and then dipping it into an aqueous growth solution. The nanocrystal solution was prepared by following the method used by Pacholski et al. [29]. 30mM NaOH (Sigma Aldrich) in methanol was slowly added to a 10mM zinc acetate dihydrate ($Zn(O_2CCH_3)_2 \cdot 2H_2O$, Sigma Aldrich) solution at 60°C over a two hour period. The stent was dipped into the nanocrystal solution and blow dried with filtered nitrogen. The coated stent was then hung in an oven at 120°C for 5 minutes to insure that the ZnO nanocrystals were well adhered to the substrate. The dipping and drying procedures were repeated several times to ensure dense and uniform nano-crystalline seeds coated on the substrates. The nano-crystalline seed coated substrate was then immersed in an aqueous mixture of 20mM zinc nitrate hexahydrate ($Zn(NO_3)_2 \cdot 6H_2O$, Sigma Aldrich,) and 20mM hexamethylenetetramine ($C_6H_{12}N_4$, Sigma Aldrich,). During the ZnO nanorod growth, the Pyrex glass flask was placed on a hot plate and maintained at 94°C for 2 hours while constantly stirring the growth solution. Afterward, the substrate was removed from the solution, thoroughly rinsed with de-ionized water and dried with nitrogen at room temperature.

Low density ZnO nanorod synthesis was achieved by modifying the nanocrystal solution with adding a surfactant, Triton X-100 (Sigma Aldrich) diluted in methanol of 0.1% by volume, in the original nanocrystal solution. The solution was placed on a vortex mixer and then agitated to create a homogeneous solution without the agglomeration of the ZnO nanocrystals. In this study, silicon substrates were used and the same nano-crystalline seed coating, drying and nanorod growth process as described earlier was employed.

Silicon dioxide was deposited with a Unaxis 790 PECVD system at 50 °C using N_2O and 2% SiH_4 balanced by nitrogen as the precursors. The deposition chamber pressure was kept at 900 mtorr and rf power of 30W was generated at 13.56 MHz. The self- bias voltage during the SiO_2 deposition was around 1 V, so that ion-induced damage is not expected to occur.

Results and Discussion

Figure 1 shows SEM images of the surface of nitinol wire before (left) and after (right) ZnO nanorod growth. The image of bare nitinol wire showed a lot of cracks and trenches, which were created during the wire fabrication through wire extrusion. As shown in the Figure 1, the ZnO nanorods grew uniformly on the surface of the nitinol surface but not in the trenches. By constantly stirring the growth solution, the ZnO nanorods could be grown inside the trenches as well. This suggests that the limiting mechanism of ZnO nanorod growth is by Zn ion diffusion. The nanorods exhibit a Wurtzite crystal structure with diameters ranging from 30 to 50 nm and approximately 500 nm in length.

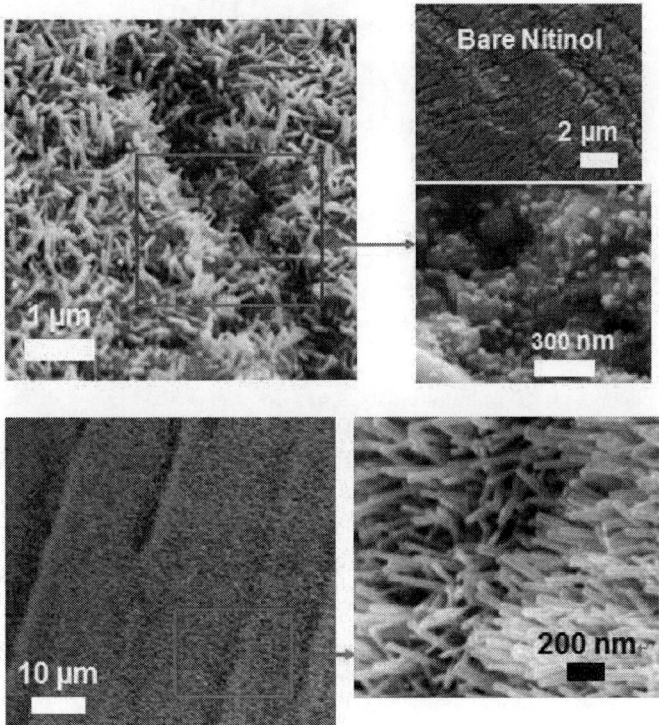

Figure 1. SEM images of (top right) bare nitinol wire (top left) ZnO nanorod growth on nitinol wire surface without agitating the zinc nitrate solution during the nanorod growth, (top insert) unsuccessful ZnO nanorod growth in the trenches on the nitinol wire, and (bottom right) ZnO nanorod growth on nitinol wire surface with agitating the zinc nitrate solution during the nanorod growth, (bottom insert) closer view of nanorods in the trenches of the nitinol wire

Figure 2 shows a photograph of a biliary plastic stent and an SEM image of the biliary stent surface and surface covered with ZnO nanorods. By constantly stirring the growth solution, ZnO nanorods were successfully grown on both the inside and outside of the stent. The crystal structure and orientation of the nanorods were identical to those grown on the nitinol wire.

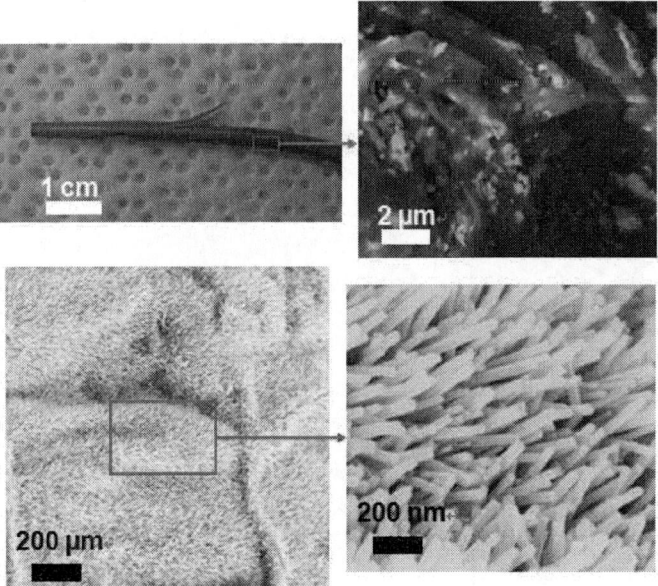

Figure 2. (top left) photograph of a plastic biliary stent. (top right) SEM image of the surface of biliary stent. (bottom left) SEM image of ZnO nanorods grown on the biliary stent surface. (bottom insert) closer view of the nanorod grown on the biliary stent surface.

Figure 3 (a) and (b) show the nanorod growths starting with diluted seed solutions of 10% and 2% from the original seed solution, respectively. The density of the nanorods was able to reduced from 210 to 140 and 90 nanorods/100µm2, for the seed solution of 10% and 2%, respectively. However, ZnO seed nanocrystals readily agglomerated for further dilution below 1% of the original nano-crystalline solutions and resulted in the formation of clustered 5 nanorods even after agitation, as shown in Figure 3 (c). On the contrary, the ZnO nanocrystal seed solutions with surfactant dispersed uniformly on the substrate surface showed no agglomeration. As shown in Figures 3 (bottom), the ZnO nanorods could be controlled with dilution of the nanocrystal seed solution from the original nano-crystalline solutions to 1%, 0.5%, and 0.2%, respectively. The density of the nanorods could be controlled from 140 to 5 nanorods/100µm2. As we mentioned previously, ZnO nanorods have been demonstrated to modulate NIH 3T3 fibroblast, umbilical vein cell, and capillary endothelial cell adhesion and survival on various substrates [28]. The distance between the adjacent nanorods is very critical for cells to assemble focal adhesions and stress fibers on nanorods. The ability to control the density of nanorods is critical for future studies of cell adhesion on the nano-materials.

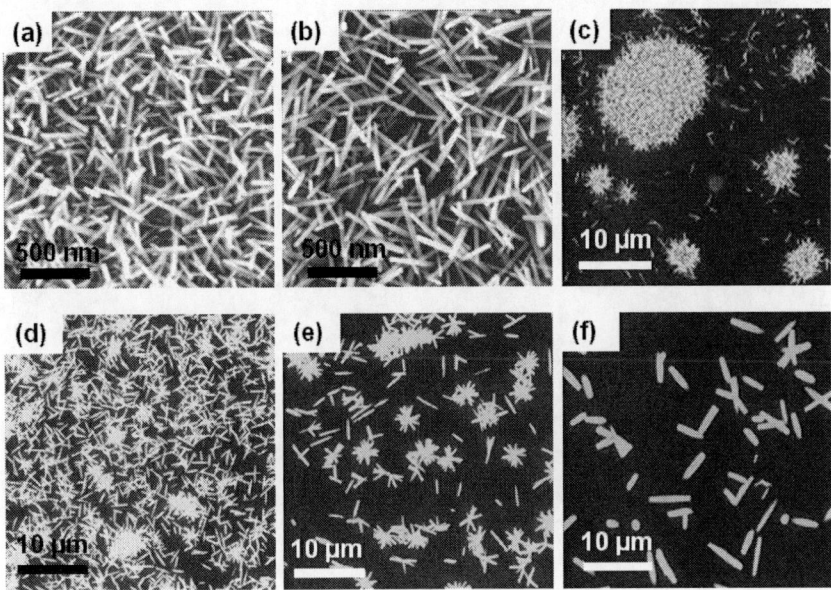

Figure 3. (top) SEM images of nanorods grown with diluted seed solutions; (a) 10%, (b) 2%, and (c) 1%. (bottom) SEM images of nanorods grown with adding surfactant in the diluted seed solutions; (d) 1%, (e) 0.5%, and (f) 0.2%.

Figure 4 compares the density of nanorods grown with and without the use of surfactant in the nanocrystal solution. Both curves show that the density of nanorods decreased according to the concentration of seed solution. The insert in Figure 4 shows the enlarged view of Figure 4 below 1 % of nanocrystal seed solution with the use of surfactant.

The nanorod density could be reduced by using diluted nanocrystal seed solution. As mentioned previously, without the assistance of surfactant, non-uniform of nanorods and formation of clustered nanorods were observed for nanorod growth using the nanocrystal seed solution concentration below 1 %.

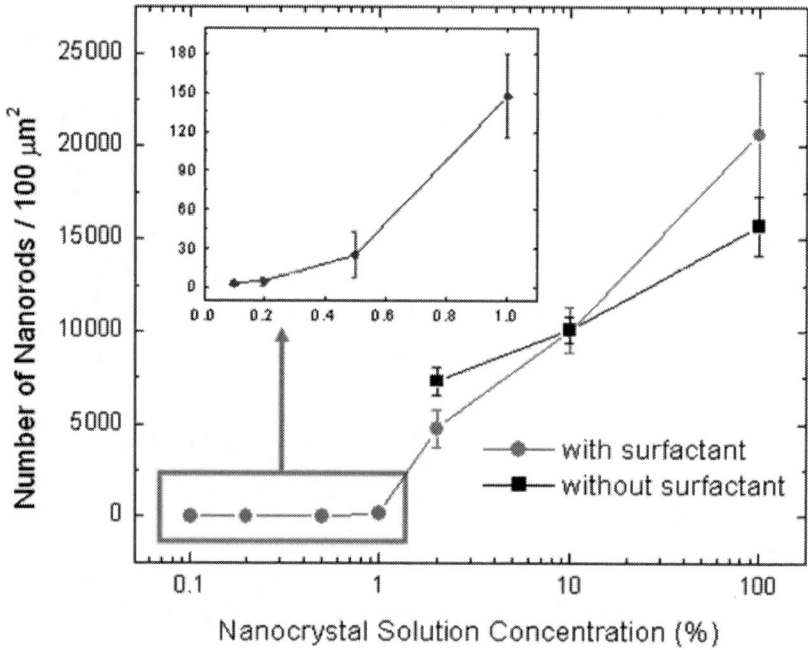

Figure 4. Nanorod densities with and without surfactant during the growth as a function of the seed density

Figure 5 (a) shows a TEM image of ZnO nanorods coated with 100 Å of SiO_2. The SiO_2 coated ZnO nanorods were released from the substrate by sonicating in methanol and then transferred on to a TEM grid. The diameter of the nanorod measures approximately 40 nm. The area that is surrounding the nanorod is slightly darker than the background indicating the presence of the SiO_2 layer. The elemental composition of the nanorods is shown in the EDX dot spectrum result shown in Figure 5 (b). There is a peak for silicon, along with zinc and oxygen, indicating the presence of silicon dioxide and ZnO. The carbon reading is from the methanol solvent.

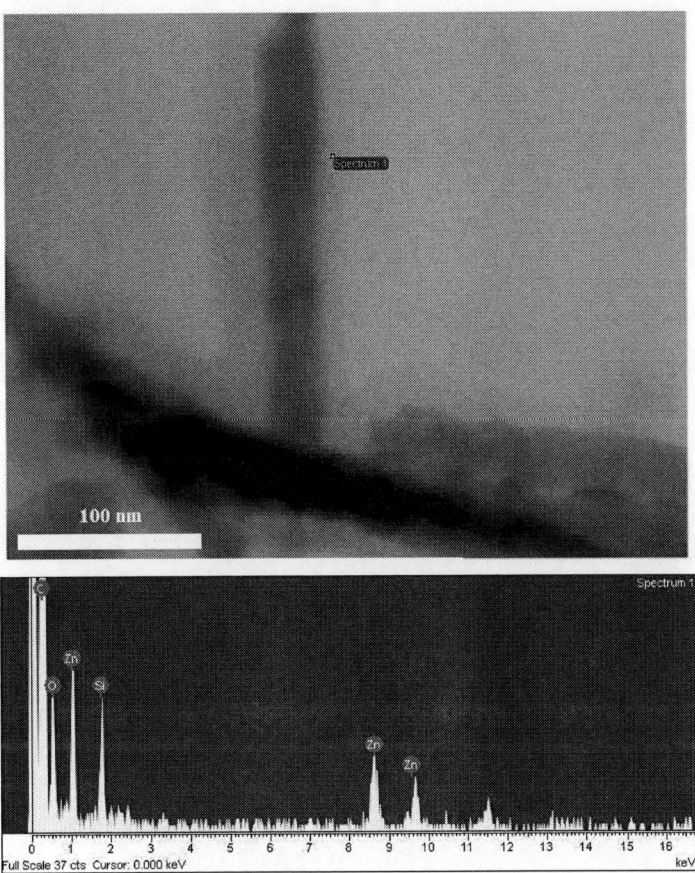

Figure 5: (a) TEM image of SiO_2 coated ZnO nanorods, (b) dot spectrum of nanorod

The EDX elemental compositions line spectra overlapped with the TEM image of the SiO_2 coated ZnO nanorod are presented in Figure 6 (left). Figure 6 (right) illustrates zinc, silicon, and oxygen concentration distribution across the nanorod and its surrounding area. The line spectrum shows that zinc is concentrated on the nanorod whereas silicon and oxygen are present in all areas of the spectrum. Importantly, oxygen is more abundant towards the center of the nanorod because of the presence of both SiO_2 and ZnO. This indicates that SiO_2 is covering the nanorods in all directions.

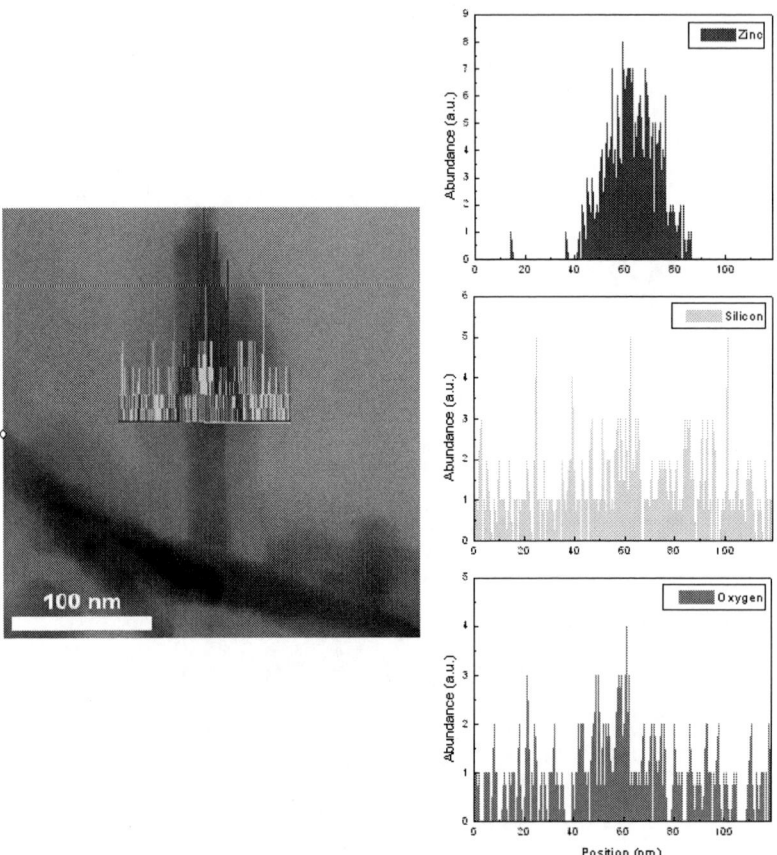

Figure 6: TEM image of SiO$_2$ coated ZnO nanorods with line spectrum results, distribution of zinc, distribution of silicon, and distribution of oxygen.

To ensure the SiO$_2$ layer deposited continuously without any areas that exposed bare ZnO, electrical conductance measurements and wet chemical etching experiments were conducted on the bare and SiO$_2$ coated ZnO nanorods samples. For the electrical conductance measurements, a thin metal layer composed of the gold and nickel bilayer structure was deposited on the nanorod samples in circular patterns as shown in Figure 7(a). Then, -5 to 5 V of bias voltage was swept across two separate circular patterns. ZnO is generally a very good electrical conductor in its as-grown state, showing high n-type electrical conductivity [1-4]. Therefore, if any nanorods were not covered in SiO$_2$, the ZnO nanorod sample would be very conductive. The results shown in Figure 7(b) indicate that there is minimal current, in the nA range, flowing through the SiO$_2$ coated ZnO nanorod sample. For the wet etching experiment, the ZnO nanorods were etched away in 1:200 HCl:H$_2$O solution in less than 1 min, consistent with the known etching properties of ZnO[32]. By sharp contrast, there was no etching observed on the SiO$_2$ coated ZnO nanorod sample even after 5 mins of sample exposure to HCl solution. This is a strong indication that the ZnO was completely encapsulated by the SiO_2. This approach looks promising for use of ZnO nanorods in controlling cell behavior in applications such as tissue engineering scaffolds, biomedical implants and implanted drug delivery devices. The SiO_2 coating reduces concerns about toxicity issues from Zn ions in these applications.

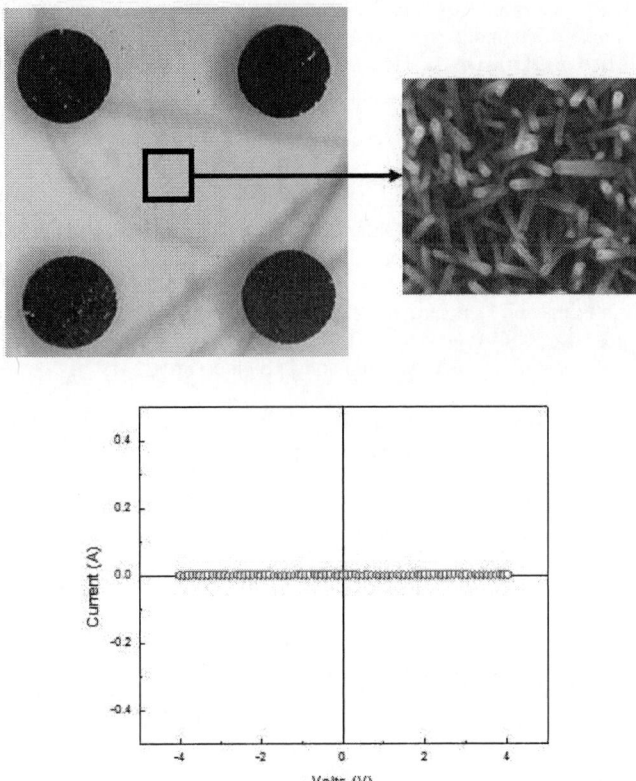

Figure 7: (a) optical microscope image of circular metal patterns and SEM of ZnO nanorods, (b) current flowing between metal patterns.

Summary and Conclusions

In summary, we have demonstrated a low temperature hydrothermal method to grow ZnO nanorods on nitinol wire and biliary stents. SEM images showed uniform highly dense nanorod arrays covering the substrate. Also, the control of density of nanorods was demonstrated by diluting the ZnO nanocrystal solution with methanol and a surfactant. In addition, low temperature PECVD was used to conformably coat SiO_2 on hydrothermally grown ZnO nanorods. These results show that nanorod monolayer can be grown on stents currently used for palliation of tumors. The promise of utilizing ZnO nanorods to control cell adhesion on implantable devices, thereby increasing their durability, patient comfort and minimizing the need for frequent stent exchanges and thus reducing hospital stays and the expense involved.

Acknowledgments

The work at UF is partially supported by ONR Grant N000140710982 monitored by Igor Vodyanoy, Army Research Office under grant no. DAAD19-01-1-0603 monitored by Dr. M. Gerhold, NSF (DMR 0700416, Dr. L. Hess), and the State of Florida, Center of Excellence for Nano-Bio Application.

References

1. Tsukazaki, A. Ohtomo, T. Onuma, M. Ohtani, T. Makino, M. Sumiya, K. Ohtani, S.Chichibu, S.Fuke, Y.Segawa, H.Ohno, H.Koinuma and M. Kawasaki, Nat Mater. 4, 42 (2005).
2. "ZnO Bulk, Thin Films and Nanostructures", Ed. C. Jagadish and S.J. Pearton (Elsevier, Oxford, UK, 2006).
3. S. J. Pearton, D. P. Norton, K. Ip, Y. W. Heo and T. Steiner, Progr. Mater. Sci .50,293 (2005).
4. Z.L. Wang, "ZnO Nanostructures", in ZnO Bulk, Thin Films and Nanostructures , ed. C. Jagadish and S.J. Pearton, Elsevier, Oxford, 2006
5. S. Ashok Kumar and Shen-Ming Chen, Analytical Letters, 41, 141 (2008).
6. W. I. Park, J. S. Kim, G. C. Yi, and H. J. Lee, Adv. Mater.17, 1393(2005).
7. M. S. Arnold, Ph. Avouris, Z. W. Pan and Z. L. Wang, J. Phys. Chem., B 107, 659 (2003).
8. H. D. Xiong, W. Wang, Q. Li, C. A. Richter, J. S. Suehle, W.-Ki Hong, T. Lee, and D. M. Fleetwood, Appl. Phys. Lett. 91, 053107 (2007).
9. S. Kang, H. T. Wang, F. Ren, S. J. Pearton, T. E. Morey, D. M. Dennis, J. W. Johnson, P. Rajagopal, J. C. Roberts, E. L. Piner, and K. J. Linthicum, Appl. Phys. Lett. 91, 252103 (2007).
10. H. Chu, B. S. Kang, F. Ren, C. Y. Chang, Y. L. Wang, S. J. Pearton, A. V. Glushakov, D. M. Dennis, J. W. Johnson, P. Rajagopal. J. C. Roberts, E. L. Piner, and K. J. Linthicum, Appl. Phys. Lett. 93, 042114 (2008).
11. S. A. Kumar and S. M. Chen, Analytical Letters, 41 (2008) 141–158.
12. Y.-Y. Noh, X. Cheng, H. Sirringhaus, J. I. Sohn, M. E. Welland, and D. Joon Kang, Appl. Phys. Lett. 91, 043109 (2007).
13. W.-K. Hong, D.-K. Hwang, I.-K. Park, G. Jo, S. Song, S.-Ju Park, T. Lee, B.-J. Kim, and E. A. Stach , Appl. Phys. Lett. 90, 243103 (2007).
14. L. Luo, B. D. Sosnowchik and L. Lin, Appl. Phys. Lett. 90, 093101 (2007).
15. S. N. Cha, J. E. Jang, Y. Choi, G. A. J. Amaratunga, G. W. Ho, M. E. Welland, D. G. Hasko, D.-J. Kang and J. M. Kim, Appl. Phys. Lett. 89, 263102 (2006).
16. 15. S. Ju, D. B. Janes, G. Lu, A. Facchetti and T. J. Marks, Appl. Phys. Lett. 89, 193506 (2006).
17. H.T. Wang, B.S. Kang, F. Ren, L.C. Tien, P.W. Sadi, D.P. Norton, S.J. Pearton, J. Lin, Appl. Phys. Lett. 87, 0243503-1-3 (2005).
18. B. S. Kang, Y. W. Heo, L.C. Tien, D. P. Norton, F. Ren, B.P. Gila, S.J. Pearton, Appl. Phys. A 80, 1029–1032 (2005).

19. P.-C. Chang, Z. Fan, C.-Jen Chien, D. Stichtenoth, C. Ronning, and J. Grace Lu, Appl. Phys. Lett. 89, 133113 (2006).
20. Y.W. Heo, D.P. Norton, L.C. Tien, Y. Kwon, B.S. Kang, F. Ren, S.J. Pearton and J.R. LaRoche, Mat. Sci. Eng: R: Reports, 47, 1(2004).
21. J. Song, S. Baek, and S. Lim, Physica B 403,1960 (2008).
22. Q Ahsanulhaq, A. Umar, and Y. B. Hahn, Nanotechnology 18,115603 (2007).
23. W. Y. Su, J. S. Huang, and C. F. Lin, J. Crystal Growth 310, 2806 (2008).
24. B. S. Kang, S. J. Pearton, and F. Ren, Appl. Phys. Lett. 90,083104 (2007).
25. H. Zhang, J. Feng, J. Wang, and M. Zhang, Materials Lett., 61, 5202 (2007).
26. H. L. Cao, X. F. Qian1, Q. Gong, W. M. Du, X. D. Ma, and Z. K. Zhu, Nanotechnology 17, 3632 (2006).
27. Z. Li, X. Huang, J. Liu, and H. Ai, Materials Lett. 62, 2507 (2008).
28. Abhilash Sugunan, H. Warad, and M. Boman, J. Sol-Gel Sci. Techn. 39, 49 (2006).
29. C. Pacholski, A. Kornowski, and H. Weller, Angew. Chem. Int. Ed. 41, 7 (2002) .
30. J. Lee, B. S. Kang, B. Hicks, T. F. Chancellor, Jr., B. H. Chu, H. T. Wang, B. G. Keselowsky, F. Ren, and T. P. Lele, Biomaterials 29, 3743 (2008).
31. S. F. Brinkman and W. D. Johnston, Arch. Environ. Contam. Toxicol. 54, 466 (2008).
32. J.J. Chen, S. Jang, F. Ren, Y. Li,H.S. Kim, D.P. Norton, S.J .Pearton, A. Osinsky, S.N.G. Chu and J.F. Weaver, J.Electron. Mater. 35,516(2006).

158

CHAPTER 5

COMPOUND SEMICONDUCTOR MATERIALS AND DEVICES POSTER SESSION

ECS Transactions, 19 (3) 161-165 (2009)
10.1149/1.3120697 ©The Electrochemical Society

Luminescent Properties of M-Al$_2$O$_4$:Eu^{2+} (M: Ca, Sr) Thin Films Prepared on Sapphire Substrates

Jun Seong Lee, Choong Ki Lee, and Young Jin Kim

Department of Materials Science and Engineering, Kyonggi University, Suwon 443-760, Korea
E-mail: yjkim@kyonggi.ac.kr

M-Al$_2$O$_4$:Eu^{2+} (M: Ca, Sr) thin films were deposited on c-plane sapphire substrates by rf magnetron sputtering, and growing behaviors and luminescent properties were investigated. As-deposited films were amorphous-like, but they were transformed to the epitaxial films by annealing at high temperatures. The epitaxial orientations and growing behaviors of CaAl$_2$O$_4$ and SrAl$_2$O$_4$ films were different each other due to the crystallographic distinction. PL spectra exhibited a blue and a green emission for CaAl$_2$O$_4$:Eu^{2+} and SrAl$_2$O$_4$:Eu^{2+}, respectively.

Introduction

Eu^{2+} doped alkaline earth aluminates, M-Al$_2$O$_4$:Eu^{2+} (M: Ca, Sr), have been recognized as excellent phosphor materials due to the prominent luminescent properties. Eu^{2+} ions in the luminescent materials generally show a broad band emission due to the electronic transitions between 4f^7 ground state and 4f^{6}5d^1 excited state. These radiative transitions are highly affected by the crystal field of the surrounding ions. Accordingly the structure of the host crystal contributes to the variation of the emission wavelength. For example, Eu^{2+} ions doped CaAl$_2$O$_4$ (CA) and SrAl$_2$O$_4$ (SA) exhibit a blue and a green emission, respectively (1-3).

The crystallographic structures of CA and SA are same: monoclinic (β-tridymite type). But the space groups are P2$_1$/c for CA and P2$_1$ for SA, respectively. [AlO$_4$] tetrahedral is constructed by a three-dimensional framework, and each oxygen surrounds M^{2+} ions (4). Alkaline earth aluminates contain three M^{2+} sites: M (I) is nine-coordinated and the others, M (II) and M (III), are six-coordinated by oxygen atoms. Eu^{2+} ions are preferentially substituted for nine-coordinated M^{2+} sites rather than for two six-coordinated ones, because larger spaces are demanded for the Eu^{2+} ions. Detail crystallographic parameters are shown in Table. I.

TABLE I. Crystal parameters.

Parameters	CaAl$_2$O$_4$	SrAl$_2$O$_4$
Crystal system	Monoclinic	Monoclinic
Space group	P2$_1$/c	P2$_1$
a (Å)	8.6980	8.4470
b (Å)	8.0920	8.8160
c (Å)	15.2080	5.1630
Melting point	1606 ℃	1960℃

Compared with powder phosphors, some advantages of thin film type enable them to be applied to electronic displays including thin film electroluminescent display and field emitting display. However, the luminescent efficiency of thin film phosphors is lower than that of powder phosphors due to the internal reflectance at the smooth film surface and the absorption of the substrates (5-6). One way for the improvement of the luminescent efficiency is an epitaxial growing, because defects and impurities, which exist at grain boundaries of poly-crystalline films, distort the crystal field surrounding the activators and generate non-radiative traps at a band gap (7).

Since the energy transitions of Eu^{2+} ions are very sensitive to the surroundings, the crystallographic distinction between CA and SA causes the different PL (Photoluminescence) properties as mentioned above. Furthermore it can be speculated that growing behaviors of thin film are possibly dissimilar.

In this work, Eu^{2+} doped SA thin films were prepared on c-plane sapphire substrates by rf magnetron sputtering, and the growing behaviors and PL properties were compared with those of CA films that were obtained in our previous work (8).

Experiment

$M-Al_2O_4:Eu^{2+}$ (M: Ca, Sr) thin films were prepared on c-plane sapphire substrates by rf magnetron sputtering system. Sputtering targets (2″ diameter) were fabricated by sintering the stoichiometric mixtures of $MCO_3-Al_2O_3-Eu_2O_3$ at 1500 °C for 3 hours under air atmosphere. $O_2/(Ar+O_2)$ ratios in sputtering gas varied in the range of 10 – 50 % at the constant working pressure of 5 mTorr. As-deposited thin films were post-annealed in the electric tube furnace at 900 °C for 3 hours under 5 % H_2 (95 % N_2) atmosphere.

The crystallographic structure of the films were measured by X-ray diffractometer (XRD, PHILIPS X'pert) θ-2θ and ϕ-scan with CuKα radiation (λ = 1.5406 Å). Scanning angle and speed were 20 - 70° and 5°/sec, respectively. Field emission scanning electron microscope (FESEM, JEOL JSM-6500F) was used to observe surface morphology and to measure the thickness. The luminescent spectra were measured by PL system (PSI DARSA-5000) with a xenon light source (250 W) at a room temperature.

Results and Discussion

XRD patterns of CA:Eu^{2+} (1 mol%) and SA:Eu^{2+} (10 mol%) films after post-annealing are shown in Figure 1. Both as-deposited films were amorphous-like, but they were epitaxially recrystallized on c-plane sapphire through atomic rearrangements after annealing at 900 °C. However, their orientations were different from each other, showing out-of-plane orientations of CA(020)//sapphire(0001) and SA(200)//sapphire(0001), respectively. This indicated that the crystallographic distinctions between SA and CA (Table I) caused the different epitaxial directions. Despite the same monoclinic structure, the lattice constant c of CA is about three times as long as that of SA, while the lattice constant a and b are a little distinctive.

Corresponding PL excitation and emission spectra are shown in Figure 2, exhibiting apparently different results. For CA films a blue emission band at 445 nm was achieved under 325 nm excitation, while the spectra of SA films showed a green emission at 535 nm under 340 nm excitation. These optical emission and excitation ranges are almost same with those of the powders that have been previously reported. The emissions generated from the energy transition of Eu (I) ions that were preferentially substituted for M (I) sites, but not from those of Eu (II) and Eu (III) ions, which corresponded to M (II) and M(III), respectively. Eu (I) sites are surrounded by nine oxygen atoms, and the respective distances between Eu and O in CA and SA structure are different, leading to the distinct luminescent properties.

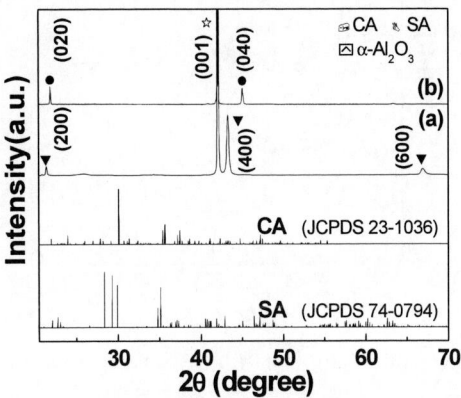

Figure 1. XRD patterns of post-annealed (a) SA films and (b) CA films deposited on c-plane sapphire substrates: $O_2/(Ar+O_2) = 0.5$.

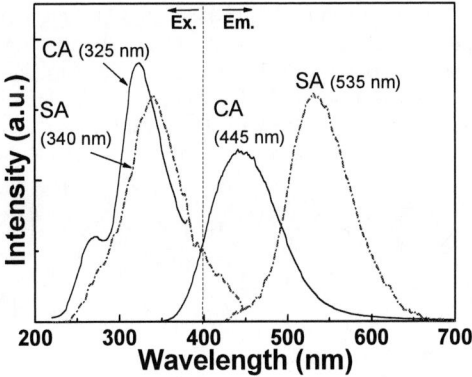

Figure 2. The excitation and emission spectra of (a) SA:Eu^{2+} and (b) CA:Eu^{2+} films after post-annealing.

PL intensities of post-annealed SA:Eu^{2+} ($\lambda_{em} = 535$ nm) and CA:Eu^{2+} ($\lambda_{em} = 445$ nm) films prepared as a function of the oxygen partial pressure, $O_2/(Ar+O_2)$, are shown in Figure 3. PL spectra could not be measured in as-deposited amorphous-like films, because crystal fields for the luminescence were not completely constructed. PL intensity of SA:Eu^{2+} films decreased with the increase of the oxygen partial pressure, but that of CA:Eu^{2+} films increased. This revealed that the oxygen partial pressure during sputtering process kinetically affected the structure of growing films, and also contributed to the compositional variation of the films.

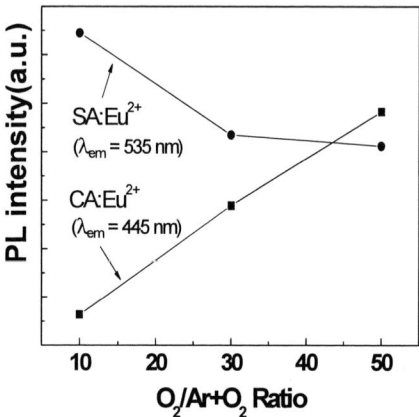

Figure 3. PL intensity of (a) SA:Eu^{2+} and (b) CA:Eu^{2+} as a function of oxygen partial pressure.

PL variations closely correlate with the crystallinity of films. It is known that the higher crystallinity leads to the higher PL intensity. FWHM (Full width of half maximum) values of XRD (400) peak of SA:Eu^{2+} films and (040) peak of CA:Eu^{2+} films as a function of the oxygen partial pressure are shown in Figure 4, which reversely coincide with the PL variation in Figure 3. The highest PL intensity corresponded to the smallest FWHM value, which meant the highest crystallinity. For SA films FWHM increased with increasing the oxygen partial pressure, while for CA films it decreased. It is not clear why the inclinations of FWHM variation as a function of the oxygen partial pressure are in the opposite way at present. But it can be speculated that the reverse tendency is ascribed to the different material properties between SA and CA.

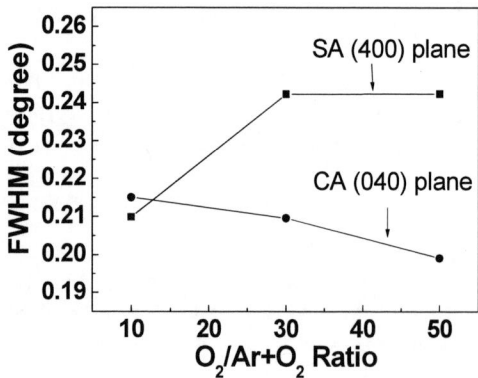

Figure 4. FWHM values of (a) (400) SA:Eu^{2+} and (b) (040) CA:Eu^{2+} as a function of the oxygen partial pressure.

Figure 5 shows the SEM micrographs (020) CA:Eu^{2+} epitaxial films and (200) SA:Eu^{2+} epitaxial films deposited at $O_2/(Ar+O_2)$ = 0.5 and 0.1, respectively, where the lowest FWHM values were obtained in each film. The grain size of CA films was significantly larger than that of SA films due to the distinction of the structural parameters and a melting point.

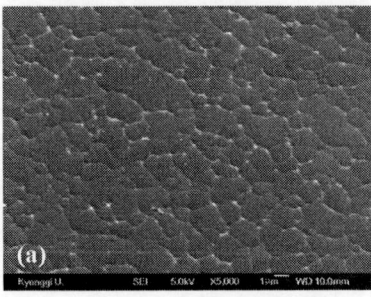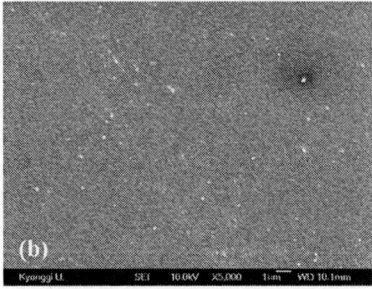

Figure 5. SEM micrographs of (a) CA:Eu^{2+} and (b) SA:Eu^{2+} at the $O_2/(Ar+O_2)$ ratio of 0.5 and 0.1, respectively.

Conclusion

Epitaxial M-Al$_2$O$_4$:Eu^{2+} (M: Ca, Sr) thin films could be achieved on c-plane sapphire substrates by post-annealing as-deposited ones, respectively. The epitaxial orientations and growing behaviors of CaAl$_2$O$_4$ and SrAl$_2$O$_4$ films were different from each other due to the crystallographic distinction. Related PL spectra exhibited a blue and a green emission for CaAl$_2$O$_4$:Eu^{2+} and SrAl$_2$O$_4$:Eu^{2+}, respectively. The variation of PL and the crystallinity closely correlated with the oxygen partial pressure in the sputtering gas.

Acknowledgments

This work was supported by the Korea Science and Engineering Foundation (KOSEF) grant funded by the Korea government (MOST). (No. R01-2005-000-10530-0).

References

1. J. Holsa, H. Jungner, M. Lastusaari, and J. Niittykoski, *J. Alloys Compd.*, **323**, 326 (2001).
2. T. Aitasalo, J. Holsa, H. Jungner, M. Lastusaari, and J. Niittykoski, *J. Alloys Compd.*, **341**, 76 (2002).
3. Y. J. Park and Y. J. Kim, *Mat. Sci. Eng. B* **146(1-3)**, 84 (2008).
4. W. Höekner and H. K. Müller-Buschbaum, *J. Inorg. Nucl. Chem.*, **38**, 983 (1976).
5. S. L. Jones, D. Kumar, K. G. Cho, R. Singh, and P. H. Holloway, *Displays* **19**, 151 (1999).
6. Y. J. Kim, S. M. Chung, Y. H. Jeong, and Y. E. Lee, *J. Vac. Sci. Technol.* **19** 1095 (2001).
7. Y. E. Lee, D. P. Norton, and J. D. Budai, *Appl. Phys. Lett.* **74**, 3155 (1999).
8. C. K. Lee and Y. J. Kim, *J. Cryst. Growth*, **311**, 904 (2009)

Observation of self-limiting regime in the atomic layer deposition of ZnO films using nitrous oxide as the oxygen supply

Ping-Han Chung[1], Hung-Wei Lai[1], Yen-Ting Lin[2]*, Kuo-Yi Yen[3], Chung-Yuan Kung[1], and Jyh-Rong Gong[1,2,3]

[1]Institute of Optoelectronic Engineering, National Chung Hsing University, Taichung 402, Taiwan, R.O.C.
[2]Institute of Nanoscience, National Chung Hsing University, Taichung 402, Taiwan, R.O.C.
[3]Department of Physics, National Chung Hsing University, Taichung 402, Taiwan, R.O.C.

We report the observation of self-limiting process in the atomic layer deposition (ALD) of zinc oxide (ZnO) films on (0001) sapphire substrates at low temperatures using diethylzinc (DeZn) and nitrous oxide (N_2O). I was found that a monolayer-by- monolayer growth regime occurred in a range of DeZn flow rates from 5.7 to 8.7 μmole/min. Furthermore, the temperature window for the self-limiting process of the ALD-grown ZnO films was also observed ranging from 290 to 310 □. The transmission and absorption spectra of the ZnO films prepared in the self-limiting regime show good optical characteristics with tramsmittance being more than 80% in the visible light region. Experimental results indicate that ZnO films grown in the self-limitng regime all exhibit improved materials characteristics and thickness uniformity.

Introduction

Zinc oxide (ZnO) has attracted much attention owing to its potential applications in optoelectronics in recent years. Compared with GaN and ZnSe, ZnO shows special advantages including having a large exciton binding energy of 60meV [1, 2] which is larger than those of GaN (25meV) and ZnSe (22meV) [3]. Moreover, ZnO is a direct wide band gap semiconductor with high melting temperature and stable chemical behavior. Recently, several techniques such as metalorganic chemical vapor deposition (MOCVD) [4], molecular beam epitaxy (MBE) [5], radio-frequency (RF) sputtering [6], vapor-liquid-solid epitaxy (VLSE) [7], pulsed laser deposition (PLD) [8] and atomic layer deposition (ALD) [9] have been conducted to grow ZnO films. Unlike the conventional growth techniques, ALD relies on alternate supply of precursors to achieve an ultimate control of ultra-thin film deposition in a monolayer–by– monolayer fashion.

We have previously reported the preparation of ZnO films at 600□ by alternate supply of diethylzinc (DEZn) and nitrous oxide (N_2O) [9]. However, the growth rate of ZnO film is far beyond one monolayer per growth cycle. To explore the regime of self-limiting process of ZnO growth, reduction in deposition temperature is required. In this paper, we present the characteristics of ALD-grown ZnO films on (0001) sapphire substrates using diethylzinc (DEZn) and nitrous oxide (N_2O) at reduced temperatures. It was found that a self-limiting process window was present at 300□. The ZnO films grown in the monolayer – by – monolayer deposition regime show improved material characteristics.

Experimental

ZnO films were prepared on (0001) sapphire substrates at 250 ~ 340 °C in an inductively heated quartz reactor operated at atmospheric pressure by ALD using DeZn and N_2O. The (0001) sapphire substrates were solvent-cleaned, hot-etched in an HCl : H_3PO_4 (3:4) solution for 5 min, rinsed with de-ionized water, and blow-dried with N_2 gas before loading. A typical ZnO film was deposited for 1200 cycles after cleaning the sapphire substrate at 1000 °C.

To evaluate ZnO film growth rate, cleaved ZnO/sapphire samples were investigated by using a HITACHI field emission scanning electron microscope (FESEM) to observe film thicknesses. Transmission and absorption measurements were conducted to explore the transmission and absorption spectra of ZnO films using a xenon lamp as the light source. Four-point probe equipment was employed to reveal the resistivities of ZnO films.

Results and discussion

A series of ZnO films were prepared under various DeZn flow rates to explore the monolayer-by-monolayer deposition regime. In this case, the ZnO growth rate was evaluated by dividing the thickness of a ZnO film to the number of ALD growth cycles. Figure 1 shows a typical plot of ZnO growth rate at 300 °C versus DEZn flow rate. Note that a self-limiting process window is present in the DEZn flow rate ranging from 5.72 to 8.77 μmol/min. Beyond the saturated growth regime, the ZnO growth rates are more than one monolayer per cycle for DeZn flow rates being larger than 8.77 μmol/min, while they are less than one monolayer per cycle for DeZn flow rates being smaller than 5.72 μmol/min.

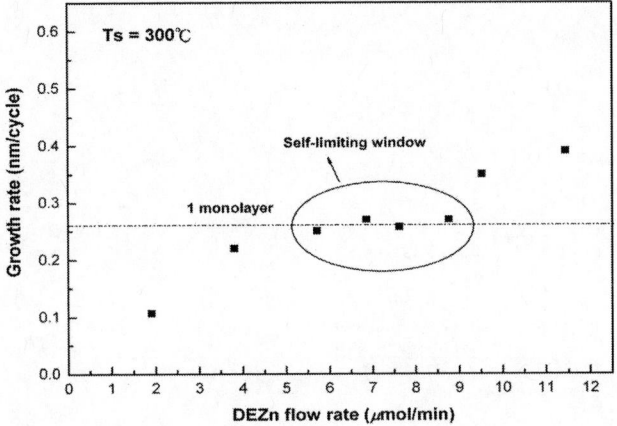

Fig. 1 The growth rate of ZnO films with different DEZn flow rates

Figs. 2(a)~ 2(d) are schematic diagrams showing the way DEZn and N$_2$O are introduced alternatively into the grown surface for the ALD of ZnO using purified nitrogen (N$_2$) as the carrier gas. The surface kinetics of ALD process are tentatively described by the following four steps. As shown in Fig.2(a), DEZn molecules are admitted to the grown surface with the Zn radicals being chemically adsorbed on the atomic sites available on the oxygen-terminated surface. Surplus DEZn molecules are physically adsorbed and will be wiped out by N$_2$ purge in the next step [Fig.2(b)]. As revealed in Fig.2(c), N$_2$O molecules are then fed to the Zn-terminated surface with the oxygen radicals being chemically adsorbed, followed by N$_2$ purge to remove excess N$_2$O molecules [Fig.2(d)].

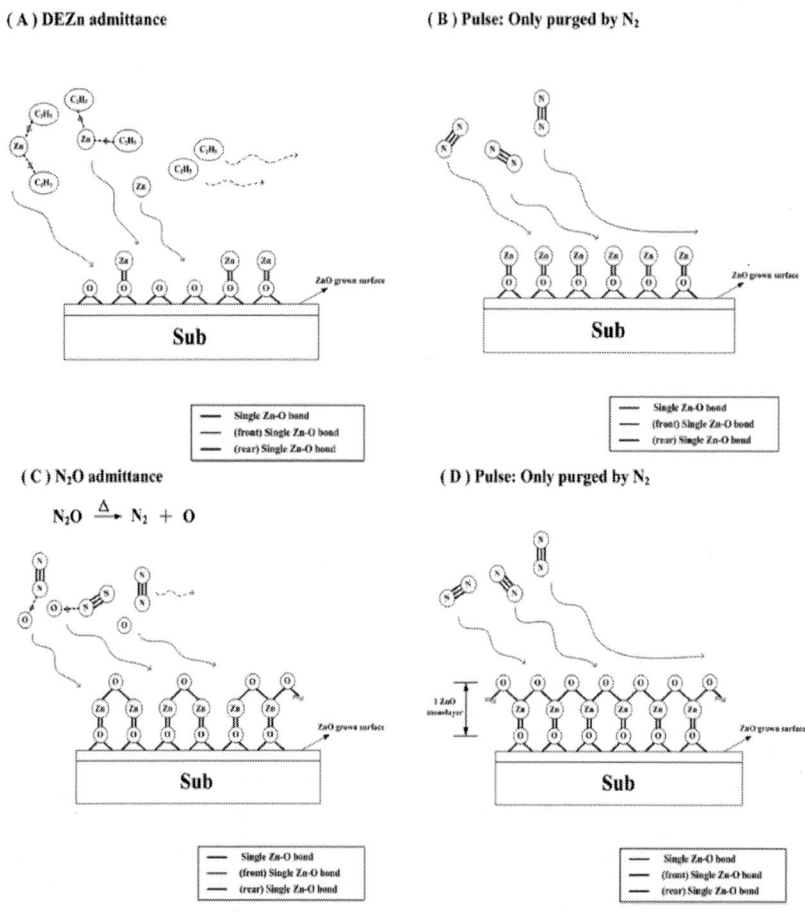

Fig. 2 The schematic of a ZnO compound monolayer grown by ALD using DEZn and N_2O as the precursors with purified nitrogen (N_2) as a carrier gas

The dependence of the ZnO growth rate on deposition temperature was also investigated in this study. Figure 3 depicts a plot of ZnO growth rate versus deposition temperature. In this case, the admittances of DeZn and N_2O remain at 7.63 and µmol/min, respectively. Although the saturated process window for deposition temperature is limited, a monolayer-by-monolayer deposition does occur at substrate temperatures between 290 and 310°C. However, the ZnO growth rate falls below one monolayer per growth cycle at substrate temperatures being less than 290 °C. It is considered that the reduced dissociation rate of N_2O molecules at low temperatures can not support enough surface coverage of atomic sites on the grown surface.

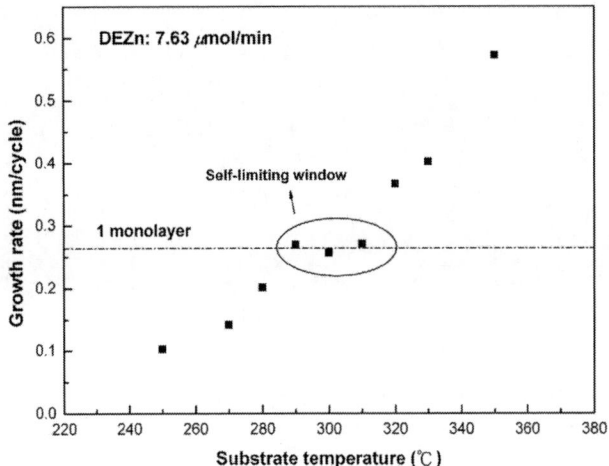

Fig. 3 The growth rates of ZnO films at various substrate temperatures

Figure 4 exhibits plots of optical transmission of the ALD-grown ZnO films prepared within the self-limiting regime. It appears that all the ZnO films exhibit optical transmittances being more than 80% in the visible spectra. According to the absorption spectroscopy, the optical absorption coefficient (α) is defined as

$$I = I_o \exp(-\alpha t) \quad ,$$

where I_o and I are the intensities of incident and transmitted light, respectively, and t is the thickness of the measured ZnO film. Also, for a direct semiconductor, $(\alpha h v)^2$ is linearly proportional to $(h v - E_g)$. Thus, one can obtain a rough estimation of energy gap (E_g) of the semiconductor material by extrapolating the linear portion of the $(\alpha h v)^2$ versus photon energy (hv) plot.

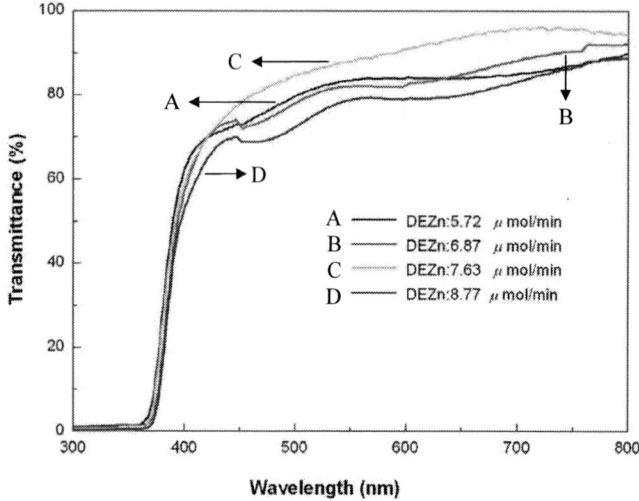

Fig. 4 Optical transmittance of ZnO films grown at 300 ▢ with the DEZn flow rates being within self-limiting window

Figure 5 shows plots of $(\alpha h v)^2$ versus hv for ZnO films grown under the self-limiting DEZn flow rates. All the plots exhibit almost the same $(\alpha h v)^2$ versus hv linear dependence with its extrapolation showing a cut-off value of ~3.26 eV which roughly represents the energy gap of ZnO at room temperature(RT).

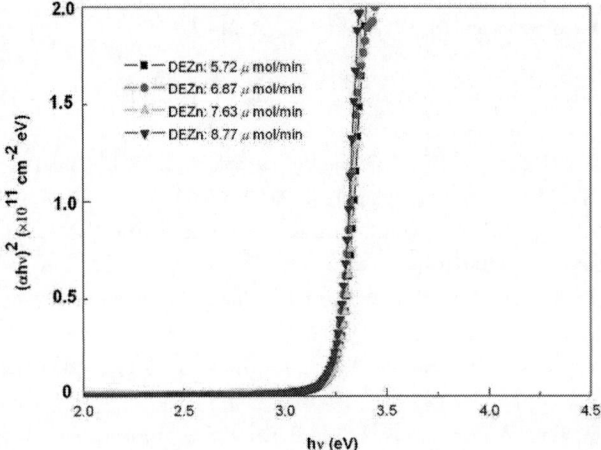

Fig. 5 Plots of $(\alpha h\nu)^2$ versus photon energy (hν) for ZnO films grown under different DEZn flow rates

Conclusions

The preparation of ZnO films at low temperatures was conducted by ALD using nitrous oxide as oxygen source. It was found that monolayer-by-monolayer deposition occurred at certain DeZn flow rates and deposition temperatures. A mechanism is proposed to interpret the presence of self-limiting window for the ALD-grown ZnO films. The ZnO films grown in the self-limitng regime were observed to exhibit improved materials characteristics and thickness uniformity.

Acknowledgment

This work was supported in part by the National Science Council of Taiwan under Contract number NSC97-2112-M-005-004-MY3.

References

1. K. Tamura,A. Ohtomo, Y. Osaka, T. Makino, Y. Segawa, M. Sumiya, S. Fuke, H. Koinuma, and M. Kawasaki, J. Cryst. **214/215**, 59 (2000).
2. Y. Chen , H. Ko, S. Hong, T. Yao, and Y. Segawa, J. Cryst. **214/215**, 87 (2000).
3. X. Liu, X. Wu, H. Cao, and R.P.H. Chang, Jour. Appl. Phys. **95**, 15 (2004)
4. S. Liu, J.J. Wu, Mater. Res. Soc. Symp. Proc. **703** 241 (2002).
5. H. Tampo, A. Yamada, P. Fons, H. Shibata, K. Matsubara, K. Iwat, S. Niki, Appl. Phys. Lett. **84**, 4412 (2004).
6. Xuhu Yu, Jin Ma, Feng Ji, Yuheng Wang, Xijian Zhang, Chuanfu Cheng, Honglei Ma, Applied Surface Science **239** (2005) 222
7. M.H. Huang, Y.Wu, H. Feick, N. Tran, E. Weber, P. Yang, Adv. Mater.**13** (2001)113.
8. A. Sasaki, W. Hara, A. Matsuda, N. Tateda, S. Otaka S. Akiba Appl. Phys. Lett. **86**, 231911 (2005).
9. P.Y. Lin, J.R. Gong, P.C. Li, T.Y. Lin, D.Y. Lyu, D.Y. Lin, H.J. Lin, T.C. Li, K.J. Chang, W.J. Lin, J. Cryst. Growth **310** , 3024 (2008).

SECTION 2

PROCESSES AT THE SEMICONDUCTOR SOLUTION INTERFACE 3

CHAPTER 6

ELECTROCHEMICAL DEPOSITION OF SEMICONDUCTORS AND THIN FILMS

ECS Transactions, 19 (3) 179-187 (2009)
10.1149/1.3120699 ©The Electrochemical Society

Towards Sustainable Photovoltaic Solar Energy Conversion: Studies Of New Absorber Materials

P.J. Dale[a] A. Løken[b], L.M. Peter[b], and J.J. Scragg[b]

[a] Laboratoire Photovoltaïque, Université du Luxembourg, 41 Rue du Brill, L-4422 Belvaux, Luxembourg

[b] Department of Chemistry, University of Bath, Bath BA2 7AY, United Kingdom

> The implementation of terawatt photovoltaic solar energy conversion will place new demands on materials supply and environmental impact. As a consequence, the search for sustainable photovoltaic materials that combine low cost with low toxicity and low energy manufacturing processes is becoming increasingly important. This paper examines the preparation and properties of two emerging indium-free photovoltaic absorber materials, Cu_2ZnSnS_4 and Cu_3BiS_3. Electrochemical routes to fabrication of absorber layers are considered, and characterization methods based on photoelectrochemistry and electrolyte contacts are discussed.

Introduction

The European Union set 2020-2030 as the target for photovoltaic (PV) generated electricity to become competitive with conventional electricity generation. The EU also agreed a binding target of a 20 % share of renewable energies in overall EU energy consumption by 2020. With the prospect of large scale deployment of terrestrial PV becoming realistic, issues of sustainability and costs of raw materials for device manufacture are assuming greater importance. It is becoming clear that there could be issues of long-term sustainability in terms of cost and availability for thin film PV technologies based on CdTe and $Cu(In,Ga)Se_2$ (CIGS). The search for alternative sustainable absorber materials, as the key component inside thin film photovoltaic devices, is therefore timely.

Two promising candidates containing only abundant non-toxic elements are being studied in our laboratory: Cu_2ZnSnS_4 (CZTS) (1-4) and, more recently, Cu_3BiS_3. CZTS can be considered as an analogue of $CuInS_2$ (CIS) obtained by replacing In(III) by Zn(II) and Sn(IV) in a 50:50 ratio. It is a p-type semiconductor with a direct bandgap in the range 1.45 – 1.6 eV (5-7). CZTS has been prepared by in-line vacuum sputtering of Cu, SnS and ZnS followed by annealing in a hydrogen sulfide atmosphere (5,8,9), and photovoltaic devices with these films have achieved AM 1.5 efficiencies of up to 6.7% (10). Cu_3BiS_3 films have been made by solid state reaction of a bismuth overlayer with a film of CuS deposited from a chemical bath (11) and by reactive sputter deposition of CuS and Bi on heated silica substrates (12). The material appears to have a direct forbidden bandgap at 1.4eV. We are unaware of any reports of photovoltaic devices fabricated with Cu_3BiS_3.

In the present work, electrodeposition was chosen as a low cost and scalable alternative to vacuum sputtering. Electrodeposition of precursor layers followed by thermal annealing has been demonstrated as a suitable method for fabricating for CdTe solar cells (13) and CIGS cells (14). Photoelectrochemical methods for characterization of absorber films have been employed in our laboratory for the characterization of CIS films prepared by an electrochemical deposition/annealing route and by sputtering (15). In the present work, electrolyte contacts containing Eu(III) as an electron (minority carrier) scavenger were used in measurements of the external quantum efficiency of the absorber film under conditions in which a Schottky barrier is formed. The results suggest that both materials should be investigated further as potential candidates for the absorber layer in heterojunction solar cells.

Experimental

Soda-lime glass substrates (25 mm x 10 mm) coated with a 1 μm radio frequency magnetron sputtered layer of molybdenum were cleaned ultrasonically in detergent, distilled water, ethanol and isopropanol and dried under flowing nitrogen. The deposition area (10 mm x 10 mm) was masked off with PTFE tape.

Cu|Sn|Zn layer stacks were deposited on the Mo substrates using a conventional 3-electrode cell with a platinum counter electrode and Ag|AgCl reference electrode. Depositions were carried out potentiostatically at room temperature (without stirring) using an Autolab 20 potentiostat. Solutions were prepared using milliQ water and metal salts of 4N purity or higher. The amounts of each metal deposited were controlled by monitoring the deposition charge. Bright and strongly adherent copper layers were deposited from an alkaline solution containing 1.5 M NaOH, 50 mM $CuCl_2$ and 0.1 M sorbitol[12] at -1.14 V vs. Ag|AgCl. Tin layers were then deposited on the copper film at -1.21 V vs. Ag|AgCl using an alkaline solution, which contained 2.25 M NaOH, 55 mM $SnCl_2$ and 0.1 M sorbitol. The final zinc layer was deposited at -1.20 V vs. Ag|AgCl from 0.15 M $ZnCl_2$ buffered to pH 3 using Hydrion buffer. The electroplated samples were washed in milliQ water and dried under nitrogen.

Cu|Sn|Zn stacked layers were annealed in a sulfur atmosphere using a quartz tube furnace. The metallic precursor films and an excess of sulfur (5N, Alfa Aesar) were loaded into a graphite container, which was inserted into the furnace tube. The samples were heated initially at 100 °C under vacuum to remove traces of water. The furnace tube was then backfilled with argon to a pressure of 1 bar and heated at 40 °C min^{-1} to a final temperature of 550 °C, which was maintained for two hours to allow the metals to react fully with sulfur. The tube was then purged with dry nitrogen and the samples removed after they had cooled naturally to room temperature.

Cu_3Bi alloy layers (1-2 microns thick) were deposited on the 10 x 10 mm Mo substrates from a solution containing 9 mM $Bi(NO_3)$, 30 mM $CuSO_4.xH_2O$, 2 M NaOH, 0.2 M sorbitol at -0.75 V vs. Hg|HgO The alloy layers were then annealed in S vapor at 450 – 500 °C for 30 minutes under flowing N_2.

The morphology of the metal precursor films and the CZTS films was examined using a JEOL JSM6310 scanning electron microscope (SEM). X-ray diffraction measurements were carried out with a Philips PW1820/00 diffractometer (λ = 1.54 Å). Localized compositional analysis by energy dispersive X-ray spectroscopy (EDS) was performed in a FEI Quanta 200 environmental SEM fitted with an Oxford X-ray analyzer at selected positions across the samples. Plating efficiencies and mean film compositions were determined by dissolving the precursor or CZTS film in HNO_3 and analyzing the solutions using atomic absorption spectroscopy (AAS).

Photoelectrochemical measurements were made of the absorber layers using 0.2 M $Eu(NO_3)_3$ (pH 2.3) as an electron scavenging redox electrolyte. The photocurrent responses of the films were measured under potentiostatic control using a 3-electrode cell with platinum wire counter and Ag|AgCl reference electrodes. Transient photocurrents were generated using chopped illumination from a high intensity white light-emitting diode. Spatially-resolved photocurrent measurements of the CZTS film were made by mounting the sample and electrodes on a computer-controlled motorized X-Y stage, which was illuminated with 1 mm diameter spot of monochromic light (470 nm) chopped at 27 Hz. A Stanford 850 lock-in amplifier was used to measure the photocurrent. XY mapping was carried out by rastering the sample in millimeter steps. A similar setup was used to measure external quantum efficiency of the semiconductor films, with the incident photon flux was calibrated using standardized silicon and germanium photodiodes traceable to NBS standards

Results and Discussion

Cu_2ZnSnS_4

CZTS films were prepared by treatment of the metal precursor films in sulfur vapor using a range of Cu/(Sn+Zn) target ratios and a Zn/Sn target ratio of 1. The actual values of these ratios were determined by AAS. The CZTS layers produced were around 0.5 microns thick and appeared gray and non-reflective with some evidence of lateral non-uniformity. The films showed good adhesion to the Mo substrate. Cross-sectional SEM suggested the presence of an intermediate MoS_2 layer between the Mo substrate and the CZTS. Figure 1 shows the morphology of the annealed CZTS film that showed the best photoresponse (cf. Figure 4). It appears to consist of small but well-defined crystallites with an average grain size of less than 0.5 microns. EDS measurements showed that the Zn component varied across the films, which were generally Zn-poor at the centre and Zn-rich at the edges (this may explain the lateral non-uniformity observed visually).

Figure 1. Top view SEM image of annealed Cu_2ZnSnS_4 film grown on molybdenum-coated glass.

Figure 2 shows the indexed XRD pattern for an annealed CZTS film. The film evidently consists primarily of CZTS, with some SnS_2. The lattice parameters (a = 0.544 nm, b = 0.542 nm, c = 1.089 nm) calculated from the X-ray diffraction pattern were matched to JCPDS card 26-0575: a = b = 0.5434 nm, c = 1.0848 nm. The XRD pattern shows no evidence of the presence of Cu_xS, but two peaks assigned to SnS_2 can be clearly seen. The ZnS content of the films is difficult to ascertain since the lattice parameters of ZnS (sphalerite) are almost identical to those for CZTS.

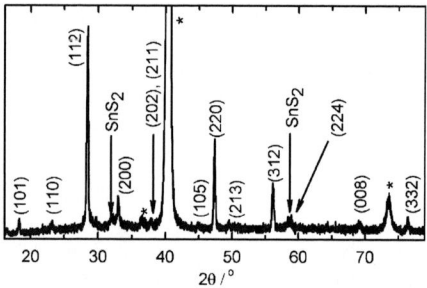

Figure 2. X-Ray diffraction pattern of Cu_2ZnSnS_4 film. The main peaks are assigned to the $CuZnSnS_4$ phase. The presence of some SnS_2 is indicated by the two labelled peaks. The unassigned peaks marked (*) arise from the molybdenum substrate.

The annealed CZTS films were characterized using a Eu(III) electrolyte to scavenge electrons reaching the surface of the illuminated semiconductor. The p-type photoactivity of the annealed films is evident from the cathodic photocurrent response seen when a flashing white LED is used to illuminate the sample during the recording of the linear scan voltammogram shown in Figure 3. The photocurrent onset is at around +0.1 V vs Ag|AgCl, and the photocurrent saturates at potentials more negative than -0.25 V. EQE spectra were therefore recorded at a potential of -0.3 V.

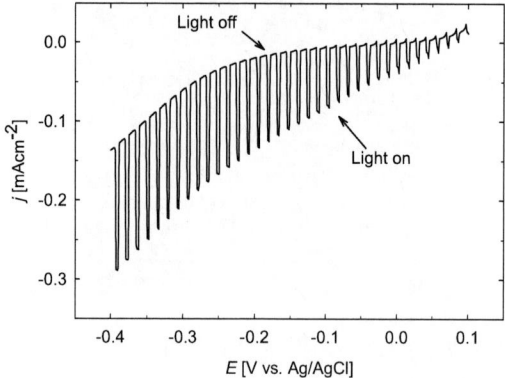

Figure 3. Voltammogram of Cu_2ZnSnS_4 film in 0.2 M Eu(III) nitrate under pulsed illumination from a white LED. The cathodic photocurrent indicates p-type behavior.

The external quantum efficiency (EQE) spectra of 4 different CZTS samples measured at -0.3V vs. Ag|AgCl in 0.2 M Eu(NO₃)₃ are compared in Figure 4. The highest external quantum efficiency photocurrent response is seen for sample B, which has a Cu/(Zn+Sn) ratio of 0.86 and a Zn/Sn ratio of 1.36. Sample D, which has a higher Cu/(Zn+Sn) ratio of 1.07 and a Zn/Sn ratio of 1.38 shows almost no response. Sample C, which has a Cu/(Zn+Sn) ratio of 0.91 and a Zn/Sn ratio of 1.30 is almost as good as sample B, whereas sample A, which has a Cu/(Zn+Sn) ratio of 0.71 and a Zn/Sn ratio of 1.30 gives a lower EQE.

The EQE data near the onset energy were analyzed using the relationship expected for a direct transition. The bandgap was found to be 1.49 eV, which is close to the optimum for terrestrial solar energy conversion. The shape of the EQE spectra suggests that the samples are rather highly doped and have a short electron diffusion length. Clearly it will be important to optimize the doping level and increase the minority carrier diffusion length if the material is to be used in the fabrication of efficient solar cells.

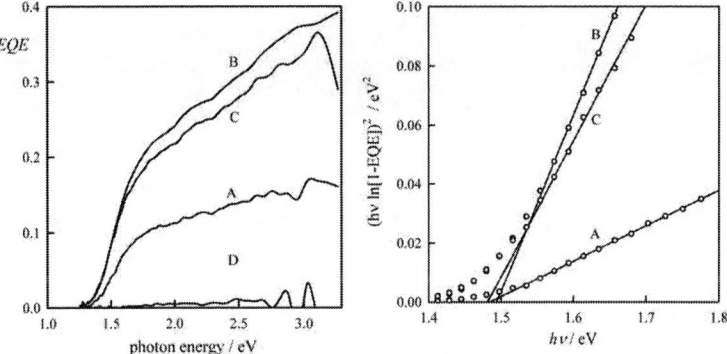

Figure 4. External quantum efficiency spectra of 4 different Cu₂ZnSnS₄ films, measured at -0.3V vs. Ag|AgCl in 0.2 M Eu(NO₃)₃. See text for details of stoichiometry. The plots on the right hand side were used to estimate the bandgap of the material.

The uniformity of the photocurrent response of a 2 μm CZTS film was examined using the X-Y scanning system. A typical result is shown in Figure 5. EDX analysis of similar samples shows that the highest photocurrent occurs where the ratio Cu/(Zn+Sn) is around 0.5, and the Zn/Sn ratio is greater than unity. Further work is in progress to relate the response more precisely to local variations in composition. Compositional uniformity is essential for larger area devices, and efforts are being made to achieve this by controlling mass transport during electrodeposition.

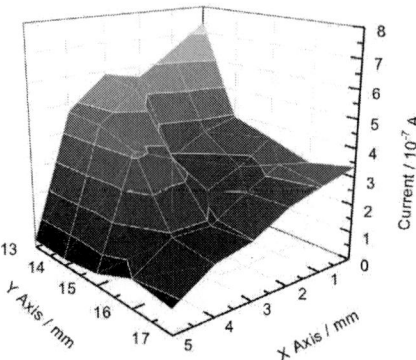

Figure 5. Spatially-resolved photocurrent response of a 2 μm thick CZTS film measured at -0.3 V vs. Ag|AgCl. Illumination wavelength 470 nm.

Cu_3BiS_3

Recently our attention has turned to Cu_3BiS_3, another potential absorber layer. Work is at an early stage, and only preliminary results are reported here. The morphology of the films produced by sulfidization of Cu-Bi alloy is illustrated by the SEM pictures in Figure 6. The film is uniform and adherent with some voids that can be seen on the cross-sectional SEM. The composition of the film is clearly very close to that expected as shown by the good match of the XRD pattern in Figure 7 to the pattern for Wittichenite.

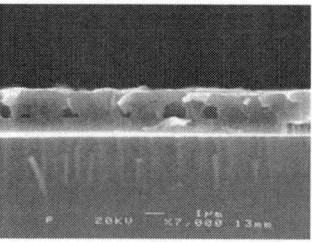

Figure 6. Top and cross-sectional SEM views of Cu_3BiS_3 film.

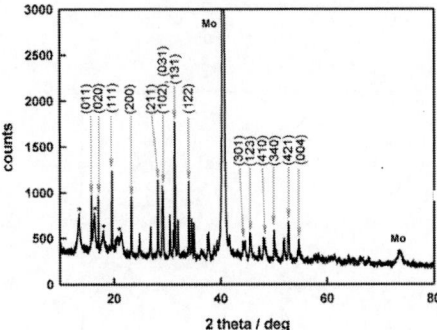

Figure 7. XRD pattern of annealed Cu_3BiS_3 film. The lines match closely with those for Wittichenite (PDF # 431479). Due to the complexity of the pattern, only the most intense peaks have been labeled. The large peak at ~40° is from the Mo substrate. Peaks marked (*) arise from the nylon sample holder.

The photocurrent response of the Cu_3BiS_3 was tested using Eu(III) electrolyte. Figure 8 illustrates the response to chopped illumination from a white LED. The cathodic photocurrent response is typical for a p-type semiconductor.

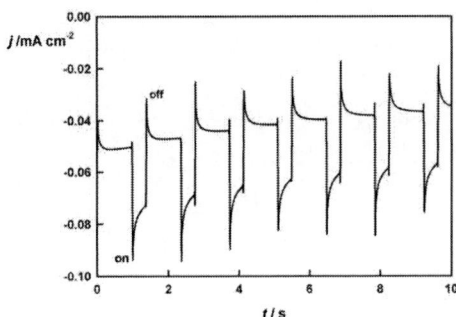

Figure 8. Transient photocurrent response measured at -0.3 V vs Ag|AgCl

The EQE spectrum of the best Cu_3BiS_3 film is shown in Figure 9. The response is smaller than that of the best CZTS film, but it is clear that the photocurrent onset wavelength is around 1000 nm, which corresponds to 1.24 eV. Work is now in progress to improve the quality of the Cu_3BiS_3 films and to enhance their photoresponse. The objective is to fabricate new solar cells using Cu_3BiS_3 as the absorber layer combined with a suitable n-type top contact.

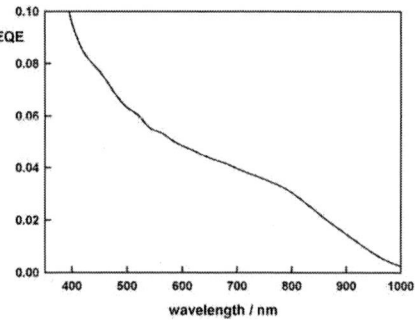

Figure 10. EQE spectrum of Cu_3BiS_3 film.

Acknowledgments

This work has been supported by the UK Engineering and Physical Sciences Research Council (EPSRC) as part of the SUPERGEN Consortium *Photovoltaic Materials for the 21st Century*. The authors thank Ian Forbes and Guillaume Zoppi for helpful discussions and collaboration in device fabrication.

References

1. J.J. Scragg, P.J. Dale, L.M. Peter, *Electrochemistry Communications,* **10**, 639-642 (2008).

2.. J.M. Scragg, P.J. Dale and L.M. Peter, G. Zoppi and I. Forbes. *Physica Status Solidi, (a).***245**, 1772-8 (2008).

3. J. J. Scragg, P.J. Dale, L.M. Peter. *Thin Solid Films,* **517**, 2481-2484 (2009).

4. $Cu_2ZnSnSe_4$ thin film solar cells produced by selenization of magnetron sputtered precursors. G. Zoppi, I. Forbes, R.W. Miles, P.J. Dale, J.J Scragg and L.M. Peter. *Progress in Photovoltaics* (in press).

5. K. Jimbo, R. Kuimura, T. Kamimura, S. Yamada, W.S. Maw, H. Araki, K. Oishi, H. Katagiri. *Thin Solid Films,* **515**, 5997 (2007).

6. K. Ito, T. Nkazawa, Jpn. *J. Appl. Phys. Part 1,* **27**, 2094 (1988).

7. T.M. Friedlmeier, N. Wiser, T. Walter, H. Dittrich, H.W. Schock in *Proceedings of the 14th European PVSEC,* Barcelona, Spain, p 1242 (1997).

8. H. Katagiri, N. Ishigaki, T. Ishida, K. Saito, *Jpn. J. Appl. Phys. Part 1* **40**, 500 (2001).

9.. H. Katagiri, N. Sasaguchi, S. Hando, S. Hoshino, J. Ohashi, T. Yokota, *Sol. Energy Mater. Sol. Cells,* **49**, 4017 (1997).

10. H. Katagiri, K. Jimbo, S. Yamada, T. Kamimura T, W.S. Maw, T. Fukano, T. Ito, T. Motohiro. Appl. Phys. Express, 1, DOI: 10.1143/APEX.1.041201 (2008)

11. V. Estrella, M.T.S. Nair, P.K. Nair, *Semicond. Sci. Technol.*, **18**, 190 (2003).

12. N.J. Gerein, J.A. Haber, *Chem. Mater.*, **18**, 6297 (2006)

13. D. Cunningham, M. Rubcich, D. Skinner, *Prog. Photovoltaics,* **10**, 159 (2002).

14. D. Lincot, J.F, Guillemoles, S. Taunier, D. Guimard, J. Sicx-Kurdi, A. Chaumont, O. Roussel, O. Ramdani, C. Hubert, J.P. Fauvarque, N. Bordereau, L. Parissi, P. Panheleux, P. Fanouillere, N. Naghavi, P.P Grand, M. Benfarah, P. Morgensen, O. Kerrec. *Sol. Energy*, **77**, 725 (2004).

15. .J. Dale, A.P Samantilleke, G. Zoppi, I. Forbes and L.M. Peter, *J. Phys. D: Appl. Phys.* **41**, 085105 8 pp. (2008).

ECS Transactions, 19 (3) 189-195 (2009)
10.1149/1.3120700 ©The Electrochemical Society

Electrochemical Growth of CuInSe₂ Compounds on Polycrystalline Mo Films Studied by Raman and Impedance Spectroscopy

E. Chassaing[a], P.-P. Grand[a], E. Saucedo[a], A. Etcheberry[b], and D. Lincot[a]

[a]Institute of R&D on Photovoltaic energy, (IRDEP, UMR 7174 EDF-CNRS-ENSCP)
6 quai Watier, 78401 Chatou Cedex, France
[b]Lavoisier Institute, (IREM, UMR 8637 - CNRS-UVSQ), 45 avenue des Etats-Unis, 78035 Versailles Cedex, France

> The growth process of Cu-In-Se compounds was investigated by in-situ electrochemical techniques and ex-situ XPS and Raman spectroscopy. The growth can be divided in three well separated steps. In a first stage, the Cu concentration is high, leading to the formation of a Cu-rich binary phase, Cu_2Se, which acts as a precursor for Se(IV) reduction into elemental Se(0). When the surface concentration of elemental Se(0) is high enough, a new surface phase is formed with a stoichiometry close to $CuSe_2$. It acts as a catalytic site for indium incorporation. This results in the formation of $CuInSe_2$ and Cu-poor chalcopyrite phases. The proportions of the different phases evolve during the growth. This complex mechanism is in agreement with the compositional, structural and electrical properties of the samples as a function of the thickness.

Introduction

$CuInSe_2$ compounds (CIS) are very promising materials for thin layer photovoltaic devices (1-3). Though their electrochemical deposition has been largely investigated, the nucleation and growth mechanism is still a matter of discussion (4-8). In addition to the main chalcopyrite phase, the deposited layer contains additional phases and may exhibit some departure from stoichiometry (9, 10).

In previous work we investigated the first stages of growth on Mo substrate and showed that the deposition process was complex involving surface reactions (11, 12). In this work, the electrochemical growth of the Cu-In-Se layers is studied, using both *in-situ* electrochemical techniques, such as current transient, quartz-crystal microbalance and impedance spectroscopy and *ex-situ* surface analysis, such as XPS and Raman spectroscopy.

Experimental

The solutions were prepared from analytical grade purity chemicals dissolved in high purity water (18.2 MΩ cm). They contained copper sulfate (10^{-3} mol L⁻¹), selenous acid ($1.7 \ 10^{-3}$ mol L⁻¹), indium sulfate ($6.0 \ 10^{-3}$ mol L⁻¹) and a supporting sulfate electrolyte, with a pH close to 2. The electrolytes were deaerated by argon bubbling prior the experiment and an argon pressure was maintained over the solutions during the experiment.

A three electrode cell with a 0.5 L capacity, was used at room temperature. The reference electrode was a saturated mercurous sulfate electrode (MSE, E° = 0.64V/NHE).

189

The anode was a Pt coil. The substrate was a 0.3 to 1 cm² disc electrode made of glass covered by sputtered Mo layer (500-600 nm) provided by Saint-Gobain Recherche, rotating at 100 rpm. Prior each experiment the Mo substrate was cleaned in 25 % ammonia solution for 5 mn, rinsed under flowing high purity water.

A Biologic VSP2 potentiostat was used for the electrochemical investigation. The electrodepositions were carried out at a given potential of –0.9V/MSE for different quantities of electricity up to 5 C cm⁻² to vary the thickness from 0.1 μm to 3 μm. Electrochemical impedance measurements were performed during deposition under potentiostatic conditions in the frequency range 100kHz-10Hz. The perturbation signal was 10-15 mV.

The chemical analysis of the deposited layers was carried out by means of X-ray fluorescence (XRF, Fischerscope). X-ray photoelectron spectroscopy measurements were performed using a Thermo electron VG-ESCALAB 220i XL system. High-resolution XPS conditions have been used with a constant analyser energy mode with 8 or 20 eV as pass energy (12).

Raman microprobe measurements were made using a HR800-UV Horiba-Jobin Yvon spectrometer coupled with an Olympus metallographic microscope (frequency cut-off at 42 cm⁻¹), and using the red line of an He-Ne laser (λ = 632.8 nm) as excitation light, due to the minimisation of thermal effects achieved with this wavelength on the thermally-sensitive CuInSe₂. In such conditions, the penetration depth of the laser is estimated in 250 nm (16).

Results

Current density response and quasi-rest potential evolution during the growth of CIS

Fig 1 shows a typical current density response recorded with a rotating disc electrode (100 rev mn⁻¹) during the deposition at –0.9V/MSE, which is the standard potential for the CIS deposition in our process (11, 12).

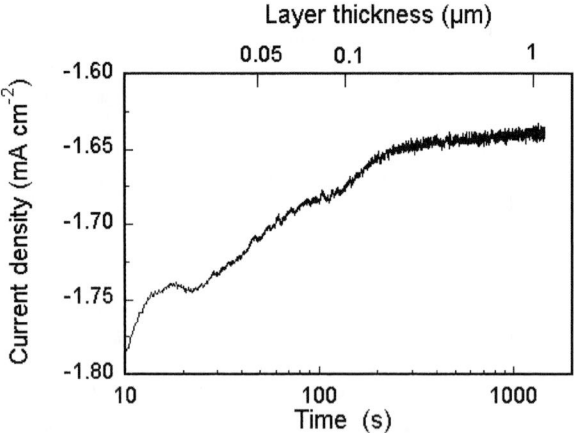

Figure 1. Influence of the deposition time at –0.9V/MSE on the current density response and on the composition of the layers (XRF analysis) (disc electrode rotating at 100 rpm).

The magnitude of the current density decreases rapidly at the beginning of the deposition. Coalescence of the layer was to occur after about 50-100 s (12). The curve shows several shoulders, indicating a multistep growth process. After about 300 s the current density tends to a nearly constant value, whose magnitude depends on the hydrodynamic conditions. A Koutecky-Levich behaviour is observed as a function of the square root of the electrode rotation, indicating that the deposition process is partly controlled by the mass transfer of the Cu(II) and Se(IV) species (13).

Electrochemical Impedance Spectroscopy

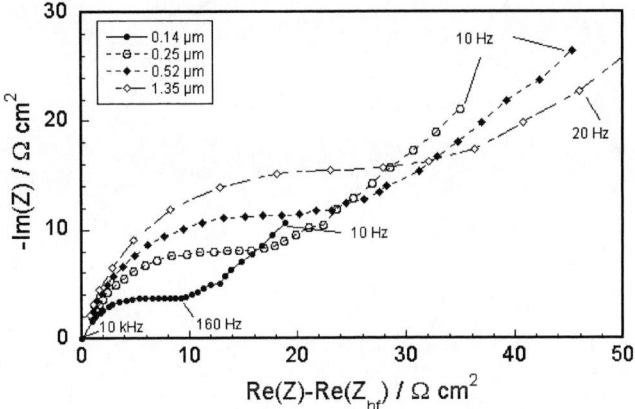

Figure 2. Nyquist plots recorded during the deposition at –0.9V/MSE (RDE: 100 rpm).

To get information on the evolution of the kinetics during the growth, the impedance was recorded during the deposition at – 0.9V, every 600 seconds. However, to perform short time measurements only the high-frequency range (100 kHz to 10 Hz) was investigated. Figure 2 shows that the Nyquist plots evolve markedly during the growth of the layers. The diagrams were fitted with a Randles equivalent circuit, with a "Constant Phase Element (CPE)" behaviour, using Biologic software. The CPE exponent of the high-frequency capacitive loop increases markedly at the very beginning of the deposition process, from about 0.7 to 0.85, even after the deposit has coalesced. This strong evolution of the CPE exponent at the beginning of the growth may be related to the composition evolution. The equivalent capacitance does not change markedly with deposition time, it is equal to $15 \pm 5 \ \mu F \ cm^{-2}$. Figure 2 shows that the charge transfer resistance, R_{ct}, increases with deposition time, indicating that the electrochemical reaction is increasingly more difficult. The $R_{ct}.j$ product of the charge transfer resistance and the current density, increases from 20 to 85 mV.

Composition Evolution

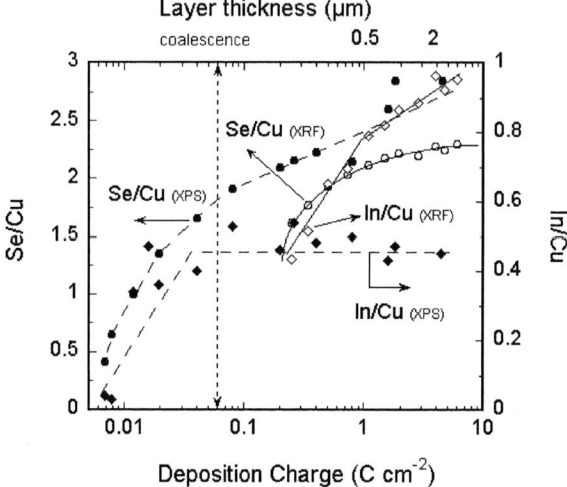

Figure 3. Influence of the deposition charge on the composition of the Cu-In-Se layers
(a) XPS analysis (plain symbols), (b) XRF analysis (open symbols)
Curves 1: Se/Cu atomic ratio, Curves 2: In/Cu atomic ratio

Figure 3 shows the evolution of the composition as a function of layer thickness. The composition changes very rapidly at the beginning of the growth. The first nuclei are Cu-rich Cu-Se nuclei, the selenium content is quite low. We showed that indium starts to incorporate when the Se/Cu atomic surface ratio becomes close to unity (11, 12). With increasing layer thickness, the indium content increases whereas the copper content decreases.

In addition, XPS showed that the surface composition is markedly different from the bulk composition. Bulk composition is close to $CuInSe_2$ stoichiometry whereas the Se/Cu surface ratio is higher on the surface and the In/Cu ratio is lower (11, 12). In addition XPS spectroscopy shows a progressive evolution in the Se3d peak position as a function of the deposition time: at the very beginning of the deposition, a single Se environment corresponding to Se(-II) oxidation state is observed. With increasing deposition time, the shape and the position of the peak evolves. At least, two or three different environments, corresponding to an intermediate oxidation state, between –II and zero, and a small amount of Se(0). This clearly indicates an evolution of the phase composition of the layers with thickness.

Raman spectroscopy

Raman spectroscopy also evidences the evolution of the phase composition with growth (14-16). Figure 4 shows, in addition to the predominant $CuInSe_2$ chalcopyrite phase, with a characteristic Se-Se A1 mode at 175 cm⁻¹, the presence of additional compounds. The shoulder at 160 cm⁻¹ is attributed to the so-called "ordered-vacancy-compound" (OVC), a Cu-poor chalcopyrite with an average composition close to

$CuIn_3Se_5$ or $CuIn_5Se_8$ (15, 16). At 240 cm^{-1}, Se-Se mode in elemental Se is observed and at 260 cm^{-1} the Se-Se mode in a Cu-Se binary phase. The OVC content increases in the first growth stages and then slowly decreases. Conversely, elemental Se decreases continuously with thickness as was also observed by selective stripping in Na_2S solutions, while the amount of binary Cu-Se compounds is constant up to the last stage where it strongly increases (16).

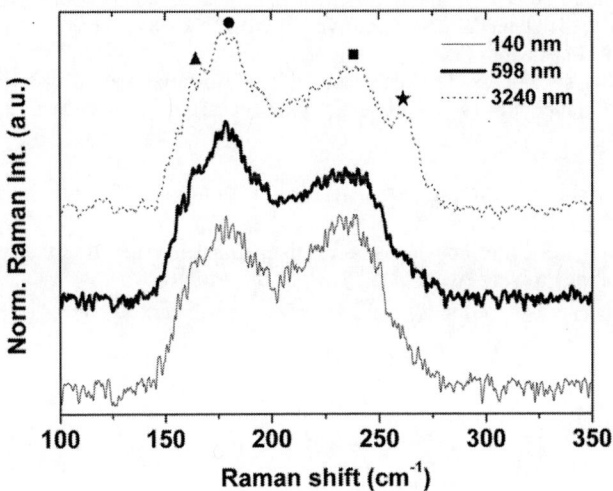

Figure 4. Raman spectra of CIS layers grown for different times, (triangle: OVC phase, circle: A1 mode of CIS chalcopyrite phase, square: elemental Se mode and star: Cu_xSe binary mode) (16).

Discussion

All these results show that the electrochemical growth of Cu-In-Se compound is a complex process leading to a multiphase structure. The first layers have a relatively high resistance and according to XPS analysis, they consist mainly in binary Cu-Se phase, in which Se is essentially as Se(-II), i.e. to a stoichiometry close to Cu_2Se (8,15).

The first step would then be the formation of binary Cu-Se phase according to:

$$2Cu(II) + Se(IV) + 8e^- \rightarrow Cu_2Se \qquad [1]$$

Although Se is a more electropositive element, it does not deposit easily on Mo substrate at room temperature. However, Se(IV) can be reduced on the previously formed Cu_2Se phase (4, 8, 11, 12) according to reaction [2]:

$$Se(IV) + 4e^- \rightarrow Se(0)_{ads} \qquad [2]$$

This reaction leads to adsorbed elemental Se on the surface. When a sufficient amount of Se(0) is present, it is consumed and a new surface phase is formed with a stoichiometry close to $CuSe_2$.

$$Cu_2Se + 3 \ Se(0)_{ads} -> 2 \ [CuSe_2]_{surf} \qquad [3]$$

This surface phase was shown to provide active sites for indium incorporation, as already observed at the potential incorporation threshold of indium (11, 15). However, this phase cannot be clearly characterized since it is amorphous and its Raman signal is close to that of elemental Se, however it was shown to have a very specific electrochemical behaviour (8).

The In(III) species would then react with this catalytic surface phase leading to chalcopyrite compound (with possibly some departure from exact stoichiometry) according to:

$$[CuSe_2]_{surf} + In(III) + 3e^- -> CuInSe_2 \qquad [4]$$

An important effect of this deposition is the permanent renewal of the active sites on the surface of the growing layer (17).

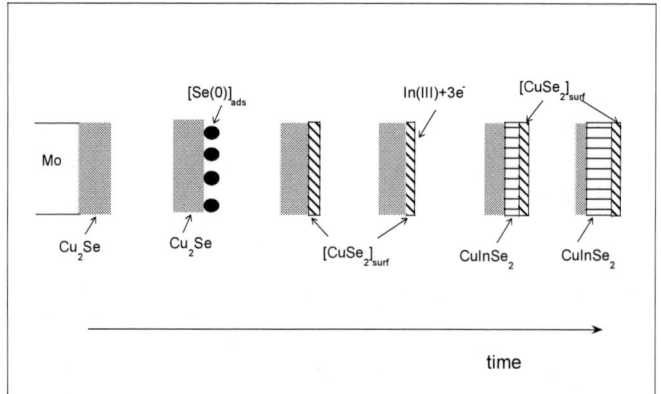

Figure 5. Schematic representation of the $CuInSe_2$ electrochemical growth.

This complex mechanism is schematized in Figure 5. The reaction path developed accounts for the main experimental results. In particular it explains the excess surface concentration of Se found by XPS and the increasing indium content with thickness.

Conclusion

The electrochemical growth of Cu-In-Se compound is a complex process leading to a multiphase structure. The first layers consist mainly of binary Cu_xSe phase (x close to 2), grown on Cu nuclei. Se is essentially as Se(-II). This Cu_2Se phase acts as precursor for the incorporation of the reduction of Se(IV) species leading to a new surface phase which catalyses indium incorporation. In addition to chalcopyrite predominant phase, secondary phases are formed, characterized by Raman spectroscopy, which shows the presence of OVC, Cu_xSe and Se(0) phases.

The amount of these secondary phases seems predominant for the final performance of the photovoltaic devices. By contrast, the presence of elemental Se is mandatory for the adequate crystallization process needed for improving the structural and electrical characteristics of the CuIn(Se,S)$_2$ absorber layer.

References

1. R. K. Pandey, S. N. Sahu, S. Chandra, in *Handbook of semiconductor electrodeposition*, Marcel Dekker, Inc., New York, 1996
2. T. E. Schlesinger, in *Electrodeposition of Semiconductors in Modern Electroplating*, Edited by M. Schlesinger, M. Paunovic, J. Wiley and Sons Inc., (2000) p 585
3. D. Lincot, J.F. Guillemoles, S. Taunier, D. Guimard, J. Sicx-Kurdi, A. Chaumont, O. Roussel, O. Ramdani, C. Hubert, J.P. Fauvarque, N. Bodereau, L. Parissi, P. Panheleux, P. Fanouillere, N. Naghavi, P.P. Grand, M. Benfarah, P. Mogensen, and O. Kerrec,.*Solar Energy*, **77**, 725 (2004).
4. K.K. Mishra and K. Rajeshwar, *J. of Electroanalytical Chemistry*, **271**, 279 (1989).
5. M. Oliveira, M. Azevedo and A. Cunha, *Thin Solid Films*, **405**, 129 (2002).
6. M.E. Calixto, K.D. Dobson, B.E. McCandless, and R.W. Birkmire, *Thin-Film Compound Semiconductor Photovoltaics*, **865**, 431 (2005).
7. J. Kois, O. Volobujeva and S. Bereznev *Physica Status Solidi (c)* **5**, 3441 (2008).
8. E. Chassaing, O. Ramdani, P.-P. Grand, J. -F. Guillemoles, D. Lincot, *Phys Stat. Solidi*, **5**, 3445 (2008).
9. M.E. Calixto and P.J. Sebastian, *Solar Energy Materials and Solar Cells*, **63**, 335, (2000).
10. C. Guillen and J. Herrero, *Solar Energy Materials and Solar Cells*, **43**, 47 (1996).
11. E. Chassaing, O. Roussel, O. Ramdani, P.-P. Grand, B. Canava, A. Etcheberry, J. -F. Guillemoles and D. Lincot, *SOTAPOCS 46 and Processes at the Semiconductor/Solution Interface 2*, L. Chou, D. Buckley, P. Chang, A. Etcheberry, C. O'Dwyer, M. Overberg, M. Yoshimoto Ed., PV 6-2, p. 577, The Electrochemical Society Proceedings Series, Pennington, NJ (2007).
12. O. Roussel, O. Ramdani, E. Chassaing, P. P. Grand, M. Lamirand, A. Etcheberry, O. Kerrec, J. F. Guillemoles and D. Lincot *J. of the Electrochemical Society*, **155**, D141 (2008).
13. L. Thouin, S. Massaccesi, S. Sanchez, and J. Vedel, *J. of Electroanalytical Chemistry*, **374**, 81 (1994).
14. O. Ramdani, J. F. Guillemoles, D. Lincot, P. P. Grand, E. Chassaing, O. Kerrec and E. Rzepka, *Thin Solid Films*, **515**, 5909, (2007).
15. V. Izquierdo-Roca, J. Alvarez-Garcia, L. Calvo-Barrio, A. Perez-Rodriguez, J.R. Morante, V. Bermudez, O. Ramdani, P.P. Grand, and O. Kerrec, *Surface and Interface Analysis*, **40**, 798 (2008).
16. E. Saucedo, C.M. Ruiz, E. Chassaing, J.S. Jaime-Ferrer, P.P. Grand, G. Savidand, and V. Bermudez, *J. of Applied Physics*, (submitted).
17. E. Chassaing, P.-P. Grand, A. Etcheberry and D. Lincot, Electrocrystallization mechanism of Cu-In-Se compounds for solar cell applications investigated by impedance spectroscopy, this meeting, # 1449 symposium I4.

Electrochemical growth gold buffer layer on H-Si(111) surfaces and their applications

P. Allongue, F. Maroun, H. Jurca, R. Cortès, P. Prod'homme, N. Tournerie

Physique de la Matière Condensée, Ecole Polytechnique, CNRS, 91128 Palaiseau, France.

This paper reviews our recent work about gold electrodeposition on H-terminated Si(111). It is shown that Au(111)/Si(111) epitaxial layers are grown and that the film morphology can be varied according to the deposition conditions (potential and solution pH). At pH = 14 (cyanide solution) selective gold nucleation at the substrate monatomic step edges is observed. A homogeneous nucleation is obtained at pH 4 (chloride solution). In the former case 3D growth occurs and in the latter case ultra smooth buffer layers are obtained. Application of these substrates to formation of magnetic nanostructures with tuneable properties is demonstrated.

Introduction

Fundamental studies on single crystals give insights into processes on the atomic scale and are invaluable in applied nanoscience (e.g. catalysis, spintronics etc.). However in numerous studies a buffer layer, i.e. a thin textured polycrystalline layer deposited on a substrate (glass, silicon, sapphire etc.), is used instead of a single crystal. Buffer layers have the strong advantage that the acquired knowledge may be easily transferred to technological devices. Among buffer layers, gold one are very widely used in different fields of research. Textured gold buffer layer may be obtained by evaporation or sputtering on glass, mica or sapphire. Electrochemical growth is an alternative method which is reputed to be a versatile method to grow with a fast rate thin films on substrates with a complex shape. It is also an invaluable method to grow epitaxial ultrathin films, in particular magnetic ones [1, 2]. One may further deposit self-ordered nanostructures on structured surfaces [3]. Gold electrochemical growth has been studied in the past by various groups. Oskam and Searson have in particular shown that Au grows epitaxially on Si(100) and form interesting Schottky barriers [4-6]. Studies of Au homoepitaxy on Au(100) [7] and Au(111) [8] using in situ using STM and in situ X-ray scattering show that the process is essentially a layer by layer process.

This paper is a brief review of our recent work about gold electrodeposition on well defined H-terminated Si(111) [9, 10]. It is shown that Au(111)/Si(111) epitaxial layers are grown and that the film morphology can be varied according to the deposition conditions (potential and solution chemistry). The main parameter controlling the nucleation seems to be the mechanism and the rate of hydrogen evolution reaction (HER). At pH = 14 (gold cyanide solution) selective gold nucleation at the substrate monatomic step edges is observed. A homogeneous nucleation is obtained at pH 4 (gold chloride solution). In the former case 3D growth occurs and in the latter case ultra smooth buffer layers are obtained. Applications of these well defined substrates to formation of magnetic nanostructures with tuneable properties are also demonstrated.

Experimental

The silicon samples were cut from 10 mΩ.cm - 10 Ωcm (111) wafers (n-type, P doped) with a miscut angle 0.2° precisely oriented towards <11-2> to obtain the long range staircase structure shown in the inset of Fig. 2a [11]. The etching included the following steps: cleaning the sample in (98% H_2SO_4)/(30% H_2O_2) 2:1 mixture, thorough rinsing with ultra pure water (UPW), immersion for 15 min in 40% NH_4F + 50 mM $(NH_4)_2SO_3$ and a final rinse in UPW.

An ohmic contact was then formed by applying an InGa eutectic layer on the rear face of the sample before mounting it on a rotating electrode (1700 rpm). The sample lateral edges were protected by an electrolytic scotch tape (Struers) to expose only the well-defined silicon surface to the solution.

Two gold plating solutions were prepared from bi-distilled water and ultra-pure chemicals: (i) 10^{-4} M $HAuCl_4$ solution with 0.1M K_2SO_4 as supporting electrolyte, whose pH was adjusted to pH ~ 4 by adding 1mM H_2SO_4; and (ii) 10^{-4} M $KAu(CN)_2$ + 2×10^{-4} M NaCN solution at pH~14 (2M NaOH). For all potentials used in this study, Au is essentially deposited under mass transport limitation at a rate of ~0.2 MLs^{-1} in both solutions. The choice of the deposition potential is detailed below. Deposition was performed at a constant potential using a three-electrode electrochemical cell connected to a potentiostat. All potentials are hereafter quoted against the mercury sulfate electrode (MSE). A gold wire served as counter electrode. The sample was pre-polarized before immersion to immediately initiate metal deposition at the desired potential. At the end of the deposition time, the sample was quickly removed from the solution, rinsed with UPW and dried with nitrogen.

The morphology of the deposits was characterized by contact mode AFM (Molecular Imaging microscope, Phoenix, USA) in a nitrogen atmosphere using Si_3N_4 cantilevers (Nanoprobes, spring constant 0.12 Nm^{-1}). Out of plane ($\theta - 2\theta$) and grazing incidence or ψ scans ($\theta \sim 0.6°$) X-ray diffraction characterizations were performed using a home-built four-circle diffractometer with a Kα1 Cu source (λ = 1.5405 Å) (for more details see Ref. [10]).

Gold electrodeposition on silicon

Electrochemistry: Figure 1 presents typical current-potential curves of a n-Si(111) electrode in the gold cyanide solution at pH = 14 (filled circles) and in the gold chloride solution at pH = 4 (open squares). Since we are using a rotating disk electrode (RDE), each electrochemical reaction gives rise to a current plateau when its kinetics becomes limited by the diffusion of the ions towards the electrode surface [12]. At pH = 14 the cathodic current onset is $U < -1.1$ V and the current it reaches a first plateau for $U \sim -1.2$ V (~ -0.08 mA/cm^2), which corresponds to Au deposition ($Au(CN)_2^- + e^- \rightarrow Au + 2\ CN^-$) [13]. A second less well-defined plateau is found for $U \sim -1.5$ V (~ -0.16 mA/cm^2), most probably corresponding to O_2 reduction reaction ($O_2 + 4\ e^- + 2\ H_2O \rightarrow 4\ OH^-$) [12]. The cathodic current finally increases abruptly for $U < -1.6$ V, due to the water decomposition reaction ($H_2O + e^- \rightarrow \frac{1}{2}\ H_2 + OH^-$) [14]. The nucleation onset potential of Au on Si, was found to ca. -1.5V, in agreement with Oskam and Searson [6].

At pH = 4 (Fig. 1, open squares), the voltammogram (scan rate 10 mV/s) presents three well defined negative current plateaus for $U < 0.2V$. The first one, in the potential range -0.4 V $< U < 0$ V (-0.14 mA/cm^2), corresponds to Au deposition ($AuCl_4^- + 3$ e$^- \rightarrow$ Au + 4 Cl$^-$). The second plateau (-1 V $< U < -0.4$ V; current density ~ -0.4 mA/cm^2) corresponds to the O_2 reduction reaction ($O_2 + 4$ e$^- + 2$ $H_2O \rightarrow 4$ OH$^-$). The third plateau (-1.85 V $< U < -1.55$ V; current density ~ -4 mA/cm^2) corresponds to H_2 evolution (2 H$^+$ + 2 e$^- \rightarrow H_2$). Finally, for $U < -1.85V$, water decomposition occurs ($H_2O + e^- \rightarrow \frac{1}{2}$ $H_2 +$ OH$^-$).

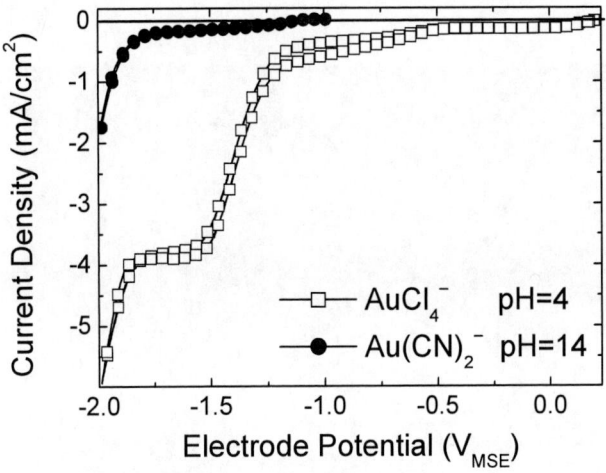

Figure 1: (a) Voltammograms (scan rate 10 mV/s) of n-type Si(111) in 10^{-4} M HAuCl$_4$ pH = 4 solution (open squares) and in 10^{-4} M KAu(CN)$_2$ pH = 14 solution (filled circles), in the rotating electrode configuration. (b) Evolution of the Au coverage as a function of the deposition time, for different deposition conditions at pH = 14 and pH = 4.

Nucleation and growth modes: Figure 2 presents AFM images of gold layers deposited on a quasi perfect vicinal H-Si(111) surface (image 1a). All steps on this surface are 3.1 Å, which corresponds to the distance between two consecutive Si(111) planes. The step to step distance is 100 nm, which is in close agreement with a 0.2° miscut angle. It should be noted that obtaining such an anisotropic etching requires a dissolution kinetics that is 7 orders of magnitudes smaller on terraces than at steps [3, 15].

Deposition of an *average* gold thickness $t_{Au} = 10$ ML from the cyanide solution (Fig. 1b) results in the formation of 3D islands that are well aligned along rows separated by ca 100 nm. It was verified that the islands positioned the substrate monatomic steps [10]. The gold islands have a flat top and a mean lateral dimension 40 nm and are 25 nm high. This deposit was obtained by applying a double step of potential: the first step at $U = -2V$ for 4s, to control nucleation along the silicon steps, followed by a second step of potential ($U = -1.5$ V for 100s) to grow the initial islands. At this potential no new nuclei are formed. This new deposition procedure reduces the island size dispersion with respect to our previous work [10]. The *linear* density of nuclei along one silicon step is $\sim 2 \times 10^5$

nuclei/cm, which is the maximum density attainable. The nucleation density is therefore 2×10^{10} cm^{-2} because the average step to step distance is 100 nm (this sample has a miscut angle 0.2°). It should be noted that this nucleation density may in principle be increased by increasing the miscut of the substrate so as to reduce the step spacing. Shorter nucleation potential steps leave vacancies in the island rows (this suggests that nucleation is still progressive at -2V). Longer pulses leads to some nucleation on the Si(111) terraces. As a whole the gold islands are assembled in a sort of rectangular array since the average island – island distance is 50 nm along the same step and 100 nm between two silicon steps.

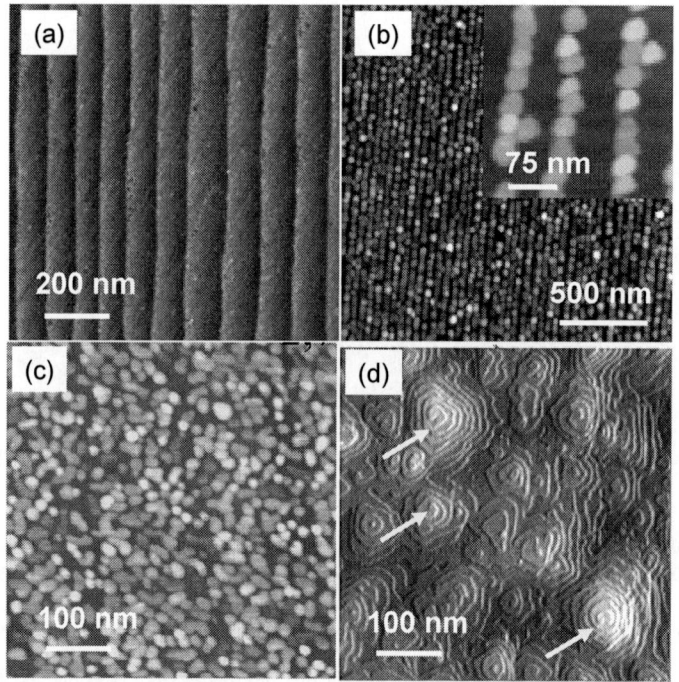

Figure 2: AFM images of: (a) bare H-Si(111) surface; (b) 10 ML thick Au film grown from the gold cyanide bath (pH = 14) at U = - 2V$_{MSE}$, (c-d) 2.5 ML and 30 ML thick films grown from the gold chloride bath (pH = 4) at U = - 2V$_{MSE}$. Arrows in (d) outline screw dislocation.

Deposition of gold from a chloride solution leads to quite different nucleation and growth modes at -2V [9]. In Fig. 2c, which corresponds to an *average* thickness t_{Au} = 2.5 ML, the surface density of islands is ~4 10^{11} islands/cm^2, which is much greater Fig. 2b. Given the surface coverage, the islands in Fig. 2c are ~8 ML high (ca 2 nm) and their average diameter 40 nm. In contrast to Fig. 2b, the islands are now remarkably uniformly distributed over the Si(111) terraces. Upon growth, the nm size islands expand essentially laterally because coalescence occurs around t_{Au} = 15 ML. Above this thickness, a layer by layer growth occurs which leads to ultra smooth Au layers (Fig. 2d, t_{Au} = 30 ML). All steps in this image are close to expectation for Au(111), i.e. steps are ~2.3 Å high. Some screw dislocations are visible in the image. However no grain boundary was resolved.

We will report elsewhere that we observe a strong correlation between the final morphology of the gold deposit (grown in the chloride solution) and the partial current related to proton reduction also referred to as hydrogen evolution reaction (HER). Namely, performing deposition at potentials more positive than -1V, i.e. at a potential where HER is minimized, gives rise to smooth layers which however retain a granular structure resulting from the nucleation stages (the nucleation stage is very similar to the one shown in Fig. 2c). This outlines the critical influence of the rate of HER and/or the electric field across the interface on the mobility of the Au adatoms on Au(111).

<u>Structure and epitaxial relationship of Au layers</u>: The crystallographic orientation of layers deposited as in Fig. 2 was characterized by XRD. The $\theta - 2\theta$ scans (not shown) of the two kinds of deposits are consistent with (111) gold planes parallel to the Si(111) surface (d_{111} = 2.35 Å). In the case of the deposit performed from the chloride bath, we also observe numerous Kiessig fringes, which are a confirmation that the layer thickness is highly uniform on the long range. This is consistent with the smooth topography observed by AFM. The periodicity of fringes gives a film thickness in close agreement with expectations.

The $\psi-$ scans (Fig. 3) are more appropriate than $\theta - 2\theta$ scans to determine the epitaxial relationship. They reveal 6 sharp peaks: the 3 peaks marked with a star correspond to (-111), (1-11) and (11-1) diffracting planes with the epitaxial relationship Au(111)<11-2> || Si(111)<11-2>. The three other peaks correspond to the same family of diffracting planes rotated by 180° with respect to the above epitaxial relationship. Looking into more details, one notices that the 6 peaks have almost the same intensity for the deposit performed from the cyanide solution (Fig. 3a). AFM images at low island density confirm that the Au(111) islands can be aligned with the silicon lattice or rotated by 180°. In the case of the chloride solution (Fig. 3b), the two families of peaks have a different intensity. The one related to rotated Au(111) are more intense. This observation was highly reproducible.

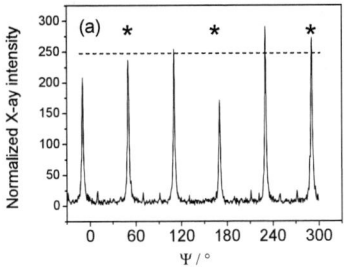 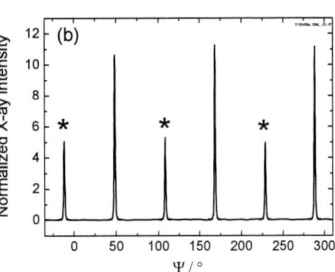

Figure 3: $\psi-$ scan of a 10 ML Au film deposited at pH = 14 (a) and at pH = 4 (b). the 3 peaks marked with a star correspond to (-111), (1-11) and (11-1) diffracting planes with the epitaxial relationship Au(111)<11-2> || Si(111)<11-2>. The three other peaks correspond to the same family of

diffracting planes rotated by 180° with respect to the above epitaxial relationship.

Nucleation and Growth mechanism:
In the ultra high vacuum (UHV), metal nucleation and growth on clean 7x7 Si(111) is very often 2D because this surface is highly reactive [16]. On a low reactive surface such as the H-terminated Si(111) surface, metal deposition generally leads to 3D growth due to a smaller surface energy [16]. In some cases it is debated whether the H monolayer is segregated or not. As discussed below, the electrochemical interface offers new routes to nucleation and growth modes.

As a whole XRD and AFM imaging confirm that Au(111) layers are grown for the two solutions. The small peak FWHM (~1°) in Fig. 3, which is related to in plane mosaicity (rotation of film lattice with respect to substrate lattice), is indicative of a strong epitaxy or a strong interface energy. This point and the pH dependence of nucleation may be explained by the electrochemical model below, in which we account for the fact that hydrogen evolution reaction (HER) takes place in parallel with gold deposition at $U = -2V_{MSE}$.

On a n-type H-Si(111) electrode the HER is promoted by the accumulation of conduction band electrons (e_{CB}^-) and its mechanism depends on pH on the molecular scale [17]:

At pH 4:
$$Si\text{-}H + H^+ e_{CB}^- \rightarrow Si\bullet + H_2 \qquad (1a)$$
$$Si\bullet + H^+ e_{CB}^- \rightarrow Si\text{-}H \qquad (1b)$$

At pH 14:
$$Si\text{-}H + H_2O\ e_{CB}^- \rightarrow Si\bullet + H_2 + OH^- \qquad (2a)$$
$$Si\bullet + H_2O + e_{CB}^- \rightarrow Si\text{-}H + OH^- \qquad (2b)$$

In these reactions, Si• corresponds to an intermediate state where the H atom has been desorbed. The exact structure of this intermediate is certainly more complex than described since an unsaturated Si• surface atoms must be highly unstable [17]. In the presence of Au species, they may be also the stage of the following electrochemical reduction reaction corresponding to the creation of Si-Au bonds:

$$Si\bullet + AuCl_4^- + 3e_{CB}^- \rightarrow Si\text{-}Au + 4Cl^- \qquad (3a)$$
$$Si\bullet + AuCN_2^- + 2e_{CB}^- \rightarrow Si\text{-}Au + 2CN^- \qquad (3b)$$

According to this description, the initial Au-Si sites act as preferential nucleation centers. A similar mechanism was inferred for electroless metal deposition on silicon in fluoride containing solutions [18]. Hence, in this scheme, Au nucleation is driven by HER induced H-desorption. This hypothesis is supported by many observations.
(i) At high pH the Si• sites are in fact mostly created at the step edges at pH because the water molecule can more easily split (it may freely adopts the necessary conformation than on (111) terraces). This explains the selective Au nucleation along the steps (Fig. 2b). (ii) At reverse at low pH the HER is much less site dependent because it involves

protons. As a consequence the Si• sites are evenly distributed at steps and on terraces, which explains the homogenous Au nucleation in the chloride solution (Fig. 2c).

After nucleation, the HER is still playing an important role during the growth. Next to the rim of Au islands, the H-monolayer is probably progressively removed from the silicon surface by a mechanism similar to the one described by reactions (1) or (2). Such a mechanism of progressive segregation of the H-monolayer increases the interface energy, in agreement with the strong epitaxy observed by XRD (Fig. 3). This mechanism applies however only to *small enough* islands. The process is indeed unable to promote a perfect 2D growth. This is particularly obvious at pH = 14: in this case the lateral growth of the gold island is quite limited and it appears difficult to cover the silicon surface unless one deposits a rather thick gold film. This limitation stems from the fact that the electrochemical transfer entirely occurs on the Au islands above a critical size [19, 20]. Hence above a critical island size the HER rate is considerably reduced at the island rim, which hinders lateral growth because the removal of the H-monolayer is less efficient. At pH = 4, the lateral growth of the nuclei seems enough to cover the surface after deposition of 15 ML. This is because islands have a smaller lateral size than at pH = 14 and also because the HER rate is several orders of magnitude larger than at pH = 14. Both factors allows developing HER at a sufficient rate close to the islands. The surface may therefore be rapidly covered by gold islands (the vertical growth rate is 3-5 times smaller than the lateral growth rate).

Magnetic nanostructures grown on Au/Si(111) buffer layers:

As mentioned in introduction, buffer layers are often used in the field of nanomagnetism. One key property of magnetic nanostructure is the magnetic anisotropy, because it plays a key role in hard disk technology [21]. It is therefore highly desirable being able to adjust the magnetization vector in done specific direction of space. In nanostructures difference sources of anisotropy coexist:
(i) The magnetocrystalline anisotropy, as in bulk materials, which depends on the crystallographic structure.
(ii) The dipolar anisotropy (also called shape anisotropy), which is responsible for the fact that a thin is generally in plane magnetized).
(iii) The interface anisotropy, which is induced by the symmetry breaking of the crystal as well as changes in local atomic environment of the ferromagnetic atoms at the surface or interface (coordination number, nature of chemical bonds etc. affect the density of state and thereby the magnetic properties). Elastic strains also induce anisotropy. The interface anisotropy may become dominant in ultrathin layers because the ratio [atoms at the surface / atoms in the films] increases. This is why ultrathin magnetic films sandwiched between two non magnetic layers often present *perpendicular* magnetization anisotropy (PMA) below a critical thickness of only few atomic layers [21].

In this section we use the above Au(111)/Si(111) substrates to illustrate that one may vary the magnetization orientation of ultrathin cobalt deposits (few atomic planes in thickness) by simply changing the gold film morphology. STM studies of electrodeposition of cobalt, from a diluted Watts solution (mainly composed of sulphate salts and of pH ~3), on flat Au(111) have shown that growth is 2D [22] and recent XRD characterizations have confirmed that Co(0001) is grown [23]. Very good epitaxial growth is achieved. These layers can be capped with an Au layer which is also in epitaxy with Co(0001). We have also grown Au/Co/Au/Co/Au(111) and checked that the same epitaxial relationship is kept in the second repeat [23].

Rectangular array of Au/Co/Au dots: Templated Co electrodeposition atop the gold clusters in Fig. 2b can be achieved in a specific window of potential. This enables to cover the Au islands with cobalt and avoid Co deposition on silicon [3]. This specificity does not exist in a UHV approach [24].

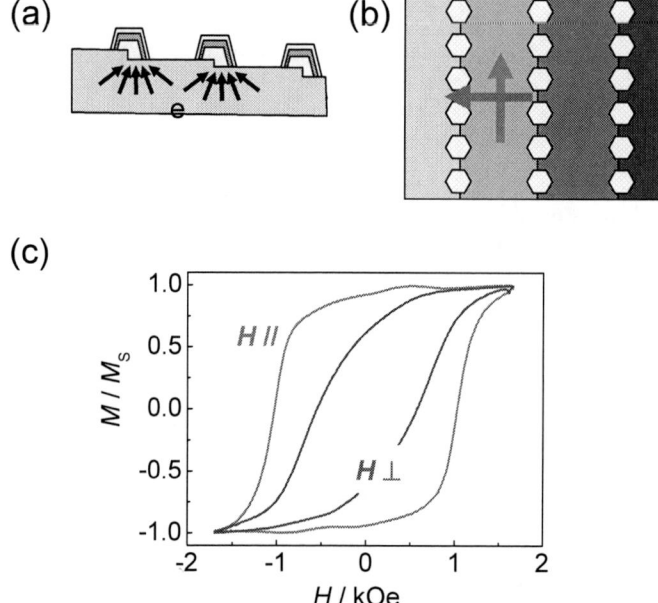

Figure 4: (a-b) Side and top view of the Au/Co/Au(111) dot array. In (b) the grey scale is the one used in AFM imaging. Lower terraces are darker. (c) Magnetization curve (MOKE). The field is applied parallel (*H//*) or perpendicularly (*H⊥*) to the silicon step direction. The larger coercive field measured with *H//* indicates that the easy axis of magnetization is parallel to the silicon steps.

Within this approach we obtain the island structure sketched in Fig. 4a (side view). Fig. 4b is a top view of the idealized array structure. For an array with 20 ML of cobalt per Au islands the normalized magnetization curves shown in Fig. 4c (magneto optical Kerr effect, MOKE) reveal that the dots are ferromagnetic at room temperature. On a macroscopic scale, the array presents a well defined easy axis of magnetization in the plane of the sample because the coercive field is larger when applying the field parallel to the silicon steps than when applying it transverse to the silicon steps: the easy axis of magnetization is clearly along the dense rows of dots. It should be emphasized that a 20

ML Co layer deposited on a flat Au(111) layer is also in plane magnetized but presents no easy axis. The present observations are therefore attributed to magnetic coupling between dots. The latter is stronger along one row (dot – dot distance is ca 50 nm along one step) than between two rows (the spacing distance is 100 nm in our case). Consistent with this conclusion, we also note that remnant magnetization is much greater when the field is parallel to the steps.

Ultrahin Au/Co/Au layers and multilayers:
In this case we studied in great details magnetization curves of Au/Co/Au(111)/Si(111) recorded (Fig. 5a). These evidence, with increasing Co thickness, a progressive evolution of the shape of hysteresis loops (HL) strictly square loop → nearly reversible loop when the field is applied perpendicularly to the sample surface. A square HL means that the sandwich layers present a perpendicular magnetization anisotropy induced by the Co/Au interface. To characterize the strength of anisotropy one generally plots the thickness dependence of remnant magnetization (M_R) to saturation magnetization (M_S) as shown in Fig. 5b [25]. Recalling that M_R/M_S is equal to $\cos(\theta)$, with θ the angle between the magnetization vector and the surface normal, such a plot better evidence the spin reorientation transition of the magnetization. It also helps quantifying the anisotropy and compare different systems. It is generally accepted that the strength of PMA is given by a critical thickness t^* for which $\theta = 45°$ or $M_R/M_S = \cos \theta = 0.707$. In the case of Fig. 5, t^* = 8-10 ML, which is a thickness comparable to state of the art sandwich layer grown in the UHV [26]. Electrodeposited layers present therefore quite interesting performances.

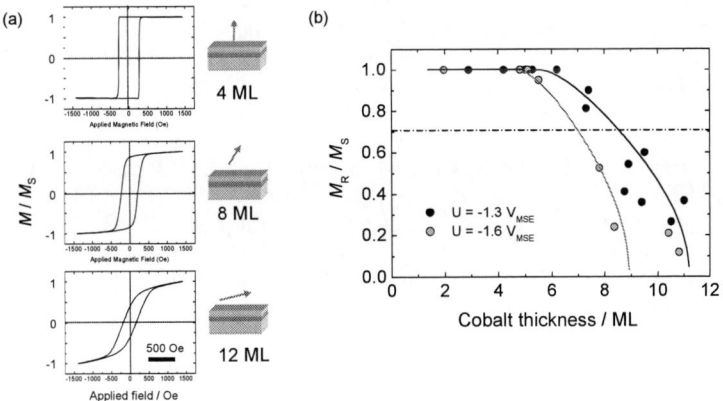

Figure 5: (a) Thickness dependence of magnetization curves recorded with the field applied perpendicularly to the surface (PMOKE). The cobalt thickness is given next to each curve. Note the progressive change of the hysteresis loop. (b) Thickness dependence of $M_R/M_S = \cos \theta$. The strength of PMA is defined by the critical thickness t^* for which q = 45°. In the present case t^* = 8 – 10 ML depending on the applied potential.

Conclusions:

Electrodeposition of Au on Si(111) allows one to grow high quality epitaxial Au(111) layers with quite different morphologies. Such buffer layers can be used to grow magnetic nanostructures (dots and ultrathin films) which present interesting magnetic properties. These substrates offer also the possibility to study in situ the magnetic properties and play with surface chemistry. Other potential applications of these substrates will be presented at the conference.

Acknowledgments

This work was partially supported by the Région Ile-de-France in the framework of C'nano IdF. C'Nano-IdF is the nanoscience competence center of Paris Region, supported by CNRS, CEA, MESR and Région Ile-de-France.

References

[1]   P. Allongue and F. Maroun, in Electrocrystallization and Nanotechnology (G. Staikov, ed.), Wiley - VCH, Weinheim, 2006, p. 217
[2]   P. Allongue and F. Maroun, *Curr. Opin. Solid State Mat. Sci.*, 10, 173 (2006)
[3]   P. Allongue and F. Maroun, *J. Phys.: Condens. Matter*, 18, S97 (2006)
[4]   G. Oskam, J. G. Long, A. Natarajan, and P. C. Searson, *J. Appl. Physics D*, 31, 1927 (1998)
[5]   G. Oskam and P. C. Searson, *J. Electrochem. Soc.*, 147, 2199 (2000)
[6]   G. Oskam and P. C. Searson, *Surf. Sci.*, 446, 103 (2000)
[7]   K. Krug, J. Stettner, and O. M. Magnussen, *Phys. Rev. Lett.*, 96, 246101 (2006)
[8]   W. Polewska and O. M. Magnussen, *Surf. Sci.*, 601, 4657 (2007)
[9]   P. Prod'homme, F. Maroun, R. Cortes, and P. Allongue, *Appl. Phys. Lett.*, 93, 171901 (2008)
[10] M. L. Munford, F. Maroun, R. Cortes, P. Allongue, and A. A. Pasa, *Surf. Sci.*, 537, 95 (2003)
[11] M. L. Munford, R. Cortes, and P. Allongue, *Sens. Mater.*, 13, 259 (2001)
[12] A. J. Bard and L. R. Faulkner, *Electrochemical Methods: Fundamentals and Applications, 2nd Edition*, John Wiley & Sons, New York, 2001.
[13] C. H. Shue, S. L. Yau, and K. Itaya, *J. Phys. Chem. B*, 108, 17433 (2004)
[14] J. O. M. Bockris and A. K. N. Reddy, *Modern Electrochemistry 1*, Plenum/Rosetta, New York, 1970.
[15] J. Kasparian, M. Elwenspoek, and P. Allongue, *Surf. Sci.*, 388, 50 (1997)
[16] K. Oura, V. G. Lifshits, A. A. Saranin, A. V. Zotov, and M. Katayama, *Surf. Sci. Reports*, 35, 1 (1999)
[17] P. Allongue, V. Costa-Kieling, and H. Gerischer, *J. Electrochem. Soc.*, 140, 1018 (1993)
[18] P. Gorostiza, M. A. Kulandainathan, R. Diaz, F. Sanz, P. Allongue, and J. R. Morante, *J. Electrochem. Soc.*, 147, 1026 (2000)
[19] P. Allongue and E. Souteyrand, *J. Electroanal. Chem.*, 362, 79 (1993)
[20] P. Allongue, E. Souteyrand, and L. Allemand, *J. Electroanal. Chem.*, 362, 89 (1993)

[21] M. T. Johnson, P. J. H. Bloemen, F. J. A. d. Broeder, and J. J. d. Vries, *Reports on Progress in Physics*, 59, 1409 (1996)

[22] P. Allongue, L. Cagnon, C. Gomes, A. Gundel, and V. Costa, *Surf. Sci.*, 557, 41 (2004)

[23] P. Prod'homme, F. Maroun, R. Cortes, P. Allongue, J. Hamrle, J. Ferre, J. P. Jamet, and N. Vernier, *J. Magn. Magn. Mater.*, 315, 26 (2007)

[24] A. Fleurence, G. Agnus, T. Maroutian, B. Bartenlian, P. Beauvillain, E. Moyen, and M. Hanbucken, *Appl. Surf. Sci.*, 254, 3147 (2008)

[25] L. Cagnon, T. Devolder, R. Cortes, A. Morrone, J. E. Schmidt, C. Chappert, and P. Allongue, *Phys. Rev. B*, 63, 104419 (2001)

[26] P. Beauvillain, A. Bounouh, C. Chappert, R. Megy, S. Ould-Mahfoud, J. P. Renard, P. Veillet, D. Weller, and J. Corno, *J. Appl. Phys.*, 76, 6078 (1994)

208

ECS Transactions, 19 (3) 209-219 (2009)
10.1149/1.3120702 ©The Electrochemical Society

Electrophoretic Deposition of Spherical and Rod-Shaped Nanocrystals into Close Packed Superlattices

S. Ahmed[a], C. A. Barrett[a], C. O'Sullivan[a], A. Sanyal[a], H. Geaney[a], A. Singh[a], R.D. Gunning[a] and K. M. Ryan[a]*

[a] Department of Chemical and Environmental Sciences and Materials & Surface Science Institute, University of Limerick, Ireland
* Corresponding Author to whom correspondence should be addressed,
kevin.m.ryan@ul.ie

> Ordered nanocrystal assemblies are an exciting new class of materials where electronic, optical or magnetic properties can be dictated by tuning the position and ligand environment of each nanocrystal structural unit. Device integration in both nanoelectronics and nanophotonics relies on the ability to locate these ensembles with nanoscale accuracy exactly where they are needed. In this paper we overview our recent work in the area of electrophoretic deposition of charged nanocrystals from toluene.(1, 2) The systems discussed are twofold 1: gold nanocrystals (8 nm) which are positively charged and assembled from a toluene solution with unprecedented selectivity into lithographic channels and 2: CdS nanorods (100 nm × 8 nm) which are negatively charged and assembled into perpendicularly oriented nanorod superlattices over centimeter squared areas suitable for nanorod solar-cells.

Introduction

Electrophoretic deposition is a well understood process whereby charged particles in a fluid medium can be induced to migrate to an electrode under the influence of an electric field. On reaching the electrodes the particles undergo discharge and attach to the surface as a coating layer. The most common solvent is water and the process has been commercially applied for large scale coatings and in paints where non uniform particles can be conformally distributed on a surface. (3-8) The technique has been successfully employed to form coatings with metal, silica, metal oxide and polymeric particles from colloidal suspensions of particles typically < 30 μm in size (9). An inherent disadvantage of the aqueous based systems is the concominant hydrolysis of water which occurs when voltage is discharged across the solution (10). Electrophoretic deposition from organic solvents is less well understood and while micron and sub micron sized particles can be deposited from organic solvents, the technique usually requires low concentrations of charged particles and hence a high volume of solvents in addition to high voltages which is unattractive commercially (11, 12). While electrophoretic deposition has been applied successfully as a bulk coating technique with all known material classes its potential as a controllable process for directed assembly of size and shape controlled nanocrystals from solution has not been significantly exploited (13, 14).

Colloidal nanocrystals are an exciting new material with size tunable electronic and optical properties (15). The most commonly studied system are pseudo-spherical (0D) nanocrystals of semiconductors which are synthesized by the pyrolysis of organometallic

209

precursors in hot long chain alkyl phosphonic surfactants (16). The surfactant is bi-functional in the reaction, limiting the size of each nanocrystal by preventing clustering during growth and further acting as a capping ligand allowing nanocrystals to be subsequently solubilized in high concentrations in organic solvents (17). At high temperatures (> 573 K) and high monomer concentrations (usually achieved by injection), each nanocrystal nucleates and grows in size equally and simultaneously until all the monomer in the reaction is used. Nanocrystals of a specific diameter e.g 3 nm can be obtained by reducing the temperature and hence quenching the growth after a specific time interval. More complex shapes such as nanorods and tetrapods can be obtained by introducing surfactant combinations which selectively passivate different growth facets (18). Such colloidal nanocrystals have found application where discrete particles can be used as biolabels, single electron devices and quantum cellular automata (19-22). However, more interesting applications are possible when the nanocrystals are assembled into supercrystals from solution. In particular if size polydispersity is less than 5%, pseudo spherical nanocrystals hexagonally close pack into assemblies where each nanocrystal is separated from its nearest neighbor by organic ligands (23). These assemblies represent a new form of nanocrystal solid where electronic and optical properties are not defined by the bulk material characteristics but by the size, interparticle separation, and ligand environment of each nanocrystal structural unit (24). This was shown to particular effect in PbSe nanocrystal assemblies by Beecher et al. where an insulator to metal transition was achieved in the material simply by manipulating the interparticle spacing (25). More recent work has shown the potential of close packed assemblies as a new material class capable of functioning as single electron devices or plasmon routers for photonics (26, 27). Nanorods can also assemble into close packed superlattices if both polydispersity in length and diameter is obtained (28-30). The most interesting geometry is where each nanorod is vertically aligned in the superlattice allowing applications in aligned nanocrystal solar-cells and field emission devices (31, 32).

The challenge with current methods of nanocrystal assembly from solution is twofold; in discrete on-chip applications methods are needed to assemble the nanocrystals exactly where they are needed while in applications such as displays and photovoltaics, large area assembly is desirable with order on the nanoscale. Current methods towards achieving large area coverage mainly rely on drop casting or controlled solvent evaporation and are not amenable either to selectivity or complete order throughout the deposit (28, 29, 33-35). A degree of selectivity is attainable when AC fields applied across patterned electrode contacts allow trapping of discrete particles, however the process usually results in particle accumulation on and around the electrodes in addition to infilling the gap (36).

Herein we overview our recent work in the electrophoretic assembly of nanocrystals from solution for device application where unprecedented selectivity and layer by layer coverage are obtained by finely controlling this process. Gold nanocrystals are selectively assembled into lithographic channels back etched from SiO_2 onto highly doped silicon precisely in areas occupying less than 1 billionth of the area under study (1). The assemblies are hexagonally close-packed and corralled exactly by the lithographic features allowing this technique to function as a reproducible method for on-chip integration. The method successfully integrates top-down (e-beam lithography) and bottom-up fabrication (self assembly) creating on-chip arrays where the unique electronic and or optical/plasmonic properties of metal particle assemblies can be exploited. The use of electrophoretic deposition as a controllable method of uniquely organizing anisotropic nanocrystals (nanorods) is further demonstrated. The phosphonic acid capped nanorods are found to be inherently negatively charged and migrate to the positive electrode under the influence of an electric field to form close packed assemblies over centimeter scale areas where remarkably all nanorods are vertically aligned. The nanorod assembly work is the first report of controlled electrophoretic deposition with anisotropic particles and has significant implications for field emission devices, nanorod based solar-cells and linear polarized emission devices.

Experimental

A stable solution of gold nanocrystals (nc) in toluene was prepared following a method similar to that described by Fink *et al* (37) and the CdS nanorods in this study were synthesized using the pyrolysis route (38). For electric-field assisted deposition of gold nanoparticles, 1.5 ml of gold-nanoparticles /toluene solution was further diluted with 25 ml of toluene (1.5 ml solution, 1.7718×10^{-3} g of gold which is equivalent to 0.36×10^{-3} mol dm^{-3} gold which is equivalent to 8.85% (w/v)). For electrophoretic deposition of CdS nanorods, nanorods are further re-dispersed in a toluene solution (5% w/v). Electric field deposition was carried out at room temperature.

For direct deposition of gold nanoparticles on carbon coated copper TEM grid, grids are fixed in the cathode and an electric field of 50 V was applied while the assembly was completely immersed in solution and deposition was carried out for 6 s. During electric-field assisted deposition into trenches, electrodes are completely immersed in a deposition bath and a negative potential of 200V was applied to the patterned wafer for 180 s. DC voltage was applied using a high voltage power supply unit (Griffin 5 kV EHT, Essex, UK). During Electrophoretic deposition of CdS nanorods, electrodes are completely immersed in a deposition bath and a positive potential of 200V was applied to the substrate using a high voltage power supply unit (TECHNIX SR-5-F-300, S/N: BU08/04971). In all cases the voltage was monitored using Black star 3225 MP multimeter.

The surface topography and depth profiles of patterned wafers were characterized by a Veeco- Enviroscope atomic force microscopy (AFM) in contact mode using a gold coated silicon nitride probe. The SEM micrographs were acquired using high resolution scanning electron microscopy (Hitachi S-4800 HRSEM). Gold nanocrystal size (~8 nm) and CdS nanorod dimensions was characterized using JEOL 2011 transmission electron microscope (TEM).

Conductivity measurements of tetraoctyl ammonium bromide (TOAB) and gold nanocrystal containing toluene were carried out using the same power supply as above but current was monitored using Keithley 610A Electrometer which is capable to measure 10^{-11} amps with full scale deflection.

Results and Discussion

Figure 1a shows a TEM image of a typical gold nanoparticles synthesized using tetraoctyl ammonium bromide [(C$_8$H$_{17}$)$_4$NBr] surfactant which has been drop-cast onto a TEM grid. The typical particle size is 8 nm. Figure 1b shows a high resolution TEM image showing lattice fringing corresponding to 0.23 nm typical of (111) crystalline gold. The gold nanocrystals are expected to be coated with the covalently bound tetraoctyl ammonium bromide ligands with a loosely coordinated sphere of bromine ions. Our studies on these surfactant passivated gold nanocrystals in organic solvents under the influence of a DC electric field showed the selective migration of the particles preferentially to the cathode indicating an overall positive particle charge.

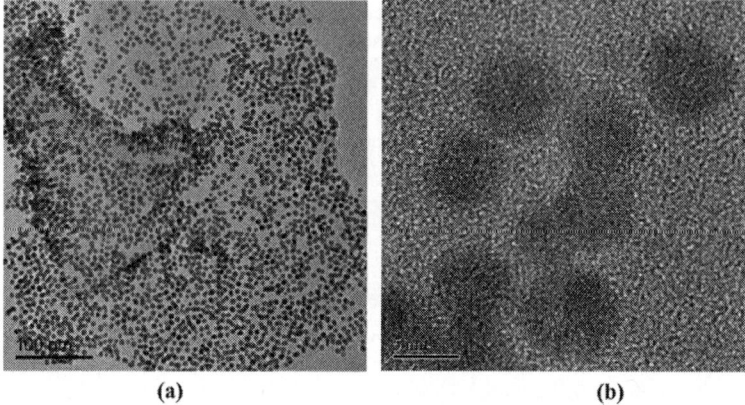

(a)   (b)

Figure 1: (a) TEM image of as synthesized gold nanoparticles (b) HRTEM image of with lattice fringes evidencing gold nanoparticles are crystalline.

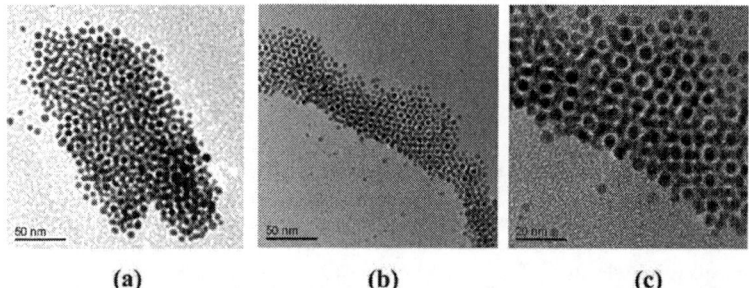

(a)   (b)   (c)

Figure 2: TEM micrograph (a) shows ~8 nm gold nanoparticles deposited on TEM grid using two parallel aluminum electrodes where negatively biased aluminum electrode has an arrangement to clamp a carbon coated copper TEM grid. TEM micrograph (b) and further magnified image (c) shows ~5 nm gold nanoparticles deposited on TEM grid using the same parallel aluminum electrode set-up as was used for ~8 nm gold nanoparticle deposition from toluene medium.

Figure 2 shows further TEM images where in figure 2(a) the carbon coated copper TEM grid was fixed on the cathode and an electric field of 50 V was applied while the assembly was completely immersed in solution. The particles deposited on the grid layer by layer in close packed assemblies. The characteristic pattern shown in figure 2c occurs when the final depositing layer of nanocrystals sit in the interstitial spaces of nanocrystals in the underlying layer. In contrast, when the TEM grid was clamped to the positive electrode no deposition occurred. The electrophoretic approach to nanocrystal deposition clearly showed an enhancement of particle ordering compared to drop-cast deposition. To investigate the selectivity of this process, we fabricated using electron beam lithography square and rectangular trenches from 30 nm SiO_2 grown on p-type silicon.

Figure 3(a) shows HRSEM images of the trench patterns showing dimensions ranging from 150 nm × 150 nm to 850 nm × 150 nm. The trenches were back etched using electron beam lithography from 30 nm SiO_2 thermally grown on p-type (100) silicon wafer. Consequently, the bottom of trench contains conducting bare p(100) silicon. Figure 3(b) shows AFM topography and the corresponding depth profiles confirming the average trench depth at 25 nm. The etch process created a deeper trench in larger area channels and in most cases left a residual non-uniform deposit of SiO_2 at the trench edges. This substrate was clamped to the negative electrode, and the assembly immersed in a solution of gold nanoparticles in toluene under an applied electric field of 200 V for 10 minutes. Figure 4 shows SEM images of the trenches after gold nanoparticle infill. The gold nanoparticles only deposited in the conducting channels at the trench base with no deposition occurring on the insulating SiO_2 layer. The extent of this selectivity was remarkable as in this experiment the area occupied by the trenches was 1×10^{-7} % of the total wafer die.

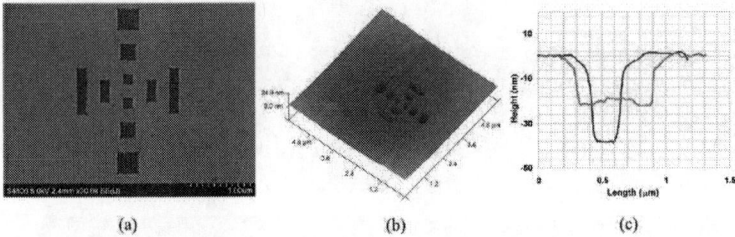

(a)   (b)   (c)

Figure 3. (a) HRSEM images of the trench patterns showing dimensions of ranging from 150 nm × 150 nm to 850 nm × 150 nm. (b) AFM topography and (c) the corresponding depth profiles showing the average trench depth at 25 nm.

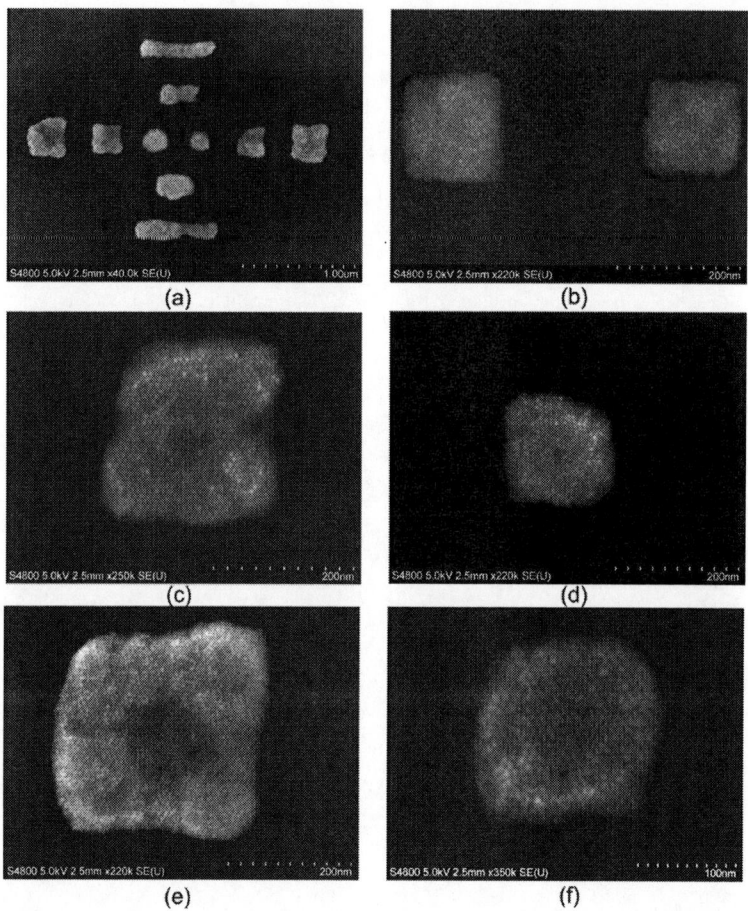

Figure 4. HRSEM micrographs of the trenches (a-f) after gold nanoparticle infill by electrophoretic deposition from toluene. The gold nanoparticles only deposited in the conducting channels at the trench base with no deposition occurring on the insulating SiO_2 layer.

The nanocrystals assemble in a layer by layer process as indicated by step wise features in the deposits which were only previously observed in electrophoretic deposition of sub-micron colloids. The gold nanocrystals are corralled exactly in the channels with the extent of filling determined by the deposition time. Sustained deposition results in outgrowth of the assemblies beyond the trench edge once the thickness of the deposit becomes greater than 30 nm. The assemblies show a high degree of order, organizing into hexagonal close packed layers directly from solution. The occurrence of such a high degree of order is interesting in this case as nanoparticles polydispersity in the as synthesized sample is ~ 10%. In all reported methods on evaporation assisted assembly of nanocrystals, polydispersity in particle size less than 5 % was needed for close-packing to occur.

The mechanism of deposition under the influence of the field is thought to be responsible for this phenomenon. As shown in Fig.3, the patterned substrate consisting of insulating and conducting regions is essentially a dielectric layer obstructing the field lines between the two electrodes. The field lines move to the path of least resistance bending towards the conducting trench base and are the conduit pathways for particle migration resulting in highly selective particle deposition. However, each positively charged particles will experience repulsion from its neighboring particles as accumulation occurs at the features allowing the particle to find its preferred position and hence lowest energy consideration in the assembly before locking in and discharging.

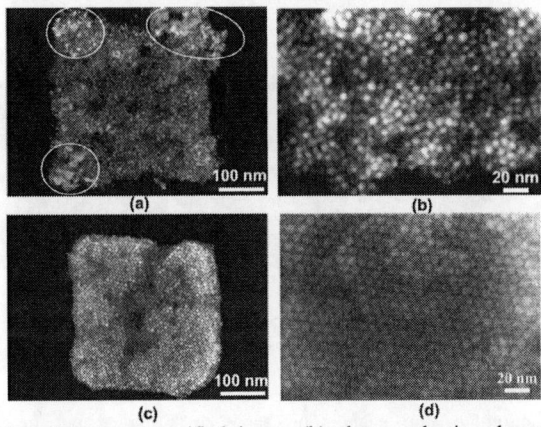

Figure. 5. (a) and further magnified image (b) shows selective deposition of gold nanoparticles with excess surfactant into trenches without ordered assembly. (c) and further magnified image (d) shows selective and ordered assembly of gold nanoparticles into the trenches.

Surfactants play a key role in the order assembly of gold nanoparticles. Excess surfactant had a negative influence on the particle ordering Fig. 5a-b in comparison to the washed and cleaned nanoparticles ordered assembly Fig. 5c-d. Although particles are selective to the channel, random ordering of the particles does not occur if pre-cleaning of the particles through re-precipitation is not carried out. The excess surfactant is evidenced by insulator charging under SEM inspection and co-deposited excess surfactants are also shown by circular marks.

Nanorod Assembly by Electrophoretic deposition

Shape anisotropy in nanorods (1D) compared to (0D) nanocrystals creates a barrier to close packing with rods usually aggregating into rafts parallel to the surface with little or no long range order. However, recently methods have been developed whereby close packed nanorod assemblies can be generated where each rod in the assembly is oriented normal to the substrate interface. The methods required to achieve vertical alignment are largely aspect ratio dependent with external electric fields required for nanorods with aspect ratios > 6 and spontaneous vertical alignment occurring for their lower aspect ratio analogues. The size of these domains can vary from as small as a few nanorods to no larger than several micron[2] when assembled on a solid surface such as an electron microscopy grid or a semiconductor wafer.

Figure 6 (a-c) shows HRSEM images of CdS nanorods (10 nm × 30 nm) which have been drop cast from solution. The rods aggregate both parallel and perpendicular to the substrate in random clusters ranging from as few as 10 nanorods up to micron sized areas. Assemblies are only obtained from the evaporation of dilute solutions of nanorods in toluene, with the evaporation of more concentrated dispersions resulting in several hundred multilayer of rod aggregation with no positional or orientational order. It is known that in the evaporation of a liquid with dissolved solids on a surface, the receding solvent line randomly pins resulting in the accumulation of particles. This evaporation mechanism accounts for the random distribution of the nanorod assemblies on the TEM grid. However, this approach is not amenable to large area surface coverage of vertically aligned nanorods which is a barrier to their device application.

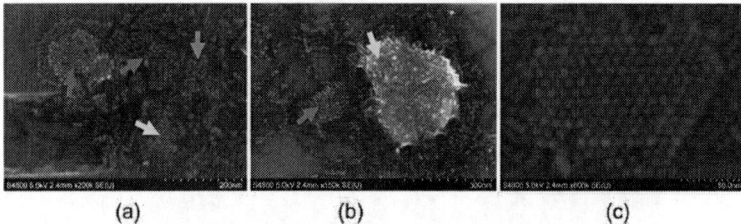

(a)   (b)   (c)

Figure 6. (a) to (c) shows short range order in vertically aligned nanorod domains obtained from drop-cast solution as the evaporating solvent randomly pins at successive points on the surface.

The limitations of the evaporation approach prompted our investigation of electrophoretic deposition as a potential route to large area nanorod assembly. The assembly from solution with the electrophoretic process without the requirement for evaporation as a driving force for assembly is a considerable advantage. Here, the CdS nanorods are capped with long chain alkyl phosphonic acid surfactants and were found to be negatively charged as determined by their zeta potential of -51.48 ± 1.44 mV. In a similar set up to the gold particle studies, a substrate was clamped to the positive electrode, in this case a Si/30-nm SiO_2 wafer, gold coated silicon wafer or ITO coated glass and a voltage of 200 V was applied for 5 minutes. Remarkably, the deposition forms a monolayer of close-packed nanorods where each rod is vertically aligned with respect to the substrate. Figure 7 (a-b) shows vertically aligned CdS nanorods on Si/SiO_2 and gold coated silicon wafer respectively. The nanorods are typically vertically aligned on the substrate with ± 15° variation to the normal.

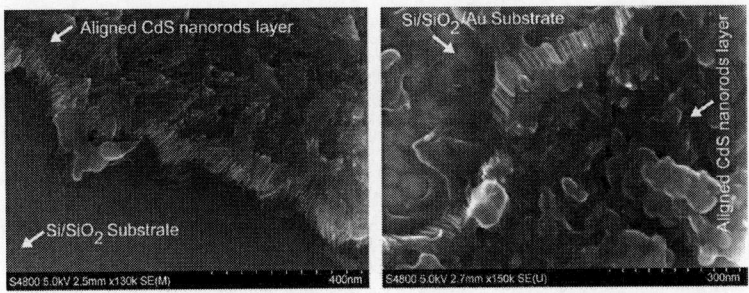

(a)   (b)

Figure 7 (a) and (b) shows vertically aligned CdS nanorods on to the Si/SiO$_2$ and Si/SiO$_2$/gold wafer respectively.

The deviation of the aligned nanorods from the normal could occur during cracking of the deposited layer when tensile stress develops at the nanorod/substrate interface during dewetting. Figure 8(a) shows an electrophoretically deposited (EPD) vertically aligned nanorod layer which shows significant cracking on solvent evaporation. This cracking can be avoided by removing/exchanging the long chain (ODPA+TOPO) surfactant with a short chain pyridine surfactant. Figure 8(b) shows a crack-free EPD layer of nanorods after mild stripping and/or exchange of (ODPA+TOPO) by the short chain ligand pyridine (10 µL pyridine injected in 20 ml toluene solution in which 1 mg CdS nanorods was dispersed). The inset zoomed image (scale bar 100 nm) shows ~ 75% vertically aligned nanorods in this assembly. When the long chain surfactant is striped-off by refluxing with pyridine the nanorods start to grow due to the Ostwald ripening and thus the degree of polydispersity increases which is detrimental to the nanorod alignment.

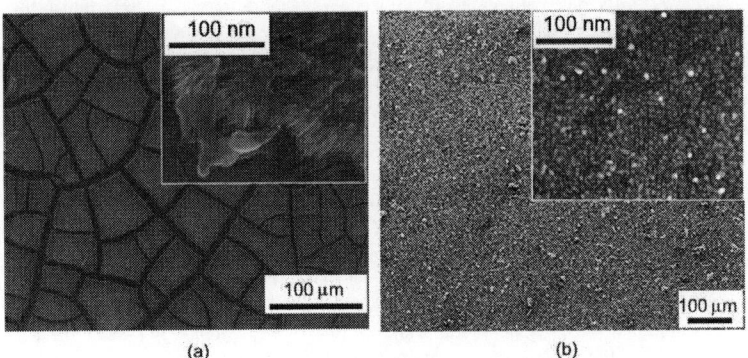

(a)   (b)

Figure 8. A macro-scale SEM image (scale bar = 100 micron) of electrophoretically deposited nanorods (a) with significant crack propagation after dewetting and (b) without cracking after pyridine stripping. Inset HRSEM image in (a (scale bar = 100 nm) shows nanorods are vertically aligned with some deviation from its normal in this deposited layer. Inset in (b) shows ~75% nanorods are vertically aligned.

While the degree of vertical alignment in pyridine stripped rods is reduced, these structures are more suitable for device application for example in nanorod solar cells. In the nanorod assemblies with long chain alkyl surfactants, the surfactant ligands interdigitate, creating an electrically insulating spacer between each nanorod pair. In contrast the pyridine stripped rods allow charge transfer at the nanorod interface and have found application in hybrid nanorod solar-cells when blended with poly-3-hexyl thiophene (31). The novel route to complete vertical alignment described here will potentially allow significant increases in power conversion efficiencies over randomly aligned cells as the rod orientation provides directional pathways for charge transport to the electrodes. Efforts to control the pyridine stripping to minimize polydispersity are currently under investigation.

Conclusion:
The use of controlled electrophoretic deposition from toluene as a reproducible method for localizing ordered and close-packed nanocrystal assemblies in precise locations on-chip is demonstrated. This method which integrates bottom up assembly and top down fabrication is highly desirable as it allows for the creation of novel devices utilizing the unique properties of nanocrystal assemblies as a single step in the multi-step processes of semiconductor manufacture. Further work on electrophoretic deposition of perpendicularly aligned nanorod arrays over large areas is also demonstrated. This process allows large deposits to be formed from solution which consist of single monolayers of vertically aligned and close-packed nanorods. Such structures are of significant interest for nanorod based solar-cells and field emission devices where the unique axial orientation of the nanorods and close packed geometry is required.

Acknowledgements
The work was principally supported by Science Foundation Ireland (SFI) under the Principal Investigator Program, Contract No. 06/IN.1/I85. The authors thank Professor Noel Buckley for access to Hitachi high resolution scanning electron microscopy S4800, Zeta PALS (zeta potential measurement instrument) and useful discussion.

References:

1.   S. Ahmed and K. M. Ryan, *Advanced Materials*, **20**, 4745 (2008).
2.   K. M. Ryan, D. Erts, H. Olin, M. A. Morris and J. D. Holmes, *Journal of the American Chemical Society*, **125**, 6284 (2003).
3.   X.-C. Guo and P. Dong, *Langmuir*, **15**, 5535 (1999).
4.   A. L. Rogach, N. A. Kotov, D. S. Koktysh, J. W. Ostrander and G. A. Ragoisha, *Chem. Mater.*, **12**, 2721 (2000).
5.   M. Semmler, E. K. Mann, J. Ricka and M. Borkovec, *Langmuir*, **14**, 5127 (1998).
6.   M. Giersig and P. Mulvaney, *Langmuir*, **9**, 3408 (1993).
7.   M. Giersig and P. Mulvaney, *J. Phys. Chem.*, **97**, 6334 (1993).
8.   O. O. Van der Biest and L. J. Vandeperre, *Annual Review of Materials Science*, **29**, 327 (1999).
9.   L. Besra and M. Liu, *Progress in Materials Science*, **52**, 1 (2007).
10.   J. Zhao, X. Wang and L. Li, *Materials Chemistry and Physics*, **99**, 350 (2006).

11. K. Moritz and E. Müller, *Journal of Materials Science*, **41**, 8047 (2006).
12. B. Ferrari, R. Moreno, P. Sarkar and P. S. Nicholson, *Journal of the European Ceramic Society*, **20**, 99 (2000).
13. N. Chandrasekharan and P. V. Kamat, *Nano Lett.*, **1**, 67 (2001).
14. S. Banerjee, S. Jia, D. I. Kim, R. D. Robinson, J. W. Kysar, J. Bevk and I. P. Herman, *Nano Letters*, **6**, 175 (2006).
15. A. P. Alivisatos, *Science*, **271**, 933 (1996).
16. Z. A. Peng and X. Peng, *J. Am. Chem. Soc.*, **123**, 183 (2001).
17. Y. Yin and A. P. Alivisatos, *Nature*, **437**, 664 (2005).
18. L. Manna, E. C. Scher and A. P. Alivisatos, *J. Am. Chem. Soc.*, **122**, 12700 (200).
19. D. Wang and C. M. Lieber, *Nat Mater*, **2**, 355 (2003).
20. Y. Cui, U. Banin, M. T. Bjork and A. P. Alivisatos, *Nano Lett.*, 1519 (2005).
21. D. L. Klein, R. Roth, A. K. L. Lim, A. P. Alivisatos and P. L. McEuen, *Nature*, **389**, 699 (1997).
22. A. Khitun and K. L. Wang, *Superlattices and Microstructures*, **37**, 55 (2005).
23. C. B. Murray, C. R. Kagan and M. G. Bawendi, *Science*, **270**, 1335 (1995).
24. H. Mattoussi, A. W. Cumming, C. B. Murray, M. G. Bawendi and R. Ober, *Physical Review B*, **58**, 7850 (1998).
25. P. Beecher, A. J. Quinn, E. V. Shevchenko, H. Weller and G. Redmond, *Nano Letters*, **4**, 1289 (2004).
26. M. Salerno, J. R. Krenn, A. Hohenau, H. Ditlbacher, G. Schider, A. Leitner and F. R. Aussenegg, *Opt. Commun.*, **248**, 543 (2004).
27. S. Chen, R. S. Ingram, M. J. Hostetler, J. J. Pietron, R. W. Murray, T. G. Schaaff, J. T. Khoury, M. M. Alvarez and R. L. Whetten, *Science*, **280**, 2098 (1998).
28. K. M. Ryan, A. Mastroianni, K. A. Stancil, H. Liu and A. P. Alivisatos, *Nano Lett.*, **6**, 1479 (2006).
29. S. Ahmed and K. M. Ryan, *Nano Lett.*, **7**, 2480 (2007).
30. C. O'Sullivan, S. Ahmed and K. M. Ryan, *JournalofMaterialsChemistry*, **18**, 5218 (2008).
31. W. U. Huynh, J. J. Dittmer and A. P. Alivisatos, *Science*, **295**, 2425 (2002).
32. Y. Xia, P. Yang, Y. Sun, Y. Wu, B. Mayers, B. Gates, Y. Yin, F. Kim and H. Yan, *Advanced Materials*, **15**, 353 (2003).
33. L. Carbone, C. Nobile, M. DeGiorgi, F. D. Sala, G. Morello, P. Pompa, M. Hytch, E. Snoeck, A. Fiore, I. R. Franchini, M. Nadasan, A. F. Silvestre, L. Chiodo, S. Kudera, R. Cingolani, R. Krahne and L. Manna, *Nano Lett.*, **7**, 2942 (2007).
34. A. C. Balazs, T. Emrick and T. P. Russell, *Science*, **314**, 1107 (2006).
35. C.-C. Kang, C.-W. Lai, H.-C. Peng, J.-J. Shyue and P.-T. Chou, *ACS Nano*, **2**, 750 (2008).
36. R. J. Barsotti Jr, M. D. Vahey, R. Wartena, Y.-M. Chiang, J. Voldman and F. Stellacci, *Small*, **3**, 488 (2007).
37. J. Fink, C. J. Kiely, D. Bethell and D. J. Schiffrin, *Chem. Mater.*, **10**, 922 (1998).
38. Z. A. Peng and X. Peng, *J. Am. Chem. Soc.*, **123**, 183 (2001).

220

ECS Transactions, 19 (3) 221-225 (2009)
10.1149/1.3120703 ©The Electrochemical Society

Spontaneous deposition of metallic Pt onto n-InP: an electroless process

A. Etcheberry, , C. Mathieu, M. Bouttemy, J. Vigneron, P. Tran-Van, A-M Gonçalves

[a] Institut Lavoisier de Versailles UMR CNRS818045ave des Etats Unis 78035 Versailles
France

InP surfaces are treated in K_2PtCl_4 dissolved in pure water. A
spontaneous electroless process is observed with an interesting
residual Pt(0) coverage and a correlated InP oxidation. The
demonstration of this effect is given by high resolution XPS.
Modifications are also detected using capacitance measurements
on the interface.

Introduction

Interfacial electrochemistry on semiconductors presents a lot of specific aspects,
compared to the responses on noble or semi-noble metallic surfaces. The understanding
of surface evolutions or deviations, on a lot of particular processes contributes to the
fundamentals of the often specific semiconductor electrochemistry. The more or less
important reactivity of the semiconductor surface, during the "electrochemical process",
is a very interesting aspect among plentiful of original behaviors. Surface reactivity
during a process depends strongly on both the nature of the semiconductor and the redox
systems implied in the overall process. In the case of surface reactivity, the steady state
situation can strongly differ, depending on Si, Ge, III-V, II-VI, large or narrow gaps are
considered. The surface evolution can happen, at open circuit potential or in a given
potential range for which oxidation or reduction global currents are observed. In this way,
a particular phenomenon, is the spontaneous mechanisms, happening at open circuit
potential. The more famous are the spontaneous or "electroless" etching processes that
imply very complex interfacial implication for the upper layers of the semiconductor
lattice (1). Others are the processes that could happen spontaneously when low
concentration of metallic salts, are present in the solution. Interaction, possible resulting
surface pollution or deposition feature are very interesting point as well as fundamental
aspects or also technological ones. However these effects are difficult to study and often
observed with rather fluctuations. In all case the characterization of the surface evolution
is an important question. A lot of procedures are able to provide information about the
interfacial interaction. If the surface chemistry is strongly changed with sufficient
stability, to undergo the sample emergence, *ex situ* characterization can be performed.
Among them XPS is a quantitative approach to describe accurately the nature of the
modification. Another way is an *in situ* characterization with electrochemical methods
that allows a comparison between the initial response and the modified one.
In this paper we focus on the behavior of a clean InP surface dipped in a neutral solution
of a 1-2 mM of K_2PtCl_4. The aim of this work was initially devoted to performed a
comparison with the behavior of a passivated InP with a monolayer of polyphosphazene
film which functionalization by Pt(II) is expected. First observations performed by XPS,

clearly shown that an unexpected strong transformation, was observed on a bare surface and we decided to look accurately at this surprising effect. Surprising effect because, only electrochemical deposits with more or less marked side transformation has been reported and accurately described on p-InP (2) and n-GaAs (3). Moreover these experiments were performed with $K_2 PtCl_6$ salt. Concerning K_2PtCl_4 nothing is reported on semiconductor electrochemistry. A recent work (4) using this salt, performed on semi-noble metals mentioned adsorbed salt phenomenon without reduced metal. Metallic sites are clearly observed only after annealing.

Experimental

Experiments were performed on n^+-InP. Surfaces are oriented (100). Before each experiment surface are chemo-mechanically polished in methanol- Br_2 solution rinsed in methanol then dipped into 2M HCl just before working. So, almost oxide free surfaces are studied. Solution for interaction are 1-3mmole L^{-1} in $K_2Pt Cl_4$ salt, ultra pure water was used systematically N_2 purged before InP was dipped inside. Electrochemical experiments was performed in a classical three electrode configuration, in HCl 1M or 0,1M solutions on clean or treated surface of InP. In this paper only impedance results are presented. These measurements were performed with a 2273 PAR system. Before XPS analysis sample are emerged from solution and dipped in ultra pure water, dried and charged in the introduction chamber. A Thermo VG ECALAB 220i XL is used as XPS spectrometer. Monochromatic X-AlK$_\alpha$ rays are used for excitation. Detection is performed perpendicularly to the surface. Constant analyser energy mode is used for the detection, pass energy are 20 or 8eV providing high or ultra high energy resolution spectrum. Thermo Avantage program is used to treated experimental results.

Results and discussion

Immersion of a clean InP surface in ultra pure DI water solution containing a 1-3 mM K_2PtCl_4 salt undergoes a strong transformation easily detected by XPS. Indeed, the unbuffered solution, is in a pH range for which an oxidation process of InP gives rise to the growth of very stable oxide phases which composition governs the behavior of the interface [5, 6]. The detection of any oxidation of InP is so very easy to do. Moreover, if something related to a Pt interaction happens, its detection is easy because XPS sensibility factor for the Pt4f core level is high. So low concentration or low coverage can be detected

The figure1 presents the XPS response of an initially clean surface submitted to the Pt(II) solution interaction. In the same figure, are presented the Pt4f and the In3d 5/2 levels. In3d5/2 presents a typical double feature structure that can be fitted using four contributions. The shape of this peak demonstrates that the initial InP surface has been significantly oxidized during its immersion inside the Pt(II) solution. Note that nothing as comparable is obtained, if the solution is a pure oxygen free water solution. The P_{2p} spectrum no presented here contained the same information with a classical P_{ox} component close to 133eV and another one centered toward 129eV that is typical of the InP response. As for In_{3d} we can say that it is typical of an oxidized InP surface only covered by a thin layer since for P as In the InP matrix components are perfectly detected. If we remark that the intensity of the O_{1s} signal sensibly increased on treated sample we can conclude that a thin oxide on InP is systematically obtained during immersion in the Pt(II) solution. We must remark that the high sensitivity of XPS allows this affirmation.

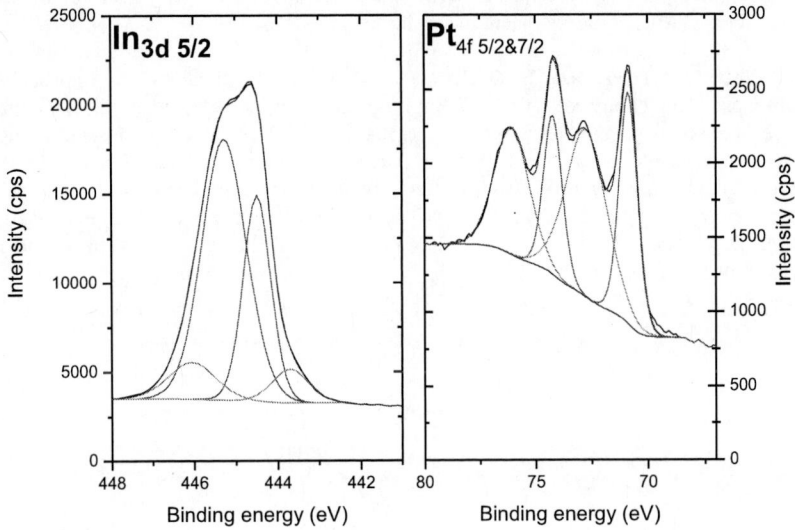

Figure1: Pt_{4f} and $In_{3d\ 5/2}$ core level registered on an InP modified surface. Fitting of the peaks is also presented in the same picture.

The second important XPS feature is the specific shape of the Pt_{4f} signal. We observed always more or less established, a triple feature structure that indicates that Pt phase was always present despite the final pure water rinsing. So clearly a Pt interaction was demonstrated then a fitting procedure can describe accurately the corresponding chemical configuration. Pt $_{4f}$ can be described with a double structure for each $Pt_{4f5/2}$ and $Pt_{4f7/2}$ component. One at low binding energy is unambiguously associated to a metallic Pt(0) phase and the other one at high binding energy can be interpreted as associated to a Pt oxide phase or to a strongly adsorbed $PtCl_4$ phase. Atomic balance shows that significant Cl signal is detected and can agree with a combined situation. Note that similar observation has been mentioned in the literature (4). So the very important point is that Pt(0) spontaneous deposited species are detected, implying a reduction of an amount of Pt(II) in interaction with the surface which origin must be explained.

Clearly a double correlated mechanism happens: oxidation of InP and Pt (0) deposit probably associated to a correlated or uncorrelated Pt(II) salt adsorption.°

This complex phenomenon can be only explained by an electroless process which happens with a side oxidation of the InP surface correlated to a Pt(II) partial reduction.

This is a typical electroless process happening progressively and it seems to saturate after 1hours of immersion. The resulting surface characterized on the basis of XPS results is only partially covered by Pt components. The oxidized InP surface is obviously uncovered even a global ultra thin Pt coverage cannot be totally excluded at this time.

So this *ex situ* XPS characterization allows a precise description of the resulting modified chemistry which in figure 1 & 2 corresponds to the steady state described by a very heterogeneous surface mixing oxidized site and metallic ones. To be exhaustive we mentioned that the balance with the oxidized P component shows that a significant In oxide enrichment, is present onto the surface. We see again a typical situation already obtained for specific anodic treatment in similar medium (5,6).

This modified surfaces, are then characterized in acidic medium by impedance measurement. The choice of this medium is governed by the fact that this medium dissolves the oxides on InP and we can expect that only the Pt phases or more interesting the Pt(0), stay onto the surface. In fact the situation is more complex because the Pt balance of the surface Pt can change differently from one sample to another. So the situation is more complex than the one described after the treatment into Pt(II) unbufferred solution. However we can often observe the evolution of the Mott Schottky plots as shown in the figure 2.

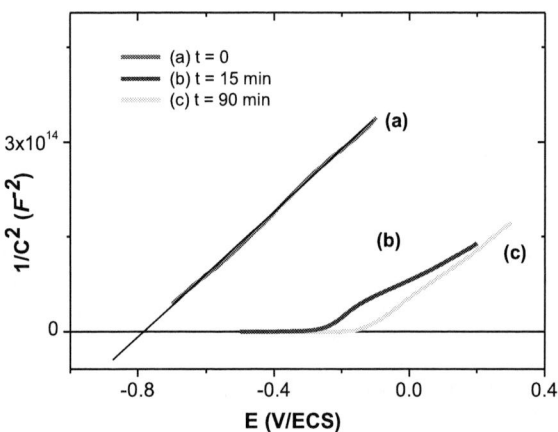

Figure 2. Evolution of the Mott-Shottky plots, f = 1kHz, in 1M HCl aqueous solution, after different immersion times (t = 15 min, 90 min) of n-InP sample into a deoxygenated neutral aqueous solution of K_2PtCl_4 (3 mmol.L^{-1}) rinsed and dipped in acidic solution.

Obviously a clear shift of the curves is obtained on treated surfaces. As observed by XPS the modification seems consequent for relatively low immersion time. At this time as the surface must be described as an inhomogeneous covering the interpretation of the effect is not evident. We can however points out that significant positive shift of the hydrogen evolution can be also observed. It depends on the sample but is always higher than 0.3 V and sometime as high as 0.5V.

Conclusion

This work points out a very interesting feature : the spontaneous deposition of Pt metallic particles on InP in Pt(II) solution. This result is original because literature on semi-noble metals mentions only adsorbed salt phenomenon without reduced metal (4) for the same experimental solutions. This process implies that a complex electroless process happens.

XPS analysis and C(V) curves suggest that an heterogeneous surface is created with dispersed Pt(0) particles. At this time we are far from a well established procedure to obtain dispersed metallic site with an equal precision as the one obtained by electrodeposition process but the way is interesting because of the local electronic exchange it supposes.

References

1. P. H. L. Notten, J. E. A. M. Van Den Meerakker, J. J. Kelly. *Etching of III-V semiconductors* Elsevier(1991)
2. A. Heller, E. Aharon-Shalom, W.A. Bonner, B. Miller, *J. Am. Chem. Soc.*, **104**, 6942 (1982)
3. P. Allongue, E. Souteyrand, *J. Vac. Sci. Technol.*, **B5(6)**, 1644 (1987)
4. S. Manandhar, J.A. Kelber, *Electrochimica Acta,* **52,** 5010 (2007)
5. N. Simon, N.C. Quach, A.M. Gonçalves, A. Etcheberry, *J. Electrochem. Soc.* **154** H340 (2007)
6. A. Etcheberry, N.C. Quach, N. Simon, I. Gérard, *ESC Trans* **6(2***)* 469 (2007)

226

ECS Transactions, 19 (3) 227-234 (2009)
10.1149/1.3120704 ©The Electrochemical Society

Characterisation of Self-Assembled Monolayers on Germanium Surfaces via NEXAFS

M. Lommel [a], F. Reinhardt [b], M. Kolbe [b], B. Beckhoff [b], M. Müller [b], P. Hönicke [b], B.O. Kolbesen [a]

[a] Institute for Inorganic and Analytical Chemistry, Goethe University, Frankfurt / Main, Germany
[b] Physikalisch-Technische Bundesanstalt (PTB), Abbestraβe 2-12, Berlin, Germany

> The formation of self-assembled monolayers (SAMs) by suitable organic molecules with appropriate anchor groups on semiconductor surfaces may be used either to probe both the chemical state and the quality of the surface or to achieve surface passivation. Molecules with thiol anchor groups are able to bond to hydrogen-terminated germanium surfaces (Ge-S bond). We have prepared SAMs of alkylthiols with trifluoroacetate head groups on germanium. The germanium surface prior to and after SAMs formation has been characterised by Near-Edge X-Ray Absorption Fine Structure spectroscopy (NEXAFS) with synchrotron radiation in the PTB laboratory at BESSY II. We succeeded in assigning S-NEXAFS peaks to the Ge-S bond formed during self-assembly and were able to distinguish between different fluorine species via F-NEXAFS.

Introduction

Germanium has gained interest in recent years as a promising material for high-performance microelectronic devices. However, unlike silicon Ge does not form a robust passivation layer on its surface – germanium oxide is water-soluble [1]. Since the surface preparation of germanium is neither well understood nor developed, the controlled preparation of an oxide-free completely H-terminated surface which is a prerequisite for SAM formation of alkylthiols turned out to be a major challenge. Several approaches have been studied [2]. The best results for H-termination of germanium so far have been obtained with a 40 s dip in HF. The immediate immersion of the HF treated germanium into a 1-mmol solution of thiol leads to self-assembly of thiol monolayers. Grazing Incidence X-Ray Fluorescence (GIXRF) [3-6] measurements have shown the distance of the fluorine head groups from the surface to be 1.4 nm, which corresponds to a bonding angle of the Ge-S bond of about 47.5°, as published elsewhere [7]. The duration of exposure to the solution is a crucial factor as too long a treatment may lead to the formation of molecule clusters.

Experiment

All wafer pieces were treated the same way. A clean and untreated piece of a p-doped germanium (100) wafer was dipped in 50% HF for 40 s, blown dry with N_2 and immersed immediately in a 1mmol solution of 11-mercaptoundecyl-trifluoroacetate (99%, Asemblon, USA) (Fig.1) in water free dichloroethane (DCE). All work was performed in a glove box at room temperature to provide oxygen and water free conditions. After the

227

exposure to the thiol each piece was rinsed with DCE for 15 s and mounted onto the sample holder. To minimize contact to air and humidity, no sample was exposed to air for more than 60 s during the transfer from the glove box to the UHV irradiation chamber. All measurements were performed in the laboratory of the Physikalisch-Technische Bundesanstalt (PTB) at the electron storage ring BESSY II where two well-characterized beamlines, a plane grating monochromator (PGM) beamline for undulator radiation and a four-crystal monochromator (FCM) for bending magnet radiation, served as excitation sources for reference-free X-ray spectrometry at grazing incidence conditions [4-6, 8]. All samples were mounted vertically, i.e. perpendicular to the plane of the storage ring, thus taking advantage of minimal scattering probabilities for the exciting radiation when detecting element-specific fluorescence radiation in the same plane at an angle of 90° with respect to the propagation direction of the incident beam. NEXAFS, GIXRF and TXRF (Total-reflection X-Ray Fluorescence) measurements were performed under these conditions.

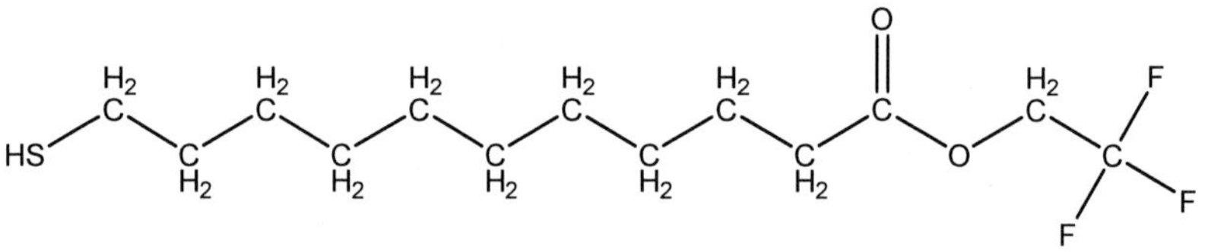

11-mercaptoundecyl-trifluoroacetate

Figure 1: The thiol (HS-) group is the "anchor group" with the sulphur bonding to the germanium. The long chain is the motor of self-assembly. The "head" group, here – $OCOCF_3$, facilitates detection of the monolayer. The distance of sulphur from fluorine in the molecule is 1.8 nm

NEXAFS Measurements of the Fluorine Head Group

The fluorine NEXAFS spectra of samples immersed for a period not longer than 1 h in a solution of 11-mercaptoundecyl-trifluoroacetate (Fig. 2) show – in addition to the main fluorine head group peak at 692 eV [Peak #5] – a small pre-peak at 685 eV [Peak #1] which disappears with longer immersion times.

This pre-peak could arise from the interaction of fluorine with the H-terminated germanium surface. It could also indicate the presence of residual HF from the preceding treatment, as the same peak at 685 eV [Peak #1, Fig. 4] is also present in spectra recorded immediately after the HF treatment. In reference-free TXRF measurements performed according to the method of Beckhoff et al. [8] the existence of HF on germanium surfaces treated as described earlier have been confirmed (Table I).

TABLE I: Amount of fluorine on germanium (reference free TXRF)

carried out preparation	amount of fluorine		maximum SAMs
	(ng/cm^2)	(atoms/cm^2)	coverage on Ge
after dipping in conc. HF for 40s	4.5	1.43×10^{14}	./.
after immersing in thiol solution for 1min	4.1	1.30×10^{14}	7%
after immersing in thiol solution for 1h	4.7	1.50×10^{14}	8%
after immersing in thiol solution for 1d	9.1	2.90×10^{14}	15%

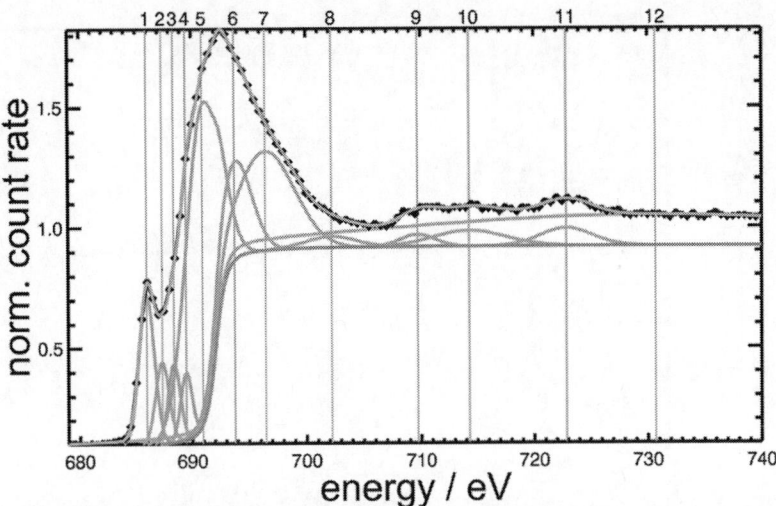

Fig. 2: A deconvolved fluorine NEXAFS spectrum of self-assembled monolayers of 11-mercaptoundecyl-trifluoroacetate on a germanium substrate after immersion in the thiol solution for 1 min.

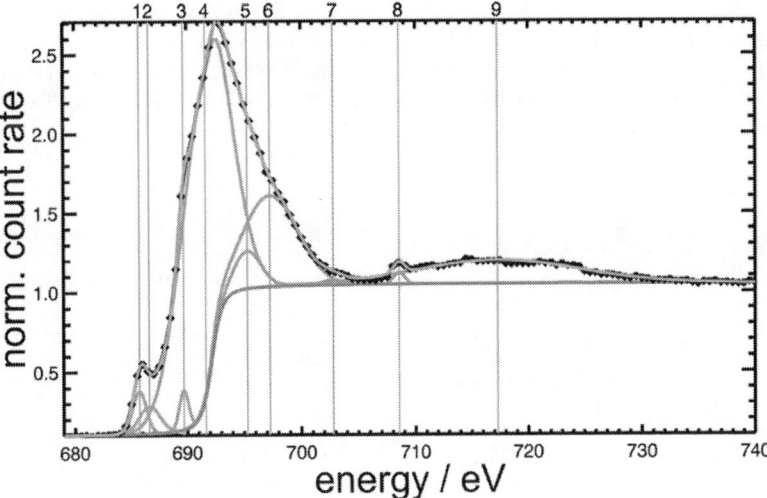

Fig. 3: A deconvolved fluorine NEXAFS spectrum of self-assembled monolayers of 11-mercaptoundecyl-trifluoroacetate on a germanium substrate, after immersion in the thiol solution for 1h.

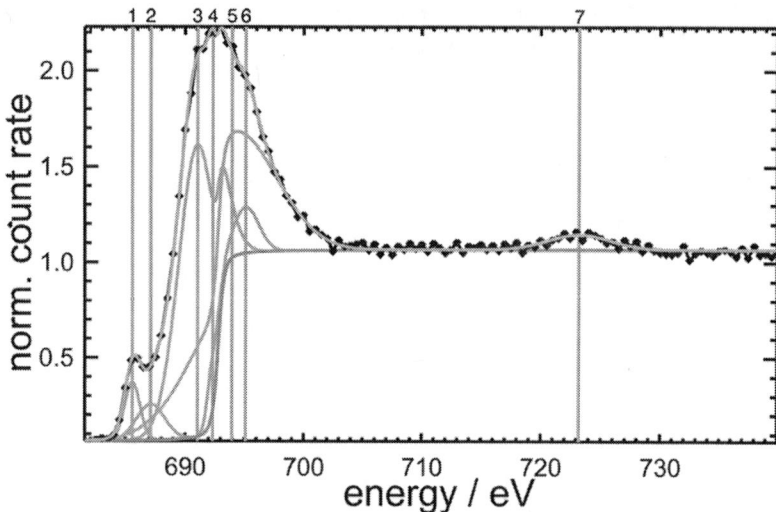

Fig. 4: A deconvolved fluorine NEXAFS spectrum of a germanium substrate dipped in HF (50%) for 40 s and then measured immediately

Molecules in the thiol solution first interact with the substrate by physical adsorption (Fig. 5) and "lie" on the substrate surface before self-assembly proceeds. At this stage, both the sulphur anchor group and the fluorine head group may interact with germanium, with which they are in close contact. With longer immersion times small groups of molecules form S-Ge bonds whereby the alkyl chain with the fluorine head group is drawn away from the surface.

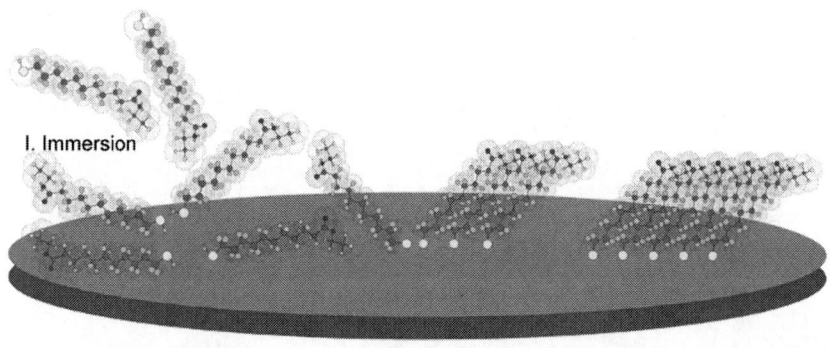

Fig. 5: Schematic sequence of self-assembly. After immersion (I.) the molecules are at first physically adsorbed (II). Chemisorption (III) commences with the formation of S-Ge bonds whereupon the carbon chain with the fluorine head group withdraws from the surface until the self-assembly is complete (IV.).

Nevertheless one has to take into account that especially the relative intensity of the pre-peaks with respect to the step function representing the absorption edge in both spectra (Fig. 2 and Fig. 4) differ significantly. The increase in intensity of the fluorine pre-peak at 685 eV [Peak #1, Fig. 2] may be ascribed to the presence of the alkyl thiol, as well as to residual HF which could still remain on the surface after the short immersion time. The gradual disappearance of this peak [Peak #1, Figs. 3 and 6] together with the increase in intensity of the main fluorine peak at 692 eV [Peak #4, Figs. 3 and 6] suggest that, with longer immersion times, the HF leaves the germanium surface and passes into the solution and secondly, the process of self assembly has reached the stage where the fluorine head group is no longer in contact with the germanium surface.

It is assumed in Table I that amount of fluorine determined by TXRF has to be assigned to both HF and the head group of the thiol molecule for the immersion times of 1min and 1h. As the sample dipped in HF is immersed in a polar solvent we suggest that the amount of HF present on the sample surface decreases with time as evidenced by the decrease in intensity of the pre-peak in the NEXAFS spectra [Peak #1 & #2, Figs. 2 & 3, respectively 4] which eventually disappears [Peak #1 & #2, Fig. 6] so that all the fluorine measured on the substrate should arise from "upright" thiol molecules. As three atoms of fluorine correspond to one molecule of mercaptoundecyl-trifluoroacetate 2.90×10^{14} atoms should represent one third of this sum in thiol molecules per cm². Nevertheless it is not straight-forward to deduce a SAMs coverage rate from these data. Therefore additional GIXRF measurements are to be performed for further evidence.

The choice of a fluorine head group was influenced by the fact that it was favourable, if not essential for some of the methods used to characterize monolayer formation. GIXRF [5] for instance would have been rather difficult without it. The difficulties involved are that the fluorine from HF used for the hydrogen termination of the surface interferes with the determination and that, under unfavourable conditions, trifluoroacetate is easily removed from the molecule during measurements.

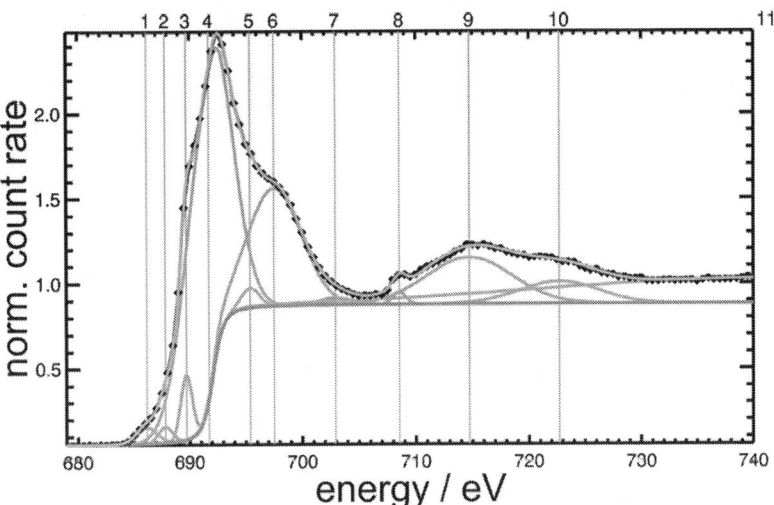

Fig. 6: A deconvolved fluorine NEXAFS spectrum of self-assembled monolayers of 11-mercaptoundecyl-trifluoroacetate on a germanium substrate, after immersion in the thiol solution for 1d.

NEXAFS Measurements of Sulfur Anchor Group

A second attempt to characterize the self-assembled monolayers was made with sulphur NEXAFS. Sulphur is the anchor group with which the molecule is bonded to the germanium. A NEXAFS of Ge did not prove useful as the Ge peak was that of the Ge substrate and therefore too large in comparison to the molecules on the surface for them to show up in the spectra.

The germanium was exposed to the thiol solution for one day (Fig. 7, dotted line). The NEXAFS spectrum shows two distinct groups of peaks: The peaks around 2470 eV correspond to a sulphur S^{II} and peaks around 2480 eV to an oxidised state of sulphur (S^{IV}), [9, 10]. These two states may be due to the preparation of the sample the conditions for which were not ideal. This split in the 11-mercaptoundecyl-trifluoroacetate peak at 2470 eV (2471 and 2473 eV) of approx. 2 eV is evidence for a change of the chemical surroundings of the sulphur. In comparison to this we measured as well the unbonded sulphur anchor group - H_2C-S-H (solid line), its peak is at 2473 eV as reported by Yagi et al [11]. Furthermore, we obtained a spectrum of a sulphur layer on germanium, the only bonds existing should be Ge-S, and this peak is at 2470.5 eV (dot and dash line). The split peak includes both peaks, which leads us to the conclusion that in the studied sample a carbon-sulphur-germanium (-H_2C-S-Ge) bond has been formed, indicating self - assembly.

To confirm this, further measurements have to be carried out, especially for very short exposures (one minute and one hour), as one would expect that with increasing exposure time, i.e. when more - H_2C-S-Ge bonds get formed, the peak at 2471 eV increases, too.

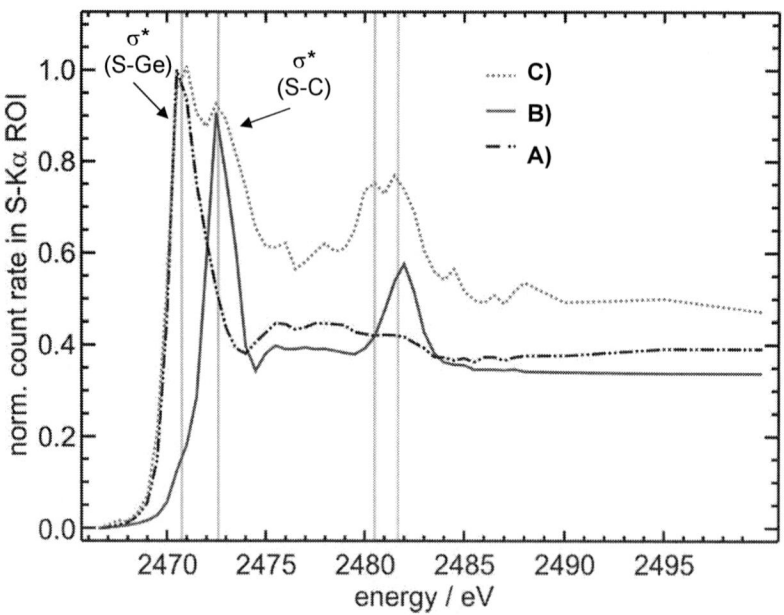

Fig. 7: Sulphur NEXAFS spectra of three different sulphur species. A) Sulphur layer on germanium (Ge-S); B) Sulphur in an alkylthiol not bonded to germanium (H-S-CH$_2$-R); C) self-assembled monolayers of 11-Mercaptoundecyl-trifluoroacetate on germanium substrate, exposure to the thiol: 1 day (Ge-S-CH$_2$-R)

AFM studies of the different stages of formation and growth of SAM islands did not provide similar clear evidence. AFM images (Fig. 8) show that reproducible results are not easy to achieve. Whether this is caused by measuring or preparation techniques has to be determined in further investigations.

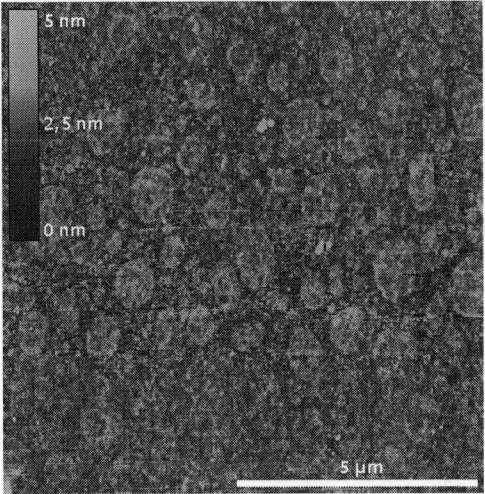

Fig. 8: AFM image: islands of self-assembled monolayers of 11-mercaptoundecyl-trifluoroacetate on a germanium substrate, after immersion in the thiol solution for 12h.

Conclusion

NEXAFS is a powerful tool for studying the chemical state of self-assembled monolayers with sulphur anchor groups and different head groups. In combination with other techniques such as GIXRF and reference-free SR-TXRF we have not only determined the amount of fluorine on the germanium substrate but were also able to distinguish between the fluorine of the HF treatment prior to self-assembly and the fluorine of the head group of the monolayer. Furthermore, the NEXAFS of sulphur and fluorine provide information concerning the progress of self-assembly.

Acknowledgments

We thank Umicore AG and M. Meuris and P. Mertens from IMEC for providing the germanium wafers, S. Sioncke and C. Fleischmann for the Ge-S NEXAFS results and Y. Filbrandt for helpful discussions around this publication. This work was supported by the European Commission - Research Infrastructure Action under the FP6 "European Integrated Activity of Excellence and Networking for Nano and Micro-Electronics Analysis" - Project number 026134(RI3) ANNA.

References

1. J. He, Z.-H. Lu, S. A. Mitchell, D. D. M. Wayner, J. Am. Chem. Soc. **120** (11) 2660 (1998)
2. M. Lommel, E. Mankel, B. O. Kolbesen; ECS Transactions **11** (20) 83-90 (2008)
3. M. Krämer, Dissertation, Dortmund (2007)
4. B. Beckhoff, J. Anal. At. Spectrom. **23**, 845 (2008)
5. B. Pollakowski, B. Beckhoff, F. Reinhardt, S. Braun, and P. Gawlitza, Phys. Rev. B **77**, 235408 (2008)
6. P. Hönicke, B. Beckhoff, M. Kolbe, S. List, T. Conard, H. Struyff, Spectrochim. Acta B **63**, 1359 (2008)
7. M. Lommel, P. Hönicke. M. Kolbe, M. Müller, F. Reinhardt, P. Möbus, E. Mankel, B. Beckhoff, B.O. Kolbesen; "Preparation and characterization of self-assembled monolayers on germanium surfaces"; *Solid State Phenomena,* **145-146,** 169 (2009)
8. B. Beckhoff, R. Fliegauf, M. Kolbe, M. Müller, J. Weser, G. Ulm, Anal. Chem., **79**, 7873 (2007)
9. Y. Takahashi, Y. Kanai, H. Kamioka, A. Ohta, H. Maruyama, Z. Song, H. Shimizu, Environ. Sci. Technol. **40**, 5052 (2006)
10. X. Liu, C.-H. Jang, F. Zheng, A. Jürgensen, J. D. Denlinger, K. A. Dickson, R. T. Raines, N. L. Abbott, F. J. Himpsel, Langmuir **22**, 7719 (2006)
11. S. Yagi, K. Matsumura, Y. Nakano, E. Ikenaga, S.A. Sardar, J.A. Syed, K. Soda, E. Hashimoto, K. Tanaka, M. Taniguchi; Nucl. Instr. and Meth. in Phys. Res. B **199** 244 (2003)

ECS Transactions, 19 (3) 235-243 (2009)
10.1149/1.3120705 ©The Electrochemical Society

**Electrochemical Stability of a Novel Inorganic Protective
and Functionalizable Monolayer onto InP**

O. El Ali[a], A.-M. Gonçalves[a], N. Mézailles[b], C. Mathieu[a],
P. Le Floch[b] and A. Etcheberry[a]

[a] Lavoisier Institute CNRS UMR 8180, Versailles 78035 Cedex- France.
[b] Laboratoire "Hétéroéléments et Coordination" École Polytechnique,
CNRS UMR 7653,- Palaiseau 91128 cedex - France

A new route of InP passivation was clearly ruled by an anodic electrochemical process in liquid ammonia as a relevant nonaqueous solvent. The coupling approach using both galvanostatic control and XPS analyses evidenced that this first protective film upon InP was definitely associated to one coating monolayer of general formula $[-(H_2N)P=N-]_n$. The electrochemical stability of the complete covering thin film was tested during hydrogen evolution in HCl aqueous solution. In spite of the high cathodic charge, unexpectedly no cathodic decomposition of the coated InP semiconductor was detected by XPS analyses. The electrical conductivity of this protective film and its electrochemical stability under negative overvoltage was then reported. In contrast to a bare InP, the electroless immersion of the protected InP in K_2PtCl_4 solution did not lead to the formation of metallic Pt cluster onto the substrate. On the contrary, the presence of amino groups in the film offered a unique opportunity for Pt (+II) grafting. A hybrid inorganic structure was proposed from the N/Pt atomic surface ratio, slightly higher than 2.0. The stability of the covered InP substrate as well as the ultra thin passivated film upon functionalization were also demonstrated from XPS analyses.

Introduction

To control the electrical surface properties and to make the semiconductor useful for electronic devices, surface passivation treatments were often required (1). For example, hydrogenation was a conventional wet method to passivate the defect states which could be detrimental to device performances. For instance, in the case of silicon, hydrogenated surface was however virtually immune against ageing effects. Even if the silicon surface had been properly hydrogenated, this extremely reactive semiconductor could be oxidized after being stored in air for several hours (2, 3). On III-V semiconductors, the surface chemistry was however more complicated than on silicon, and plain oxidation or hydrogenation of the surface could not lead to chemically stable defect-free surfaces, and efficient passivation was therefore required (4, 5). Concerning InP semiconductor, common wet methods to stabilize the surface chemistry involved inorganic sulfite with $(NH_4)_2S$ or Na_2S (6), organothiolated self-assembled monolayers (7, 8) or ruthenium deposition (9, 10). Even if an improvement of InP electrical properties was observed, the main drawback of these passivated treatments was the air surface ageing which led inescapably to an uncontrolled native oxides evolution.

235

We have already published that we are able to passivate electrochemically InP in acidic liquid ammonia (NH$_3$ *liq.*). This passivation was ascribed to an anodic wave (5-10 mC.cm^{-2}) obtained during the first scan by cyclic voltametry in acidic NH$_3$ *liq.* (11). The InP surface was indeed covered by a thin film involving " P-N " terminations. A specific " P-N " surface bond chemistry was therefore associated to a phosphinimidic amide-like electrodeposition involving both InP and solvent oxidation according the fellowing equation (11, 12):

$$InP + 9NH_3 + 9h^+ \rightarrow \text{« } H_2N\text{-}P\text{=}NH \text{ »} + 6NH_4^+ + In^{3+} + \frac{1}{2}N_2 \qquad [1]$$

Stability tests of the " P-N " film have been also performed. The passivated film remained unchanged even after ageing in air for one year at room temperature (11). The InP air immunity required a complete recovering of surface sites by the " P-N "film. As a result, InP active sites were not more available for native oxide evolutions on the surface. In this work, the thickness of the " P-N "film was resolved by the combined approach using galvanostatic control, *in-situ* capacity measurements and XPS analyses. The aim of this paper was also to determine the electrochemical stability of the " P-N " film as well at open circuit potential (V$_{OC}$) using an oxidizing solution (K$_2$PtCl$_4$) as under negative over-potential during hydrogen evolution. These results were systematically compared to those obtained onto an unprotected InP surface.

Experimental

n-InP wafers with a (100) orientation were purchased from InPact Electronic Materials, Ltd, with a doping density of 10^{18} cm^{-3}. The wafers were cut into small squares (≈ 0.1 cm^2). Prior to use, semiconductors were chemomechanically polished with a solution of bromine in methanol (2%), and rinsed with purest methanol and dried under an argon stream (13, 14). In order to remove residual oxides, samples were dipped for a few minutes in 2 M HCl just before the experiment (15, 16).

The electrochemical set-up was a classical three-electrode device, using a 283 EG&G potentiostat-galvanostat with a linear potential scan of 20 mV/s. Ammonia condensation, from gaseous ammonia ('electronic grade' from Air Liquide), was provided by a glass column assembly and required a low operating temperature under atmospheric pressure (17). An electrochemical cell filed up with 150 cm^3 of NH$_3$ *liq.* was maintained at 223 K in a cryostat. The acidic medium was obtained by addition of 0.1M NH$_4$Br (purest available quality from Aldrich). All potentials were measured *vs.* a silver reference electrode (SRE) (18). The anodic galvanostatic control was performed at 10 µA.cm^{-2} under illumination using an optical fiber with a tungsten lamp. In-situ capacitance measurements *vs.* potential, C(V), were used to probe the surface chemical evolution during the anodic galvanostatic control. C(V) were carried out in the dark after each galvanostatic control, in NH$_3$ *liq.*, with a linear potential scan of 20 mV/s. The sinusoidal modulation frequency was set at 11kHz and the peak-to-peak amplitude at 10 mV. After the anodic treatment in NH$_3$ *liq.*, InP was rinsed at open circuit in purest liquid ammonia (without conducting salt). The sample was then transferred toward an XPS analyzer using a specific procedure avoiding any air contamination.

The chemical stability of the passivated film, was already established as well in the open air as in HCl (2M) aqueous solution (11). The film stability allowed the sample transfer in aqueous solution. In 0.1M HCl aqueous solution, a galvanostatic control at -100 μA.cm^{-2}, was used to test the electrochemical stability of the coated film against hydrogen evolution. C(V) were carried out in the dark after each galvanostatic control with a linear potential scan of 20 mV/s. The potential was measured against a saturated calomel electrode (*ECS*). The sinusoidal modulation frequency was set at 1k Hz and the peak-to-peak amplitude at 10 mV. After a cumulated charge of 0.3 C.cm^{-2}, the sample was rinsed at open circuit in purest water and transferred to the XPS analyzer. The electrochemical stability of the coated film was also tested at open circuit in contact with an oxidized solution of potassium tetrachloroplatinate [K$_2$PtCl$_4$] in a desaerated neutral aqueous medium. The chemical evolution of the surface was also examined by XPS.

Results

" P—N " film: a coated monolayer

In acidic liquid ammonia a galvanostic treatment at 10μA.cm^{-2} was applied onto a freshly deoxidized *n*-InP. The evolution of the interfacial potential was reported on the figure 1.

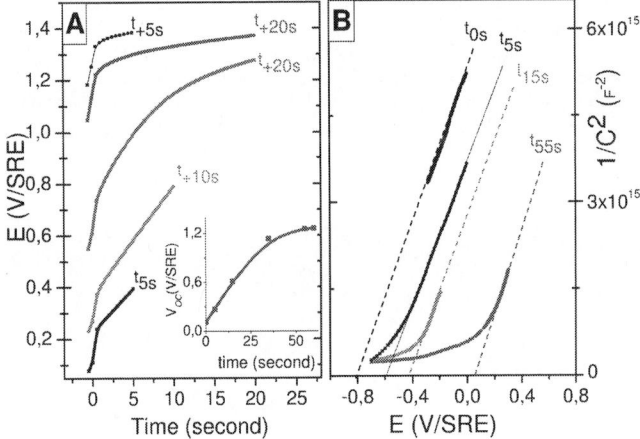

Figure 1. Sequential galvanostatic control at 10μA.cm^{-2} in acidic liquid ammonia (223K) for a total duration of 60 seconds (Q $\approx$ 0.6 mC.cm^{-2}).
A- Resulting *in-situ* evolution of the interfacial potential during these sequential galvanostatic treatments. Inset: corresponding *in-situ* evolution of the open circuit portential (V$_{oc}$).
B- Resulting *in-situ* evolution of the Mott-Shottky plots, f = 11kHz.

Before the galvanostatic treatment, the open circuit potential (V_{oc}) of n-InP in acidic NH$_3$ *liq.* stabilized at +0.1 V *vs.* SRE. After each galvanostatic treatment, the V_{oc} gradually increased from +0.1 V to +1.3 V *vs.* SRE (see figure 1A and inset). After a total duration of 60 s, an anodic charge lower than 1mC.cm^{-2} (Q $\approx$ 0.6 mC.cm^{-2}) went through the InP surface but a potential variation higher than one volt was involved at the interface. After each galvanostatic treatment, the interfacial modification was widely followed by the Mott-Shottky plots evolution (see figure 1B). Initially, without galvanostatic control, the flat band potential of the bare InP surface was determined from the Mott-Shottky plot. Right from the start of the galvanostatic control (t =5s), the Mott-Shottky plots drastically changed (Fig. 1B). An infinite capacitance involved in almost 700 mV was observed at the end of the galvanostatic treatment (Q $\approx$ 0.6 mC.cm^{-2}). This phenomenom was related to a classical flat band potential shift since the same Mott-Shottky plot, keeping nearly the same slope, was recovered after a positive over-voltage of 700 mV.

After the galvanostatic control, InP electrode was removed from NH$_3$ *liq.* toward XPS analyzer using a transfer procedure avoiding any air contamination. In comparison to a bare surface, atomic surface ratios were strongly modified since the ratio (In/P)$_{total}$ decreased from 1.1 to close to 0.6. This ratio (1.1 to 1.2) was typically observed onto a deoxidized surface of InP <100> (15, 16). Consequently, the ratio between the new energy phosphor contribution (P$_{133ev}$) and the one issued from the matrix (P$_{matrix}$) strongly increased to a value close to 0.8. This result was crucial since the phosphorus enrichment due to the P$_{133ev}$ contribution was undoubtedly related to the galvanostatic control in NH$_3$ *liq.* As a constant result, the ratio between the corrected areas of N$_{1s}$ and P$_{133\,eV}$ was always close to 2 (see figure. 2). It was noteworthy that these atomic ratios had been already reported during the electrochemical passivation of InP in acidic NH$_3$ *liq.* This passivation had been however ascribed to a larger anodic charge (5-10 mC.cm^{-2}) obtained during the first scan by cyclic voltametry in acidic NH$_3$ *liq.* (11). As a consequence the total indium contribution was only related to the InP response, never to the passivating film. The nitrogen chemical contribution on the passivating film was clearly evidenced by XPS.

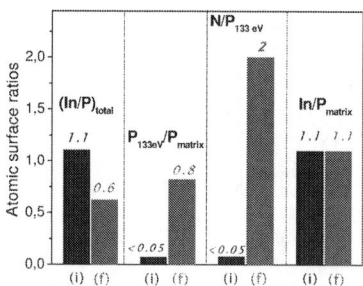

Figure 2. Evolution of atomic surface ratios from XPS spectra surface (In$_{3d}$, P$_{2p}$ and N$_{1s}$) obtained before (I) and after (II) the galvanostatic control at 10 μA.cm^{-2} in acidic NH$_3$ *liq.* on n-InP at T = 223 K, [NH$_4$Br] = 0.1 M.

Despite the lower anodic charge involved during the galvanostatic control ($Q \approx 0.6$ mC.cm^{-2}), the same chemical composition of the surface was obtained after the anodic wave by cyclic votammetry (11). The anodic wave required an anodic charges nearly ten times higher than those involved in the galvanostatic control. Since no chemical deviation was detected from the XPS analyses, we can conclude that an anodic charge around 1 mC.cm^{-2} was enough to passivate entirely InP sample. We could assume that the large excess of the anodic charge, observed by cyclic voltammetry, might result from parasitic electrochemical reaction such as ammonia oxidation. To confirm our assumption the anodic charge was reduced ($Q \approx 0.3$ mC.cm^{-2}). This charge was obtained by the same galvanostatic control at $10\,\mu$A.cm^{-2} but for a duration divided by two ($t \approx 30$ s). After the galvanostatic control at lower anodic charge ($Q \approx 0.3$ mC.cm^{-2}), InP electrode was proprely removed from NH$_3$ *liq.* toward XPS analyzer (see figure 3)

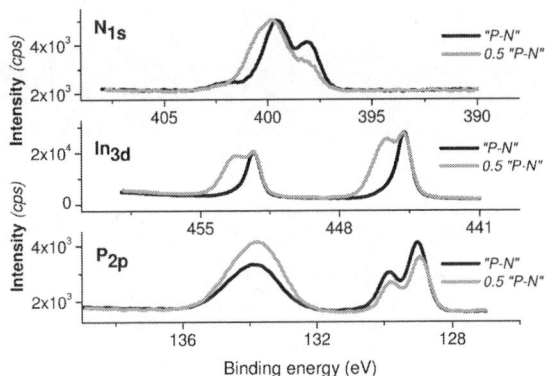

Figure 3. Evolution of XPS spectra surface (N$_{1s}$, In$_{3d}$ and P$_{2p}$) obtained in acidic NH$_3$ *liq.* on *n*-InP at T = 223 K, [NH$_4$Br] = 0.1 M. A complete covering film (black line) and partial covering film (grey line) were provided by the same galvanostic control at $10\,\mu$A.cm^{-2} but the partial covering film required a time duration divided by two. Both samples were removed from liquid ammonia without being sheltered from air.

As reported in the figure 3, in contrast to the complete covering film, a widening of the indium peak (In $_{3d}$) was clearly observed for a lower anodic charge. The enlargement at higher energy for both indium peaks can suggest the oxidization of the sample. This assumption was confirmed by both oxidation increase of the P$_{2p}$ and N$_{1s}$ components. As we could expected, from the lower anodic charge ($<<$ 1mC.cm^{-2}), the InP surface was partially covered by the passivated film. The surface was therefore not entirely protected by the " P-N " film. The sample oxidation from air, was obviously the direct consequence of the partial protection of the InP surface.

Assuming that a charge around 1 mC.cm^{-2} was involved for the formation of the passivating film according a 9 charges mechanism (see equation [1]), we concluded that one monolayer of " P-N "termination covered entirely InP surface.

" P—N " film: electrochemical stability against H_2 evolution

The electrochemical stability of the monolayer " P–N " film was tested during the hydrogen evolution in HCl (0.1M) aqueous solution under a galvanostatic mode. A galvanostatic control at -100 µA.cm^{-2} was used to test the electrochemical stability of the coated film against hydrogen evolution. Mott-Shottky plots were carried out in the dark after each galvanostatic control. These plots were compared to those obtained onto a bare surface in HCl (0.1M) aqueous solution (see figure 4).

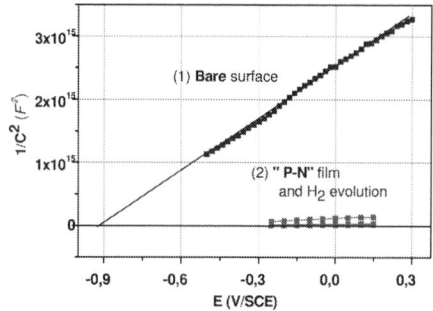

Figure 4. Comparison of the Mott-Shottky plots, f = 11kHz, in HCl (0.1M) aqueous solution.
(1) Onto a bare surface (unprotected surface, without anodic treatment in NH$_3$ *liq.*)
(2) Onto a " P-N " film (after anodic treatment in NH$_3$ *liq.*). The Mott-Shottky plots were nearly superimposed even after a cathodic charge around 0.3 C.cm^{-2}.

Like in NH$_3$ *liq.* (fig 1.B), an infinite capacitance was observed onto a " P-N " film in aqueous solution (Fig 4.(2)). In spite of the high cathodic charge ($\approx$ 0.3 C.cm^{-2}), the initial Mott-Shottky plots (bare InP surface) was not recovered (Fig 4.(1)). The hydrogen evolution occured through the " P-N " film without apparently altering it. The electrical conductivity property of this protective film was then reported for the first time.

After a cumulated charge of nearly 0.3 C.cm^{-2}, the sample was rinsed at open circuit in purest water and transferred to the XPS analyzer (see figure 5). In spite of the high cathodic charge, unexpectedly no cathodic decomposition of the coated InP semiconductor was detected by XPS analyses. Nevertheless, we could note a slight oxidation of the P$_{2p}$ component. This thin oxidation was probably related to the modification of the N$_{1s}$ component at low energy. Even if a chemical transformation of the coated film can be however assumed, its protective power was still efficient since no enlargement at higher energy for both indium peaks was detected.

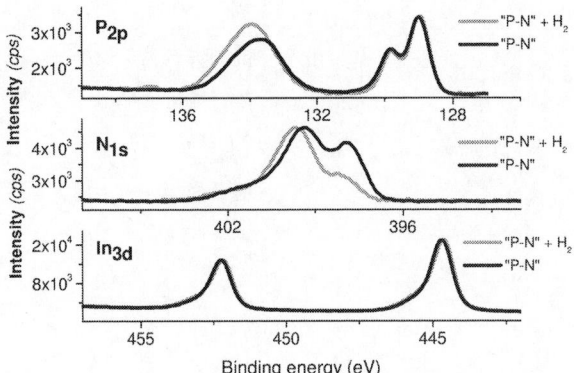

Figure 5. The XPS spectra surface (P_{2p}, N_{1s} and In_{3d}) of a complete covering" P-N "film obtained onto n-InP in acidic NH_3 *liq.* at T = 223 K, by anodic treatment. These spectra were compared to those obtained with " P-N "film sample after hydrogen evolution by cathodic galvanostatic treatment (≈ 0.3 C.cm^{-2}) in (0.1M) HCl aqueous solution.

" P—N " film: electrochemical stability against an oxidized solution at open circuit

The electrochemical stability of the monolayer " P–N " film was also tested during the electroless immersion of the coated InP in an oxidized solution of potassium tetrachloroplatinate [K_2PtCl_4]. The neutral aqueous solution was properly desaerated at open circuit. After few hours of immersion, the coated sample was rinsed by purest water and the chemical evolution of the surface was again examined by XPS. In contrast to a bare surface, no oxidation of the surface was observed in spite of the fact that the " P-N " film was a long time in contact with an oxidized solution. As a result, in contrast to an unprotected InP surface, no platinum electrodeposition was therefore detected on the " P-N " film. Furthermore, it was also well established that platinum grafting on surfaces was easily revealed by XPS analyses even at low coverage. Moreover, the different oxidation states of platinum (+0, +II, +IV) were defined by well separated chemical shifts (19-22). The presence of amino groups in the film, trough the -P-N-P- chains, offered a unique opportunity for Pt (+II) grafting. The Pt (+II) coordination sphere was therefore maintained upon the " P-N " film. From the N/Pt atomic surface ratio, slightly higher than 2.0, an idealized view of this designed hybrid architecture upon the monolayr " P-N " film onto InP matrix was proposed in figure 6.

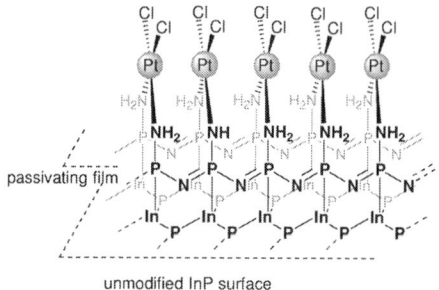

unmodified InP surface

Figure 6. Proposed structure for the hybrid obtained after grafting of [PtCl₂] fragments through reaction with a solution of [K₂PtCl₄] upon the protected " P-N "film at open circuit.

Conclusion

In this work, the mono-layer " P-N "film was attested from combined techniques using galvanostatic control, in-situ capacity measurements and XPS analyses. An anodic charge around 1 mC.cm^{-2} was then enough to passivate entirely InP sample in acidic liquid ammonia. We thus report herein the electrochemical stability of the " P-N " film as well at open circuit potential (V_{OC}) using an oxidizing solution (K₂PtCl₄) as under negative over-potential during hydrogen evolution in 0.1M HCl aqueous solution. These results were systematically compared to those obtained onto an unprotected InP surface. Whereas the bare InP surface was drastically damaged, the protective power of the " P-N " film was clearly established by the systematic lack of enlargement at higher energy for both indium peaks. We thus report herein the first chemical method to control, at a molecular scale, the functionalization of these amino groups upon the protective film. New opportunities for further high-speed optoelectronic applications are offered from these new functional molecular grafting. Physical characterizations by Angular Resolution X-ray Photoelectron Spectroscopy analyses (ARXPS) are now in progress.

References

1. R. Cohen, G. Ashkenasy, A. Shanzer, D. Cahen, Encyclopedia of Electrochemistry, (Eds: A. J. Bard, M. Stratmann), Vol. 6 Semiconductor Electrodes and Photoelectrochemistry, (Eds: S. Licht), WILEY-VCH Verlag GmbH, Weinheim, Germany, 2002.
2. N. Sieber, B. F. Manter, Th. Seyller, J. Ristein, Ley, *Diamond and Related Materials.*, **10**, 1291 (2001).
3. G. S. Higashi, Y. J. Chabal, G. W. Trucks, K. Raghavachari, *Appl. Phys. Lett.*, **56**, 656 (1990).

4. M. C. Traub, J. S. Biteen, B. S. Brunschwig, N. S. Lewis, *J. Chem. Soc.*, **130**, 955 (2008).
5. A. Shaporenko, K. Adlkofer, L. S. O. Johansson, M. Tanaka, M. Sharnikov, *Langmuir*. **2003**, *19*, 4992.Y. H. Jeong, S. K. Jo, B. H. Lee, T. Sugano, IEEE Electron Device Letters. 16 (1995) 109.
6. Y. H. Jeong, S. K. Jo, B. H. Lee, T. Sugano, *IEEE Electron Device Lett.* **1995**, *16*, 109.
7. M. Schvartzman, V. Sidorov, D. Ritter, Y. Paz, J. Vac. Sci. Technol. 21 (2003) 148.
8. M. Schvartzman, V. Sidorov, D. Ritter, Y. Paz, *Semicond. Sci. Technol.* **2001**, *16*, L68.
9. D. N. Bose, Y. Ramprakash, S. Basu, Mater. Lett. 88 (1989) 364.
10. D. N. Bose, J. N. Roy, S. Basu, Mater. Lett. 2 (1984) 455.
11. O. Seitz, C. Mathieu, A-M Gonçalves, M. Herlem, A. Etcheberry, J. Electrochem. Soc. 2003, 150, E461.
12. A-M. Gonçalves, C.Mathieu, M. Herlem, A. Etcheberry, Electrochem. Acta. 2001, 46, 2835.
13. Y. Robach, A. Gagnaire, J. Joseph, E. Bergignat, G. Hollinger, Thin Solid Films. 62 (1988) 81.
14. I. Gérard, N. Simon, A. Etcheberry, Appl. Surf. Sc. 175 (2001) 734.
15. N. Simon, N.C. Quach, A-M. Gonçalves, A. Etcheberry, J. Electrochem. Soc. 154 (2007) H340.
16. N. C. Quach, I. Gérard, N. Simon, A. Etcheberry, Phys. Stat. Sol (c). 3 (2003) 1033.
17. J. Jander, in Anorganische und allgemeine Chemie in flüssigen Ammoniak. Part I, Friedr. (Eds. Vieweg & Sohn), Braunschweig, 1966.
18. D. Guyomard, M. Herlem, C. Mathieu, J. L. Sculfort, J. Electroanal Chem. 216 (1987) 101.
19. T. F. Nolan, A. A. Gewirth, *Phys. Chem.*, **6**, (2004) 1310.
20. M. Manolova, V. Ivanova, D.M. Kolb, H. G. Boyen, P. Ziemann, M. Büttner, A. Romanyuk, P. Oelhafen, *Surf. Sci.*, **590**, (2005) 146.
21. C. K. Pal, S. Chattopadhyay, C. Sinha, A. Chakravorty, *Inorg. Chem.*, **33**, (1994) 6140.
22. S. Adachi, H. Kawaguchi, *J. Electrochem. Soc.*, **128**, (1981) 1342.

ECS Transactions, 19 (3) 245-272 (2009)
10.1149/1.3120706 ©The Electrochemical Society

Optimization of PbSe Nanofilms formation by Electrochemical Atomic Layer Deposition (ALD)

Dhego O Banga[+], Youn-Geun Kim[+], Stephen Cox[++], Uwe Happek[++], and John L Stickney[+]*

[+]Department of Chemistry
[++]Department of Physics and Astronomy
University of Georgia, Athens, GA 30602-2556, USA

Optimization studies of lead selenide (PbSe) nanofilm formation using electrochemical Atomic Layer Deposition (ALD) are reported here. IV-VI compounds semiconductors, such as the lead chalcogenides (PbSe, PbTe and PbS), have narrow band gaps and crystallize in the cubic rock salt structure. They are of interest for their optical and electronic properties, which make them useful in thermoelectric device structures, infrared sensors, and photovoltaics. PbSe has the narrowest band gap of the lead chalcogenides, 0.26 eV at room temperature. PbSe deposits were formed using an ALD cycle on Au substrates, one atomic layer at a time, from separate solutions, containing Pb or Se ions. Single atomic layers were formed using surface limited reactions, referred to as underpotential deposition (UPD). UPD is a phenomenon where an atomic layer of one element deposits on a second at a potential prior to that needed to form that element in a bulk form. The deposition cycle consisted of the alternated UPD of Se and Pb. In the optimized process, Se and Pb deposition potentials were ramped up from -0.3 V and -0.25 V respectively to -0.04 V over the first 20 cycles, and then held constant for the remaining cycles. Coverages near one monolayer in each cycle were indicated by the coulometry for the reduction of both Se and Pb. Electron probe microanalysis (EPMA) indicated a uniform and stoichiometric deposit, with a Se/Pb ratio of 1.02. X-ray diffraction patterns showed peaks matching the rock salt structure, with a preferential (200) orientation for the as formed deposits. Infrared reflection absorption measurements of the PbSe films formed with 100, 200, 250 and 300 cycles indicated strong quantum confinement. STM suggested conformal film growth on the Au on glass substrate.

Introduction

The interest in lead chalcogenides (PbS, PbSe and PbTe) and their alloys stems from their importance in applications, in crystalline and polycrystalline forms, as detectors of IR radiation, IR resistors, thermoelectric devices and more recently as IR emitters and solar control coatings(1). Lead chalcogenides exhibit quite unique properties relative to other semiconductors such as the positive temperature coefficients (dE_g/dT) of the minimum energy gap E_g, the narrow direct band gap which decreases with hydrostatic pressure and increases with temperature and the large dielectric constants(1). IV-VI compound semiconductors have small and relatively equal electron and hole masses compared to III-V or II-VI compound semiconductors which result in larger confinement effects for nanostructures, that is they have a large Bohr radius(2, 3). The Bohr radius of PbSe is ~ 46 nm, suggesting that strong quantum confinement effects should still be evident when confining dimensions are as large 50 nm (4, 5). In the present report, very strong band gap shifts, from 0.26 eV, are reported for nanofilms of PbSe.

Other methods used to grow PbSe thin films have included molecular beam epitaxy (6-8), chemical vapor deposition (9), sonochemistry (10), sonoelectrochemistry (11), microwave heating (12), codeposition (4, 13-20) and electrochemical atomic layer deposition (ALD) (3) also known as electrochemical atomic layer epitaxy (EC-ALE) (21).

Extensive work on the growth of compound semiconductor and metal nanofilms have been carried out by the Author's group and others using electrochemical atomic layer deposition (ALD) (22-30, 31-36). Electrochemical ALD makes use of underpotential deposition (UPD) (37-40), a phenomenon where an atomic layer of one element deposits on a second, at a potential prior to that needed to form the element in bulk form. The result is a surface limited reaction: deposition limited by the surface area of the deposit. The basis of the electrochemical ALD consists of forming compound monolayers using UPD in a sequence, referred to here as a cycle, which can be repeated as many times as needed to form more compound monolayers. In ALD, materials can be grown layer by layer with atomic level control (21).

Electrochemical ALD has been used successfully to grow various compounds such as CdTe (41, 42), CdS (22, 43), ZnSe (44), ZnS (43), CdS / HgS superlattices (45), PbS (46), PbSe (3), PbTe (47), PbSe / PbTe superlattices (32), GaAs (48, 49), InAs (50-52), InSb (51) and superlattices of InAs / InSb (51). In this report, which is an extension of previous work on PbSe deposition by this group(3), attempts to optimize the formation PbSe nanofilms on Au on glass and Au on mica substrates will be presented. The first new cycles involved reductive deposition of both Se and Pb at constant potentials for the duration of the deposition process. The second new ALD cycles involved the inclusion of a Se reductive stripping step, at a controlled potential to strip off excess Se, while both Se and Pb deposition potentials were maintained constant throughout the deposition process. The last new ALD cycle involved ramping positively both Se and Pb potentials over the first 20 cycles, before holding them constant for the remaining cycles.

Experimental

An automated thin layer flow electrodeposition system described previously (Electrochemical ALD L.C.)(42, 47, 53) consisting of a series of solution reservoirs, computer controlled pumps, valves, a thin layer electrochemical flow cell and a potentiostat was used in this study to grow thin films of PbSe. Minor changes were made to the reference electrode compartment and the auxiliary electrode. A simple O-ring was used to hold the reference electrode instead of the Teflon compression fitting, providing a better seal, and a Au wire embedded into the Plexiglas flow cell was used as the auxiliary electrode, instead of the ITO glass slide, which has proven more durable. The electrochemical flow cell was designed to promote laminar flow. The working electrode was Au on glass, the auxiliary was a Au wire, and the reference electrode was Ag/AgCl (3M NaCl) (Bioanalytical systems, Inc., West Lafayette, IN). The cell volume was about 0.3 mL and solutions were pumped at 50 mL min^{-1}. The system was contained within a nitrogen purged Plexiglas box to reduce the influence of oxygen during electrodeposition.

Substrates used in the study consisted of 300 nm thick vapor deposited gold on glass microscope slides. Prior to insertion into the vapor deposition chamber, the glass slides were cleaned in HF, and rinsed with copious amounts of deionized water from a Nanopure water filtration system (Barnstead, Dubuque, IA), fed from the house distilled water system. Deposition began with 3 nm of Ti, as an adhesion layer, followed by 300 nm of Au deposition at around 250 °C. The glass slides were then annealed in the deposition chamber at 400 °C for 12 hours, while the pressure was maintained at 10^{-6} Torr.

Reagent grade or better chemicals and deionized water were used to prepare the solutions. Three different solutions were used in this study. The lead solution consisted of 0.5 mM Pb(ClO$_4$)$_2$ (Alfa Aesar, Ward Hill, MA), pH 3, with 0.1 M NaClO$_4$ (Fischer Scientific, Pittsburgh, PA) as the supporting electrolyte. The selenium solution was made of 0.5 mM SeO$_2$ (Alfa Aesar, Ward Hill, MA), pH 5, buffered with 50.0 mM sodium acetate and 0.1 M NaClO$_4$ (Fischer Scientific, Pittsburgh, PA) as the supporting electrolyte. A pH 3 blank solution of 0.1 M NaClO$_4$ (Fischer Scientific, Pittsburgh, PA) was used as well. NaOH and HClO$_4$ (Fischer Scientific, Pittsburgh, PA) were used to adjust the pH values of all the solutions in this study.

The electrochemical ALD cycle program used to grow PbSe consisted of the following: the Se solution was flushed in to the cell for 2 s at the chosen potential, and then held quiescent for 15 s for deposition. The blank solution was then flushed through the cell at the same potential for 3 s, to rinse out excess HSeO$_3^-$ ions. The Pb solution was then flushed through the cell for 2 s at the Pb deposition potential, and held quiescent for 15 s for deposition. Blank solution was again flushed through the cell for 3 s to rinse out excess Pb^{2+} ions from the cell. This process was referred to as one cycle, and was repeated 100 times to form each of the deposits described in this paper.

Microanalyses (EPMA) of the samples were performed with a Joel JXA-8600 electron probe. Glancing angle X-ray diffraction patterns were obtained on a Scintag PAD V diffractometer, equipped with a 6" long set of Soller slits on the detector to improve resolution in the asymmetric diffraction configuration. Absorption measurements were performed using a variable angle reflection rig in conjunction with a Bruker 66v FTIR spectrometer equipped with a Si detector. STM measurements were taken with a Princeton Research Instruments Nanoscope.

Results and Discussion

Figures 1 and 2 display the cyclic voltammograms for the deposition of Se and Pb on the Au on glass substrates; which were used to determine initial deposition potentials for the ALD cycle. Many studies of the electrochemical deposition of both Se and Pb have been carried out by this and other groups(3, 24, 32, 44, 47, 54-62). From previous studies by this group, Se deposition is known to be kinetically slow (57, 59, 60). Four CVs of Au in the Se solution, each to an increasingly lower potential and each to just past a different reduction feature, are shown in Figure 1. Distinct reduction features were observed at 0.6 V, 0.1 V, and -0.25 V, with a sharp increase in reduction current at -0.45 V, corresponding Se reduction to selenide and hydrogen evolution (Fig. 1(b)). The first reductive feature, around 0.6 V, was reduction of small amounts of Au oxide, formed at the most positive potentials. Previous studies of Se deposition (57, 59, 60) showed that selenite reduction did not result in deposition at an underpotential, as suggested by the term UPD. However, if UPD is considered to denote formation of a surface limited amount, then the reduction peak at -0.05 V (an overpotential) could be considered UPD. The peak occurs at an overpotential because of the slow kinetics for $HSeO_3^-$ reduction. The reductive features (Figure 1a) at -0.05 V display all the characteristics of surface limited deposition, except that it occurs at an overpotential, since the formal potential for $HSeO_3^-$/Se appears to be 0.4 V. When the scan was reversed at -0.3 V (Figure 1a), the only oxidative feature occurs at around 0.6 V, positive of the formal potential, and as the more negative the scan (-0.025 V to -0.3 V), the larger the 0.6 V peaks becomes. It is felt by these authors that all this reduction (to -0.3) is associated with surface limited deposition (Figure 1a). When the scan was reversed at much more negative potentials as (Figure 1b), a new oxidative feature was evident, just positive of 0.4 V, apparently the oxidation of bulk Se. It is known from other studies that near -0.5 V, a different process begins, reduction of Se to selenide (Se^{2-}), a soluble species which diffuses away. So there are really three processes contributing to the increase in reduction near -0.5 V, bulk Se formation, selenide formation, and hydrogen evolution. The clear difference between bulk Se and UPD, evident in Figure 1, prompted use of -0.3 V for Se atomic layer deposition, which is well positive of the potential corresponding to bulk Se formation (-0.45 V). Some bulk Se will still deposit at -0.3 V, given that the formal potential is 0.4 V, but the kinetics appear slow enough that if little time is given for deposition, some small fraction of a monolayer of bulk at most will form.

Figure 2 depicts Au CVs in the Pb^{2+} solution. A slowly increasing reduction current was observed as the negative going scan (started at 0.25 V) progressed, then producing a characteristic Pb^{2+} reduction peak (reversible) at -0.25 V, reminiscent of Pb UPD on a Au(111) electrode (58, 63-68). The sharp increase in current negative of -0.45 V, and displaying hysteresis suggesting nucleation and growth, corresponds to bulk Pb deposition. In the corresponding positive going scan, there was a small sharp peak for bulk Pb oxidation at -0.47 V, as well as a large broad peak around -0.38 V, suggesting oxidation of Pb from a Pb/Au alloy (69). Based on the features in Figure 2, -0.25 V was selected for use as the starting potential for Pb atomic layer formation in the ALD cycle.

From the CVs for Se and Pb, the ALD cycle shown in Figure 3 was devised. The initial step was filling the cell with the $HSeO_3^+$ solution for 2 s, followed by depositing Se for 15 s with no flow and then rinsing the cell with blank for 3 s, all at -0.3 V. This was followed by filling for 2 s with the Pb^{2+} solution, depositing Pb for 15 s with no flow, and then rinsing the cell for 3 s, all at -0.25 V. Each cycle was intended to form a compound monolayer or single compound bilayers of PbSe on (2 x 2) cm^2 electrode surface.

Figure 4 displays the charges in monolayers (ML), where a monolayer is defined as one depositing atom for each surface atom, or 1.35×10^{15} atoms/cm^2, assuming the geometric area of the substrate is composed of Au(111), hexagonal surface of fcc Au. The substrate will not be completely (111), and there will be some roughening relative to an atomically flat surface. However, such numbers give a relative idea of the amounts of depositing elements. During the first cycle, the Pb deposition was much larger than expected. Previous studies of the II-VI compound CdTe suggested that after deposition of a chalcogenides layer on Au, the first metal deposition results in deposition both on top of the chalcogenide and below it, as the Cd has a strong affinity for the Au surface. If such a process is at work here, after one cycle of deposition (Figure 3), there should be an atomic layer of Pb on the Au, then Se, followed by another layer of Pb.

It is notable that the coverages for the first 3 cycles are below 1 ML. On the other hand, the coverages increase dramatically up to 2 ML, about twice the amount needed to form one compound bi-layer. It has been observed by this group that more than the amount needed to form a compound bi-layer is generally too much, as excess deposition suggests bulk formation, and it is not clear if bulk can be converted quantitatively into the desired compound. Thus when more than that needed to form the bi-layer is deposited; it is probable that surface roughening will result. From Figure 4, it is clear that too much is being deposited, although it starts out reasonably, over the first 20 cycles the coverages increase to about 2 times too much. This particular trend was also observed for the formation of PbTe using electrochemical ALD (47).

An explanation suggested for these changes (3, 70, 71) was that there was a drop of some of the applied potential across a growing space charge layer (SCL), or Schottky barrier, between the Au electrode and the growing semiconductor thin film. It seems, however, unrealistic that a space charge layer fully developed after only 20 ALD cycles

(72). Reasons for the changes in coverage as the cycle number increased are not yet understood, though it may have to do with deposition and growth kinetics, as similar behavior is frequently observed for gas phase ALD studies (73-75). Deposits formed under the conditions described above appeared overgrown and roughened, when inspected using optical microscope, as expected based on the excess deposited amounts. Given the voltammetry shown in Figures 1 and 2, the chosen potentials should have worked. However, such choices were based on deposition on Au surfaces, not on deposition on the compound.

Se electrodeposition is generally kinetically slow (57, 59, 60). The kinetics for PbSe might be as well, beginning slowly, and increasing as more facets of the compound are formed, possibly explaining the overgrowth observed above. Au on glass or mica substrates , such as used in this experiment, are polycrystalline surface with steps and defects on and between 50 nm clusters, with only small Au (111) terraces; as shown in Figure 5. The energetics for deposition on such defect sites and terraces may be slower then for deposition on atomically flat Au(111) planes. Furthermore the lattice mismatch between the substrate, say Au(111), and the PbSe deposit will increase dislocations and defects as the deposit grows. As the defects grow, the surface area increases, and the amounts deposited increase. The Se/Pb ratios for these deposits were 1.1 from EPMA analysis, indicating an excess of Se in the deposit, which may also increase the deposit roughness.

Another ALD cycle was devised to circumvent the excessive Se deposition observed with increasing cycle number, when the Se and Pb potentials were held constant. An extra step, designed to reduce excess Se to a soluble selenide species (22, 76), was added to the cycle as shown in Figure 6. The ALD cycle started with rinsing the Se solution into the cell for 2 s, where it was held static for 15 s of Se deposition, both at -0.3 V. The cell was then flushed with blank solution for 3 s at -0.45 V. After that, the Pb solution was flushed into the cell for 2 s and held without flow for 15 s for deposition, all at -0.35 V. This was followed by another blank rinse for 3 s at -0.45 V, referred to here are the Se reductive stripping step. The inclusion of the extra Se stripping step at a much more negative potential was intended to remove excess Se, as at this potential bulk Se was expected to reduce to HSe^-, a soluble species which diffused away, leaving only a Se atomic layer. Figure 7 shows the current time traces for three ALD cycles of PbSe, where current for Se reductive stripping is clearly depicted. However, the deposit grown using this cycle appeared thin, suggesting that too much Se was removed during the stripped step, leaving little Se to react with the Pb ions. EPMA of the deposit suggested only ¼ of the amount expected, around 6 atomic % of both Se and Pb which resulted in atomic ratio of ~ 1. Upon inspection under the microscope, the deposits appeared too thin for 100 cycle-run.

A third ALD cycle was then devised where the potentials were shifted positive from cycle to cycle, for the first 20 cycles, before constant potentials were applied for the remaining cycles. The use of a sequence of potential shift for the first 10 to 20 cycles has been use in previous studies to grow compounds such as PbSe (3), CdTe (42), InAs (51), and others , in order to achieve constant cycle to cycle growth and stoichiometry ratio of

1.0. Other than in the case of PbTe (47) where the potential shift for Te was positive for the first 20 cycles, the potential shifts have been negative. A series of experiments were carried out using different functional forms for calculating the sequence of positive shifts of both Se and Pb potentials, in order to optimize this process. Figure 8 shows the potential steps selected for the first 20 ALD cycles, which resulted in the most consistent deposited amounts. For that deposit, the Se potential began at -0.3 V, while the Pb potential started at -0.25 V, and both were stepped as in Figure 8, over 20 cycles, to a "steady state" value of -0.04 V, which was used for all subsequent cycles. A set of current time traces using this ALD cycle is shown in Figure 9. Figure 10 shows the coverages in compound ML, from coulometry, for all 100 cycles, formed by stepping the potential positive for the first 20 cycles. There was some scatter for the first couple cycles, but for the rest of the run, the coverages were essentially equal, and just under one compound ML a cycle. In general, it is better for a cycle to be low than to be high. More than a compound ML in a cycle is likely to result in roughening of the deposit, while less will only result in a slower deposit, but not harm the quality. Stepping the potentials avoided the gradient in coverage observed over the first 20 cycles seen in Figure 4, where the same potentials were used for the whole deposit. No significant increase in coverage or charge was observed when both Se and Pb deposition times were increased from 15 to 30 s, confirming the surface limited nature of the chosen potentials for Se and Pb, as was observed for PbTe (47). EPMA results for the 100 cycles PbSe deposit formed using ramps for Se and Pb potentials, from -0.3 V and -0.25 V respectively to -0.04 V, are shown in Table 1. The deposit was homogeneous from inlet to outlet, and the Se/Pb ratio was within a couple % of 1.

Figure 11 displays the glancing angle ($1°$ from the sample plane) X-ray diffraction patterns of a 100 cycle PbSe deposit, formed with potential steps for both Se and Pb, from -0.3 V and -0.25 V to -0.04 V. The sample was not annealed. As expected, peaks for the rock salt structure were for the (111), (200), (220), (311), (222), (400), (331) and (420) planes of PbSe [JCPDS 20-0494], along with polycrystalline Au substrate peaks, were clearly evident in the X-ray diffraction pattern. The X-ray diffraction pattern exhibited a preferential (200) peak. No elemental peaks for Se or Pb were evident in the XRD pattern.

Figure 12 displays a series of absorption spectra for PbSe electrodeposited nanofilms formed with different numbers of cycles, where both the Se and Pb potentials were stepped positively from -0.3 V and -0.25 V to -0.04 V over the first 20 cycles. Bulk PbSe has a direct bandgap of 0.26 eV. The band gaps of the nanofilms formed in this report, however, were strongly blue shifted, as expected given that their thicknesses were similar to the Bohr radius for PbSe (~ 46 nm). When samples with different cycle numbers (100, 200, 250 and 300 cycles) were deposited, the extent of the blue shift resulting was clearly evident as attested by Figure 12. Despite the large value of PbSe bandgap of around 0.26, the blue shifts of 0.7, 0.46, 0.38 and 0.36 eV were observed for 100, 200, 250 and 250 cycles respectively. A considerable amount in spectrum of the 100 cycle deposit attesting that its absorption is blue shifted out of the spectrometer's range.

A PbSe nanofilm formed by stepping the potentials for both Se and Pb from -0.3 V and -0.25 V, respectively, to -0.04 V was grown on a template stripped Au (TSG) substrate ref. The TSG was formed by taking a Au film deposited on mica, and covering it with epoxy. The mica was then defoliated, leaving the Au film on the epoxy, and with the surface, previously in contact with the ultra flat mica surface, exposed. EPMA analysis of the sample indicated a similar composition to that grown on Au on glass. Figure 13 displays STM images of a 100 cycle deposit formed on Au on mica substrate, which showed 80 nm, approximately hexagonal terraces, unlike those displayed in Figure 5. In both cases (Figures 5 & 13), three different spots were measured, resulting in similar images. Roughness (RMS) values of around 4.0 nm were observed for both the Au surface before and the PbSe deposit surface, suggesting the PbSe films formed using electrochemical ALD did not result in a rougher deposit. However, the terraces shown in Figure 13 were more hexagonal and flatter than the TSG substrate used, indicating that the ALD PbSe films showed improved crystallinity in Figure 13, relative to the Au substrate, where the crystallites were more rounded. In addition, the grain size of the PbSe film in Figure 13 appears relatively larger, compared with the bare gold (Figure 5). The X-ray diffraction pattern of the PbSe grown on Au on mica is displayed in Figure 14. The rocksalt structure of PbSe is once again evident in the pattern for the as formed deposit, given the peaks for the (200), (111), (220) and (311) reflections. There was little indication of the Au substrate, besides a small (111) peak since a glancing angle of 0.1° was used for this pattern. As it was the case for the ALD cycle PbSe film grown on Au on glass, no elemental peaks for either Se or Pb were present in the XRD pattern attesting to the quality of the deposit.

Conclusions

This paper presented various electrochemical ALD cycle methods of which the one containing the ramps of both Se and Pb for the first 20 cycles resulted in the optimal and successful growth of smooth epitaxial PbSe films on both Au on glass and Au on mica substrates after the inspection under the microscope. Coulometry and charges confirmed coverage of near or below a ML/cycle. Quantum confinement effects were visible in the samples grown at 300 cycles or less as observed from the absorption spectra. EPMA results confirmed that the deposits were homogeneous and stoichiometric. XRD revealed the films had a rocksalt structure and a preferential (200) orientation in deposits on both Au on glass and Au on mica substrates. STM images of PbSe samples showed clusters of improved crystallinity and similar roughness to the ones seen on Au on glass.

Acknowledgements

Support from the National Science Foundation is gratefully acknowledged.

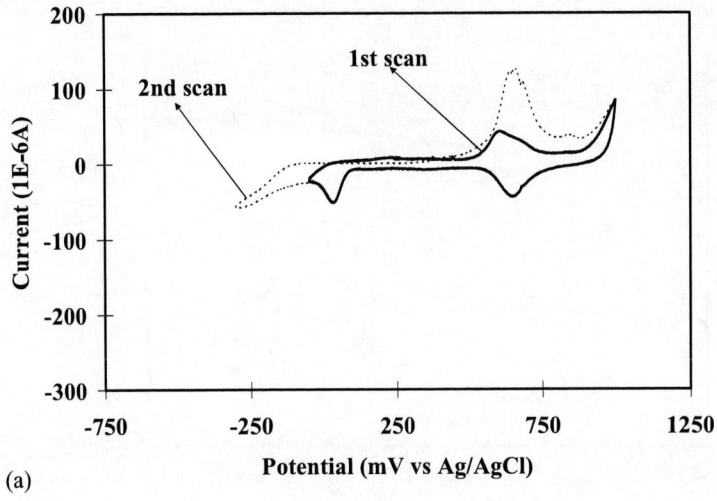

(a)

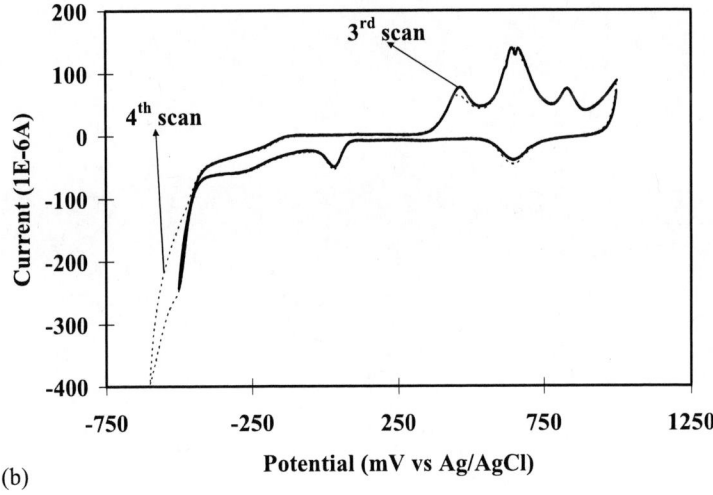

(b)

Figure 1. Cyclic voltammograms (a) and (b) of Au electrode in 0.5 mM $HSeO_3^+$ to successive decreasing potentials, pH 5, (electrode area: 4 cm^2, scan rate: 10 mV s^{-1}).

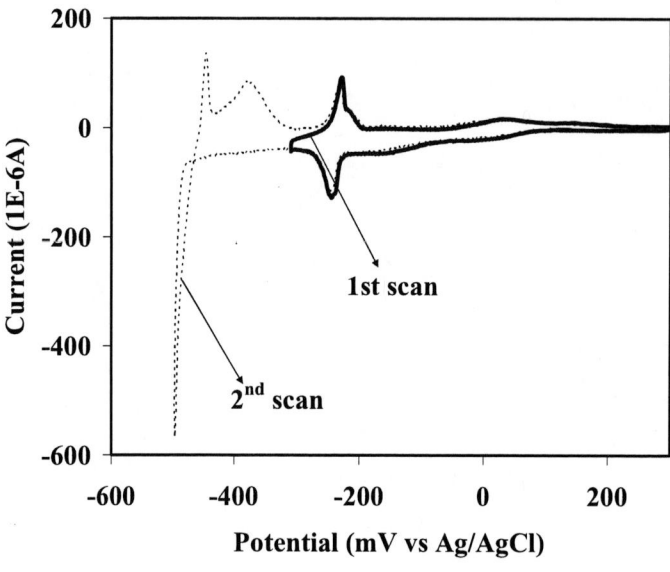

Figure 2. Cyclic voltammogram of Au electrode in 0.5 mM Pb^{2+}, pH 3, (electrode area: 4 cm^2, scan rate: 10 mV s^{-1}).

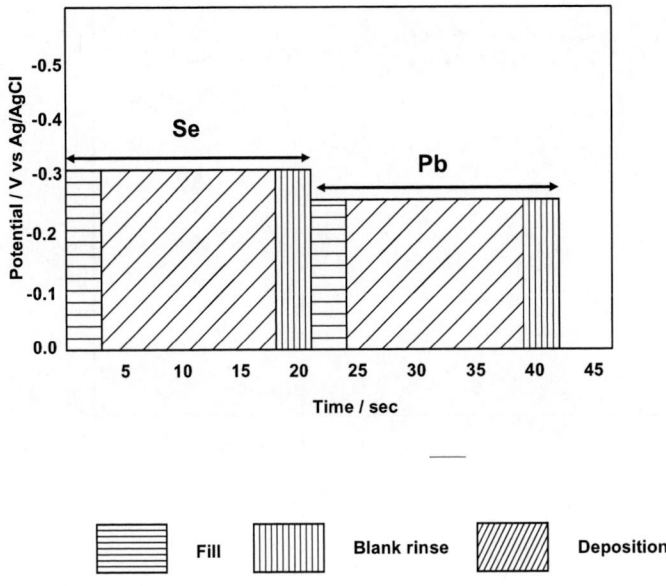

Figure 3. PbSe ALD deposition program with both Se and Pb potentials held constant at -0.3 and -0.25 V respectively.

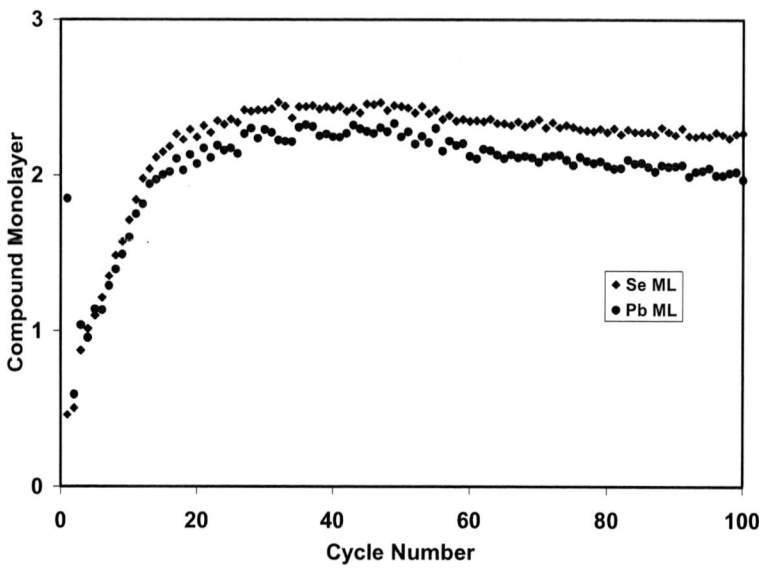

Figure 4. Deposition currents (coverage) of 100 ALD cycles of PbSe with both Se and Pb potentials held constant at -0.3 V and -0.25 V respectively. Electrode area: 4 cm^2.

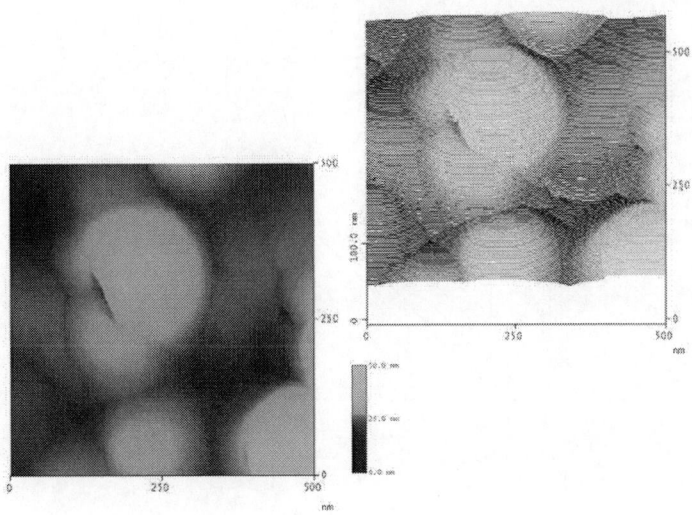

Figure 5. STM image of template stripped Au (500 nm² image). Electrode area: 4 cm².

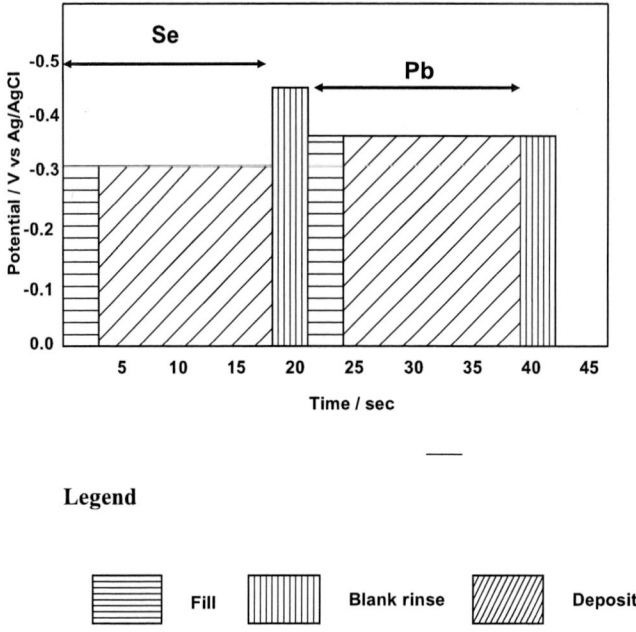

Legend

Fill   Blank rinse   Deposition

Figure 6. PbSe ALD deposition program with both Se and Pb potentials held constant including a Se reductive stripping step.

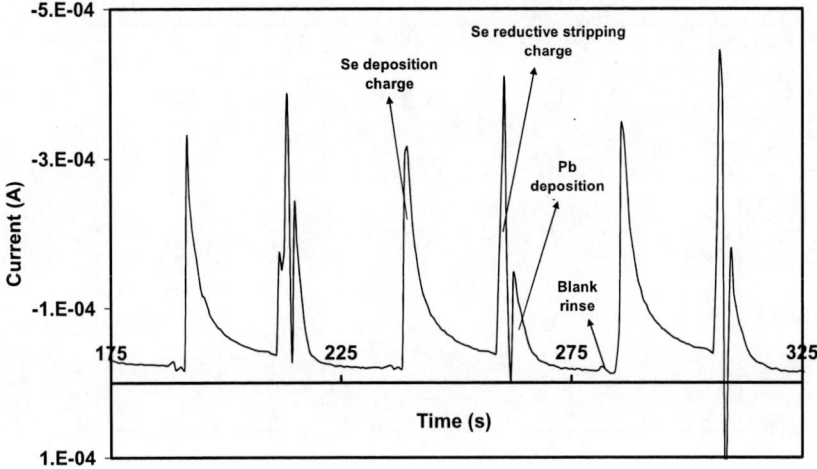

Figure 7. Current time traces of 100 cycles of PbSe formed by electrochemical ALD with the inclusion of the Se reductive stripping step. Electrode area: 4 cm^2.

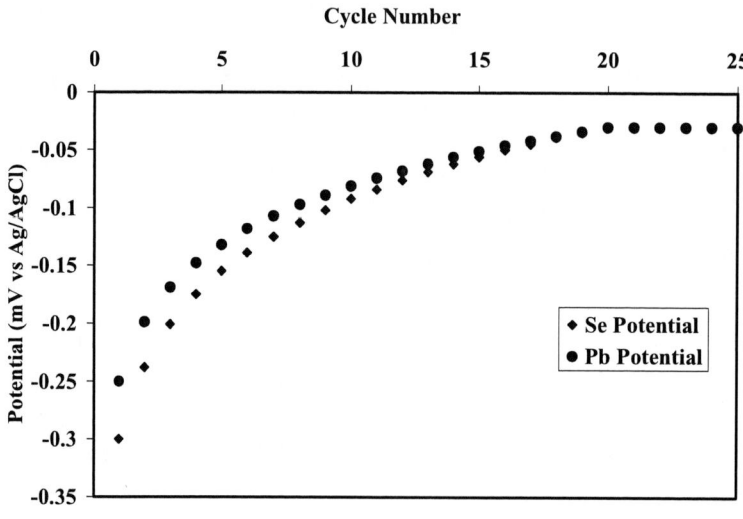

Figure 8. PbSe ALD cycle showing Se and Pb deposition potentials being shifted positive from -0.3 V and -0.25 V respectively to the steady state -0.04V. Electrode area: 4 cm^2.

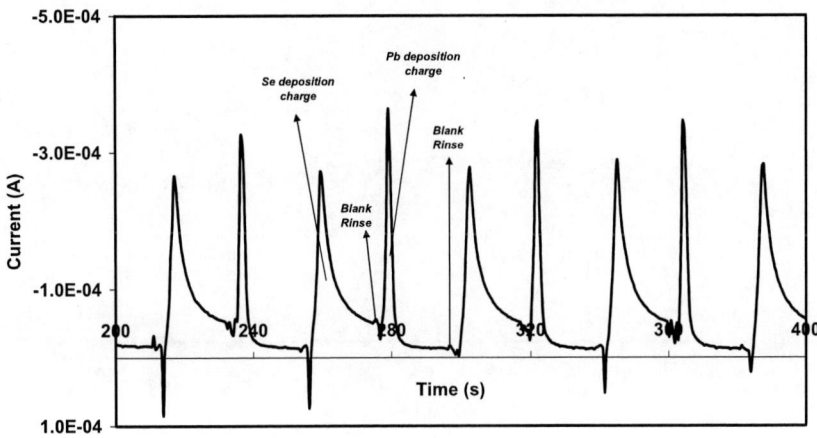

Figure 9. Current time traces of 100 cycles of PbSe formed when Se and Pb potentials were positively shifted from -0.3 V and -0.25 V respectively to the steady state -0.04 V. Electrode area: 4 cm².

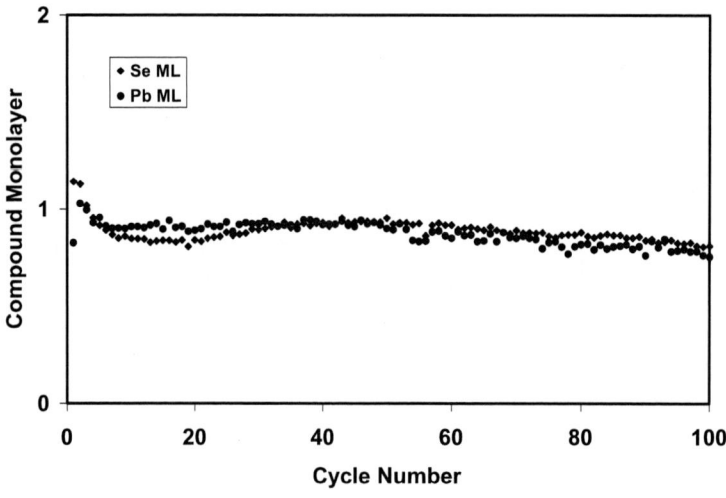

Figure 10. Deposition currents (coverage) of 100 ALD cycles of PbSe formed when both Se and Pb potentials were positively shifted from -0.3 V and -0.25 V respectively to -0.04 V. Electrode area: 4 cm^2.

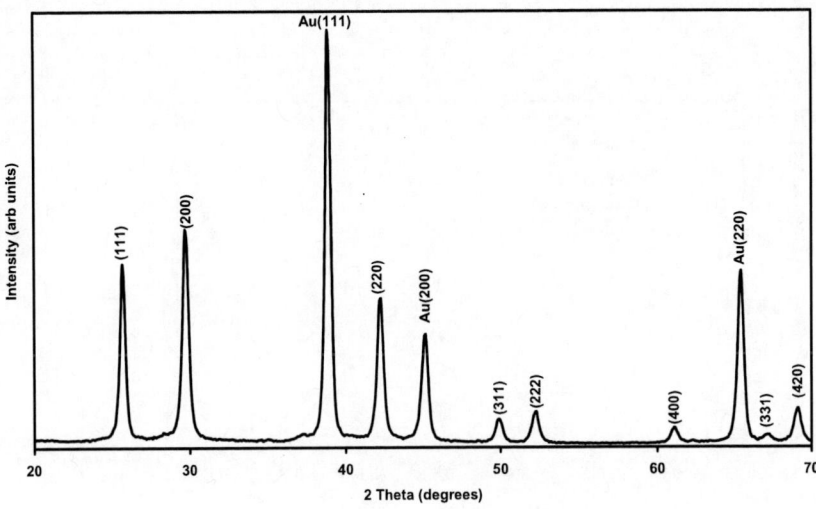

Figure 11. X-ray diffraction of 100 ALD cycles of PbSe thin film formed when both Se and Pb potentials were positively shifted from -0.3 V and -0.25 V respectively to -0.04 V. Angle of incidence is 1°, Cu Kα source. Electrode area: 4 cm^2.

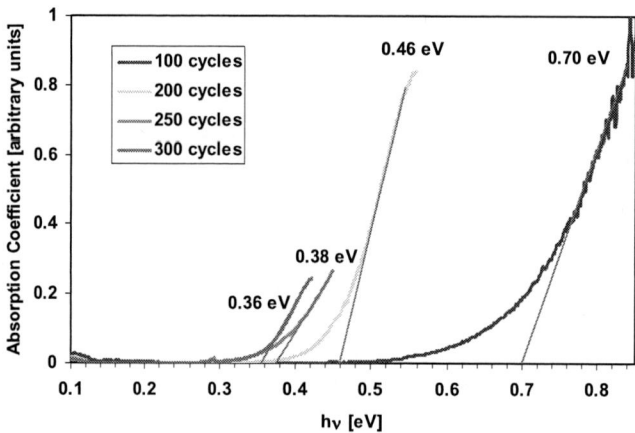

Figure 12. Absorption spectra of a) 100 ALD cycle, b) 200 ALD cycles, c) 250 ALD cycles and d) 300 ALD cycles of PbSe thin films formed when both Se and Pb potentials were positively shifted from -0.3 V and -0.25 V respectively to -0.04 V. Electrode area: 4 cm^2.

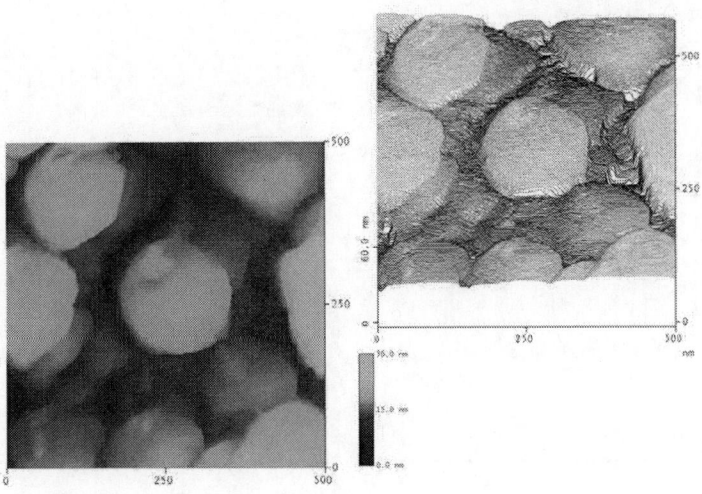

Figure 13. STM image (500 nm^2 image) of 100 ALD cycles of PbSe thin film on template stripped Au formed when both Se and Pb potentials were positively shifted from -0.3 V and -0.25 V respectively to -0.04 V. Electrode area: 4 cm^2.

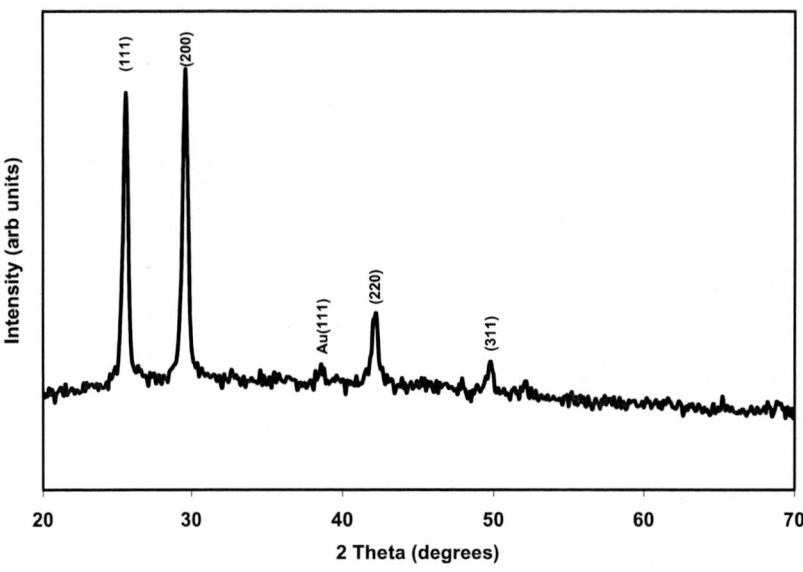

Figure 14. X-ray diffraction of 100 ALD cycles of PbSe thin film on Au on Mica formed when both Se and Pb potentials were positively shifted from -0.3 V and -0.25 V respectively to -0.04 V. Angle of incidence is 0.1°, Cu Kα source. Electrode area: 4 cm².

Table I. Electron Probe Microanalysis (EPMA) Data of 100 ALD cycle PbSe film formed when both Se and Pb potentials were positively increased from -0.3 V and -0.25 V to steady state -0.04 V. Electrode area: 4 cm^2.

PbSe Sample	Se Atomic %	Pb Atomic %	Se / Pb Ratio
Inlet	18.78	18.06	1.04
Middle	20.03	19.86	1.01
Outlet	20.27	19.94	1.02

References

1. S. Kumar, H. K. Zishan, M. A. M. Khan and M. Husain, *Current Applied Physics*, 5, 561 (2005).

2. G. Allan and C. Delerue, *Physical Review B*, 70, (2004).

3. R. Vaidyanathan, J. L. Stickney and U. Happek, *Electrochimica Acta*, 49, 1321 (2004).

4. Z. Hens, D. Vanmaekelbergh, E. S. Kooij, H. Wormeester, G. Allan and C. Delerue, *Physical Review Letters*, 92, (2004).

5. F. W. Wise, *Accounts of Chemical Research*, 33, 773 (2000).

6. V. Mathet, P. Galtier, F. Nguyenvandau, G. Padeletti and J. Olivier, *Journal of Crystal Growth*, 132, 241 (1993).

7. P. J. McCann, X. M. Fang, W. K. Liu, B. N. Strecker and M. B. Santos, *MBE growth of PbSe/CaF2/Si(111) heterostructures*, p. 1057, (1997).

8. F. Nguyenvandau, V. Mathet, P. Galtier, G. Padeletti, J. Olivier, D. G. Crete and P. Collot, *Materials Science and Engineering B- Solid State Materials for Advanced Technology*, 21, 317, (1993).

9. M. J. Bierman, Y. K. A. Lau and S. Jin, *Nano Letters*, 7, 2907 (2007).

10. B. Li, Y. Xie, J. Huang and Y. Qian, *Ultrasonics Sonochemistry*, 6, 217 (1999).

11. J. J. Zhu, S. T. Aruna, Y. Koltypin and A. Gedanken, *Chemistry of Materials*, 12, 143 (2000).

12. J. J. Zhu, O. Palchik, S. G. Chen and A. Gedanken, *Journal of Physical Chemistry B*, 104, 7344 (2000).

13. L. Beaunier, H. Cachet, R. Cortes and M. Froment, *Electrochemistry Communications*, 2, 508 (2000).

14. Z. Hens, E. S. Kooij, G. Allan, B. Grandidier and D. Vanmaekelbergh, *Nanotechnology*, 16, 339 (2005).

15. D. K. Ivanou, E. A. Streltsov, A. K. Fedotov and A. V. Mazanik, *Thin Solid Films*, 487, 49 (2005).

16. D. K. Ivanou, E. A. Streltsov, A. K. Fedotov, A. V. Mazanik, D. Fink and A. Petrov, *Thin Solid Films*, 490, 154 (2005).

17. D. K. Ivanov, C. K. Poznyak, N. P. Osipovich and E. A. Strel'tsov, *Russian Journal of Electrochemistry*, 40, 1044 (2004).

18. Y. A. Ivanova, D. K. Ivanou and E. A. Streltsov, *Electrochimica Acta*, 53, 5051 (2008).

19. K. W. Li, X. T. Meng, X. Liang, H. Wang and H. Yan, *Journal of Solid State Electrochemistry*, 10, 48 (2006).

20. H. Saloniemi, T. Kanniainen, M. Ritala, M. Leskela and R. Lappalainen, *Journal of Materials Chemistry*, 8, 651 (1998).

21. J. L. Stickney, T. L. Wade, B. H. Flowers, R. Vaidyanathan and U. Happek, *Encyclopedia of Electrochemistry*, 1, 513 (2003).

22. L. P. Colletti, B. H. Flowers and J. L. Stickney, *Journal of the Electrochemical Society*, 145, 1442 (1998).

23. T. E. Lister, L. P. Colletti and J. L. Stickney, *Israel Journal of Chemistry*, 37, 287 (1997).

24. T. E. Lister and J. L. Stickney, *Formation of the first monolayer of CdSe on Au(111) by electrochemical ALE*, p. 153, (1996).

25. B. M. Huang, L. P. Colletti, B. W. Gregory, J. L. Anderson and J. L. Stickney, *Journal of the Electrochemical Society*, 142, 3007 (1995).

26. J. L. Stickney, *Abstracts of Papers of the American Chemical Society*, 210, 297 (1995).

27. L. B. Goetting, B. M. Huang, T. E. Lister and J. L. Stickney, *Electrochimica Acta*, 40, 143, (1995).

28. L. B. Goetting, B. M. Huang, T. E. Lister and J. L. Stickney, *Abstracts of Papers of the American Chemical Society*, 208, 115 (1994).

29. T. E. Lister, R. D. Herrick and J. L. Stickney, *Abstracts of Papers of the American Chemical Society*, 208, 128 (1994).

30. L. Colletti, D. Teklay, S. Thomas, E. M. Wilmer and J. L. Stickney, *Abstracts of Papers of the American Chemical Society*, 208, 131 (1994).

31. B. M. Huang, L. B. Goetting and J. L. Stickney, *Abstracts of Papers of the American Chemical Society*, 208, 134 (1994).

32. R. Vaidyanathan, S. M. Cox, U. Happek, D. Banga, M. K. Mathe and J. L. Stickney, *Langmuir*, 22, 10590 (2006).

33. X. Zhang, X. Z. Shi and C. M. Wang, *Journal of Solid State Electrochemistry*, 13, 469 (2009).

34. W. Zhu, J. Y. Yang, D. Zhou, C. J. Xiao and X. K. Duan, *Electrochimica Acta*, 53, 3579 (2008).

35. W. Zhu, J. Y. Yang, D. X. Zhou, C. J. Mao and X. K. Duan, *Journal of Electroanalytical Chemistry*, 614, 41 (2008).

36. W. Zhu, J. Y. Yang, X. H. Gao, S. Q. Bao, X. A. Fan, T. J. Zhang and K. Cui, *Electrochimica Acta*, 50, 4041 (2005).

37. D. M. Kolb and H. Gerischer, *SS*, 51, 323 (1975).

38. D. M. Kolb, Przasnys.M and Gerische.H, *Journal of Electrochemical Society*, 54, 25 (1974).

39. O. M. Magnussen and R. J. Behm, *Journal of Electroanalytical Chemistry*, 467, 258 (1999).

40. F. Moller, O. M. Magnussen and R. J. Behm, *Electrochimica Acta*, 40, 1259 (1995).

41. V. Venkatasamy, N. Jayaraju, S. M. Cox, C. Thambidurai, U. Happek and J. L. Stickney, *Journal of Applied Electrochemistry*, 36, 1223 (2006).

42. B. H. Flowers, T. L. Wade, J. W. Garvey, M. Lay, U. Happek and J. L. Stickney, *Journal of Electroanalytical Chemistry*, 524, 273 (2002).

43. M. Innocenti, G. Pezzatini, F. Forni and M. L. Foresti, *Journal of Electrochemical Society*, 148, C357 (2001).

44. G. Pezzatini, S. Caporali, M. Innocenti and M. L. Foresti, *Journal of Electroanalytical Chemistry*, 475, 164 (1999).

45. A. Gichuhi, B. E. Boone and C. Shannon, *Langmuir*, 15, 763 (1999).

46. T. Oznuluer, I. Erdogan, I. Sisman and U. Demir, *Chemistry of Materials*, 17, 935 (2005).

47. D. O. Banga, R. Vaidyanathan, X. H. Liang, J. L. Stickney, S. Cox and U. Happeck, *Electrochimica Acta*,53, 6988, (2008).

48. I. Villegas and J. L. Stickney, *Journal of Vacuum Science & Technology A*, 10, 3032 (1992).

49. I. Villegas and J. L. Stickney, *Journal Electrochemical Society*, 139, 686 (1992).

50. M. Innocenti, F. Forni, G. Pezzatini, R. Raiteri, F. Loglio and M. L. Foresti, *Journal of Electroanalytical Chemistry*, 514, 75 (2001).

51. T. L. Wade, R. Vaidyanathan, U. Happek and J. L. Stickney, *Journal of Electroanalytical Chemistry*, 500, 322 (2001).

52. T. L. Wade, L. C. Ward, C. B. Maddox, U. Happek and J. L. Stickney, *Electrochemical and Solid-State Letters*, 2, 616 (1999).

53. V. Venkatasamy, N. Jayaraju, S. M. Cox, C. Thambidurai and J. L. Stickney, *Journal of the Electrochemical Society*, 154, H720 (2007).

54. R. R. Adzic, J. Wang, C. M. Vitus and B. M. Ocko, *Surface Science*, 293, L876 (1993).

55. M. Aindow and J. P. G. Farr, *Transactions of the Institute of Metal Finishing*, 70, 171 (1992).

56. M. Alanyalioglu, U. Demir and C. Shannon, *Journal of Electroanalytical Chemistry*, 561, 21 (2004).

57. B. M. Huang, T. E. Lister and J. L. Stickney, *Surface Science*, 392, 27 (1997).

58. Y. G. Kim, J. Y. Kim, C. Thambidurai and J. L. Stickney, *Langmuir*, 23, 2539 (2007).

59. T. E. Lister, B. M. Huang, R. D. Herrick and J. L. Stickney, *Journal of Vacuum Science & Technology B*, 13, 1268, (1995).

60. T. E. Lister and J. L. Stickney, *Journal of Physical Chemistry*, 100, 19568 (1996).

61. M. K. Mathe, S. M. Cox, B. H. Flowers, R. Vaidyanathan, L. Pham, N. Srisook, U. Happek and J. L. Stickney, *Journal of Crystal Growth*, 271, 55 (2004).

62. M. O. Solaliendres, A. Manzoli, G. R. Salazar-Banda, K. I. B. Eguiluz, S. T. Tanimoto and S. A. S. Machado, *Journal of Solid State Electrochemistry*, 12, 679 (2008).

63. A. S. Bondarenko, G. A. Ragoisha, N. P. Osipovich and E. A. Streltsov, *Electrochemistry Communications*, 7, 631 (2005).

64. B. Y. Chang, E. Ahn and S. M. Park, *Journal of Physical Chemistry C*, 112, 16902 (2008).

65. B. E. Conway and J. C. Chacha, *Journal of New Materials for Electrochemical Systems*, 7, 231 (2004).

66. N. Dimitrov, R. Vasilic and N. Vasiljevic, *Electrochemical and Solid State Letters*, 10, D79 (2007).

67. M. Seo and M. Yamazaki, *Journal of the Electrochemical Society*, 151, E276 (2004).

68. G. R. Stafford and U. Bertocci, *Journal of Physical Chemistry C*, 111, 17580 (2007).

69. M. P. Green and K. J. Hanson, *Surface Science*, 259, L743 (1991).

70. J. Y. Yang, W. Zhu, X. H. Gao, S. Q. Bao, M. Fan, X. K. Duan and J. Hou, *Journal of Physical Chemistry B*, 110, 4599 (2006).

71. W. Zhu, J. Y. Yang, J. Hou, X. H. Gao, S. Q. Bao and X. A. Fan, *Journal of Electroanalytical Chemistry*, 585, 83 (2005).

72. S. M. Sze, *Proceedings of the Ieee*, 69, 1121 (1981).

73. J. W. Elam, C. E. Nelson, R. K. Grubbs and S. M. George, *Thin Solid Films*, 386, 41 (2001).

74. J. W. Elam, A. Zinovev, C. Y. Han, H. H. Wang, U. Welp, J. N. Hryn and M. J. Pellin, *Thin Solid Films*, 515, 1664 (2006).

75. F. H. Fabreguette, Z. A. Sechrist, J. W. Elam and S. M. George, *Thin Solid Films*, 488, 103 (2005).

76. V. Venkatasamy, M. K. Mathe, S. M. Cox, U. Happek and J. L. Stickney, *Electrochimica Acta*, 51, 4347 (2005).

ECS Transactions, 19 (3) 273-281 (2009)
10.1149/1.3120707 ©The Electrochemical Society

Morphology, Composition and Electrical properties of Thin Anodic Oxides on InP

N. Simon*, C. Decorse-Pascanut, L. Santinacci, A.M. Gonçalves, A.L. Joudrier
And A. Etcheberry

Lavoisier Institute (UMR 8081), Université de Versailles
45 avenue des Etats Unis 78035 Versailles cedex, France

Depending on the applied electrochemical parameters, various anodic films can be grown onto InP in aqueous media. In borate buffer at pH = 9, the formation of homogeneous and thin $InPO_4$-like films occurred when anodic treatment is performed with a low current density, while with higher imposed current configuration porous and thicker layers are grown. In this work, AFM measurements coupled with XPS analyses and capacitance-voltage investigations have been used to study the relationship between morphology, composition and electrical properties of different anodic oxide films performed on InP surface, with current densities ranging from 1 mA.cm^{-2} to 100 mA.cm^{-2}.

Introduction

In aqueous media, electrochemical methods can be used to control the growth of various oxide structures onto InP surfaces (1-9). Previous works have established that the local oxidation conditions are crucial; a "constrained" or "non-constrained" process will give rise to a final homogeneous or heterogeneous interfacial layer with a specific chemical composition that determined the resulting electrical properties (10, 13). At pH = 9, depending on the anodization parameters contrasted anodic layers are grown, whose composition, thickness and electrical properties, strongly rely on the applied current density (13, 14): low current density leads to a thin $InPO_4$-like layer exhibiting good electrical properties. Conversely when the oxidation is performed with high current densities, thicker In-rich oxide films without electrical blocking properties are produced. These previous works demonstrate a correlation between composition, texture and electrical properties.

In the present paper, a galvanic method has been used to prepare anodic films, with current densities ranging from 1 to 100 mA·cm^{-2} and durations varying from 1 s to a few minutes. The chemical composition and morphology of the anodic layers have been investigated by x-ray photoelectron spectroscopy (XPS) and atomic force microscopy (AFM) respectively and have been correlated to the passivating properties of InP/oxide interfaces characterized by capacitance-voltage measurements.

Experimental

Anodic oxidation was performed onto n-InP (100) wafers from INPact Inc. (Sn doped, $N_D = 10^{18}$ at·cm^{-3}). The electrodes were cleaved in 0.1 cm^2 pieces and chemo-mechanically polished with a Br_2/methanol solution and successively rinsed with pure methanol and water. In order to remove residual oxides, samples were dipped few minutes in 2 M HCl, prior to the experiments. Electrochemical experiments were

273

performed in a borate buffered Tritisol (Merck) solution, at pH = 9, using a classical three-electrode configuration with a Mercury Sulfate Electrode (E_{MSE}= + 0.65 V/SHE)) as reference and a Pt wire as counter electrode. The electrochemical cell was connected to a PAR 283 potentiostat/galvanostat and placed in the dark. Oxidation treatments have been performed by applying a constant anodic current density, j_a ranging from 1 to 100 mA·cm^{-2} for durations varying from 1 second till a few minutes. For each treatment, the durations have been adjusted in order to apply a similar anodic coulometric charge (Qa = 100 mC·cm^{-2}) to the samples.

Capacitance-measurements vs. potential (C–V), have been used to investigate the oxide passivating properties. C(V) measurements were carried out before and after oxidation treatments using a PAR 283 potentiostat and an EG&G 5210 lock-in amplifier. The perturbating ac-voltage frequency is set to 1107 Hz with amplitude of 10 mV and the dc-voltage scan rate was 10 mV/s.

XPS analysis was performed using a VG-Escalab 220 iXL, spectrometer. A focused monochromated x-ray beam (Al Kα) was used as excitation source while detection was performed using a constant analyzer energy with threshold energy of 8 or 20 eV and photoelectrons were collected perpendicularly to the surface.

Atomic force microscope was used to characterize the topography of oxide films. A Dimension 3100 microscope (Veeco Instruments) equipped with pyramidal Si$_3$N$_4$ tip was used in contact mode. Forces in the range of 100 nN were applied to image the surface. Both root mean square (R_{rms}) and averaged (R_a) roughness were calculated using the manufacturer software.

Results and Discussion

Anodic oxide growth

Oxide growth has been achieved by galvanostatic control at different current densities (j_a) ranging from 1 to 100 mA·cm^{-2}. For each treatment, the durations have been adjusted in order to apply a similar anodic coulometric charge (Qa = 100 mC·cm^{-2}) to the samples.

The evolutions of potential against oxidation duration (U–t) for each current are given in figure 1a to 1d. When j_a is applied at the SC/electrolyte interface, the potential, U, reaches instantaneously an average value that strongly depends on the current value. When an anodic current of 1 mA.cm^{-2} is imposed, the average potential difference between the SC and the reference electrode is close to 1.3 V/MSE (Fig 1a). For galvanostatic processes performed with 5, 10 and 100 mA.cm^{-2}, U respectively attains 2.5; 6 and 10.5 V/MSE. These different U-t curves suggest that the oxide growth proceeds differently according to the applied current density. The resulting in-situ capacitance-voltage measurements and ex-situ XPS and AFM characterizations have been compared for the four resulting oxidized surfaces.

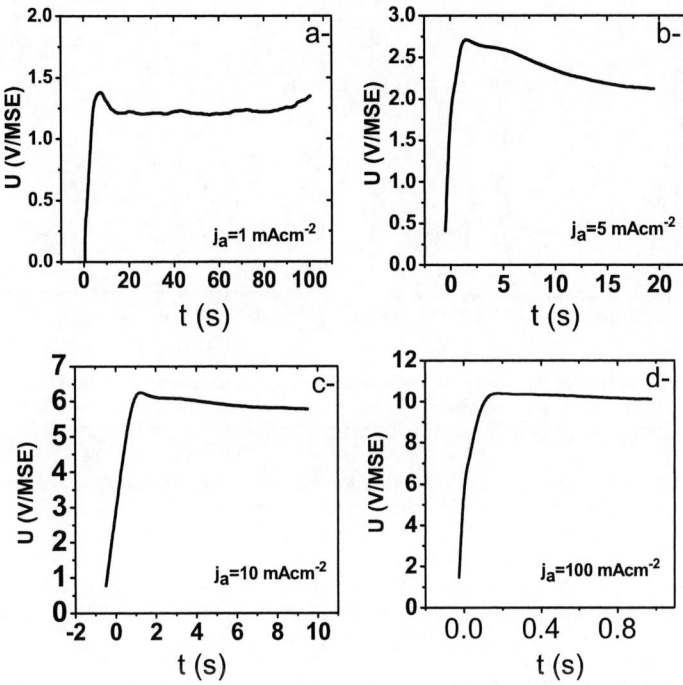

Figure 1: Potential-time evolution during anodic treatment in borate buffer (pH = 9) at j_a = 1 mA·cm^{-2} (a-); ja = 5 mA·cm^{-2} (b-), j_a = 10 mA·cm^{-2} (c-); ja = 100 mA·cm^{-2} (d-)

Chemical surface analysis: x-ray photoelectron spectroscopy (XPS)

The chemical composition of oxidized surfaces has been studied by XPS. Figure 2 shows the evolution of In$3d$ and P$2p$ photoelectron peaks after oxidation at different current densities while a same coulometric charge (100 mC.cm^{-2}) was involved.
A quantitative analysis of spectra recorded on InP oxidized in the same buffered solution (pH 9) with current densities ranging from 0.2 to 12 mA.cm^{-2} has already been described in our previous paper (13). In the present study, additionally to the InP matrix signal, after each anodic treatments, new contributions positively shifted in binding energy appear on XPS spectra. These contributions noted In_{ox} and P_{ox}, already described in ref (13), evidence the presence of oxidized phases in which In and P atoms are bounded to O atoms.

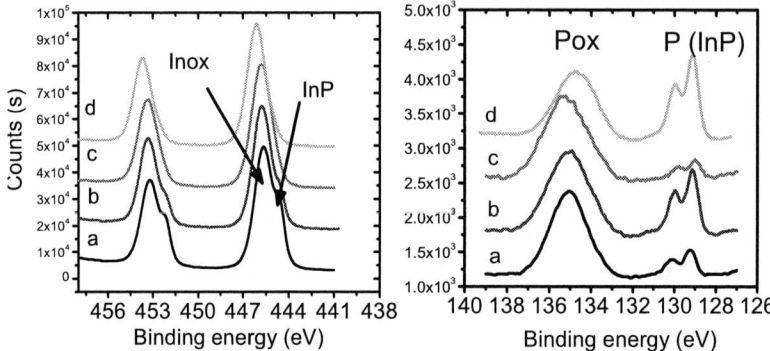

Figure 2. Evolution of P_{2p} and In_{3d} spectra recorded after a galvanostatic process in borate buffer solution at pH 9, for respectively (a-) j_a = 1 mA·cm^{-2} and 100 s ; (b-); ja = 5 mA·cm^{-2} and 20 s ;(c-), j_a = 10 mA·cm^{-2} and 10 s (d-); ja = 100 mA·cm^{-2} and 1 s.

Fitting treatments have been carried out on the four oxidized surfaces, the P_{2p} and In_{3d} levels were kept associated by constraints on InP bulk response: fixed f_{whm} for In_{InP} and P_{InP} contributions and corrected ratio $In_{InP}/P_{InP}= 1 \pm 0.1$. It appears that P_{2p} can be perfectly fitted with two contributions: one associated to the bulk, P_{InP}, and the other one to oxide, P_{ox}. For $In_{3d\ 3/2\ or\ 5/2}$, when InP bulk component, In_{InP}, is fixed, one or two additional components must be added to describe the oxide phase contribution, In_{ox}.

Table 1 gives the values of area ratios of both phosphorus contributions P_{ox}/P_{InP}. Additionally, the area ratio of Indium against Phosphorus peaks points to the evolution of the chemical composition of oxidized surfaces.

Table 1

Oxidation conditions $Qa = 100$ mA.cm^{-2}	P_{ox}/P_{InP}	In/P	dox (Å)
1 mA.cm^{-2} and 100 sec	1.4	1	20-30
5 mA.cm^{-2} and 20 sec	2.5	0.7	30-40
10 mA.cm^{-2} and 10 sec	10.8	2	70-80
100 mA.cm^{-2} and 1 sec	6.2	1	50-60

As the current density varies from 1 to 10 mA.cm^{-2}, the P_{ox}/P_{InP} ratio progressively increases (1.4 to 10.8) evidencing a larger oxide covering level. Additionally, figure 2 shows a substrate response almost masked for InP surface oxidized with 10 mA.cm^{-2} (see figure 2-c). The oxide thicknesses could be estimated from P_{ox}/P_{InP} ratios using equation [1] (15):

$$d_{ox} = \lambda_{ox}^{(P2p)}.Ln[(I_{ox}^{(P2p)}/I_{Inp}^{(P2p)}) \cdot A+1] \qquad [1]$$

The mean free path of photoelectrons coming from $P2p$ core levels inside the oxide, $\lambda_{ox}^{(P2p)}$, is assumed to be 2-3 nm. The correlation factor, A, is taken equal to 0.65 (7). d_{ox} values deduced from this simulation are given in table 1.

For applied currents lesser than 5 mA.cm^{-2}, XPS analyses evidence the growth of thin oxide layers ($d_{ox} \leq 4$ nm) with a In/P ratio close to 1, suggesting the formation of InPO$_4$-like oxide as already described for anodic layers obtained with $ja = 0.2$ mA.cm^{-2} (13). While a same coulometric charge is involved in the anodic process, a thicker Indium enriched anodic film is produced, as the current density reaches 10 mA.cm^{-2} ($d_{ox} \approx 8$ nm and In/P = 2).

Since the galvanic treatment is performed with the highest current density, 100 mA.cm^{-2} $(t_{ox} = 1s)$, the substrate response appears again (see figure 2-d). The P_{ox}/P_{InP} ratio indicates an oxide covering level less important than that corresponding to the InP surface oxidized 10 sec with 10 mA.cm^{-2}.

Electrical characterization: capacitance-voltage measurements

Previous works have shown that capacitance-voltage measurements performed, before and after the semiconductor surface oxidation, are very sensitive to oxide coverage and passivating properties (10, 13-15). Compared to the initial C^{-2}-V response (Mott-Schottky plot, MS), a flat plot (constant C^{-2} values) has been measured for a semiconductor surface covered by a thin oxide film with good electrical properties. Differently, linear Mott-Schottky behaviors with a similar or decreased slope, compared to the bare InP surface, have been attributed to inhomogeneous oxide layers exhibiting poor electrical properties, either oxide islands or thick and probably porous oxidized films.

In the present work, capacitance-voltage curves obtained before and after the anodic treatments are given on figure 3. The C^{-2}-V measurements performed on bare samples are very reproducible and give an apparent donor density close to the value indicated by the wafer provider (with $\varepsilon_{InP} = 9.3$, $N_D \approx 10^{18}$ cm^{-3}). As expected at pH = 9, the extrapolated flat band potential is approximately equal to -1.3 V/MSE. Different MS responses are, however, obtained after the four different anodic processes.

The MS representation of InP surface anodically oxidized with 1 mA.cm^{-2} for 100 sec is nearly flat and the corresponding interfacial capacitance is high and constant over almost 1 V which is very similar to that already described in reference (13) for anodisation performed during a few minutes with $j_a = 0.2$ mA.cm^{-2}. This flat Mott-Schottky plot suggests that the applied potential entirely drops in the anodic film, keeping constant the SC band bending and evidencing thus good passivating properties. As the applied current density used in the anodic process increases from 5 to 100 mA.cm^{-2}, the C^{-2}-V representation of the resulting "InP/oxide/electrolyte" interfaces progressively deviates from the flat behavior with a rising slope (see figure 3-b to 3-d), indicating different electrical behaviors for the SC/Oxide interfaces.

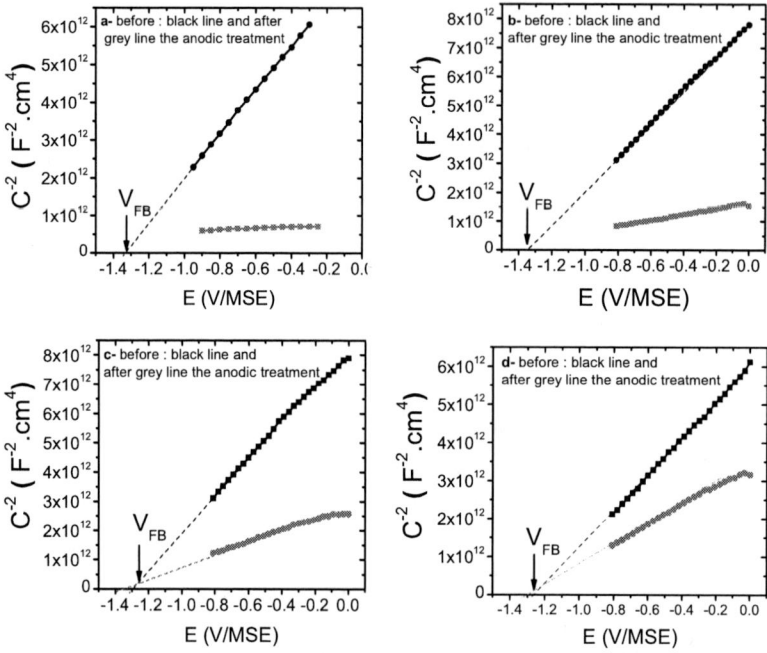

Figure 3: Mott Schottky plots on n-InP performed in borate buffer (pH = 9) with a perturbation frequency of 1107 Hz. before and after anodic oxidation at j_a = 1 mA·cm^{-2} (a-); j_a = 5 mA·cm^{-2} (b-), j_a = 10 mA·cm^{-2} (c-); ja = 100 mA·cm^{-2} (d-)

The MS plots recorded on InP surfaces oxidized respectively with 5, 10 and 100 mA.cm^{-2}, point out increasing negative slopes indicating that the SC band bending occurs while the polarization varies and becomes more and more noticeable as j_a increases. In these conditions the applied potential does not wholly drops in the oxide layer as observed for InP anodized with 1 mA.cm^{-2}. C^{-2}=$f(V)$ curves, thus suggest that the oxide films exhibit lesser passivating properties as the current density applied in the anodic process increases.

Morphological observation: atomic force microscopy

Atomic force microscopy has been used to investigate the morphology of the layers grown with the different current densities. Figure 4 shows the 3D views of InP surfaces before and after oxidation, respectively with 5, 10 and 100 mA·cm^{-2} while Table 2 reports the evolution of root mean square (R_{rms}) and averaged (R_a) roughness corresponding to the different surfaces.

The initial InP surface (Figure 4a), obtained after a chemo-mechanical polishing with Br_2-methanol, exhibits a flat and smooth morphology: R_{rms} and R_a values are close to 0.2 and 0.17 nm respectively.

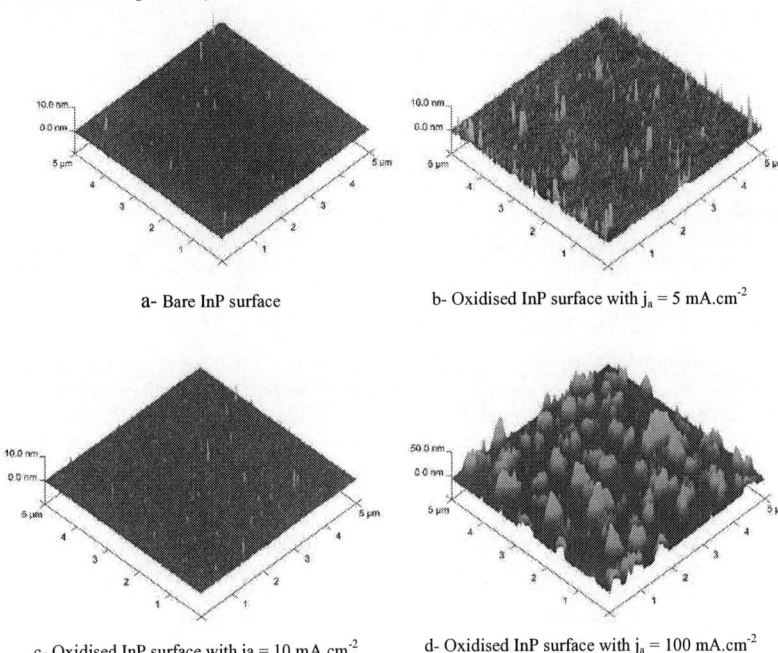

a- Bare InP surface   b- Oxidised InP surface with j_a = 5 mA.cm^{-2}

c- Oxidised InP surface with ja = 10 mA.cm^{-2} d- Oxidised InP surface with j_a = 100 mA.cm^{-2}

Figure 4: AFM 3D view of *n*-InP surface after (a) chemo-mechanical polishing with Br_2-methanol; (b) anodic oxidation in borate buffer at 5 mA·cm^{-2} for 20 s; (c) after anodic oxidation in borate buffer at 10 mA·cm^{-2} for 10 s, (d) after anodic oxidation in borate buffer at 100 mA·cm^{-2} for 1 s

Figure 4b shows the surface topography after a galvanic oxidation at 5 mA·cm^{-2}. Except some particles that can be attributed to a surface pollution occurred before the AFM measurements, a flat and homogeneous surface is observed. The roughness increases slightly compared to the bare surface: R_{rms} = 0.45 nm and R_a = 0.35 nm. Similarly a low roughness, R_{rms} = 0.45 nm and R_a = 0.33 nm, and a uniform topography are observed after the anodic treatment performed at 10 mA·cm^{-2}. Conversely, the surface morphology is strongly modified when a high current density is applied (100 mA·cm^{-2}). Numerous globular features exhibiting a mean diameter of 140 nm, a low aspect ratio (height/diameter) of 0.1 to 0.15 and a density of 4-5·10^8 cm^{-2} are formed. The overall roughness is thus strongly increased (R_{rms} = 8.01nm and R_a = 6.01nm). However the roughness measured between these clusters remains relatively low (R_{rms} = 1 nm and R_a = 0.8 nm). Taking into account the diameter and the density of these islands, it appears that only 25-30% of the overall InP surface is covered by the globular particles.

Table 2

	Initial	Oxidized (20 s, 5 mA·cm^{-2})	Oxidized (10 s, 10 mA·cm^{-2})	Oxidized (1 s, 100 mA·cm^{-2})
Rrms (nm)	0.20	0.45	0.45	8.01
Ra (nm)	0.17	0.35	0.33	6.01

Although a similar coulometric charge ($Q_a = 100$ mC·cm^{-2}) is involved in the four anodic treatments, capacitance-voltage measurements, XPS analyses and AFM examinations evidence the formation of strongly different oxidized phases.

For current density ranging from 1 to 10 mA.cm^{-2}, AFM characterization indicates that the resulting oxidized surfaces are flat and compact suggesting a 2D growth process. Conversely, when j_a is equal to 100 mA.cm^{-2}, a very heterogeneous (presence of islands) and rougher surface is detected evidencing therefore a 3D growth mechanism.

Oxide growth with $j_a = 1$ to 10 mA.cm^{-2} : While a similar 2D growth process occurs, XPS and C^{-2}-V measurements point out the formation of different oxide layers as the current density increases from 1 to 10 mA.cm^{-2}. Thicker, indium enriched and less passivating oxide films are grown as j_a is more important. This result, already described in reference (13), is completed by AFM examination suggesting that less passivating properties are related to either micro-porosities (undetectable with AFM) and/or chemical heterogeneities in the oxide film. In this current density range, although the oxide growth mechanism seems to be the same, as very flat and compact surfaces are observed, the properties of the resulting layers appear to be controlled by the galvanic conditions. Low current mode providing InPO$_4$-like thin films with good electrical properties as already reported (13), while higher current mode produce indium enriched and thicker film with poor electrical qualities. The oxide chemical composition could be the key factor for the resulting electrical properties.

Oxide growth with $j_a = 100$ mA.cm^{-2}: In this configuration, a very different oxide growth is evidenced. As a smooth surface is observed between large clusters, a 3D growth governed by a "dissolution/deposition" mechanism is suggested. XPS analysis confirms this observation, evidencing a non masked substrate response. The SC surface oxidized with the highest current seems thus partly covered with large oxide islands and partly uncovered. The very large uncovered zones (nearly 75% of the surface) allow the direct contact between electrolyte and bare InP. In these conditions, the poor electrical properties revealed by the important MS slope are related to the presence of large uncovered part of InP. The oxide morphology appears to be a crucial point governing the electrical properties of the layer grown in these conditions.

Conclusion

In this work, a successful correlation of morphological investigations to chemical and electrical studies has been presented. It has been established that anodic oxide films grown on InP in borate buffer at pH = 9 by galvanostatic methods (low and high current modes) provide very different oxide layers, whose composition, thickness, porosity and electrical properties, strongly depend on the applied current density. The 2D growth of homogeneous oxide films has been evidenced when anodic treatment is performed with a current density fewer than 10 mA.cm^{-2}, while porous heterogeneous 3D oxide layer is grown when a higher current (100 mA.cm^{-2}) is applied. The resulting electrical properties (C^2-V measurements) are completely related to the chemical (XPS analysis) and/or morphological (AFM examination) aspects of the different films. As suggested by the flat and high capacitance value measured on the modified InP/oxide/electrolyte interface, good electrical properties are related to thin and homogeneous "InPO$_4$- like" layers performed with smallest current densities (< 5 mA.cm^{-2}). Conversely, with a similar coulometric charge involved in the anodic process, as the current density is increased (> 10 mA.cm^{-2}), thick oxide films are grown with weak passivating qualities.

This work has evidenced that the poor electrical properties of thick oxides grown on InP could be associated either on a chemical composition aspect (Indium enrichment) or on a morphological aspect (very important roughness and presence of bare InP zones) Both the chemical composition of oxides and their morphology (thickness and porosity) are crucial points governing the electrical quality of InP/oxide interfaces. Depending on the galvanic conditions, one or the other of these parameters becomes decisive and govern the resulting properties.

References

1. J.F. Wager and C. Wilmsen, *J. Appl. Phys.,* **51**, 812 (1980).
2. H.L. Hartnagel, *Oxides and oxide films*, Marcel Dekker, New York and Basel, Vol. 6 (1986).
3. G. Hollinger, E. Bergignat, J. Joseph, Y. Robach, , J. Vac. Sci. Technol. A 3, 2082 (1985).
4. P. H. L. Notten, J. E. A. M. Van den Meeraker, J. J. Kelly, *Etching of III-V Semiconducors: An Electrochemical Approach*, Elsevier, Amsterdam, (1991).
5. A. Pakes, P. Skeldon, G. E. Thompson, S. Moisa, G. I. Sproule and M. J. Graham, *Corr. Sci.*, **44**, 2161 (2002).
6. T. Djenizian, G. I. Sproule, S. Moisa, D. Landheer, X. Wu, L. Santinacci, P. Schmuki, M. J. Graham, *Electrochim. Acta.*, **47**, 2733 (2002).
7. J. Joseph, A. Mahdjoub and Y. Robach, *Revue Phys. Appl.,* **24**, 189-194 (1989).
8. M.P. Besland,Y. Robach and J. Joseph, *J. Electrochem. Soc.*, **1**,140 (1993).
9. P. Louis, M.P. Besland,Y. Robach and J. Joseph, *J. Electrochem. Soc.*, **4**,142 (1995).
10. N. Simon, I. Gérard, C. Mathieu and A. Etcheberry, *Electrochim. Acta*, **47** 2625 (2002).
11. I. Gérard, N. Simon, A. Etcheberry, *Appl. Surf. Sc.*, **175**, 734 (2001).
12. N. Simon, I. Gérard, J. Vigneron and A. Etcheberry, *Thin Sol. Films*, **400**, 134 (2001).
13. N. Simon, N.C. Quach, A.M. Gonçalves and A. Etcheberry, *J. Electrochem. Soc.*, **154**, H340 (2007).
14. N. Simon and A. Etcheberry , *ECS Trans.*, **6**, 453 (2007).
15. N. Simon, L. Santinacci, C. Decorse-Pascanut, S. Jaskierowicz, A. Etcheberry, *C. R. Chimie*, **volume 11**, Issue: 9, 1030-1036 (2008).

282

ECS Transactions, 19 (3) 283-292 (2009)
10.1149/1.3120708 ©The Electrochemical Society

In-situ infrared kinetic study of multistep chemical modifications of organic monolayers at silicon surfaces

L. Touahir, A. Moraillon, S. Sam, A. C. Gouget-Laemmel, P. Allongue, J.-N. Chazalviel, C. Henry de Villeneuve, F. Ozanam

Physique de la Matière Condensée, Ecole Polytechnique, CNRS, 91128 Palaiseau, France.

The reaction of activation of acid-terminated organic monolayers grafted on (111)Si surfaces into succinimidyl-ester terminated monolayers by treatment in N-ethyl-N'-(3-dimethylaminopropyl)-carbodiimide (EDC) and N-hydroxysuccinimide (NHS) has been studied by infrared spectroscopy and its kinetics have been followed in situ. The reaction exhibits non exponential kinetics, with an initial fast rise in the appearance of succinimidyl ester followed by a long reaction tail. The origin of this behavior can be ascribed either to the existence of a fast reaction pathway involving anhydride formation, or to the rate limitation by steric hindrance effects. The subsequent amidation of the succinimidyl ester layers by a primary amine has been studied similarly. Here, the kinetics are found to be exponential, but the reaction rate exhibits a sublinear dependence on the amine concentration. This behavior is ascribed to the effects of electrostatic interactions on the adsorption of the protonated amine onto the Si surface prior to reaction.

Introduction

Using silicon as a substrate for chemical or biological sensors appears to be an attractive strategy (1). It offers the high electronic quality of silicon / organic monolayer interfaces when electronic or electrochemical transduction is sought for, and of a good stability of the silicon / molecule link owing to the chemical robustness of the non polar Si-C bond. However, since steric hindrance prevents all the surface silicon atoms from binding an organic molecule, it appears of prime importance to carefully control the quality and the packing density of these grafted organic layers in order to ensure an efficient and stable protection of the silicon surface (2). Moreover, in order to achieve a sensor, specific probes have to be immobilized on the surface, with a packing density optimized according to the specific probe-target pair chosen for the sensor. For that purpose, multi-step processes are generally required, and it is then needed to control all the successive steps of the process in order to obtain the well-defined functionalized surface required for optimum sensor performances.

A popular and versatile route for immobilization of biological probes on solid surfaces consists in grafting acid terminated organic molecules on a substrate, and then converting the acid groups into succinimidyl ester groups by reaction with N-hydroxysuccinimide (NHS) using N-ethyl-N'-(3-dimethylaminopropyl)-carbodiimide (EDC) as a coupling reagent (3). Such an activated surface can be further reacted with the primary amine of a linker attached to the biological probe in physiological conditions (4). We have already demonstrated that acid terminated alkyl chains can be grafted at a hydrogenated silicon

283

surface in one step, yielding a clean surface, with a surface concentration in acid groups which approaches 3×10^{14} cm^{-2} and an interface of high electronic quality (5). Achieving an efficient and well controlled conversion of these surfaces into succinimidyl-ester terminated surfaces needs careful optimization (6). Also, it is highly desirable to achieve the amidation reaction under well-controlled conditions, with the best immobilization yield and expending a minimum amount of biological probes. Therefore, as a part of our effort for controlling these processes, we have undertaken ex-situ and in-situ infrared studies of the activation and amidation steps aimed at clarifying the kinetic mechanisms governing these reactions.

Experimental

The silicon samples were cut from [111]-oriented n-type silicon wafers (0.2° misorientation toward [11$\overline{2}$], float zone, $\rho = 30$-40 Ωcm) and prepared according to established procedures (7). The samples were initially cleaned in "piranha", a 3:1 mixture of concentrated H_2SO_4 and H_2O_2, in order to remove all organic contaminations. This wet oxidation was followed by copious rinsing with ultrapure water. The cleaned silicon samples were then chemically etched in NH_4F (ca. 0.05 mol L^{-1} ammonium sulfite was added to the etching solution) or HF solutions to remove native oxide. The first treatment yields a hydrogenated atomically flat surface ("SiH" surface) whereas the second one yields a hydrogenated surface rough on the atomic scale ("SiH$_x$" surface). After etching, the surface was rinsed with ultrapure water provided by a Millipore station. All cleaning (H_2O_2 30%, H_2SO_4 96%) and etching (NH_4F 40%, HF 50 %) reagents were of VLSI grade and supplied by Carlo Erba. Ammonium sulfite monohydrate (92%) was purchased from Aldrich.

A one-step procedure was used to graft well-defined 10-carboxydecyl organic monolayers on hydrogenated silicon surfaces via direct photochemical hydrosilylation of undecylenic acid (5). Undecylenic acid (98%) was purchased from Acros. The neat undecylenic acid was outgassed under argon in a Schlenk tube at 100°C for 30 minutes and then cooled to room temperature under continuous argon bubbling to insert the freshly prepared H-terminated silicon sample. The Schlenk tube was then closed and exposed to UV irradiation (312 nm, 6 mW/cm^2) during 3 hours for performing the grafting. The functionalized surface was then rinsed in oxygen-free hot acetic acid (75°C) under argon during 30 min (to desorb the hydrogen-bonded undecylenic acid molecules) (5).

The EDC and NHS solutions for the activation reaction were prepared with cold water. The mixture of EDC and NHS (with the same concentration in EDC and NHS, set to 0.005 or 0.01 mol/L as indicated in the following) was deoxygenated with argon for 15 min in a Schlenk tube placed in a water bath at 15 °C. For the ex-situ studies, the freshly prepared acid-terminated alkyl surface was placed into the Schlenk tube and allowed to react under continuous argon bubbling for 90 min at 15 °C. The resulting succinimidyl-ester terminated surface was copiously rinsed with water and dried under a stream of nitrogen.

For amidation, 20 mL of a fresh solution of hexylamine in 1X PBS buffer was prepared. The pH was adjusted to 7 with a solution of HCl (2N). This solution was transferred into a degassed Schlenk tube and kept under bubbling for 15 min. The activated surface was immersed into the solution and after 10 min of bubbling the Schlenk tube was closed overnight. After amidation, the surface was copiously rinsed in water (1 min) and blown dry under nitrogen.

Activation

The controlled immobilization of amine terminated probe molecules at an acid terminated surface is performed in two steps: activation of the acid groups into succinimidyl ester groups, which are then reacted with an amino group borne by the probe molecules, yielding formation of an amide bond. The various steps of this procedure can be quantitatively followed by using infrared spectroscopy. Figure 1 shows the infrared spectra of an acid-terminated surface (Fig. 1a), an NHS-ester-terminated surface (Fig. 1b) and a surface obtained after reaction of the NHS ester groups with ethanolamine (Fig. 1c). The most salient features allowing for distinguishing one spectrum from each other are observed in the range of carbonyl vibrations, between 1500 and 1850 cm^{-1}. On acid-terminated surfaces, a single peak is observed close to 1710 cm^{-1}, corresponding to the $vC=O$ vibration mode of the carbonyl group. NHS-ester terminated surfaces are characterized by a triplet with peaks at 1745, 1785 and 1815 cm^{-1}. These vibrations essentially correspond to the antisymmetric and symmetric $vC=O$ modes of the carbonyl groups of the succinimidyl group, and to the $vC=O$ mode of the ester group, respectively (4). Finally, after amidation, two broad characteristic bands are observed at 1650 and 1550 cm^{-1}, commonly labeled amide I and amide II bands, which essentially correspond to the $vC=O$ and the δNH vibration modes of the amide group.

Many reactions can actually take place during the conversion of acid-terminated surfaces into activated-ester surfaces via an EDC-NHS treatment. On porous silicon, the concentrations in EDC and NHS in the activation solution have been found to be critical parameters in order to obtain a high activation yield and avoid the formation of unwanted residues or by products at the surface (6). Here we have investigated the kinetics of this activation process, by in-situ infrared spectroscopy, in a range of EDC and NHS concentrations close to the optimum conditions determined on porous silicon. In addition,

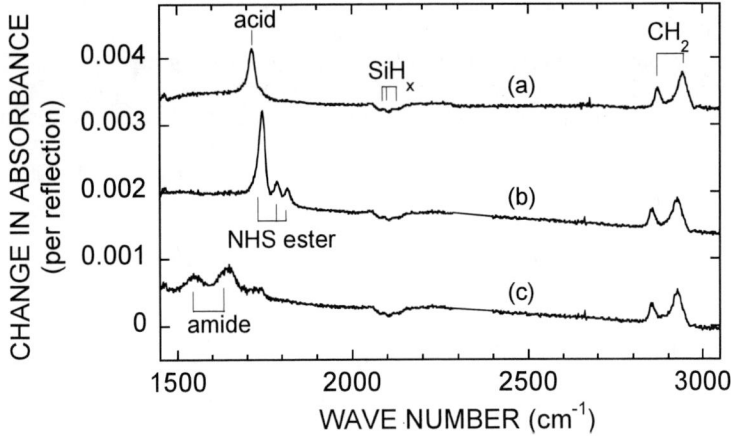

Figure 1. FTIR spectra of a SiH$_x$ surface after grafting of carboxydecyl groups (a), conversion into succinimidyl ester by reaction in a [EDC]=[NHS]=0.01 mol/L solution (b), and amidation in a 0.05 mol/L ethanolamine solution during 15 min(c).

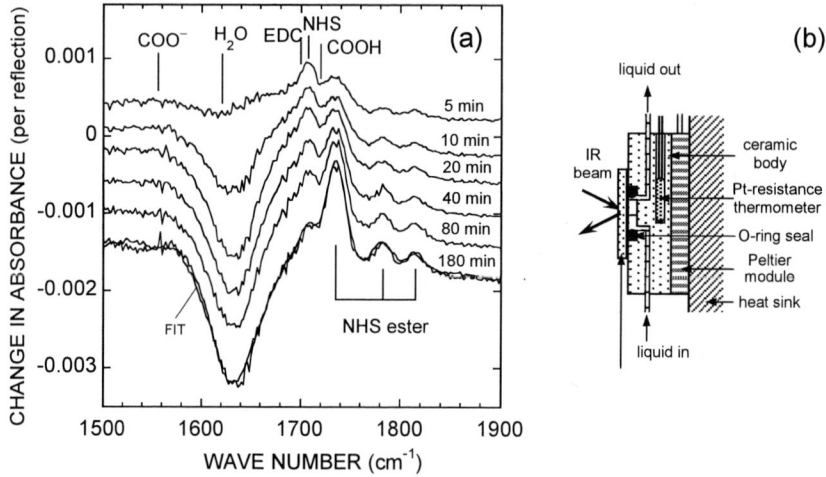

Figure 2. (a): successive in-situ spectra recorded at the various steps of the activation reaction of a SiH surface at 5°C in a [EDC]=[NHS]=0.005 mol/L solution. For each spectrum, the reaction time is indicated on the right part of the graph. In the carbonyl absorption range, various contributions are superimposed on the parallel loss in COOH absorption and gain in succinimidyl-ester absorption. (b): scheme of the IR cell.

we have used the special cell shown in Fig. 2 allowing the control of the solution temperature, because we have noticed that using a too high temperature results in the formation of unwanted products. Figure 2 shows a set of infrared spectra successively recorded during an activation treatment at 5°C in a [EDC]=[NHS]=0.005 mol/L solution. In the 1500-1900 cm^{-1} range, a complex evolution is observed. As expected, a lowering of the acid-group absorption (negative peak at ~1720 cm^{-1}) takes place (especially clearly observed in the first spectra), correlated with a progressive increase of the characteristic triplet of the succinimidyl ester. Other contributions are making the quantitative analysis of these spectra more complex. The most prominent one is water absorption (δOH_2 scissor mode of water), which results in the presence of the large and broad contribution close to 1650 cm^{-1}. Contributions from EDC (1700 cm^{-1}) and NHS (1707 cm^{-1}) molecules in solution are also discernible. Finally, a contribution from carboxylate absorption at 1550 cm^{-1} also appears in the last spectrum, the pH of the solution during the experiment making the acid groups partially ionized.

For the purpose of quantitative analysis, five pseudo-Voigt peaks, accounting for the succinimidyl-ester, carboxyl and carboxylate absorption, superimposed on a quadratic baseline and a contribution proportional to electrolyte absorption, have been used to fit the experimental spectra. An example of such a decomposition pertaining to the last spectrum of Fig. 2 is shown in Fig. 3, and the fitted spectrum is shown superimposed on the experimental one in Fig. 2. Such a quantitative analysis allows for a quantitative following of the various contributions during the experiment. Figure 4 shows the

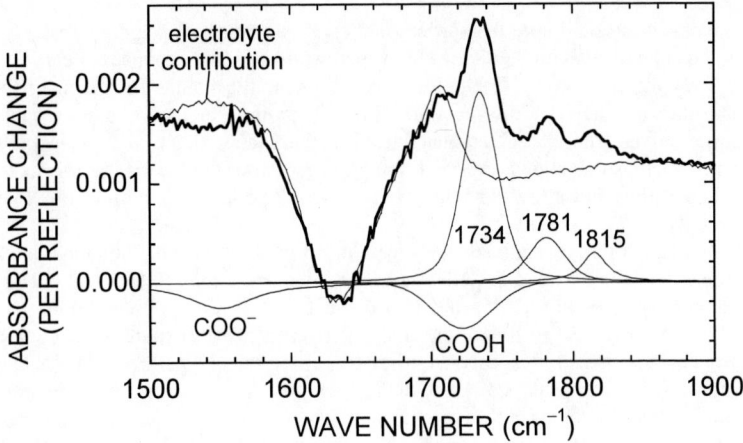

Figure 3. The various contributions used for fitting the experimental data. Five peaks account for the surface species. The electrolyte contribution is adjusted from independently recorded spectra of water and of the reactants. The resulting fit is shown superimposed to the data in Fig. 2 (bottom spectrum).

evolution of the acid and succinimidyl ester contributions as a function of reaction time. As expected, the acid loss can be paralleled with the simultaneous increase of the three succinimidyl ester peaks. Noticeably, the evolution does not appear exponential, which

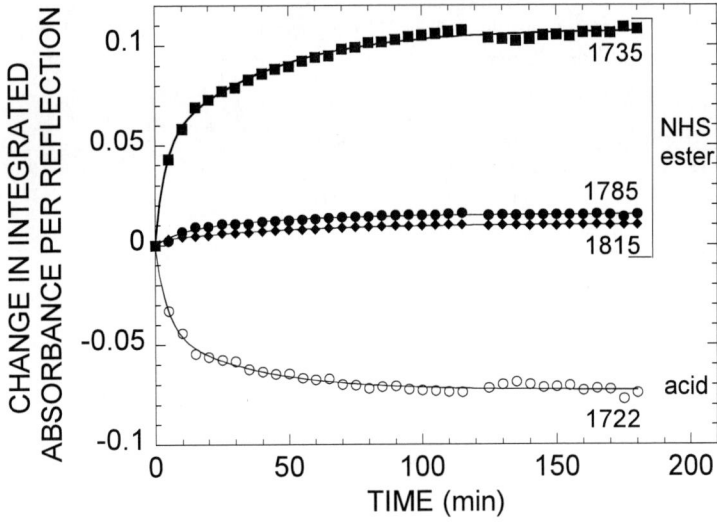

Figure 4. Evolution of succinimidyl ester and COOH adsorption as a function of the reaction time, as deduced from the fitting model sketched in Fig. 3. Notice the correlated evolution of NHS-ester and acid peaks.

would be the case for simple first-order kinetics. As a matter of fact, the transients shown in Fig. 4 can be nicely fitted either as the superposition of two exponential curves, or as a power-law evolution. In the first case, a fast exponential accounts for the fast rise of the curve in the first steps of the reaction, and the second one for the slower subsequent evolution. Given the characteristic time constants and shape of the experimental transients, such as that shown in Fig. 4, the practical duration of our experiments (usually 2 to 3 hours) does not allow for telling between the two kinds of profile (bi-exponential or power-law).

Such a quantitative analysis has been performed on results obtained in various conditions, i.e., for two distinct NHS and EDC concentrations (0.005 and 0.01 mol/L), two temperatures (5 and 15°C) and two distinct initial surface preparations (SiH and SiH_x). The experimental results of the analysis in terms of a bi-exponential evolution (i.e., the two characteristic times derived from the analysis) are gathered in Table I. The evolutions of the four peaks corresponding to succinimidyl ester and acid absorption are separately fitted to the specific model (here the superimposition of two exponential profiles). Therefore, two characteristic times τ_1 (the fastest) and τ_2 (the slowest) are determined for each of the analyzed peaks. As a matter of fact, the characteristic times for each peak are found to be close to each other, especially for τ_2, and the values given in Table I correspond to an average for the three NHS-ester peaks. The uncertainty associated with the dispersion and the accuracy of the fits of the spectra is estimated to be not larger than 10 min on τ_2. The accuracy of τ_1 is more questionable, especially in view of the time required for recording one spectrum (slightly more than two min). Therefore, the τ_1 values in Table I cannot be considered as accurately determined. It can be noticed that within experimental and analysis accuracy, the kinetics are independent of the initial surface preparation. More surprisingly, the kinetics appear to be also weakly dependent on NHS and EDC concentration (a somewhat shorter characteristic time is found when EDC and NHS concentrations are increased by a factor of two). In contrast, lowering the temperature from 15 to 5°C appears to have a measurable effect on the kinetics, the τ_1 and τ_2 values being significantly increased.

TABLE I. Characteristic times as determined by a bi-exponential fit

Surface preparation	SiH	SiH	SiH	SiHx
Temperature	15°C	15°C	5°C	15°C
[EDC]/[NHS] mM	5/5	10/10	5/5	5/5
τ_1 (min)	3	4	9	4
τ_2 (min)	50	40	80	50

The complexity of the activation reaction (multiple reaction paths) could appear to be a plausible origin for the non-exponential kinetics (6, 8). In a first step, EDC reacts with the acid groups, which yields the formation of an O-acylurea intermediate:

$$\equiv Si\text{-}(CH_2)_{10}\text{-}C\overset{O}{\underset{OH}{<}} + CH_3\text{-}CH_2\text{-}N=C=N\text{-}(CH_2)_3\text{-}N(CH_3)_2$$

$$\longrightarrow \equiv Si\text{-}(CH_2)_{10}\text{-}\overset{O}{\overset{\|}{C}}\text{-}O\text{-}C\overset{N\text{-}(CH_2)_3\text{-}N(CH_3)_2}{\underset{NH\text{-}CH_2\text{-}CH_3}{<}}$$

The O-acylurea intermediate can be transformed through three competitive reactions:

- an intramolecular rearrangement into an unreactive N-acylurea, which remains bound to the surface;
- a reaction with a neighboring surface acid group into an anhydride compound bound to the surface, which can subsequently react with NHS, yielding a succinimidyl ester bound to the surface and regenerating one surface acid group;
- a direct reaction with NHS into a surface succinimidyl ester.

Only the last two pathways yield the formation of succinimidyl ester groups, which are precisely the species whose appearance is monitored in our in-situ experiments. Therefore, a possible interpretation of the bi-exponential kinetics could be sought in terms of a competition between a fast anhydride formation and subsequent reaction with NHS, which remains the favored pathway in the first stages of the reaction as long as the surface coverage in acid groups remains high enough, and a slower direct reaction of the O-acylurea adduct with NHS. Fast anhydride formation is indeed detected when the experiment is performed in the absence of NHS.

Steric hindrance effects could also be invoked, especially in order to account for power-law kinetics. At a surface, the reactivity of a given site is likely to depend on the occupancy of the neighboring sites by bulky species. It has already been shown that when a reaction leads to the attachment at the surface of a species which might suffer from steric hindrance with its neighbors, the resulting distribution of the site reactivity as a function of the coverage in the reaction product may yield power-law kinetics (9). In the present case, two bulky groups are successively attached to the reactive site at the surface: the O-acylurea group and the succinimidyl group. Both of these groups are likely to lower the reactivity of a neighboring acid site (O-acyl urea can react with such a site and form an anhydride, but subsequent reaction with NHS will restore an acid site at a next-neighbor position of a succinimidyl-ester site), slowing down the reaction rate when the coverage in succinimidyl ester increases. Therefore, a lowering of the kinetics and a change to a power-law behavior might be accounted for by such steric hindrance effects. Noticeably, and in contrast with the case of electrochemical grafting of alkyl chains from Grignard reagents, such a mechanism would be effective even for the layers grafted on well-defined SiH surfaces, and not only for those grafted on the more disordered SiH_x surfaces.

At first sight, the weakness of the dependence of the reaction rate on EDC and NHS concentrations (close values of the time constants deduced from the fits pertaining to experiments performed at 0.005 and 0.01 mol/L) appears surprising. Such a behavior would need to be verified in an extended concentration range in order to be firmly established. Nonetheless, the present results suggest that the reaction might proceed through strong physisorption of one of the reactants. As a matter of fact, NHS adsorption has already been postulated in order to account for the limited activation efficiency on porous silicon at high NHS concentration (6). Here, we have performed complementary in-situ infrared measurements which give independent complementary indication that NHS does adsorb on an acid-terminated surface.

All in all, preliminary indications can be deduced from these in-situ infrared investigations of the activation reaction. First, NHS adsorption takes place on the acid-terminated surface, which could result in a weak dependence of the reaction kinetics on NHS concentration. Also, the kinetics of the activation reaction appear non-exponential. A fast formation of succinimidyl ester is observed during the first stages of the reaction, and a slower tailing of the reaction takes place at subsequent times. At least two effects might be invoked in order to account for this phenomenon: fast anhydride formation and reaction with NHS at low succinimidyl-ester coverage, or steric hindrance effects lowering the reactivity of acid sites at a next-neighbor position of a succinimidyl-ester reacted site. As a matter of fact, these two effects are not incompatible and may even appear closely related: the (fast) formation of anhydride is indeed more favorable at the first steps of the reaction, since it requires the presence of at least one unreacted site next neighbor of the reacting (O-acylurea modified) site. In such conditions, it might be considered that the reacting site is not much impeded (at least not fully blocked) by steric hindrance. At subsequent times, anhydride formation becomes statistically more unlikely, and steric hindrance effects plausibly become rate limiting for the formation of O-acylurea.

Amidation

The coupling of a primary amine on the activated ester surface has also been studied. This aminolysis reaction is commonly used for immobilizing biological probes on a surface; therefore, it has to remain efficient for probe (amine) concentration down to the 10^{-5} mol/L concentration range. Here, probes are mimicked by hexylamine, a short linear molecule which has been chosen in order to study the reaction kinetics in the absence of severe steric hindrance limitation. Again, infrared spectra have been recorded in situ at various stages of the reaction (here at room temperature). The experiments have been performed on mixed decyl/carboxydecyl monolayers obtained by photochemical grafting of an SiH_x surface in a mixture of decene:undecylenic acid, 1:1 in volume. The successive spectra clearly show the progressive disappearance of the succinimidyl ester from the surface (progressive build-up in the spectra of a negative triplet characteristic of succinimidyl ester). The associated build-up of the amide peaks is not conveniently seen, because of their lower intensity and, in the case of the amide I band, of the interferences with the δOH_2 vibrational band from water. Nevertheless, the intensity of the succinimidyl ester can be more easily determined than for the activation reaction, mostly because there is no interfering change of the carboxylic acid absorption. Figure 5a shows the change in the amplitude of the main succinimidyl ester peak at ~ 1745 cm^{-1} as a

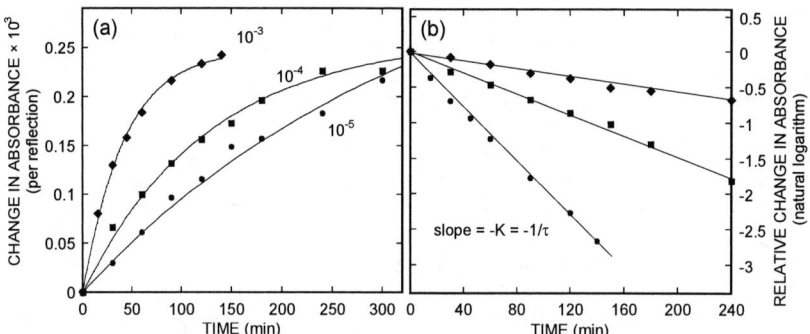

Figure 5. Plot of the change in absorbance at 1745 cm^{-1} during amidation of a succinimidyl-ester terminated surface in n-hexylamine solutions of various concentrations, as indicated in (a). The solid lines are exponential fits across the experimental points. The associated time constant τ can also be conveniently derived from the slope of a log-lin plot (b).

function of the reaction time for three amine concentrations (10^{-3}, 10^{-4} and 10^{-5} mol/L). In striking contrast with the case of the activation reaction, the kinetics are seen to be exponential, as shown in Fig. 5b. Fitting the three curves to an exponential profile yields a characteristic time for each amine concentration, namely 50 min at 10^{-3} mol/L, 130 min at 10^{-4} mol/L and 345 min at 10^{-5} mol/L (4).

In contrast with the above study of the activation reaction, it has been possible here to vary the concentration in the reacting species over an extended range. It can be seen that the characteristic time exhibits a dependence on amine concentration which is intermediate between the two limiting cases of a first-order reaction (characteristic time inversely proportional to the amine concentration) and a reaction involving an adsorbed species present at the surface at saturation coverage (characteristic time independent of the amine concentration). Here, the sublinear dependence of the amidation-reaction rate on amine concentration suggests that the reaction takes place under mixed control of amine physisorption and surface reaction. As shown in Fig. 6 (dotted line), a simple description of amine physisorption by a Langmuir isotherm approximately accounts for this dependence. However, the deviations of the model from the experimental points appear to be somewhat beyond experimental uncertainties, which calls for refining the model.

Such an improvement may be sought by remembering that amine molecules are protonated in the operating conditions. In these conditions, electrostatic interactions between adsorbed amine molecules are expected to affect the amine concentration close to the surface. Such electrostatic interactions between charged species have recently been investigated in some details by carefully studying the acid-base titration of a well-defined molecular layer bearing an acid termination. In this case, the electrostatic interactions between surface ionized groups account for the spreading of the ionization of surface acid groups over a wide pH range (10). The effect can be quantitatively modeled by identifying the extra potential due to the presence of charged species at the surface to the

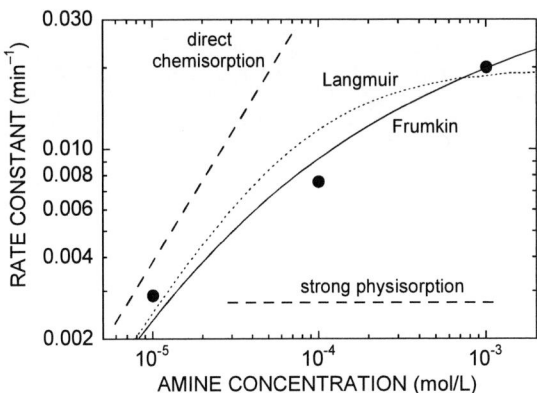

Figure 6. The three rate constants K determined from Fig. 5 exhibit a sublinear dependence against the n-hexylamine concentration. It suggests that the reaction involves physisorbed amine whose concentration is determined by an adsorption equilibrium. Accounting for electrostatic interactions among physisorbed protonated amines through a Frumkin isotherm allows for a better fit of the data than considering adsorption without interactions through a simple Langmuir isotherm.

potential drop across the Helmholtz layer. In the present case, following this procedure amounts to introducing an interaction term $e^2 N_s \theta / C_H$ in the free enthalpy of amine adsorption (with e the elementary charge, N_s the surface concentration in adsorption sites, θ the coverage in adsorbed amine, and C_H the Helmholtz capacitance) and describing the amine physisorption by a Frumkin isotherm (4). The resulting relationship between amine concentration and amidation-reaction rate constant yields a much better fit to the experimental data than previously, as shown in Fig. 6 (solid line).

Conclusion

In conclusion, in-situ infrared spectroscopy appears to bring valuable pieces of information for disentangling the mechanism of complex reactions, unreachable by ex-situ characterization of surfaces after chemical modification. For the activation of acid surfaces using the EDC/NHS treatment, the kinetics clearly appear non exponential. This behavior might arise either from the involvement of different reaction pathways as a function of the advancement degree of the reaction, or from steric hindrance effects. Longer experiments (though not easy in order to ensure stable conditions during the whole experiment) would bring valuable information for determining the asymptotic form (bi-exponential or power law) of the reaction evolution. Experiments on molecular layers in which functional chains are diluted in nonfunctional ones might also help in telling between anhydride involvement (which requires reactive neighboring sites) or steric effects (absent if dilution is high enough). The amidation reaction provides a case study in which a simple surface reaction exhibits exponential kinetics but a non trivial dependence as a function of reactant concentration, because of the existence of electrostatic interactions among the adsorbed intermediates.

References

1. R. Boukherroub, *Curr. Opin. Solid State Mater. Sci.*, **9**, 66 (2005)
2. P. Gorostiza, C. Henry de Villeneuve, Q. Y. Sun, F. Sanz, X. Wallart, R. Boukherroub and P. Allongue, *J. Phys. Chem. B*, **110**, 5576 (2006).
3. R. Voicu, R. Boukherroub, V. Bartzoka, T. Ward, J. T. C. Wojtyk and D. D. M. Wayner, *Langmuir*, **20**, 11713 (2004).
4. A. Moraillon, A. C. Gouget-Laemmel, F. Ozanam and J.-N. Chazalviel, *J. Phys. Chem. C*, **113**, 7158 (2008).
5. A. Faucheux, A. C. Gouget-Laemmel, C. Henry de Villeneuve, R. Boukherroub, F. Ozanam, P. Allongue and J.-N. Chazalviel, *Langmuir*, **22**, 153 (2006).
6. S. Sam, L. Touahir, J. Salvador Andresa, P. Allongue, J.-N. Chazalviel, A. C. Gouget-Laemmel, C. Henry de Villeneuve, A. Moraillon, F. Ozanam, N. Gabouze and S. Djebbar, submitted.
7. P. Allongue, C. Henry de Villeneuve, S. Morin, R. Boukherroub and D.D.M. Wayner, *Electrochim. Acta*, **45**, 4591 (2000).
8. N. Nakajima and Y. Ikada, *Bioconjugate Chem.*, **6**, 123 (1995).
9. S. Fellah, A. Teyssot, F. Ozanam, J.-N. Chazalviel, J. Vigneron and A. Etcheberry, *Langmuir*, **18**, 5851 (2002).
10. D. Aureau, F. Ozanam, P. Allongue and J.-N. Chazalviel, *Langmuir*, **24**, 9940 (2008).

CHAPTER 7

POROUS SEMICONDUCTORS

294

ECS Transactions, 19 (3) 295-304 (2009)
10.1149/1.3120709 ©The Electrochemical Society

Deconvolution of the Potential and Time Dependence of Electrochemical Porous Semiconductor Formation

N. Quill,[1] C. O'Dwyer,[1] R. Lynch,[1,2] D. N. Buckley[1]

Materials and Surface Science Institute, University of Limerick, Ireland

[1] *Department of Physics, University of Limerick, Ireland*
[2] Present address: *Department of Materials Science, Institute for Surface Science and Corrosion (LKO), Friedrich-Alexander University of Erlangen-Nurmberg, Germany*

> A layer of porous InP is grown beneath a thin dense surface layer when n-InP electrodes are anodized to sufficiently high potentials in aqueous KOH solutions. The shape of the linear sweep (LSV) or the cyclic voltammogram (CV) is dependent on carrier concentration. A technique is presented to deconvolute the effects of potential and time on a CV. The results obtained from this technique are used to explain the shape of the anodic current response and its relation to porous layer formation. The accuracy of the deconvolution technique is then tested by comparison to experimental results.

Introduction

The formation of porous structures during electrochemical etching of semiconductors under anodic bias has been the focus of considerable research efforts, due to the fundamental insight such studies reveal about semiconductor etching characteristics, and the potential applications of porous semiconductor structures (1-7). Although the bulk of this work has focused primarily on porous Si, a lot of attention has also been given to III-V compounds such as GaAs (8,9) and InP (10-12). There have been many models proposed (13-15) to account for the wide range of pore types, sizes and growth directions that have been observed. However, none of these models can predictably and accurately reproduce the variety of structures observed experimentally; such variations can be influenced by electrolyte type and concentration (16,17), substrate type (18), orientation (19) and doping density (20). The differences stem from the wide range of porous morphologies that are consistent with varied levels of dependence on potential, current and time, and on the relative reactivity of various crystal faces to their chemical environment. We previously reported nano-pore formation in n-InP electrodes anodized in >2 mol dm^{-3} KOH (17,21,22). These pores were shown to originate from pits etched in the surface and grow along the <111A> crystallographic directions ~(22), eventually forming a large porous domain beneath a thin (~40 nm) dense near-surface layer. The merging of the porous domains constitutes the formation of the entire porous layer.

To analyze the formation of these porous structures, linear sweep voltammetry (LSV) and cyclic voltammetry (CV) have been used. A technique to deconvolute the effects of potential and time on a CV has been developed and used to characterize the contribution/dependence of porous layer formation on potential and time during anodization.

295

Experimental

The working electrode consisted of polished (100)-oriented monocrystalline sulphur doped n-InP. An ohmic contact was made to the back of the InP sample and isolated electrically from the electrolyte by means of a suitable varnish. The electrode area was typically 0.1 cm². Two different InP wafers with different carrier concentrations, n, were used for two different series of experiments. InP anodes were used with carrier concentrations of $n = 5 - 5.6 \times 10^{18}$ cm^{-3} and a lower carrier concentration with $n = 2 - 4 \times 10^{18}$ cm^{-3}. The etch pit density of all samples used was less than 5×10^3 cm^{-2}. Anodization was carried out in aqueous 5 mol dm^{-3} KOH electrolytes. A conventional three-electrode cell configuration was used employing a platinum counter electrode and a saturated calomel reference electrode (SCE) to which all potentials were referenced. Prior to immersion in the electrolyte, the working electrode was dipped in an etchant (3:1:1 H_2SO_4:H_2O_2:H_2O) for 4 minutes and then rinsed in deionized water. All electrochemical experiments were carried out in the absence of light at room temperature.

A CH Instruments Model 650A Electrochemical Workstation was employed for cell parameter control and for data acquisition. Cleaved {011} cross-sections were examined using a Hitachi S-4800 field emission scanning electron microscope (SEM) operating at 5 kV.

Results and Discussion

A typical LSV for n-InP (carrier concentration = $5 - 5.6 \times 10^{18}$ cm^{-3}) at 2.5 mV s^{-1} in 5 mol dm^{-3} KOH is shown in Figure 1. Initially, the increase in potential corresponds to a small increase in current until a potential E_{pit} is reached; E_{pit} is the potential at which the first etch pits appear on the InP surface (23), as shown in the inset micrograph, Fig. 1a.

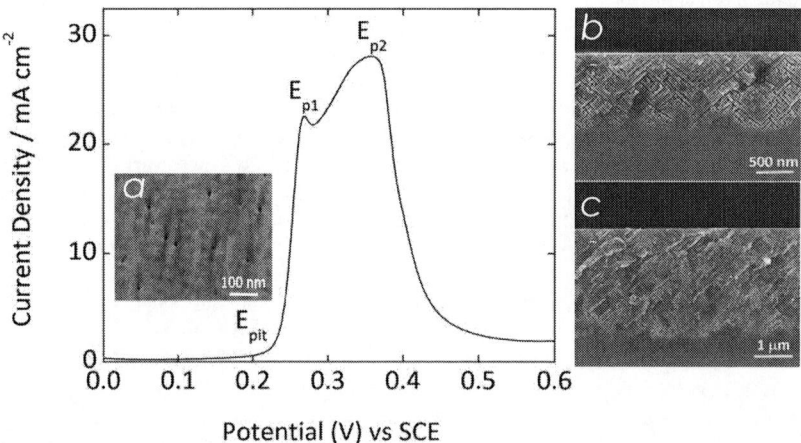

Fig. 1 Linear sweep voltammogram of an n-type InP electrode ($n = 5 - 5.6 \times 10^{18}$ cm^{-3}) in 5 mol dm^{-3} KOH at a scan rate of 2.5 mV s^{-1}. The potential was swept from 0.0 V to 0.6 V (SCE). Labelled on the plot are the pitting potential E_{pit} (0.22 V), the potential of the first current peak E_{p1} (0.27 V) and the potential of the second current peak E_{p2} (0.36 V). The SEM images were acquired after anodization to (a) E_{pit} (b) E_{p1} and (c) E_{p2}.
Deconvolution of the cyclic voltammetric response

At potentials greater than E_{pit}, the current begins to increase at a much faster rate; this current exhibits two peaks at E_{p1} and E_{p2}. In relation to the characteristics of porous layer growth observed in these samples, these current peaks correspond to merging of porous domains by growth in lateral and vertical directions, and to continued vertical growth (deepening) of the complete porous layer. This can be observed in the micrographs shown in Figs 1b and 1c. To determine to what extent the increase in current is caused by increasing potential, compared to what would be observed under potentiostatic conditions, a technique to deconvolute the effects (or contribution) of potential and time is presented below.

As the measured current I in a cyclic voltammogram (CV) is a function of both potential E and time t, we can write

$$I = I(E,t) \tag{1}$$

Therefore

$$dI = \left(\frac{\partial I}{\partial t}\right)_E dt + \left(\frac{\partial I}{\partial E}\right)_t dE \tag{2}$$

If v is the scan rate then $E = vt$ and

$$dI = \left(\frac{\partial I}{\partial t}\right)_E dt + v\left(\frac{\partial I}{\partial E}\right)_t dt \tag{3}$$

Also,

$$dI = \frac{1}{v}\left(\frac{\partial I}{\partial t}\right)_E dE + \left(\frac{\partial I}{\partial E}\right)_t dE \tag{4}$$

Thus, at any potential E_{exp} on a cyclic voltammogram, the slope of the forward and reverse curves respectively are

$$\left(\frac{dI}{dE}\right)_{FWD,E_{exp}} = \frac{1}{v}\left(\frac{\partial I}{\partial t}\right)_E + \left(\frac{\partial I}{\partial E}\right)_t \tag{5}$$

and

$$\left(\frac{dI}{dE}\right)_{REV,E_{exp}} = -\frac{1}{v}\left(\frac{\partial I}{\partial t}\right)_E + \left(\frac{\partial I}{\partial E}\right)_t \tag{6}$$

Although the functional relationship in Eqn. 1 is not necessarily the same for the forward and reverse curves, it is assumed that it is instantaneously the same at the point of potential reversal E_u. Thus it follows from Eqns 5 and 6 that

$$\left(\frac{\partial I}{\partial E}\right)_t = \frac{\left(\frac{dI}{dE}\right)_{FWD,E_u} + \left(\frac{dI}{dE}\right)_{REV,E_u}}{2} = \frac{m_f + m_r}{2} \tag{7}$$

and

$$\left(\frac{\partial I}{\partial t}\right)_E = \frac{v\left\{\left(\frac{dI}{dE}\right)_{FWD,E_u} - \left(\frac{dI}{dE}\right)_{REV,E_u}\right\}}{2} = \frac{v\left(m_f - m_r\right)}{2}$$

(8)

where m_f and m_r are the forward and reverse slopes at E_u. From Eqns 7 and 8 the effect of potential and time on a cyclic voltammogram at the point of potential reversal can be calculated simply by measuring the slopes of the current response from CVs in the forward and reverse directions.

A series of cyclic voltammograms were obtained for the higher carrier concentration samples, in which the potential was scanned from 0 V to more anodic values of the upper potential E_u over a range of scan rates (0.625 mV s^{-1} to 10 mV s^{-1}) in 5 mol dm^{-3} KOH. Typical cyclic voltammograms are shown in Fig. 2 at a single potential scan rate, where E_u is less than the first peak (E_{p1} from Fig. 1) in the LSV. It can be seen that when the potential sweep is reversed, the current increases initially until it reaches a peak, after which the current decreases non-linearly with decreasing potential.

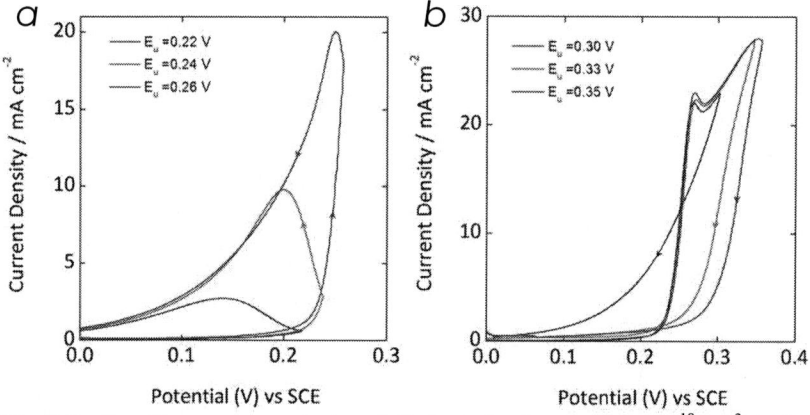

Fig. 2 Cyclic voltammograms of n-type InP electrodes ($n = 5 - 5.6 \times 10^{18}$ cm^{-3}) in 5 mol dm^{-3} KOH at a scan rate of 2.5 mV s^{-1} with (a) $E_u = 0.22$ V, 0.24V, and 0.26 V (SCE) and (b) $E_u = 0.3$ V, 0.35 V and 0.36 V (SCE).

In Fig. 2b, CVs of InP anodes with $E_u > E_{p1}$ are shown. On the reverse scan, the current densities are lower than the corresponding current densities on the forward scan and continue to decrease as the potential is decreased. Of note, we observe negligible hysteresis in CVs where $E_u = E_{p1}$, *i.e.* the reverse scan current overlays the forward scan current values.

From CVs with a range of upper potentials E_u, the slopes of the forward and reverse curves, m_f and m_r, were estimated. The resulting values are plotted in Fig. 3. The slope of the forward curve has the expected dependence for the derivative of the peaks in the cyclic voltammogram (a corresponding LSV is superimposed on Fig. 3 for comparison). The reverse slope behaves in roughly the opposite way. From the data in Fig. 3, $(\partial I/\partial E)_t$ and $(\partial I/\partial t)_E$ were estimated according to Eqns 7 and 8, and the values obtained were plotted against upper potential E_u to obtain the plot in Fig. 4.

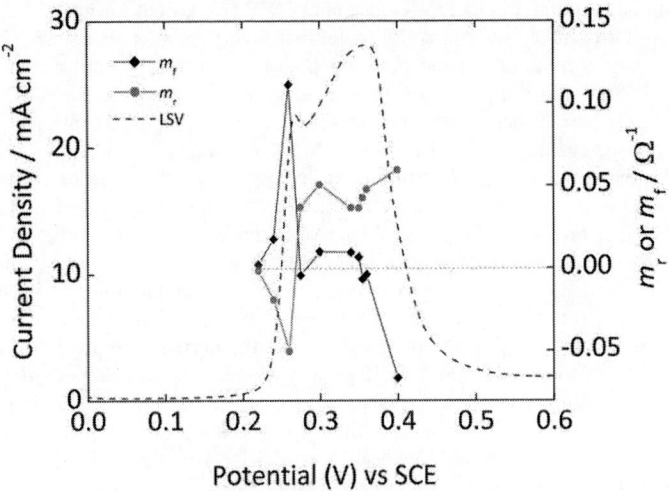

Fig. 3 Potential dependence of the slopes of the forward and reverse curves, m_f and m_r, at the upper potential limit E_u for a series of cyclic voltammograms of n-InP electrodes ($n = 5 - 5.6 \times 10^{18}$ cm^{-3}) in 5 mol dm^{-3} KOH at a scan rate of 2.5 mV s^{-1}. The corresponding LSV is also shown for comparison.

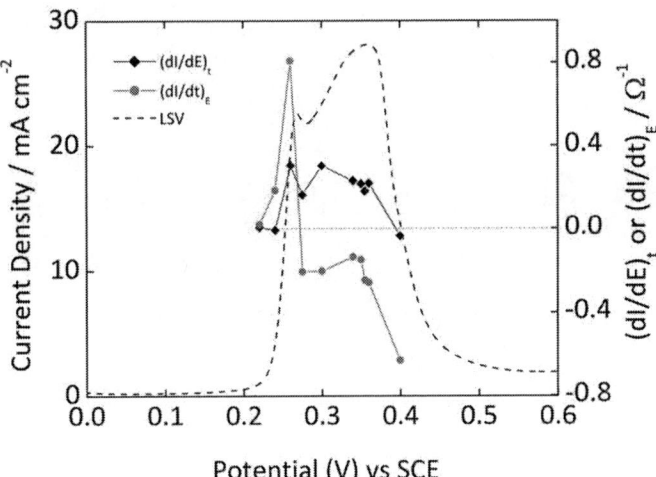

Fig. 4 Variation of $(\partial I/\partial E)_t$ and $(\partial I/\partial t)_E$ at the upper potential limit E_u, derived from the data in Fig. 3 for CVs of InP electrodes ($n = 5 - 5.6 \times 10^{18}$ cm^{-3}) in 5 mol dm^{-3} KOH at a scan rate of 2.5 mV s^{-1}. The corresponding LSV is shown for comparison. The values for $(\partial I/\partial t)_E$ as calculated by Eqn. 8, were scaled by a multiplicative factor of 400 s V^{-1} (*i.e.* the inverse of the scan rate).

It can be seen that at potentials less than 0.22 V (E_{pit}), both quantities are essentially null since the current flowing at that potential is very small (~1 mA cm^{-2}). However, $(\partial I/\partial t)_E$ rises sharply after E_{pit} and stays much larger than $(\partial I/\partial E)_t$ until the first peak, E_{p1}, is reached. Then, between E_{p1} and the second peak, E_{p2}, it is $(\partial I/\partial E)_t$ that is dominant. Finally, as the current begins to decrease, $(\partial I/\partial E)_t$ becomes approximately 0 and $(\partial I/\partial t)_E$ dominates once again.

These experiments were repeated for n-InP samples with a carrier concentration in the range $2 - 4 \times 10^{18}$ cm^{-3}. The LSV of the lower carrier concentration anode is shown as a dashed line in Fig. 5a. The most obvious difference with respect to Fig. 2 is that there is now only one current peak. It is also noted that this peak occurs at a higher potential ($E_p = 0.48$ V (SCE)) than the peaks in Fig. 2. The corresponding CVs with E_u less than the peak potential are shown for the lower carrier concentration sample in Fig. 5a. It can be seen that upon reversing the potential sweep, the current continues to increase and eventually reaches a peak, before decreasing with reducing applied potential. Figure 5b shows corresponding CVs with E_u greater than the peak potential. The current density is routinely observed to decrease for the entire reverse sweep. Clearly the behaviour of the lower doped electrodes around the peak potential is the same as the behaviour of the more highly doped electrodes at potentials close to E_{p1}.

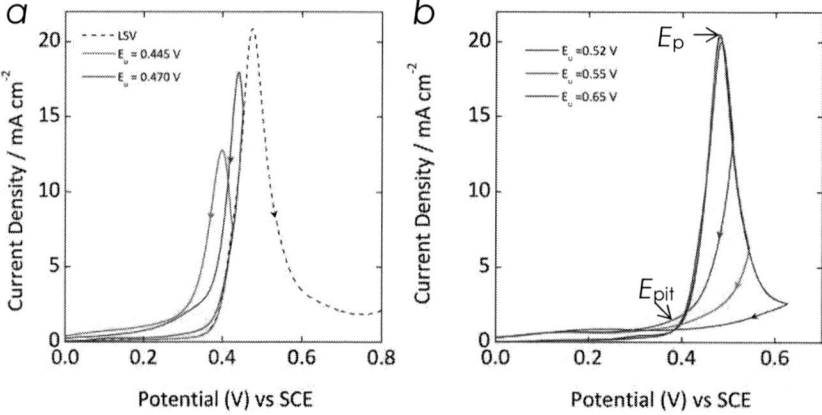

Fig. 5 Cyclic voltammograms of InP electrodes ($n = 2 - 4 \times 10^{18}$ cm^{-3}) in 5 mol dm^{-3} KOH at a scan rate of 2.5 mV s^{-1} with (a) $E_u = 0.445$ V and 0.47 V (SCE) and (b) $E_u = 0.52$ V, 0.55 V and 0.65 V. The linear potential sweep from 0.0 V to 0.8 V at 2.5 mV s^{-1} is also shown (dashed line in (a)).

As before, the slopes of the forward and reverse curves, m_f and m_r, were estimated from cyclic voltammograms (of the lower carrier concentration InP) to a range of upper potentials, and the resulting values are plotted in Fig. 6. The deconvolution technique was applied to this data and the results are plotted in Fig. 7.

It can be seen from Fig. 7 that the changing potential minimal effect on the current flow for the InP samples of lower carrier concentration. The increase in current (and also its decrease) is observed to occur under potentiostatic conditions (where the potential is stepped to values in the range E and this has been verified by experiment (21).

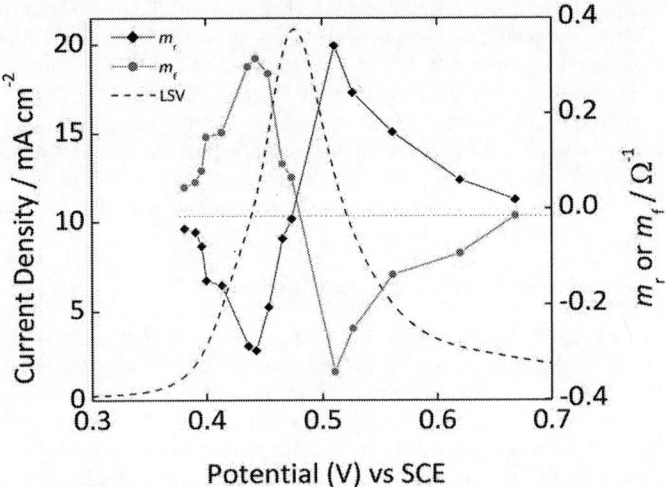

Fig. 6 Potential dependence of the slopes of the forward and reverse curves, m_f and m_r, at the upper potential limit E_u for a series of cyclic voltammograms of n-InP electrodes ($n = 2 - 4 \times 10^{18}$ cm^{-3}) in 5 mol dm^{-3} KOH at a scan rate of 2.5 mV s^{-1}. The corresponding LSV is shown for comparison.

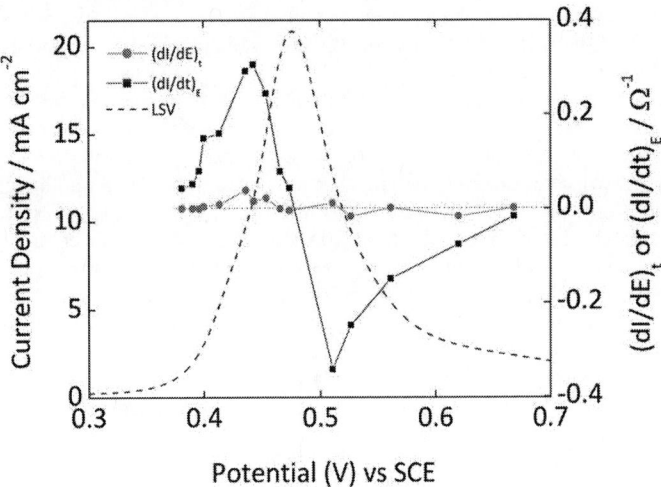

Fig. 7 Potential dependence of $(\partial I/\partial E)_t$ and $(\partial I/\partial t)_E$ at the upper potential limit E_u, derived from the data in Fig. 6 for cyclic voltammograms of InP electrodes (carrier concentration $= 2 - 4 \times 10^{18}$ cm^{-3}) in 5 mol dm^{-3} KOH at a scan rate of 2.5 mV s^{-1}. The values for $(\partial I/\partial t)_E$ as calculated by Eqn 8, were scaled by a multiplicative factor of 400 s V^{-1} (*i.e.* the inverse of the scan rate). The corresponding LSV is shown for comparison.

The principal difference between the anodic behaviour of the two InP anodes is that the higher carrier concentration sample reaches its peak in current at a much lower potential. At potentials greater than E_{p1}, the current continues to increase linearly until a second peak is reached. This potential controlled region is completely absent from the deconvoluted curve of InP samples doped at $n = 2 - 4 \times 10^{18}$ cm^{-3}. Dedicated AFM and SEM observations have shown that both the solitary peak in the low concentration sample (c.f. Fig. 5a), and E_{p1} in the higher concentration sample (c.f. Fig. 1), correspond to the merging of porous domains to form one continuous porous layer beneath the surface (23).

Relating the deconvolution technique to porous layer growth

These results can be understood in the framework of the following model for the relation between the shape of the LSV and the characteristics of porous layer growth. At potentials less than E_{pit}, almost no current flows through the electrode because of the space charge double layers that are present at the interface between electrode and electrolyte; no etching would formally occur in the case of complete carrier depletion in the InP. As the potential is increased past E_{pit}, carriers can gain enough energy to tunnel through the space charge layer resulting in the initiation of etching at the surface (21). From this point on the current begins to increase rapidly. This is due to more and more pits forming in the surface as the potential is increased and the sub-surface propagation of etching pore tips within the porous domains. This leads to an increased number of current paths and consequently an increased current density. Deconvolution of the CVs associated with porous layer growth suggests that the increase is much more significant than can be accounted for by the increasing potential alone.

The onset of the first current peak corresponds to the merging of the porous domains into a single porous layer, as shown in Fig. 1. As the potential is increased further, the current density increase cannot be accounted for by just an increase in the area it flows through. This corresponds to a limit in the number of new/existing pore formation/growth after complete domain merging. Therefore any increase in current density must be directly due to processes driven by an increase in applied potential. Even though no new domains can form, the current can still cause existing domains to grow in the vertical direction. This is what is observed after E_{p1} in the higher carrier concentration sample. The increase in depth of the fully formed porous layer occurs as the current density increases linearly with increasing potential. Only a single peak is observed for the lower carrier concentration samples under identical conditions. In addition, porous layers formed in such samples do not grow to the same depth after domain merging (~1 μm for the low concentration sample and ~3 μm for the high concentration sample). Similar behaviour is observed in the higher carrier concentration electrode after the second current peak, E_{p2}.

To assess the accuracy and applicability of the deconvolution technique to porous layer growth, a series of experiments were performed (using the higher carrier concentration samples) in which the potential was scanned from 0 V to different values of E_u. The potential E_u was then held constant in order to observe the effect of time on the electrochemical process. The slope of the current-time curve was measured to verify that the value of $(\partial I/\partial t)_E$ given by the deconvolution technique agrees with experimental observation and the results are given in Fig. 8. We observe good agreement in the data in Fig. 8, showing that the technique can produce quantitative information on independent potential and time contributions to the growth of the porous layer.

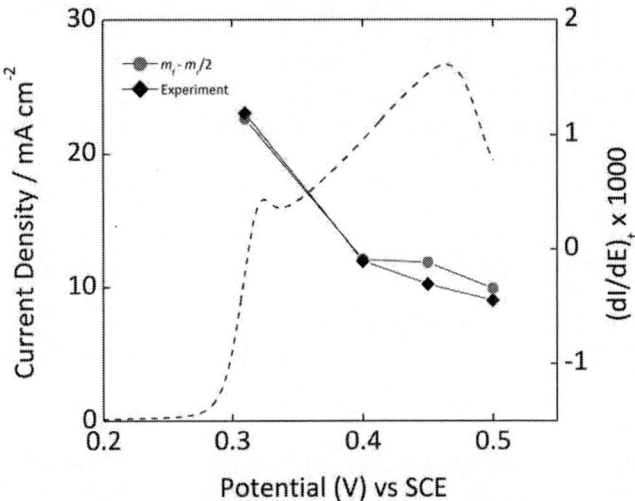

Fig. 8 Plot of the rate of change of current with respect to time at constant potential $(\partial I/\partial t)_E$ for InP anodes ($n = 5 - 5.6 \times 10^{18}$ cm^{-3}) anodized in 5 mol dm^{-3} KOH as measured by experiment and calculated by the deconvolution technique. The corresponding LSV is superimposed for comparison.

Conclusions

A technique to deconvolute the effects of potential and time on linear and cyclic sweep voltammograms was developed and used to explain the shape of the anodic response of n-InP in 5 mol dm^{-3} KOH. This technique analyzes the instantaneous value of the rate of change of current with respect to time and potential by extracting the slopes of the current profile acquired during porous layer formation. Characteristic voltammograms are observed for n-InP anodes with different carrier concentrations. For InP electrodes with $n = 5 - 5.6 \times 10^{18}$ cm^{-3}, the voltammetric response showed two distinct current peaks on the forward potential sweep; the first current peak was shown to be caused by the merging of porous domains (growing both vertically and laterally) into a continuous porous layer. In this potential region, the porous layer growth is predominatly a function of time, i.e. the porous layer would continue to grow if the potential was swept to, and held, at values in the range $E_{pit} - E_{p1}$. The 'ohmic' increase in current after the first peak was shown to be due to the vertical growth of the fully formed porous layer.

The deconvolution technique was shown to be valid at different carrier concentrations and potential sweep rates. InP anodes with $n = 2 - 4 \times 10^{18}$ cm^{-3}, only exhibited a single current peak on the forward sweep, but equally amenable to deconvolution using this technique. The deconvoluted values for $(\partial I/\partial t)_E$ and $(\partial I/\partial E)_t$ agree well with the experimental values obtainable for these quantities, showing the technique to be both qualitatively and quantitatively accurate in relating the cyclic voltammogram to the resulting porous layer formation.

References

(1) L.T. Canham, *Appl. Phys. Lett.*, **57**, 1046 (1990).

(2) H. Föll, *Appl. Phys. A*, **53**, 8 (1991).

(3) T. Holec, T. Chvojka, I. Jelínek, J. Jindřich, I. Němec, I. Pelant, J. Valenta and J. Dian, *Mater. Sci. Eng. C*, **19**, 251 (2002).

(4) R.J. Martín-Palma, J.M. Martínez-Duart, L. Li and R.A. Levy, *Mater. Sci. Eng. C*, **19**, 359 (2002).

(5) A. Matoussi, T. Boufaden, A. Missaoui, S. Guermazi, B. Bessaïs, Y. Mlik and B. El Jani, *Microelectronics Journal*, **32**, 995 (2001).

(6) A. Jain, S. Rogojevic, S. Ponoth, N. Agarwal, I. Matthew, W.N. Gill, P. Persans, M. Tomozawa, J.L. Plawsky and E. Simonyi, *Thin Solid Films*, **398**, 513 (2001).

(7) N.E. Chayen, E. Saridakis, R. El-Bahar, Y. Nemirovsky, *J. Molec. Biol.*, **312**, 591 (2001).

(8) S. Langa, J. Carstensen, M. Christophersen, H. Föll, and I.M. Tiginyanu, *Appl. Phys. Lett.*, **78**, 1074 (2001).

(9) G. Oskam, A. Natarajan, P.C. Searson and F.M. Ross, *Appl. Surf. Sci.*, **119**, 160 (1997).

(10) S. Langa, J. Carstensen, I.M. Tiginyanu, M. Christophersen and H. Föll, *Electrochem. Solid-State Lett.*, **4**, G50 (2001).

(11) S. Langa, I.M. Tiginyanu, J. Carstensen, M. Christophersen and H. Föll, *Electrochem. Solid-State Lett.* **3**, 514 (2000).

(12) R.L. Smith and S.D. Collins, *J. Appl. Phys.*, **71**, 8 (1992).

(13) H. Föll et al, *Phys. Stat. Sol. (a)*, **182**, 7 (2000).

(14) T. Unagami, *J. Electrochem. Soc.*, **127**, 476 (1980).

(15) V. Lehmann and U. Gösele, *Appl. Phys Lett.* **58**, 856 (1991).

(16) P. Schmuki, J. Fraser, C.M. Vitus, M.J. Graham, H.S. Isaacs, *J. Electrochem. Soc.*, **143**, 3316 (1996).

(17) R. Lynch, C. O'Dwyer, D.N. Buckley, D. Sutton and S.B. Newcomb, *ECS Trans.*, **2**, 131 (2006).

(18) M. Christopherson, J. Carstensen, A. Feuerhake and H. Föll, *Mater. Sci. Eng. B*, **69**, 70, 194 (2000).

(19) S. Rönnebeck, J. Carstensen, S. Ottow and H. Föll, *Electrochem. Solid-State Lett.*, **2**, 126 (1999).

(20) P. Schmuki, L.E. Erickson, D.J. Lockwood, J.W. Fraser, G. Champion, H.J. Labbé, *Appl. Phys. Lett.*, **72**, 1039 (1998).

(21) C. O'Dwyer, D.N. Buckley, M. Serantoni, D. Sutton and S.B. Newcomb, in *Proceedings of the State-of-the-Art Program on Compound Semiconductors XXXIX*, R.F. Kopf, D.N. Buckley, F. Ren, C. Monier, K. Shiojima, A.G. Baca, H.M. Ng, T.D. Moustakas, S.J. Pearton, Editors, **PV 2003-11**, p. 136, The Electrochemical Society, Proceedings Series, Pennington, NJ (2003)

(22) R. Lynch, C. O'Dwyer, D. Sutton, S.B. Newcomb and D.N. Buckley, *ECS Trans.*, **6**, 355 (2007).

(23) C. O'Dwyer, D.N. Buckley, D. Sutton, M. Serantoni and S.B. Newcomb, *J. Electrochem. Soc.*, **154**, H78 (2007).

Effects of "P-N" Terminations on the Initial Stages of Pore Growth onto *n*-InP in HCl Aqueous Solution.

A.-M. Gonçalves, L. Santinacci, N. Simon, M. Bouttemy, C. Mathieu, A. Etcheberry

Lavoisier Institute UMR 8180, University of Versailles, 78035 Versailles Cedex, France

> In this paper we report, for the first time, a galvanostatic control in 1M HCl aqueous solution onto a InP surface that had been previously entirely nitrogenated. The surface nitrogenation required an anodic electrochemical treatment in acidic liquid ammonia (NH_3 *liq.*). An homogeneous covering film with " P—N " terminations was obtained onto the InP surface properly deoxidized. This thin film is notable for its lack of air ageing and also for its chemical stability in HCl. A galvanostatic control at different current density (10 mA.cm^{-2}, 100 mA.cm^{-2} and 300 mA.cm^{-2}) in 1M HCl was used to identify the effect of this " P—N " terminations on the porosification process. In comparison to a bare surface, a contrasted evolution of the interfacial potential was clearly observed for low current density. As expected, a crystal oriented pore morphology (CO) is obtained onto a bare InP surface whereas "oscillating" current line oriented pore morphology (CLO) is kept onto a surface entirely recovered by a " P—N " film. The presence of the nitrogenated film can therefore be significant on the pore growth evolution in 1M HCl aqueous solution.

Introduction

In aqueous solutions, a wide variety of pore morphologies has been reported onto semiconductors (SC) (1, 2). Pore growths are essentially controlled by the applied potential or current density. The change from a crystal oriented pore morphology (CO) to a current line oriented pore morphology (CLO) occurred around 10 mA.cm^{-2}. In aqueous solution, whatever the pH, oxidized or hydroxides terminations could not be avoided on the surface before the porosification process. The question of the surface chemistry of the interface that undergone the dissolution is not well established with the very difficult question of an eventual specific surface layer that trigger the initiation of pores.

The control of the chemical surface provides then an original way to understand the critical initial step of SC porosification. We have already published that we are able to electrochemically passivate InP in acidic liquid ammonia (NH_3 *liq.*). This passivation was ascribed to an anodic wave obtained during the first scan by cyclic voltametry in acidic NH_3 *liq.* (3). The InP surface was indeed covered by a thin film involving " P—N " terminations. A specific "P—N" surface bond chemistry was therefore associated to a phosphinimidic amide-like electrodeposition involving both InP and solvent oxidation (3, 4):

$$InP + 9NH_3 + 9h^+ \rightarrow \text{« } H_2N\text{-}P\text{=}NH \text{ »} + 6NH_4^+ + In^{3+} + \tfrac{1}{2} N_2 \tag{1}$$

Stability tests of the " P—N " film have also been performed. The passivated film remains unchanged even after one year ageing in air at room temperature (3). The InP air immunity requires a complete recovering of the surface sites by the " P—N " film. As a result, InP active sites are no more available for native oxide evolutions on the surface. NH_3 *liq.* is then single out to perform an anodic treatment in true water free condition since oxidized terminations such as "In-O" or "P-O" can be definitively excluded on the surface. The nitrogenated film is also chemically stable since it withstands aqueous immersion at concentrated solution of 2M HCl. The high chemical stability of the nitrogenated film offers then an opportunity to study the interface response during a galvanostatic control in HCl solution. Do we observe a pore growth formation with " P—N " film onto InP surface? The aim of this paper is therefore to determine the effect of this non oxidized chemical surface on the initial stage of InP growth in HCl aqueous solution.

Experimental

n-InP wafers with a (100) orientation were purchased from InPact Electronic Materials, Ltd, with a doping density of 10^{18} cm^{-3}. The wafers were cut into small squares $(0.5 \times 0.5 \ cm^2)$. Prior to use, semiconductors were chemomechanically polished with a solution of bromine in methanol (2%), and rinsed with purest methanol and dried under an argon stream (5, 6). In order to remove residual oxides, samples were dipped for a few minutes in 2M HCl just before the experiment (7, 8).

The electrochemical set-up was a classical three-electrode device, using a 283 EG&G potentiostat-galvanostat with a linear potential scan of 20 mV/s. Pore formation were performed in analytical grade 1M HCl with a galvanostatic control of 10 $mA.cm^{-2}$, 100 $mA.cm^{-2}$ and 300 $mA.cm^{-2}$. The chemical environment of InP surface was previously controlled by cyclic voltametry in NH_3 *liq.* Ammonia condensation, from gaseous ammonia ('electronic grade' from Air Liquide), was provided by a glass column assembly and required a low operating temperature under atmospheric pressure (9). An electrochemical cell filed up with 150 cm^3 of liquid ammonia was maintained at 223 K in a cryostat. The acidic medium was obtained by addition of 0.1M NH_4Br (purest available quality from Aldrich). All potentials were measured *vs.* a silver reference electrode (SRE) (10). Current voltage curves were performed under illumination using an optical fiber with a tungsten lamp (11). After the anodic treatment in NH_3 *liq.* at open circuit, InP was rinsed in purest liquid ammonia. The sample was then transferred toward an XPS analyzer using a specific procedure avoiding any air contamination.

XPS analysis was performed on a V.G-Escalab 220 iXL spectrometer. A focused monochromated X-ray beam (Al Kα) was used for excitation. For detection, a constant analyzer energy mode was used, with pass energy of 8 or 20 eV. Photoelectrons were collected perpendicularly to the surface. Calibration of the spectrometer was done using the E902-94 ASTM procedure. Fitting procedures were performed with the Thermo VG ECLIPSE and AVANTAGE programs.

Morphology of the porous layer was characterized by SEM using a Jeol JSM 5800 tungsten filament.

Results

"P-N" film formation

Due to its high purity (electronic grade quality), liquid ammonia (NH_3 *liq.*) is able to perform anodic treatment in true water free condition. As a polar solvent, NH_3 *liq.* exhibits unique physical properties such as low viscosity allowing better conductivity than in aqueous media. Its low dielectric constant leads to ammonia molecules contributions on interfacial processes. Thus, a new interface is expected due to oxygen free surfaces and an interfacial electrochemistry ruled by ammonia molecules. In acidic liquid ammonia, we are able to passivate InP surface by anodic cyclic voltammetry. An anodic wave was systematically observed during the first scan onto a freshly polished InP surface (see figure 1). According to the low anodic charge involved in the passivating process ($\approx$ 5-10 mC.cm^{-2}) and the preceding equation (1), the passivated film might reach a few " P—N "monolayers. The thickness of this passivated film was experimentally clearly confirmed by the perfect detection of the InP matrix contribution from XPS analysis. XPS measurements allow the comparison of atomic surface ratios from InP surface before and after the formation of the " P-N " film (see figure 1 inset). After dipping the " P-N " film into 2M HCl, no evolution of " P-N " spectra was detected. As no evolution of the surface composition was depicted (similar surface ratios and constant photopeaks binding energy positions), we clearly demonstrated the chemical stability of the passivated film in aqueous (2M) HCl solution as well as in the open air. In these conditions, the transfer of the nitrogenated surface can proceed in 1M HCl.

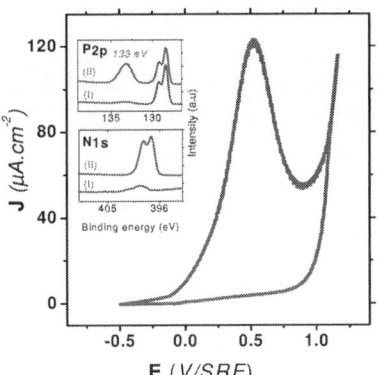

Figure 1. Formation of an anodic wave during the first current-voltage curve under illumination in acidic NH_3 *liq.* at 223 K, (1M NH_4Br), v = 20mV/s. Inset: Evolution of XPS analysis before (I) and (II) after recording the anodic wave.

Interfacial evolution in HCl

A galvanostatic control at $10 \, mA.cm^{-2}$ in 1M HCl was used to identify the effect of this " P—N " coverage (figure 2). As expected, no potential oscillations were observed onto a bare surface.

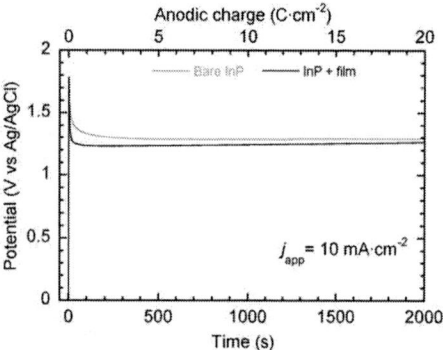

Figure 2. Comparison of the potential evolution, for a galvanostatic control at $10 \, mA.cm^{-2}$ in 1M HCl, onto a bare InP surface and a InP surface covered by a "P—N" film.

In comparison to a bare InP surface, no potential oscillations were detected onto a surface entirely recovered by a "P—N" film (figure 2). The potential evolution was however slightly higher ($\approx 100mV$) during all the galvanostatic treatment ($20 \, C.cm^{-2}$). This experimental data confirm again the thickness of the covering film. The potential evolution was slightly resisted by the "P—N" film. The lack of spectacular potential deviation may also emphasize the low electrical blocking properties of the passivated film or, in other words, the conducting electrical properties of the "P—N" film.

A current density equal to $10 \, mA.cm^{-2}$ was used since the resulting pore morphology was clearly defined in 1M HCl aqueous solution. Cross-section morphologies of the resulting porous layer was then characterized by SEM (figure 3). As expected for low current density, crystal oriented pore morphology (CO) was exhibited onto a bare InP surface (see figure 3A and the corresponding enlarged view A). For the same coulometric charge ($20 \, C.cm^{-2}$) and applied currend density ($10 \, mA.cm^{-2}$), a contrasted evolution of the pore morphology was observed for a InP surface previously coated by a "P—N" film. A wavy aspect of the dissolution front layer was indeed described (see figure 3 A') in opposition to the homogeneous dissolution front observed onto a bare InP surface (see figure 3 A) indicating that the initial stages of the porosification were inescapably perturbed by the "P—N" film. In spite of the thickness of the film, the dissolution process occurred obviously through the lowest electrical resistant area of the coated film. The wavy aspect of the dissolution front can be then associated to an electrical gradient on the "P—N" film. In the initial stage of the porosification, preferential dissolutions appeared, probably due to the inhomogeneous charge distribution involved in the "P—N" film. The wavy aspect of the dissolution layer was already been exhibited during the porosification of n-InP in acidic NH_3 $liq.$ (12, 13). In this significant non aqueous solvent, a specific "P—N" film surface was evidenced before the porosification initiation (13). The wavy aspect of the dissolution front layer seemed therefore associated to the specific "P—N" film onto n-InP surface. The resulting pore morphology was then modified since tubular pores were described through tetrahedral channels (2). For the same coulometric charge (20 C.cm^{-2}) and low applied current density (10 mA.cm^{-2}), the pore growth changes from crystal oriented pore morphology (CO) to current line oriented pore morphology (CLO) when the InP surface was coated by the "P—N" film. As a result, the "P—N" film might behave as an eventual specific surface layer that triggers the initiation of pores.

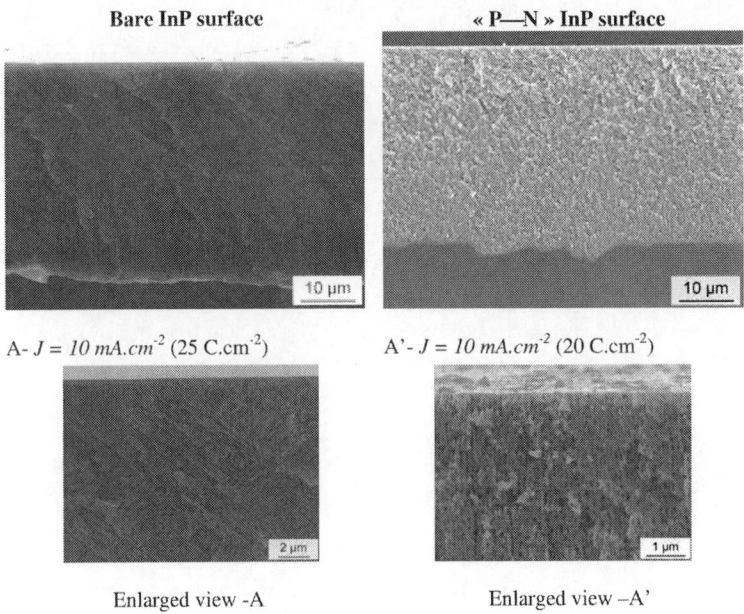

Bare InP surface   **« P—N » InP surface**

A- $J = 10\ mA.cm^{-2}$ (25 C.cm^{-2})   A'- $J = 10\ mA.cm^{-2}$ (20 C.cm^{-2})

Enlarged view -A   Enlarged view –A'

Figure 3. SEM cross sections pictures of pore morphology of n-InP surfaces obtained at low current density ($J = 10$ mA.cm^{-2}) in galvanostatic mode ($Q= 20$ C.cm^{-2}) in HCl (1M) aqueous solution. Two contrasted morphologies are ascribed according to the initial chemical state of the SC surface: a bare InP surface (right side-A) and a InP surface covered by a "P—N" film (left side-A').

The effect of the "P—N" film was also explored for higher current density. A current density equal to 100 mA.cm^{-2} and 300 mA.cm^{-2} was used since the resulting pore morphology was clearly defined in 1M HCl aqueous solution. The galvanostatic control was performed onto two different chemical surfaces of InP. The bare surface of InP was systematically used as a reference behaviour. The interfacial potential evolution onto the bare surface was reported in the figure 4. This potential evolution was compared to that observed on the InP surface entirely recovered by a " P—N " film.

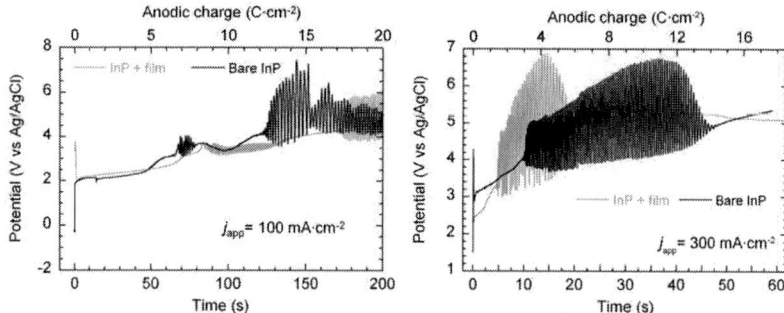

Figure 4. Comparison of the potential evolution in 1M HCl, between a bare InP surface and a InP surface covered by a "P—N" film for a galvanostatic control at 100 mA.cm^{-2}, (left side) and a galvanostatic control at 300 mA.cm^{-2}, (right side).

For a galvanostatic control at 100 mA.cm^{-2}, no contrasted evolution of the interfacial potential evolution was observed according to the chemical surface of n-InP. Concerning a bare surface, for a galvanostatic control at 100 mA.cm^{-2}, a gradual increase of the potential was evidenced from 1.8 V to 5 V (1). Nearly 3 volts of interfacial potential variation was noted onto a bare InP surface. For anodic charge higher than 13 C.cm^{-2}, oscillations with large amplitude ($\approx$ 1.5V) were exhibited. These oscillations were also combined with the high increase of the interfacial potential. Onto a " P—N " film, a high step increase of the potential up to 4V was initially evidenced and the potential rapidly stabilized and follow the potential evolution already observed onto a bare InP surface. Apart from this high step potential increase, no contrasted evolution of interfacial potential was evidenced onto a " P—N " film. This fact was also observed for higher current density (300 mA.cm^{-2}).

In order to extend the effect of the " P—N " film on the resulting pore formation, SEM pictures were reported on the figure 5. As supposed onto a bare surface, a gradual evolution of the pore morphology is evidenced (1, 2). For low anodic charge, crystal oriented pore morphology (CO) was expected to grow (see figure 3) whereas for higher anodic charge, this initial morphology switches to a current line oriented pore morphology (CLO) as it is reported on the figure 5. These morphology variations are directly correlated to the evolution of the interfacial potential (see figure 4). A wavy aspect (see enlarged view in figure 5) was indeed related to the large oscillations observed during the potential evolution onto a bare surface. Concerning the InP surface covered by the " P—N " film, no significant deviation of the pore morphology was observed apart from the wavy aspect due to the dissolution front layer (figure 5). For high anodic charge, current line oriented pore morphology was then exhibited through the " P—N " film.

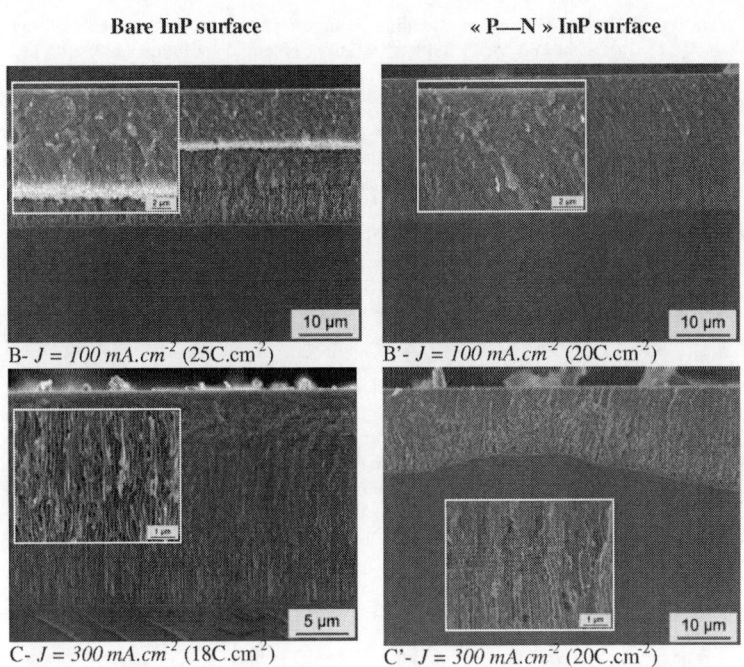

Bare InP surface	« P—N » InP surface
B- $J = 100\ mA.cm^{-2}$ (25C.cm^{-2})	B'- $J = 100\ mA.cm^{-2}$ (20C.cm^{-2})
C- $J = 300\ mA.cm^{-2}$ (18C.cm^{-2})	C'- $J = 300\ mA.cm^{-2}$ (20C.cm^{-2})

Figure 5. SEM cross sections pictures of pore morphology of n-InP surface obtained at high current density ($J = 100$ mA.cm^{-2} and $J = 300$ mA.cm^{-2}) in galvanostatic mode ($Q \approx 20$ C.cm^{-2}) in HCl (1M) aqueous solution. Apart from the dissolution front layer, no contrasted morphologies are ascribed according to the initial chemical state of the SC surface: a bare InP surface (right side) and a InP surface covered by a "P—N" film (left side).

Conclusion

In spite of the complete coverage of the surface by a passivated thin film with "P—N" terminations, the porosification of n-InP was not prevented. A singular porosification was evidenced at low current density ($J = 10$ mA.cm^{-2}) in galvanostatic mode ($Q= 20$ C.cm^{-2}) in HCl (1M) aqueous solution. The pore growth changes from crystal oriented pore morphology (CO) to current line oriented pore morphology (CLO) when the InP surface was previously coated by the "P—N" film. Since the nucleation phase seems different, an influence of the surface chemistry might be taken into account during the initial step of pore growth. The nitrogenated film is therefore significant on the pore growth. At low current density ($J = 10$ mA.cm^{-2}), the "P—N" film might behave as an eventual specific surface layer that trigger the initiation of pores.

For higher current density ($J = 100$ mA.cm^{-2} and $J = 300$ mA.cm^{-2}) in galvanostatic mode

($Q= 20$ C.cm^{-2}) in HCl (1M) aqueous solution, no significant contrasted morphology was evidenced. This fact might be correlated to the similar potential evolution observed onto both InP chemical surface. The stability of the "P—N" film at high current density is still an open question. XPS analyses at low anodic charges might help to answer.

To access to a better understanding on the effect of the "P—N" film, a partial recovering of the InP surface by the "P—N" film can be consider (14). A partial surface nitrogenation requires a previous anodic galvanostatic treatment in aqueous solution at pH9. Using a current density of 0.2 mA.cm^{-2}, the InP surface was then recovered by oxyde islands. Due to their stability in acidic NH$_3$ $liq.$, only the non oxidized parts of InP surface were recovered by the passivated " P—N " film. With a " P—N " film, another anodic parasitic process can be involved such as the oxidation of the solvent. Quantitative analyses of the dissolved products by atomic adsorption are in progress and should be helpful to understand the effect of the "P—N" film on pore growth.

References

1. L. Santinacci and T. Djenizian, *C. R. Chimie.*, **11**, 964 (2008).
2. H. Foll, S. Langa, J. Carstensen, M. Christophersen and I. M. Tiginyanu, *Adv. Mater.*, **15**, 183 (2003).
3. A.-M. Gonçalves, O. Seitz, C. Mathieu, M. Herlem, A. Etcheberry, *Electro. Chem. Comm.*, **10/2**, 225 (2007)
4. A.-M. Gonçalves, O. Seitz, A. Eb, C. Mathieu, M. Herlem and A. Etcheberry,. *ECS Trans.*, **6**(2); 461 (2007).
5. H.L. Hartnagel, Oxides and oxide films, Vol 6, Marcel Dekker, New York and Basel, (1986)
6. G. Hollinger, E. Bergignat, J. Joseph, Y. Robach, *J. Vac. Sci. Technol.*, **A3**, 2082 (1985).
7. P.H.L. Notten, J.E.A.M. Van den Meeraker, J.J. Kelly, Etching of III-V Semiconducors: An Electrochemical Approach, Elsevier, Amsterdam, (1991).
8. A. Pakes, P. Skeldon, G.E. Thompson, S. Moisa, G.I. Sproule and M.J. Graham., *Corrosion Science*, **44**, 2161(2002).
9. J. Jander, Anorganische und allgemeine Chemie in flüssigen Ammoniak. Part I, Friedr. (Eds. Vieweg & Sohn), Braunschweig (1966).
10. D. Guyomard, M. Herlem, C. Mathieu, J. L. Sculfort, *J. Electroanal Chem.*, **216**,101 (1987).
11. A.-M. Gonçalves, C. Mathieu, M. Herlem, A. Etcheberry. *J. Electroanal Chem.*, **462**, 88 (1999).
12. A.-M. Goncalves, L. Santinacci, C. David, C. Mathieu, M. Herlem, A. Etcheberry, *Physica Status Solidi A.*, **204**, (2007), 1286.
13. A-M. Gonçalves, L. Santinacci, A. Eb, I. Gerard, C. Mathieu, A. Etcheberry, *Electrochemical and Solid state Letters.*, **06** (2006), 1730.
14. A-M. Gonçalves, N. Simon, C. Mathieu, A. Etcheberry, *C. R. Chimie.*; **11**, (2008), 1037.

ECS Transactions, 19 (3) 313-319 (2009)
10.1149/1.3120711 ©The Electrochemical Society

Unexpected Dissolution Process at Porous n-InP Electrodes

L. Santinacci, M. Bouttemy, I. Gérard, and A. Etcheberry

Institut Lavoisier de Versailles (UMR CNRS 8180), Université de Versailles Saint Quentin, France

In this paper, we report an unexpected dissolution phenomenon at n-InP electrodes during porous etching in $1\,M$ HCl
under potentiostatic conditions ($5\,V$ *vs.* Ag/AgCl). Although
the scanning electron microscope examinations show the typical
tubular and regular pores that exhibit a depth proportional to
the anodic charge, further chemical analyses of the dissolved material performed after the pore formation by atomic absorption
spectroscopy reveal a dual dissolution process. An electrochemical etching exhibiting a valence of 8 and a simultaneous chemical
dissolution are evidenced.

Introduction

In the present days, one-dimensional nanostructures are intensively investigated since peculiar
properties originate directly from their specific geometries. Anodic porous semiconductors are obviously part of such nanostructures' family. Although numerous studies have been performed onto
III-V semiconductors, most of the mechanistic information report to porous etching of silicon (1).
Gravimetric investigations have, for instance, given access to the dissolution valence (z) during both
electropolishing and porous etching of Si in HF containing solutions (2). In the case of Si electropolishing, it has been found that z is constant ($z = 4$, expected value according to the valence of the Si
in the crystal) and does not depend on the applied current (j), the doping (N_D) and the electrolyte
concentration (c). Conversely, z increases with these parameters during porous etching (z varies
between 2 and 4). Furthermore, according to Lehman (3), the dissolution valence for n-type Si
during pore formation can be 4 at the pore tips and 2 at the pore walls. Many models developed
for an elemental semiconductor such as Si are not directly relevant for *III-V* compounds. Thus the
dissolution mechanism associated to the pore formation onto semiconductor such as InP is still an
issue (4). It is indeed known for InP that charge injection from adsorbed corrosion intermediaries
can assist the dissolution process of the compound semiconductor (5). A 6 or 8 charge dissolution
mechanism was determined for flat InP surfaces (6), while the etching processes have not been
considered in the case of porous layers. Without an accurate description of the dissolution process,
it appears further difficult to explain the switch of the pore growth from crystal oriented (CO) to
current line oriented (CLO) morphologies. Recently, Foll *et al.* have reported investigations carried
out by electrochemical impedance spectroscopy (EIS) that have targeted such purpose (7), however
questions are still opened.

In the present paper, we also aim to understand the pore formation mechanism. Controlled and

313

regular CLO pores were grown potentiostatically at $5\,\text{V}$ in $1\,\text{M}$ HCl with anodic charges ranging from 0.3 to $20\,\text{C·cm}^{-2}$. Instead using EIS or gravimetric measurements, atomic absorption spectroscopy (AAS) has been used to determine the amount of released indium after the porous etching. The comparison with the calculated mass using the Faraday's law allows the calculation of the dissolution valence and gives valuable information about the dissolution processes and the transport of corrosion products.

Experimental

Porous layers were grown onto Sn doped n-type InP (100) wafers from InPact Inc. ($N_D = 1 - 2 \times 10^{18}\,\text{at·cm}^{-3}$) in analytical grade $1\,\text{M}$ HCl deoxygenated and stirred by N_2 bubbling. The samples were cleaved in $4 \times 4\,\text{mm}^2$ square pieces and were degreased with methanol. Electrochemical experiments were performed using a standard three electrodes cell. Pt wire and $1\,\text{M}$ KCl silver/silver chloride electrode (Ag/AgCl, $U^\circ = 236\,\text{mV}$ $vs.$ SHE) respectively served as counter and reference electrodes. The porous layers were formed by applying a constant potential step ($U = 5\,\text{V}$ $vs.$ Ag/AgCl) using an $EG\&G$–PAR 173 potentiostat/galvanostat connected to an $EG\&G$–PAR 175 pilot in dark conditions.

Measurement of the indium concentration in the solution at the end of the polarization and its evolution against time were carried out by atomic absorption spectroscopy. Depending on the released metal quantity, a $Thermo$ $Scientific$ $M6$ atomic absorption spectrometer was used in flame ($c = 1 - 30\,\text{mg·L}^{-1}$) or furnace ($c = 50\,\mu\text{g·L}^{-1}$) modes. Background corrections to the spectra were carried out using a deuterium lamp and a Zeeman effect system for the flame and furnace modes, respectively. Calibration solutions were freshly prepared in a $1\,\text{M}$ HCl reconstituted matrix. For flame analysis the air/acetylene flux was set to $0.9\,\text{L·min}^{-1}$ and the burner position to $10.2\,\text{mm}$. Furnace experiments were performed with an extended lifetime cuvette, a working volume of 20–$40\,\mu\text{L}$ and the following optimized operating parameters : $45\,\text{s}$ drying at $110°\text{C}$, $20\,\text{s}$ ashing at $950°\text{C}$, $3\,\text{s}$ atomization at $2300°\text{C}$ and $3\,\text{s}$ cleaning at $2500°\text{C}$. Ascorbic acid and $Pd(NO_3)_2$ solutions were used as modifiers and automatically added to calibration and sample solutions by the autosampler in 1–$4\,\mu\text{L}$ proportion for a $20\,\mu\text{L}$ working volume before analysis.

Morphology of porous layer was characterized by field-emission gun scanning electron microscopy (FE-SEM) using a Hitachi S4800 microscope.

Results and Discussion

Pore growth and morphology

Figure 1a shows the typical current response and the corresponding passed anodic charge during a potential step at $5\,\text{V}$ $vs.$ Ag/AgCl. The current initially increases steeply to reach a very high current density ($j = 5.4\,\text{A·cm}^{-2}$) and decreases progressively to reach a plateau at $j \approx 1\,\text{A·cm}^{-2}$. These two phases correspond to the nucleation and growth of the pores, respectively. The inset in Figure 1a shows current oscillations that are related to the nucleation layer presented in the inset

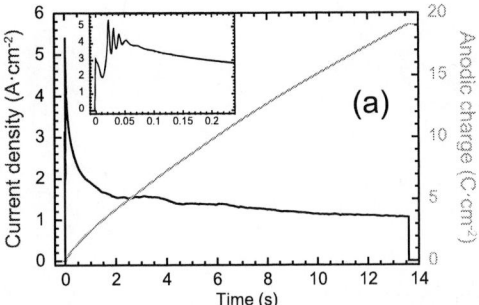
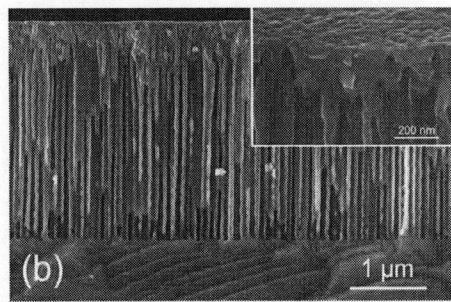

Figure 1: (a) Current density (black) and anodic charge (gray) plotted against polarization time ($U = 5\,\text{V}\ vs.\ \text{Ag/AgCl}$. The inset corresponds to an enlarged view of the initial part of the $j\ vs.\ t$ curve. (b) SEM cross section after passing an anodic charge of $2\,\text{C·cm}^{-2}$. The inset shows an enlarged view of the top layer region.

of Figure 1b. This part of the porous film is indeed irregular while the underlying section exhibits regular tubular channels that are etched during the growth phase. Under potentiostatic conditions very uniform pores are formed, almost no diameter oscillations are observed and the etching front is planar and well defined. Although the dimensions of the porous structures measured by SEM can be strongly modified by the contrast of the images, the range of the pore size can be given. The pores show a mean diameter, r_p, of 40 nm and their density, N_p, is approximately 10^{10} pores·cm^{-2}. Considering a tubular geometry, the etched volume (V_p) can be described, using Eq. 1, as a linear function of the pore depth (d_p):

$$V_p = \pi \cdot r_p^2 \cdot N_p \cdot d_p \tag{1}$$

Using the density of InP ($\rho_{InP} = 4.81$), and assuming that the quantity of released In and P are equal, the mass of dissolved indium can be deduced from Eq. 2 and is plotted on Figure 3.

$$m_{In} = (0.787 \rho_{InP} \cdot \pi \cdot r_p^2 \cdot N_p) \cdot d_p \tag{2}$$

Figure 2 presents the evolution of the thickness of the porous layer with the anodic charge. A linear behavior is observed for charges higher than $1\,\text{C·cm}^{-2}$. The linear fitting does not cross the axis at zero. This can be attributed to the different etching process between the nucleation and the growth phase. Since the top layer, with a thickness in the 400 to 500 nm range, exhibits a lower porosity, its depth increases quicker than the bottom part. Thus the slope is higher in this range of coulometric charge.

Depending on the conditions (chemical or electrochemical parameters, electrolyte nature and concentration, crystal orientation...), several electrochemical dissolution mechanisms have been proposed for InP. The generic equation, in which 6 or 8 charges are involved ($z = 6$ or 8), can be written

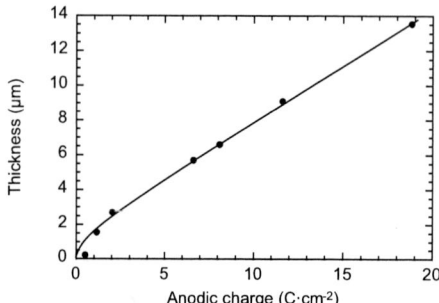

Figure 2: Evolution of the thickness of the porous layer with the anodic charge.

as follows:

$$InP + x\,h^+ \longrightarrow In^{3+} + P^{3+}\,(\text{or } P^{5+}) + y\,e^- \tag{3}$$

with $\quad x + y = z = 6 \quad$ if $\quad P^{3+} \qquad$ or $\qquad x + y = z = 8 \quad$ if $\quad P^{5+}$

Thus the comparison of m_{In}^{geo} calculated geometrically from SEM images with m_{In}^{Q} deduced from the passed anodic charges (Q) using the Faraday's law ($Q = nzF$, n: number of mol, F: Faraday's constant) shown on Figure 3 suggests the dissolution valence is close to 6. It is important to note that the error on m_{In}^{geo} is quite high. Therefore evaluation of the mass of released indium by AAS should therefore give a more relevant indication.

Chemical analysis by AAS

Since AAS detection threshold for indium is approximately hundred times higher than for phosphorus, the amount of dissolved InP was calculated from the measured In concentration. Thus it has been assumed that the detected amount of In is equal to the quantity of etched InP. The mass of indium (m_{In}) released in the electrolyte has been measured, by AAS, 30 s after the end of the polarization. Figure 3 shows the evolution of m_{In} against the anodic charge. A linear behavior is observed and the comparison of the measured m_{In} by AAS with the expected mass for a 6 or $8\,h^+$ dissolution mechanism indicates, this time, that etching occurs according to an $8\,h^+$ process.

It is noteworthy that such a large amount of matter is quickly extracted from the porous structure. In order to determine if the full amount of the etched InP is removed out the pore or not, indium concentration was followed by AAS several hours after the anodic polarization. The evolution of m_{In} is plotted against time in Figure 4 (for seek of clarity only the curves for $Q \leq 4.7$ C·cm² are shown). Surprisingly, a large amount of indium is released longtime after the end of the polarization. An almost linear increase of m_{In} is firstly observed until a plateau is reached after approximately 70 to 150 min with a high level of reproducibility.

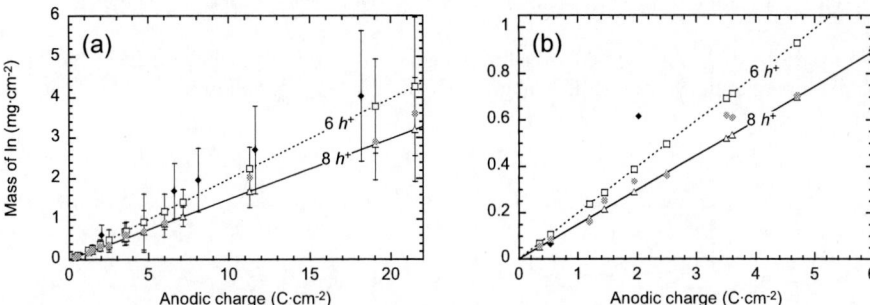

Figure 3: (a) Dissolved mass of In after the porous etching measured by AAS (•), calculated from SEM cross sections (♦) and determined from the anodic charge assuming a 6 (□) or $8h^+$ (△) mechanism, respectively. (b) Enlarged view of the low anodic charges.

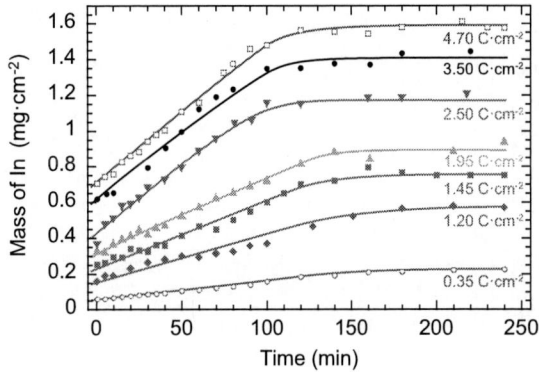

Figure 4: Evolution with time of the dissolved mass of In after the porous etching at various anodic charges (colors available online).

As seen on Figures 5a and 5b, the ratio between the initial and the final masses of dissolved indium ($R = m_{In}^f/m_{In}^i$) is high ($3 > R \geqslant 2$) and approximately constant ($R \approx 2$). Figure 5a shows that the ratio is higher for low anodic charges ($Q \geqslant 4\,C\cdot cm^{-2}$). The normalized evolution of the m_{In} with t, plotted in Figure 5b, reveals a quite uniform aspect of the curves. The plateau is however reached for shorter durations when the coulometric charge increases.

Since the amount of dissolved indium calculated using the Faraday's law (for a 8 charges mechanism) and the initial quantity measured by AAS (m_{In}^i) are in agreement, the additional quantity of released matter ($\Delta m = m_{\text{In}}^{\text{f}} - m_{\text{In}}^{\text{i}}$) can be related to two processes, in which chemical dissolution occurs:

 (i) Δm is associated to a chemical dissolution of the pore walls after the electrochemical etching;

 (ii) Δm corresponds the remaining quantity of the etched semiconductor that is progressively extracted from the pores to the solution.

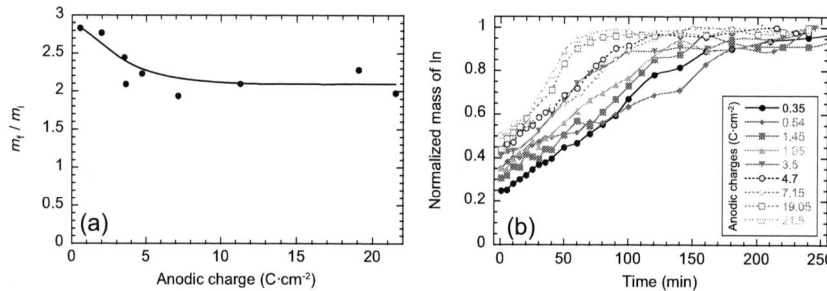

Figure 5: (a) Ratio between initial and final masses of dissolved In ($R = \frac{m_{\text{In}}^{\text{f}}}{m_{\text{In}}^{\text{i}}}$). (b) Variation of the normalized mass of In with time after the porous etching at various anodic charges (colors available online).

 Although the porous films are particularly fragile after a long immersion in the solution (*i. e.* numerous layers are lifted off during the sample rinsing and handling), SEM cross section examinations have confirmed the porous layers remain almost unchanged after the AAS concentration monitoring. It indicates therefore that no further chemical dissolution occurs while the porous layers are dipped within the electrolyte. Thus the mass of In released after the end of the polarization (Δm) also originates from InP dissolution occurring during the polarization. The effective dissolution valence (z_{eff}) of the pore formation is 4 instead of 8. This leads to consider another dissolution mechanism. It seems coupled and sequential electrochemical and chemical dissolution steps take place during the polarization. As mentioned in the introduction, it is known for Si (3) that the dissolution valence varies from 2 to 4 while it is always 4 in the case of electropolishing. Confinement of the chemical species within the pores can thus induce such variations. The linear evolution observed for m_{In} before the plateau suggests a mass transport different than a planar diffusion. Preusser *et al.* (5) have reported a current doubling effect during the photoelectrochemical etching of n-InP. They ascribed this phenomenon to an electron injection from adsorbed corrosion products to the conduction band of the semiconductor. Although such a process can occur in the present case, the electron should be collected by the external electric circuit and, then, be quantified in the measured anodic charge. Thus simultaneous dissolution must be only a pure chemical process. This phenomenon should be related to the corrosion intermediates because it does not continue unlimitedly after the porous etching. Moreover, it is known that InP can be chemically etched only in highly concentrated HCl solutions ($c_{\text{HCl}} > 5\,\text{M}$) because in such conditions HCl is not fully dissociated (6). AAS measurements were performed onto flat InP surfaces immersed 160 hours in 1 M HCl and no significant traces of dissolved indium were detected.

Conclusion

In the present work, growth of regular tubular pores under potentiostatic conditions has been performed. Except during the initiation phase (depth of 400 to 500 nm), the pore depth is proportional to the passed anodic charge. The comparison of the amount of dissolved materials measured by AAS just after the polarization and calculated using the Faraday's law suggests an 8 charge dissolution process. However, the monitoring of the c_{In} after the porosification over several hours and in the whole range of the charge shows that the total quantity of released InP is approximately doubled. Chemical analyses performed by AAS highlight therefore a new porous etching mechanism in which a dual process occurs. Both expected electrochemical and unexpected chemical dissolutions take place during the pore formation.

Acknowledgements

The authors would like to gratefully acknowledged Dr. Fouad Maroun (Ecole Polytechnique-CNRS, France) for high-resolution SEM characterizations.

References

1. L. Santinacci and T. Djenizian, *C. R. Chimie* **11**, 962 (2008).

2. V. Lehmann, *Electrochemistry of Silicon*, Wiley-VCH, Weinheim, 2002.

3. V. Lehmann, *J. Electrochem. Soc.* **140**, 2836 (1993).

4. H. Foll, S. Langa, J. Carstensen, M. Christophersen, and I. M. Tiginyanu, *Adv. Mater.* **15**, 183 (2003).

5. S. Preusser, M. Herlem, A. Etcheberry, and J. Jaume, *Electrochim. Acta* **37**, 289 (1992).

6. P. H. L. Notten and J. J. Kelly, *Etching of III-V Semiconductors: An Electrochemical Approach*, Elsevier Science & Technology, 1991.

7. J. Carstensen, E. Foca, S. Keipert, H. Foll, M. Leisner, and A. Cojocaru, *Phys. Status Solidi A* **205**, 2485 (2008).

ECS Transactions, 19 (3) 321-328 (2009)
10.1149/1.3120712 ©The Electrochemical Society

Simulating Crystallographic Pore Growth in III-V semiconductors

M. Leisner, J. Carstensen, and H. Föll

Institute for Materials Science, Christian-Albrechts-University of Kiel
Kaiserstrasse 2, 24143 Kiel, Germany

In this work, the growth of crystallographically oriented pores in
III-V semiconductors has been investigated. Based on new and
previous results a model for pore growth has been developed,
which is mainly based on a stochastic branching probability of the
pores. The stochastic nature of the model allowed to implement it
as the core for Monte-Carlo-Simulations of pore growth. The
simulations were able to reproduce the main features of
crystallographically oriented pores, like uniform pore growth in n-
type InP or the formation of domains on n-type GaAs and InP. The
model is also capable of reproducing the logarithmic growth law
for the pore depth, as well as the pore density oscillations with
depth, as recently found.

Introduction

Electrochemical pore growth on III-V semiconductors can be basically classified into
two groups according to the resulting pore morphologies. Most attention has been
focused on the so-called current-line oriented pores ("Curros"). These pores are typically
non-facetted. i.e. they have smooth pore walls with often cylindrical cross-sections and
always grow perpendicular to the equipotential lines, i.e. usually perpendicular to the
surface, independent of the substrate's crystallographic orientation (1 - 5). Curros so far
have been observed in n-type InP, as well as in GaP and Si; in GaAs no curros have yet
been found. As shown by Tiginyanu et al. (6), pores on II-VI semiconductors like CdSe
show striking similarities to pores found in III-V's and might thus also fall into this class
of pores. The second kind of pores are crystallographically oriented pores ("Crystos") (2,
7, 8). These pores always grow into <111>B directions, and often are heavily facetted
with a triangular cross-section, and pore tips and pore walls composed from {111} planes
(for details see (9)). Crysto pores have been observed in Si and Ge but particularly in n-
type InP, GaAs, and GaP, even though it is difficult to nucleate crysto pores
homogeneously on the latter two semiconductors (10). Crysto pores are rather unsensitive
to the electrolyte used. The standard electrolyte for pore etching in III-V's is based on
HCl, but similar crysto pores can be obtained for electrolytes like KOH (11, 12) or "salt
water" (6).

While a considerable amount of work has been spent on electrochemical pore growth
in III-V semiconductors, the understanding of the growth mechanisms still remains
patchy. In this work, a model for the growth of crysto pores in III-V's is developed,
which is mainly based on a stochastic process of pore branching, i.e it is of the "current
burst" type (13). This allows implementing the model with a Monte-Carlo approach in a
three-dimensional array of voxels. Results of the model are compared to dedicated
experiments undertaken for this purpose and to older results either not fully appreciated
or not fully understood in the past.

Experimental

Pore Etching

Samples consisted of single-crystalline (100)-oriented n-type InP, with a doping concentration $N_D = 10^{17}$ cm^{-3}, as well as of (100)-oriented n-type GaAs with a doping concentration $N_D = 10^{17}$ cm^{-3}. The sample size was $A = 0.25$ cm^2. Experiments have been performed in the electrochemical double-cell described in full detail in (14). Aqueous HCl with a concentration of 5 wt. % (= 1.4 M) has been used as electrolyte. All experiments have been performed under constant-current conditions at a constant temperature of $T = 20$ °C.

Monte-Carlo-Simulations

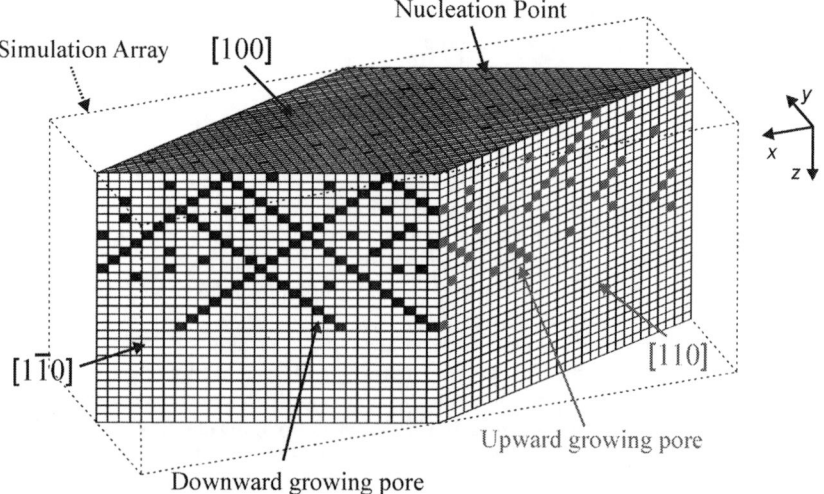

Figure 1: Schematic illustration of the simulation array. On top the nuclei or nucleation points that lead to downward growing pores are illustrated (left-hand part). All pores can branch at the tips or from a pore wall. Pores growing into the depth are shown as a succession of black squares on the front view (left-hand part), while upwardly growing pores show up as isolated back squares. On the side view (right-hand part) the situation is reverse.

Based on previous results and models from our group, in particular the current burst model for pore growth and self-induced oscillations (13), a three-dimensional model has been specifically developed that allows Monte-Carlo-Simulations of crysto pore growth in III-V semiconductors. In these simulations, an array of $(1024)^3 = 1.07 \cdot 10^9$ voxels has been used, with a mesh size set to 100 nm, corresponding to the experimentally determined typical crysto pore size. The top layer of this array is filled with nucleation seeds for the crysto pores in some distribution and density as a starting condition. The nuclei density, the depth of the nucleation layer, and the kind of nuclei distribution (random, periodic,...) can be freely chosen.

As illustrated in Fig. 1, after nucleation has taken place, the pores begin to grow one voxel into the two downward oriented <111>B directions with every iteration cycle. As observed experimentally, the pores are able to branch in two different ways. A pore can branch at the tip into the other downward oriented <111>B direction, or a new pore tip can branch out of the previously etched pore walls into one of the two upward oriented <111>B directions. The upward growing pores are then also able to branch similar to the downward growing pores.

The key point of the model is that certain different branching probabilities k_{tips} and k_{walls}, respectively, are attributed to both branching processes. In addition, an adjustable length of the pore in the order of the space charge region (SCR) width d_{SCR}, measured from its point of generation by branching will be assigned $k = 0$, i.e. it will not be able to form branches on that length because as long as the SCRs overlap the new tip cannot experience the full voltage drop. In a last primary point it is assumed that all pores stop to grow if other pores block their trajectory. In other words, pores cannot grow through the space-charge-region surrounding other pores.

The model then needs two more basic but reasonable assumptions:
1. The dissolution valence is constant.
2. The two different branching probabilities are both proportional to the current density j_{tips} at the pore tips.

This is sufficient to run the Monte Carlo simulation. Note that while the probability of branching is constant, the generation rate of pores may vary because for pores formed by branching from a pore wall it is proportional to the pore length and that increases the rate with time, while the proportionality to the current decreases both rates since j_{tips} decreases as the number of tips increases.

Results

Homogeneous Crysto Pore Structures on InP

Etching homogeneous layers of crysto pores in n-type InP can be best achieved by performing constant current experiments and applying the conditions as outlined in the experimental section. For a (comparatively low) current density $j = 0.4$ mA/cm^2 pore structures as illustrated in Fig. 2a result after an etching time of 240 min. One can easily observe the upward growing crysto pores as straight lines in this cross-sectional SEM micrograph; the downward growing pores intersect the cleavage plane and appear as triangles. The inset shows details at higher magnification. Fig. 2b shows a result from the Monte-Carlo-Simulation, which has been obtained for an adequate set of model parameters (Nucleation tip density: 0.09 μm^{-2}, randomly distibuted, current density: 0.4 mA/cm^2, starting branching probability walls: 0.003, starting branching probability tips: 0.3, starting velocity: 2.5 μm/min). Note that the triangles are shown as quadratic pixels in the simulation pictures. It can be stated, that there is a fairly good agreement of the resulting pore structures.

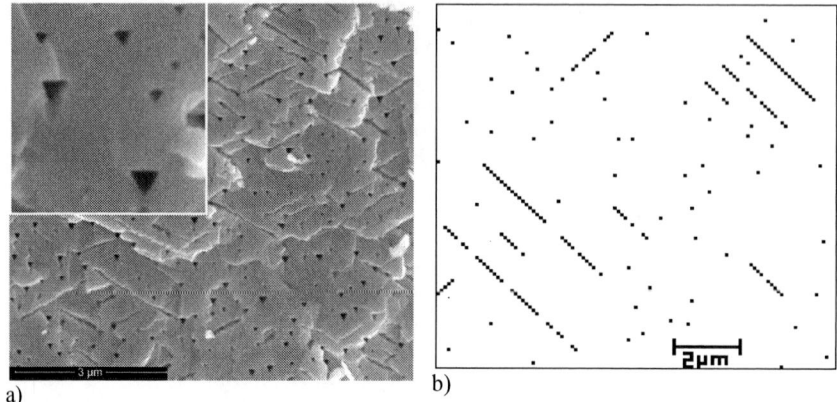

a)   b)

Figure 2: Comparison of experimental and simulated pore structures. a) Crysto pore structure obtained after etching galvanostatically at $j = 0.4$ mA/cm^2 for 240 min. b) Directly comparable crysto pore structure obtained from the Monte-Carlo-Simulation for an adequate set of parameters.

For the same conditions as before, several experiments with different etching times have been performed. The depth of the etched porous layers has been determined from SEM pictures. The resulting pore depth vs. time is plotted in Fig. 3 (black squares). A logarithmic growth law can describe the functionality:

$$d_{pore} \propto \ln t .$$ [1]

The solid line in Fig. 3 represents the result from Monte-Carlo-Simulations with one set of paramters; it matches the experimentally determined data very well. Theoretically, the logarithmic growth law is a direct consequence of the main assumptions of the model, i.e. of the proportionality of the pore branching to the current density at the pore tips (15). The fact that the simulation yields the observed growth law not only lends credibility to the model, but also confirms that the model is properly implemented in the simulations. The same line of reasoning applies to the number of pore tips as a function of time (not shown here). From the model a linear dependence is expected, which the simulation yields as well.

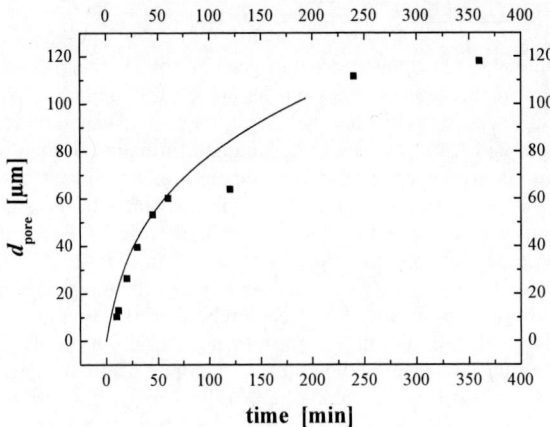

Figure 3: InP crysto pore depth as a function of etching time. Black squares represent the experimentally determined data. The line is the result from the Monte-Carlo-Simulations.

A rather surprising result has been obtained for the pore density as a function of depth. This quantity has been determined from cross-sectional SEM micrographs. The density is not monotonously decreasing, as one would intuitively expect, it is showing sinusoidal oscillations with a wavelength in the range between (6 – 10) μm, as can be seen in Fig. 4a. The Monte-Carlo-Simulations reproduced the effect partially, as shown in Fig 4b. A closer look at previous work revealed that these oscillations have already been observed, but not been recognized, in the TEM investigations published in (9).

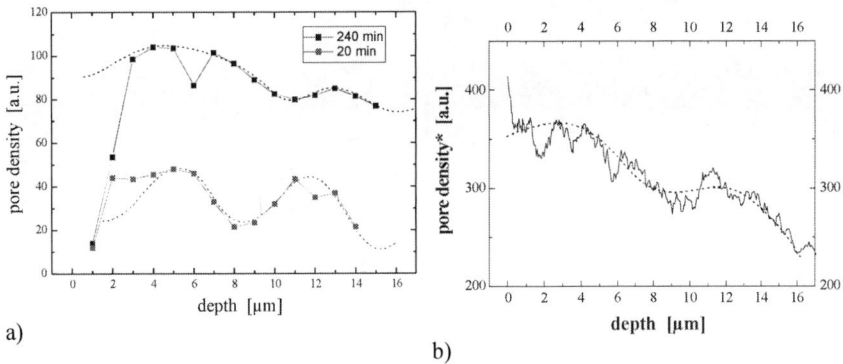

Figure 4: InP crysto pore density as a function of depth. a) Results from the dedicated etching experiments. b) Results from the Monte-Carlo-Simulations. For an explanation of the oscillatory behavior see the text.

This simulation of crysto pore growth should also be possible for GaAs and GaP, after adjusting model parameters like mesh size, current density, and in particular the nucleation.

Pore Domains on GaAs and InP

As has been shown in (2), domain structures can be obtained on n-type GaAs and InP. The formation of domains mainly relies on the nucleation conditions, i.e. a rather low nucleation density is necessary. In combination with high constant current conditions ($j =$ 80 mA/cm^2 here), it will lead to strong branching and formation of the domain structure in short times. Fig. 5a gives an example of a domain grown on GaAs as seen in a SEM surface view. The initial pore pair starts to grow in the center of the square and strong branching subsequently leads to upward growing pores, which will eventually reach the surface again, and thus form the pore domain as shown in Fig. 5a. A closer look at the domain surface reveals that 4 quadrants exist, 2 triangles with a higher pore density and 2 triangles with a lower pore density. For a comprehensive illustration of this effect see reference (2). Fig. 5b shows the corresponding result from the Monte-Carlo-Simulations(Nucleation tip density: 0.0004 μm^{-2}, randomly distibuted, starting branching probability walls: 0.5, starting branching probability tips: 0.5, starting velocity: 2.5 μm/min).. The similarity to the experimental results is striking; even the 4 triangles and the line-up of the pores in two of them are clearly visible.

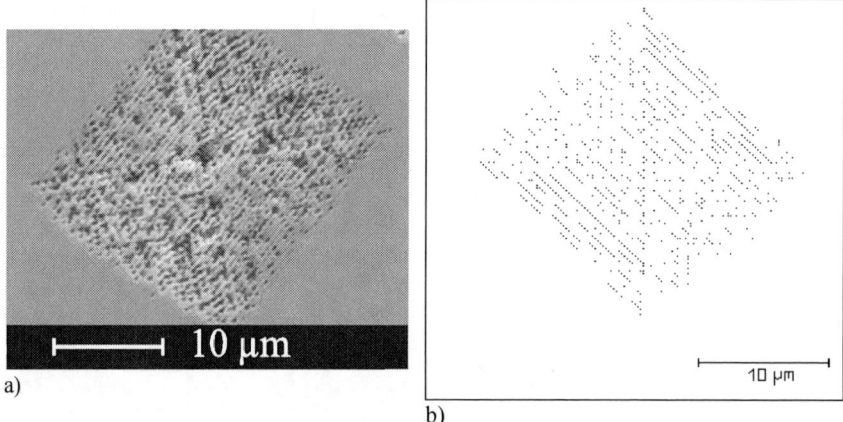

a)

b)

Figure 5: Crysto pore domains in GaAs. a) SEM surface view of a crysto pore domain. Constant current experiment: j = 80 mA/cm^2, electrolyte concentration 5 wt.% HCl aq. b) Corresponding results from the Monte-Carlo-Simulations.

Discussion

The relatively simple model, with two basic assumptions, which has been implemented into Monte-Carlo-Simulations, is able to reproduce different experimentally observed crysto pore structures on different III-V semiconductors, like homogeneous pore layers on n-type InP and the phenomenon of domain formation on n-type GaAs and InP. The main simulation parameters, used to achieve the different pore structures, are the branching probabilities at the pore tips and pore walls. The images of the simulation results shown in Figs. 2 and 5 show a very good agreement to the experimentally found pore structures observed in the corresponding SEM micrographs.

For the time dependence of the pore depth, the simulation yields the logarithmic growth law, which very well matches the experimentally found results. This is expected, since the model was designed to result in a logarithmic pore depth growth. The same holds for the number of pore tips as function of etching time, a linear relationship is observed in the simulations, as is theoretically expected from the model. Still both results strongly indicate, that the model is a valid description of crysto pore growth on III-V semiconductors.

For the pore density as function of depth, the interesting oscillations shown in Fig. 4 have been found in good agreement to the results of the simulation. The strength of a simulation is the full access to all parameters as a function of time at all positions which could never be measured in such detail. This allows an investigation of the origin of the pore density oscillation, which can be summarized as follows: The density oscillations occur, because in a certain depth a critical pore density is reached, so that no further branching is possible, since all free space between the pores is covered by space charge regions. Thus, some pore tips die out and will hence increase the current density at the pore tips. In the framework of the model this will lead to enhanced branching probabilities at the tips and walls. Since this branching can only occur at some distance to the region where the critical pore density has already been obtained, an intrinsic length scale is introduced, which will define the wavelength of the oscillations. Geometrically this length scale is defined by the (average) distance of nucleation sites at the surface and the angle of the first <111> pores growing into the depth.

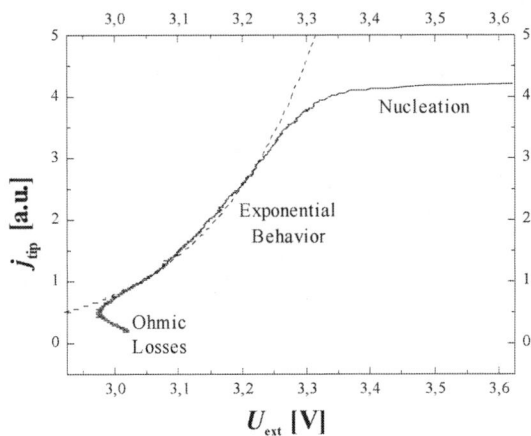

Figure 6: Pseudo IV Curve for crysto pore growth on n-type InP. The curve is constructed out of simulation ($j_{tips}(t)$) and experimental data ($U(t)$). For details see the text.

The strongest assumption in the model is that at all pore tips (up-growing, down-growing, branched) the same current density is flowing. In the following we will combine the data extracted from the simulation with the measured $U(t)$ data to check, if the assumption is valid and which further consequences follow from the assumption. Thus, dividing the externally flowing constant current by the number of pore tips, resulting from the Monte-Carlo-Simulation, will yield the time development of $j_{tips}(t)$ during an experiment. Since the time development of the voltage is directly measured in the experiments, it is possible to construct a "pseudo IV curve" by combining the simulation data ($j_{tips}(t)$) and the experimental data ($U(t)$). Fig. 6 shows the resulting "pseudo IV curve". In the main part of the etching experiment an exponential behavior is found (as indicated by the dashed line), which is a strong hint that simple Butler-Vollmer kinetics dominates the etching processes at all pore tips. In the nucleation phase the behavior deviates from the exponential behavior, which is expected, since the real nucleation is probably far more complicated than the approach in the simulations. Towards the end of the experiment the curve also deviates. A possible explanation for this phenomenon might be that the ohmic losses due to the increase in pore length have been ignored so far. To get the correct voltage at the pore tips the ohmic losses can be calculated from the simulation of the pore geometry as a function of time, i.e. the (average) pore length has to be calculated from the pore length distribution during pore growth. If these results would correct the dip in the IV curve, it would be a further strong indication for the validity of the model for crysto pore growth in III-V semiconductors.

References

1. E. Kikuno, M. Amiotti, T. Takizawa, and S. Arai, *Japan J. Appl. Phys.* **34(1, 1)**, 177 (1995).
2. H. Föll, S. Langa, J. Carstensen, S. Lölkes, M. Christophersen, and I.M. Tiginyanu, *Adv. Mater.* **15(3)**, 183 (2003).
3. H. Hasegawa and T. Sato, *Electrochim. Acta* **50**, 3015 (2005).
4. P. Schmuki, U. Schlierf, T. Herrmann, and G. Champion, *Electrochim. Acta* **48**, 1301 (2003).
5. A.M. Gonçalves, L. Santinacci, A. Eb, I. Gerard, C. Mathieu, and A. Etcheberry, *Electrochem. Solid-State Lett.* **10(4)**, D35 (2007).
6. I.M. Tiginyanu, V.V. Ursaki, E. Monaico, E. Foca, and H. Föll, *Electrochem. and Sol. State Lett.* **10(11)**, D127 (2007).
7. T. Takizawa, S. Arai, and M. Nakahara, *Japan J. Appl. Phys.* **33(2, 5A)**, L643 (1994).
8. H. Tsuchiya, M. Hueppe, T. Djenizian, P. Schmuki, and S. Fujimoto, *Sci. Technol. Adv. Mater.* **5**, 119 (2004).
9. E. Spiecker, M. Rudel, W. Jäger, M. Leisner, and H. Föll, *Phys. Stat. Sol. (a)* **202(15)**, 2950 (2005).
10. S. Langa, J. Carstensen, M. Christophersen, K. Steen, S. Frey, I.M. Tiginyanu, and H. Föll, *J. Electrochem. Soc.* **152(8)**, C525 (2005).
11. C. O'Dwyer, D.N. Buckley, D. Sutton, and S.B. Newcomb, *J. Electrochem. Soc.* **153(12)**, G1039 (2006).
12. C. O'Dwyer, D.N. Buckley, D. Sutton, M. Serantoni, and S.B. Newcomb, *J. Electrochem. Soc.* **154(2)**, H78 (2007).
13. J. Carstensen, R. Prange, G.S. Popkirov, and H. Föll, *Appl. Phys. A* **67(4)**, 459 (1998).
14. S. Langa, I.M. Tiginyanu, J. Carstensen, M. Christophersen, and H. Föll, *Electrochem. Solid-State Lett.* **3(11)**, 514 (2000).
15. M. Leisner, J. Carstensen, A. Cojocaru, and H. Föll, *ECS Trans.* **16(3)**, 133 (2008).

ECS Transactions, 19 (3) 329-345 (2009)
10.1149/1.3120713 ©The Electrochemical Society

Growth Mode Transition of Crysto and Curro Pores in III-V Semiconductors

H. Föll, M. Leisner, J. Carstensen, and P. Schauer

Institute for Materials Science, Christian-Albrechts-University of Kiel
Kaiserstrasse 2, 24143 Kiel, Germany

The formation of crystallographically oriented and current line oriented pores in n-type InP is reviewed and compared to other semiconductors in the light of some new results. A model for the formation of crystallographically oriented pores is presented that reproduced salient features of this pore type rather well. Impedance data together with their model-based evaluation are given and discussed. Some self-organization features of current line oriented pores and their possible relation to self-induced or externally triggered growth mode transitions between the two pore types conclude the paper.

Introduction

The growth mode transition between crystallographically oriented (crysto) pores to current-line oriented (curro) pores in III-V semiconductors (InP, GaP, GaAs) remains one of the most interesting phenomena in the field of electrochemically etched porous semiconductors. Even though a considerable amount of work has been performed (1 - 3), a consistent theory for the growth mode transition does not yet exist. In this work some new findings with respect to this topic in InP will be presented and discussed with regard to older results and other semiconductors.

The two pore types mentioned above may not be the only pore types in InP (or other III-V semiconductors); at least some mixtures or intermediate structures ("currystos") between crysto and curro pores exist, if not truly other types, for example if liquid ammonia is used as electrolyte (4). However, crysto and curro pores are the dominating species at least in InP, and they are found in a surprisingly large range of electrolytes including acidic electrolytes like diluted HCl (1, 5 - 8), almost pH-neutral "salt-water" (9), alkaline electrolytes like KOH (10, 11) and liquid ammonia (4). While GaAs so far produced only crysto pores, both kinds are known from GaP (1, 12), which also supports a new kind of pores with rectangular cross-section as shown by the Erlangen group (13) that seems to combine features of both crysto and curro pores. The few pore etching experiments with regard to II-VI semiconductors (9) also show typical curro pores in CdSe and ZnSe, plus yet another new pore type (three-dimensional fractal pores in ZnSe (14)).

The extensive literature to macropores in Si (for reviews see (1, 15 - 25) is almost completely centered on pores that would be termed crysto pores in this context. However, true curro pores as well as mixes have been reported, too (24): moreover, in (24) a claim was made that at least some macropores in p-type Si are actually current line pores and not crystallographically oriented pores as implicitly assumed.

While in some of the cases mentioned above only one pore type was observed, the crysto – curro issue also includes the transition between those two basic pore types either self-induced or externally triggered. Pore growth mode transitions also might be seen in a more general context (with crysto-curro transitions being a special case), in particular in

329

the light of recent findings in Si (24, 26, 27); in (27) a first attempt was made to look at this phenomenon in a systematic way.

The crysto – curro issue is thus a quite general feature of pore etching in semiconductors, and a complete understanding of the underlying mechanisms would be an important step in the ongoing attempt to understand pore etching in semiconductors. While this goal has not yet been reached, some progress in the general direction will be reported in what follows. In particular, experiments dedicated to increase the size of curro pores in InP, in-situ-FFT impedance spectroscopy, and first results from an attempt to simulate crysto pore growth will be given.

Crystallographic Pores

The distinctive features of crystallographic pores are:

- Their growth direction is strictly a crystallographic direction. In III-V semiconductors and, as far as is known, in II-VI semiconductors, so far always <111>B has been found as growth direction (with the exception of the rectangular pores in GaP that are growing in <110> directions (13)). In Si, the major growth directions of crysto pores are <100> and <113> (28, 29), for Ge <100> and <111> have been found (23).
- Pore walls consist of "stopping planes", i.e. planes that are not or only slowly dissolved and the pore thus consists of a facetted tube. On a closer look, crysto pores are actually a sequence of either tetrahedrons for <111> pore directions, i.e. for the bulk of the III-V's (30), or octahedrons for <100> directions, i.e. for Si (31) that are connected like pearls on a string. The sidewalls then consist of {111} planes. The tetrahedrons / octahedrons may show considerable overlap (32, 33) (cf. **Fig. 1a**) and at lower magnification the pore may actually appear to have smooth walls (with, e.g.,{112} pore walls (32) for <111>B pores).
- The formation mechanism of crysto pores relies on two ingredients: i) local production of the holes needed for dissolution at the pore tips by some junction breakdown mechanism (avalanche or tunneling), and ii) strong passivation of the pore walls allowing the formation of a space charge region (SCR) around the pore. The second ingredient may well be the more important one for pore formation as will become clear from the simulation model described below.
- The crysto pore size in InP in the sense of its (average) diameter is rather constant (within a factor of 2 – 3) and not, as one could expect, tied to the critical field strength needed for junction breakdown. It increases somewhat with pore depth or time, due to leakage currents causing lateral growth (32) and seems to be smaller in alkaline electrolytes (11). For pure crysto pores it is essentially a function of the size of the tetrahedrons formed and thus given by the passivation kinetics of the pore wall or rather tetrahedron walls (cf. (30, 31)), which appears to be rather insensitive to the electrolyte composition. The doping level has some influence, because it controls the size of the SCR. However, no systematic study with respect to crysto pore size has been done so far.

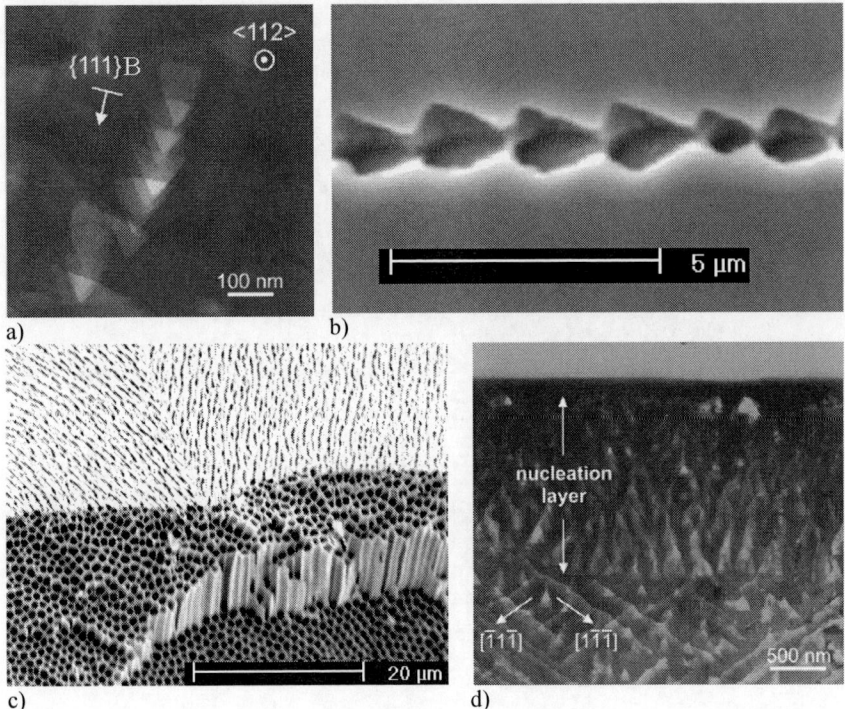

Figure 1: a) TEM picture of a sequence of overlapping tetrahedra in InP (cf. (32)). b) Crysto pore formed by a string of tetrahedra in GaAs. Nucleation layer in InP followed by a thin layer of crysto pores (not visible) and curro pores showing domains in the nucleation layer. For details see text. d) TEM picture of nucleation layer for crysto pores at high magnification (cf. (32)).

- Crysto pores branch in all 4 <111>B directions until the current density at pore tips is small enough. This leads to pore growth not only into the depth of the sample but also towards the surface.
- The nucleation of crysto pores may not be uniform and it might require some nucleation layer that is not well characterized at present, cf. **Fig. 1c, d** and (34). In the case of GaAs, nucleation of crysto pores is preceded by some lateral etching producing grooves on the surface as described in (1).
- So far it was not possible to induce nucleation by some lithographically provided mask (see below).
- There is little self-organization among crysto pores.

Recently an attempt has been made to model some of the pertinent features of crysto pores; details are given in (35, 36). A three-dimensional Monte Carlo model as illustrated in **Fig. 2** with about 1 billion voxels reproduced the experimental findings for a reasonable set of assumptions and parameters, including the pore density oscillations; for details cf. (35, 36) and forthcoming publications. Three assumptions have been made. i)

The valence of the dissolution process is constant, ii) The branching probability of a pore is proportional to the current density j_{tip} at the pore tip (with different probabilities for branching at pore tips and from pore walls). iii) Pore growth stops if the pore tip approaches a pore wall of another pore.

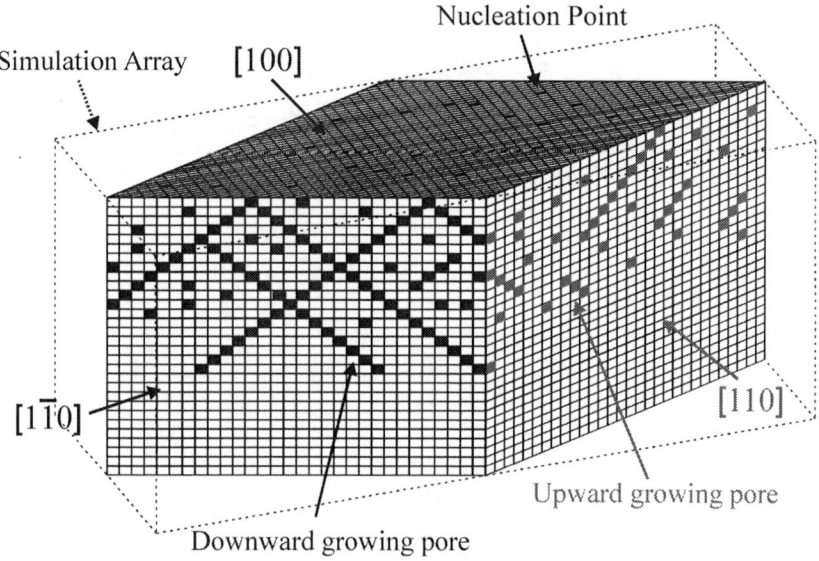

Figure 2: Simulation array of the Monte-Carlo-Simulation model. On top, the nucleation points for pores can be observed, which lead to downward growing pores (left hand side) and subsequently to upward growing pores (right hand side). A three-dimensional crysto pore structure evolves.

The model needs a nucleation scenario as input, but otherwise reproduces many features of crysto pores with good accuracy; for example the weak oscillation pore density with depth as found in (32, 35). **Fig. 3** compares two experimental results with simulations. In **Fig. 3a** examples of the crysto pore structure in (100) n-type InP obtained for relatively low current densities of 0.4 mA/cm^2 and a 5 wt. % HCl electrolyte (1.4 M), branching probability ratio: k_{tips} / k_{walls} = 100 is shown in cross-section. Downwardly branching pores show up with their triangular cross-section as shown in the inset. While experiment and simulation agree rather well, the simulation of a typical growth domain as found in GaAs (1) (and in InP) in **Fig. 3c,d** is even more striking. In this case the current density of 80 mA/cm^2 is rather high, the ratio of the branching probabilities was adjusted to k_{tips} / k_{walls} = 1.

Note that the model is essentially of the "current burst" (CB) type (37). In other words, current flow is not uniform in time but occurs in "bursts" that are governed by some probability function. The CB model implemented in this case is simpler than the full-fledged CB model employed to model oscillatory phenomena at the Si electrode (38) but maintains the essential stochastic component.

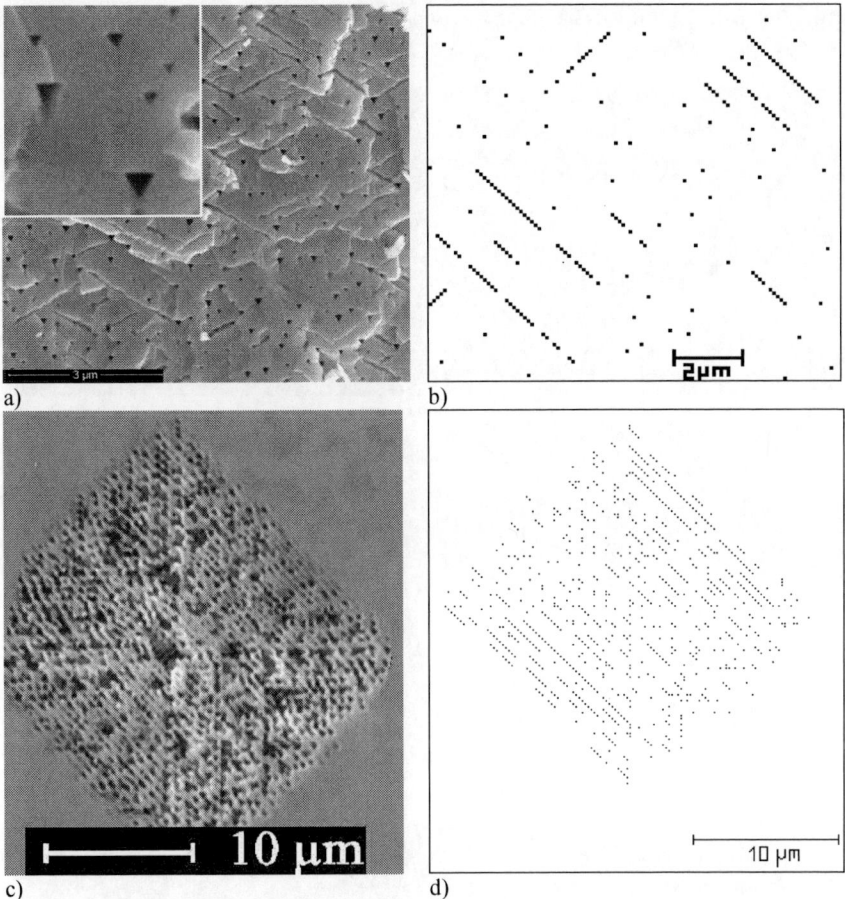

Figure 3: Comparison of observed (a) and simulated (b) crysto pore structures from a uniform (random) nuclei distribution. b) From just one nucleation point pore domains result as shown in c) for GaAs, with the simulated structure in d).

Nucleation of crysto pores in general is not well understood. While rather uniform and easy nucleation is often found in Si and InP, the opposite is true for Ge and GaP, with GaAs being a kind of intermediate case. The two examples shown in **Fig. 1c,d** for InP serve only to show that the surface layer shows nucleation structures with a texture that changes at some kind of domain boundaries, and that fully developed crysto pores might be preceded by a thin nucleation layer of heavily branched pores with no clear expression of <111> directions. On occasion (in particular at high current densities), this nucleation layer may switch to curro pores without forming "proper" crysto pores first; an example is given in **Fig. 4a**. The doping of the sample also appears to influence nucleation, While the mostly used more heavily doped InP samples show rather uniform pore structures, this is not true for lightly doped samples, **Fig. 4** gives an example. Initiating etching proved to be difficult in this case; the (curro pore) obtained in this case is reminiscent of structures in Si (39).

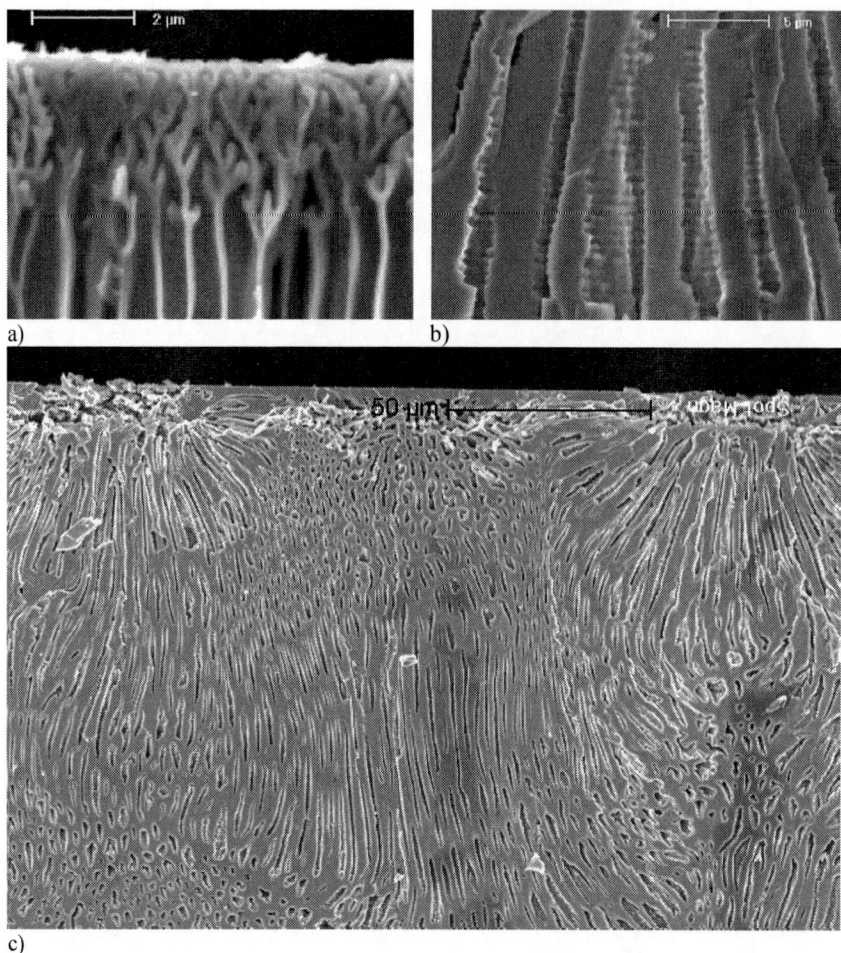

Figure 4: a) Curro pore formation without well-developed crysto pores at high current densities. b), c) Pore structure in lightly doped n-type InP ($2 \cdot 10^{15}$ cm^{-3}). c) shows an overview, b) some detail of c) at a depth around 150 μm.

Given the points made above, it cannot be expected that the initial nucleation of crysto pores would follow some lithographically defined structure. One experiment undertaken in this vein (unfortunately) proved this assumption to be correct, **Fig. 5** shows some results. The idea was to see if - against expectations - one could nucleate curro pores with the desired spacing directly without the (crysto) nucleation layer. As it turned out, this was not possible. A number of crysto pores or their predecessors nucleated at each opening in the mask (usually at the edge as can be seen from **Fig. 5a**, and while curro pores formed after about 2 μm, the original structure was completely lost. This underscores once more that curro pores need some diffusion limitation that is provided by first forming crysto pores.

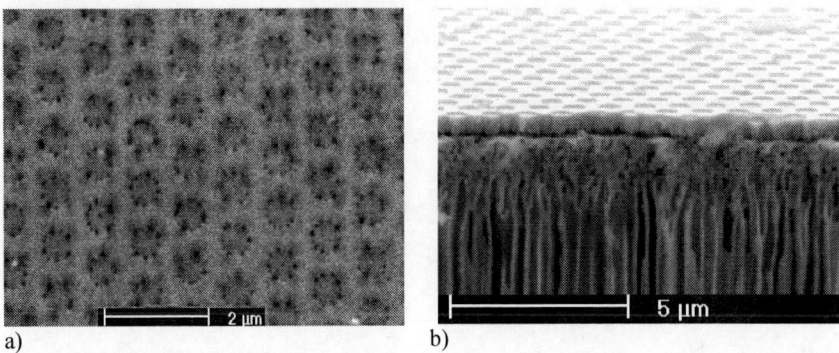

a)   b)

Figure 5: a) Nucleation of (crysto) pores at the openings of the mask; typically at the edge. b) Pore structure obtained with the mask.

Considering that crysto pore growth starts with some as yet unspecified nucleation phase and then proceeds by continued branching until the current density at the pore tip is below some limit, it is clear that crysto pore growth only achieves some "steady state" for low current densities and after some time. If a steady state cannot be achieved, a transition to curro pore occurs. In-situ FFT IS data, always averaging over many pores, must thus be expected to show some complex time behavior as long as steady state is not established.

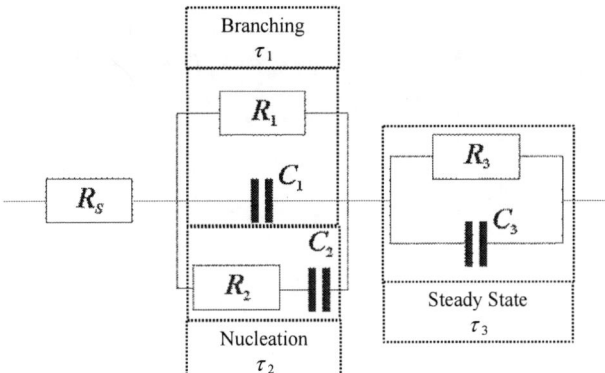

Figure 6: Equivalent circuit diagram used for fitting impedance data. The three $RC = \tau$ components are particular important for the three growth modes as indicated. R_S is the series resistance of the system.

335

The equivalent circuit in **Fig. 6** described by **Eq. 1** gives a rather perfect fit to typically more than 10000 IS spectra obtained in a crysto pore etching experiment.

$$Z(\omega) = R_S + \frac{1}{\dfrac{1}{R_1} + i\omega C_1 + \left(R_2 + \dfrac{1}{i\omega C_2}\right)^{-1}} + \frac{R_3}{1 + i\omega R_3 C_3} \qquad [1]$$

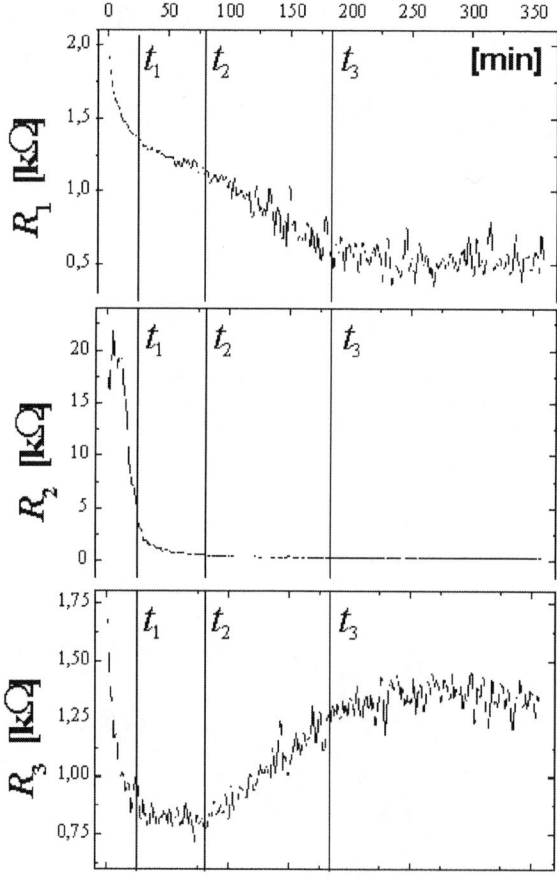

Figure 7: Time dependence of the three important resistances from the equivalent circuit shown in **Fig. 6**. Together with the other 8 curves not shown here, the 4 time regimes emerge as indicated.

In total, 7 parameters can be extracted (4 resistances and 3 capacities) as a function of time, and thus also the three time constant $\tau_i = R_i C_i$. Together with the time dependence of the potential $U(t)$ for a galvanostatic experiment, a first picture for what happens during pore growth emerges. **Fig. 7** shows just three out of the 11 curves obtained, which allow to make a distinction between four crysto growth modes (indicated by the times when they occur in **Fig. 7**): Nucleation (t_1), branching mainly at pore tips (t_2), branching mainly from pore walls (t_3) and steady state with little branching (t_4).

This interpretation agrees well with the independent results obtained form the Monte Carlo model. The next step in modeling crysto pore formation would be to identify the physico-chemical processes behind the components of the equivalent circuit, but presenting and discussing the present speculative efforts in this vein would go beyond the scope of this paper. It is clear that R_2 is only important at the beginning of the experiment and thus most likely associated to the nucleation process. R_1 decreases as branching slows down whereas R_3 increases and then stays roughly constant, indicating steady state.

Current Line Pores

The distinctive features of current line pores are:

- Growth direction is in the direction of current flow or at right angles to the equipotential planes. Under uniform etching conditions away from the sample boundaries, the current flow is perpendicular to the surface and curro pores then appear to grow in the crystallographic direction perpendicular to the surface (usually a low-indexed direction). A distinction between the two types might then be impossible without further data.

- If current flow is very inhomogeneous for whatever reason (e.g. because nucleation occurs only at some widely spaced points), very irregular structures might be obtained that are nevertheless still curro pores. **Fig. 8** gives some examples.

- The decisive intrinsic length scale is foremost the space charge region width d_{SCR}, which gives directly the distance between curro pores or the pore wall thickness. The curro pore diameter often also is in the order of d_{SCR} but may be considerably larger, too. It is sensitive to many parameters including the flow rate of the electrolyte (see below).

- The formation of curro pores relies on only one ingredient: Chemical passivation of the pore walls breaks down (or becomes sufficiently weak) because of diffusion limitation by the nucleation layer (= crysto pores). Pore growth essentially proceeds by direct dissolution but without the strongly crystallographically dependent passivation of pore walls. Pore walls are mainly stabilized by the overlapping of the space charge regions of neighboring pores typically ordered in a hexagonally closed packed arrangement.

- At least for some length of time pore growth proceeds in or close to steady-state conditions, and this is reflected by the impedance data.

- Curro pores cannot nucleate directly and always need some precursors, usually crysto pores that provide for diffusion limitation of the passivating species.

- There are many self-organized structures resulting from the interaction between curro pores via the SCR between the pores.

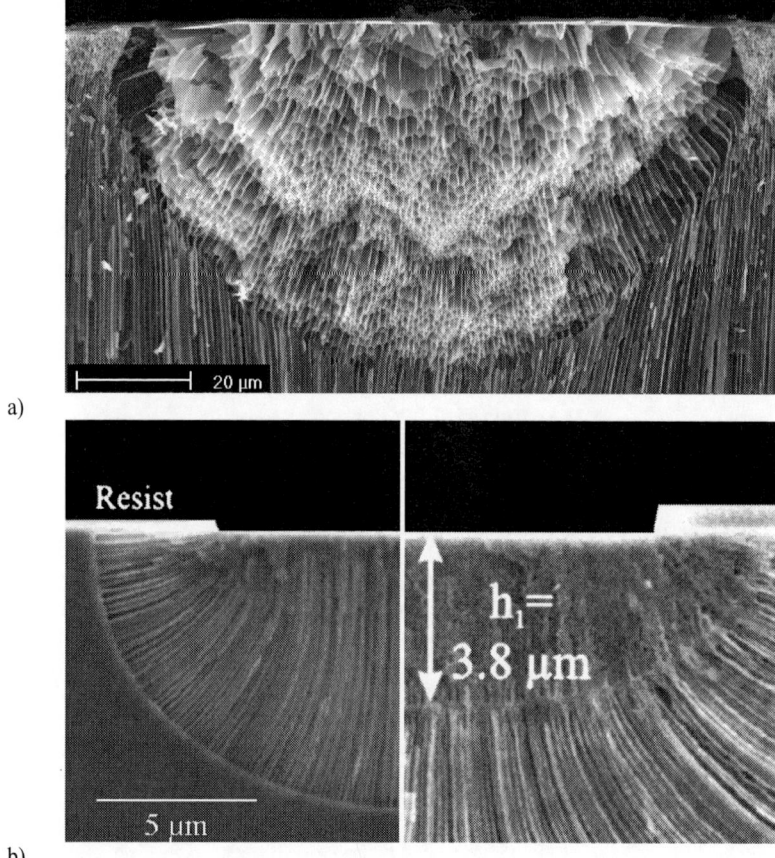

a)

b)

Figure 8: a) "Hedgehog" structure of curro pores obtained for point-like nucleation with pores growing in all directions. b) Waveguide structure (cf. (40)) obtained for linear nucleation; overview left; detail (showing initial crysto pores) on the right.

The fact that the remaining semiconductor between curro pores is an insulator allows to predict that III-V or II-VI membranes with curro pores will be good piezoelectric materials in contrast to bulk samples where the piezo voltage is short-circuited by the low resistance of the semiconductor (41).

SCR "filled" pore walls provide for a strong coupling between neighboring pores and this is the reason for the many self-organization effects observed with curro pores. Besides the formation of single curro-pore crystals and synchronized pore diameter oscillations described before in some detail (42, 43), a new effect that will be called "pore bundle oscillations" has been reported more recently (24). **Fig. 9** shows this effect on an unusually large scale. Obviously, the current density in a pore bundle oscillates in anti-phase to its neighbors, causing "antiphase" oscillations of the pore bundle diameter.

A similar effect has been observed recently in Si for single macropores and termed anti-phase oscillations in view of the fact that the diameters of single pores oscillate in "antiphase" as shown in **Fig. 9c**. Note that from a top view, the structure of the Si pores is "frustrated" (for details see (24)), since this kind of oscillations cannot be obtained in a hexagonal close-packed symmetry. This is also true for the pore bundles here, which have a one-dimensional symmetry (i.e. they look the same on any plane parallel to the one shown but different at right angles).

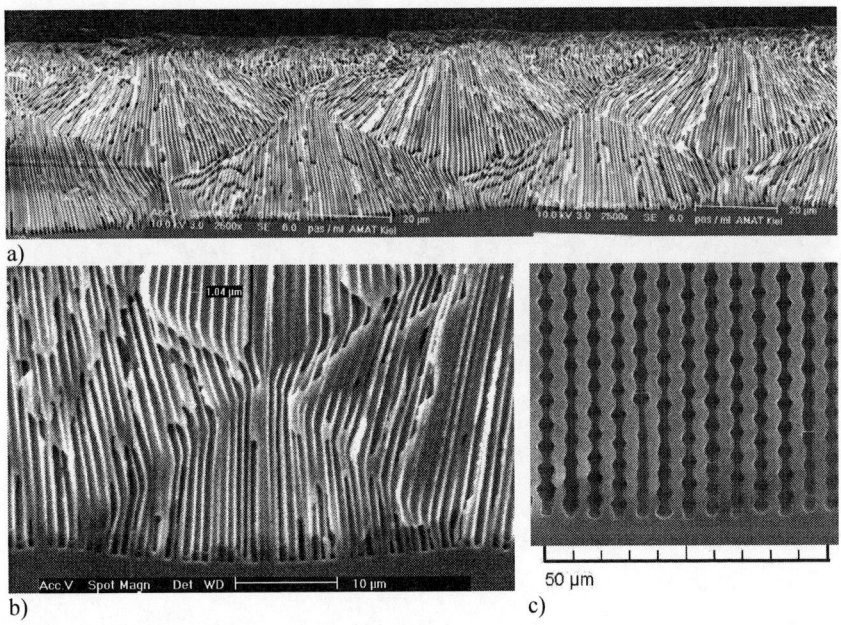

Figure 9: a) Overview of pore bundle oscillations with a wavelength of about 70 μm. b) Detail of a). c) Antiphase oscillations of macropores in Si (cf. (26)).

In attempting to understand this phenomenon it is important to realize that curro pores can neither cross each other nor branch, and this necessitates that they are always coupled: what one pore does for whatever reason determines what its neighbors must do and vice verse. For example, one curro pore cannot change its direction; this can only be done in a collective. This "explains" to some extent why a pore bundle oscillation, if started somehow, must continue for some time or depth. The puzzle is the wavelength of the oscillations shown in **Fig. 9**, since it does not scale with any of the known intrinsic length scales of the electrochemical system like d_{SCR}, critical radius of curvature etc. (cf. (1)). The experiments suggest that it is tied to inhomogeneities of the primary nucleation, possibly to the domains mentioned above, and that the pumping of the electrolyte is a decisive ingredient in this case.

The geometry of curro pores is directly tied to the width of the space charge region d_{SCR} as pointed out before, as far as the pore wall thickness is concerned, and to several parameters, as far as the pore diameter is concerned. For the relatively high-doped samples, used in most studies, pore diameters in the (50 – 200) nm region resulted. For some possible applications larger diameters are desired and an attempt was made to produce large diameter curro pores by using low-doped InP. This proved to be surprisingly difficult, however: **Fig. 4** gives an example. Excessively high voltages (> 100 V) were needed to induce current flow, nucleation was always very inhomogeneous, and often no clear pore structures could be obtained. Therefore attempts were made to increase just the diameter, while keeping the thin pore walls unavoidable in medium to high-doped samples. This approach worked to some extent, **Fig. 10** shows results.

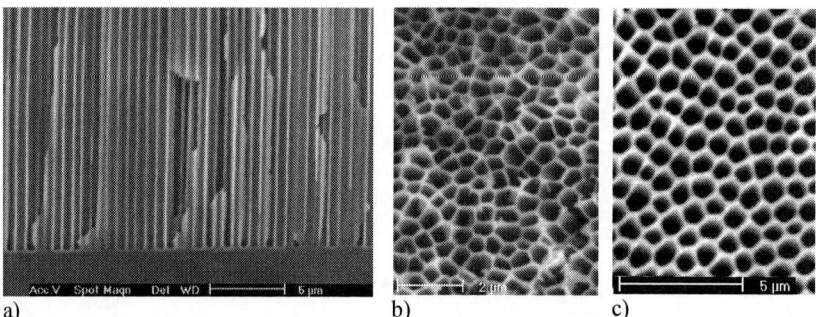

a)   b)   c)

Figure 10: a) Curro pores in InP with large diameters. a) Cross section; b), c) top view of two different samples (after removal of the nucleation layer).

The curro pore diameter proved to be a function of many parameters, in particular the inducement of uniform nucleation by, e.g., a series of voltage steps at the beginning of a (potentiostatic) experiment. The following empirical relation for the pore diameter d_{Pore} has been found for an n-InP $\{100\}$, specimen with $N_{Doping} = 5.4 \cdot 10^{17}$ cm^{-3}, processed in a 2 wt % HCL electrolyte + a variable small amount of H_2SO_4 at 20 °C.

$$d_{Pore} \propto (j \cdot c_{Ox} \cdot PS)/c_{EI} \qquad [2]$$

with j= current density (around (60 - 80) mA/cm^2, c_{Ox} = concentration of the oxidizing species (H_2SO_4), PS = pump speed (settings at the electrolyte pump), and c_{EI} concentration of the dissolving species (HCl). While this result, in particular the dependence on c_{Ox}, is in accordance with predictions from the current burst theory in particular as far as the oxidizing part is concerned (cf. (22)), it should be emphasized that the dissolution chemistry is at present not very well established. The dissolution valence, for example, has only been addressed quite recently by the Versailles group.

Impedance data obtained during curro pore growth provide a rather clear picture of what is going on; for details see (44). All data can be fitted very well by **Eq. 3**, that describes besides a general series resistance R_S the series connection of two simple RC circuits with a third R in parallel to a Warburg impedance (44).

$$Z(\omega) = R_S + \frac{R_1}{1+i\omega\tau_1} - \frac{R_2}{1+(1+i)\sqrt{\omega\tau_2}} + \frac{R_3}{1+i\omega\tau_3} \qquad [3]$$

The last (R_3) component is related to the unavoidable nucleation via crysto pore formation necessary for curro pores and thus not important in this context. The Warburg impedance describes diffusion through the length of the pore, and the R_1 term describes the reaction at the interface. **Fig. 11a** shows this relation between equivalent circuit components and physico-chemical processes schematically.

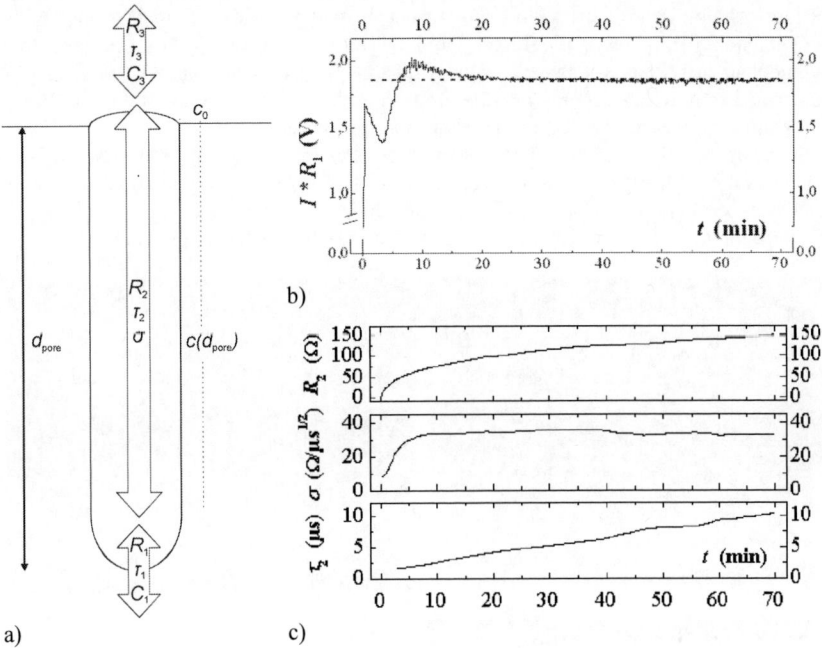

Figure 11: a) Processes in curro pore etching and their schematic relation to the components of the equivalent circuit. b) The product of the etching current and R_1 shows a nearly constant voltage in steady-state etching. Three out of 7 curves from the impedance data showing the development of R_2, the associated time constant τ_2 and the Warburg parameter σ.

R_2 can be fitted very well by

$$R_2 = 29.0\Omega \left(\frac{t}{\min} \right)^{0.4}, \qquad [4]$$

which corresponds in its functionality directly to the experimentally determined pore tip growth speed given in **Eq. 5** via $v_{\text{tip}} \propto 1/R_2$.

$$v_{\text{tips}} = 21.4 \frac{\mu m}{\min} \left(\frac{t}{\min} \right)^{-0.40} \qquad [5]$$

The foregoing consideration just serves in a spurious way to demonstrate that it is possible to move from equivalent circuit diagrams to the physico-chemical processes of interest; for details and more results the reader is referred to (44).

Intermediate Pores

It is easy to imagine some pore morphologies intermediate between crysto and curro. In Si, for example, pores that follow the current flow may consist of small tripods with "legs" extending in the crystallographic <113> direction as shown in **Fig 12a**, that stop growing at an early stage (because encountering a neighboring pore) and instead nucleate a new tripod somewhat deeper. **Fig. 12b** shows InP pores that are neither here nor there; the Versailles group has also found such "meandering" pores.

The examples given in the context of the pore nucleation (e.g. **Figs. 4a and 5b**) also show pores not quite crysto or curro. The same is true for pores found in some transition region between crysto and curro. We may conclude once more that many features of pores in semiconductors are still awaiting discovery, not to mention explanation (45).

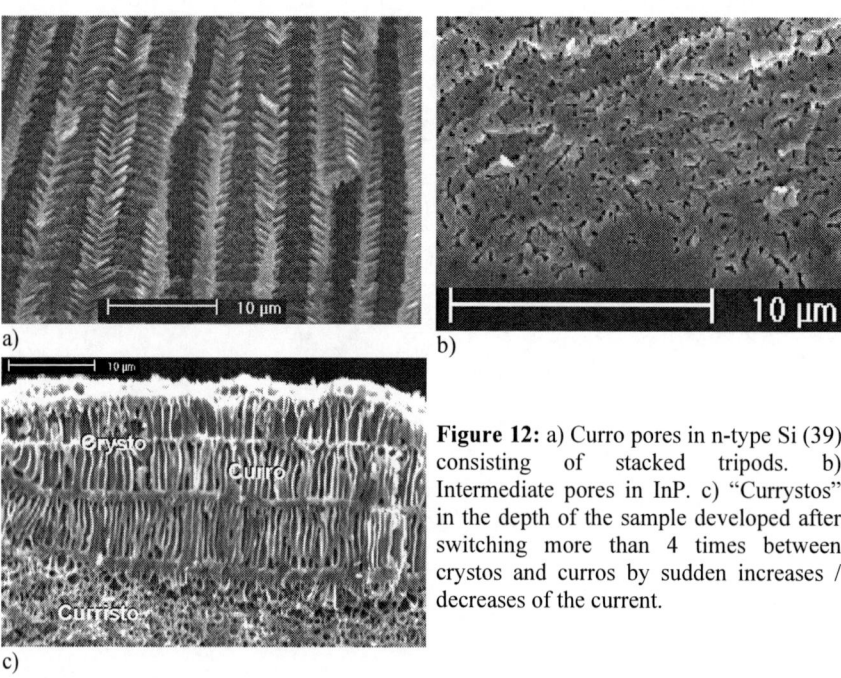

a)

b)

c)

Figure 12: a) Curro pores in n-type Si (39) consisting of stacked tripods. b) Intermediate pores in InP. c) "Currystos" in the depth of the sample developed after switching more than 4 times between crystos and curros by sudden increases / decreases of the current.

Crysto-Curro Transitions

As a general rule, a transition form crysto to curro pores occurs if the current density at the tips of the crysto pores is larger than some steady state value of the system. If left to themselves, the pores react by branching; this goes on until the current density is low enough, followed by some kind of steady-state crysto pore growth with little branching. If that is not possible because the pore density becomes too large for further branching (i.e. the pore distance is in the order of d_{SCR}), a transition to curro pores occurs. If and when such a self-induced transition occurs thus depends on many parameters, the most important one being the current density, the initial nucleation density and uniformity, and the passivation difference between tips and pore walls (expressed in the Monte Carlo model by different nucleation probabilities for a current burst at the tip or on the side wall (then initiating branching)).

It is of interest in this context that similar transitions have recently been observed in Si (27), showing that the crysto – curro issue including the growth mode transitions transcends specific semiconductor (electro)chemistry but describes phenomena on a meta level that is common to many semiconductor pore etching situations.

A sudden increase in the current density during crysto pore growth then will trigger a sharp transition from crysto to curro (and back for a sharp current decrease). This is partially due to the fact that all pores are then induced to do "the same thing at the same time", which is possible with little geometric constraints. On contrast, if the current is gradually increased, some regions may undergo the transition ahead of others, which leads to a non-uniformity that may amplify because "latecomers" find their intended path blocked by others. **Figs. 9a** and **13b** show this indirectly to some extent: pores growing with lateral components sooner or later encounter other pores and are then forced to grow downwards. Alternatively, the system may find ways to spread non-uniformities in space and time by inducing pore bundle oscillations. **Fig. 13** shows a striking example.

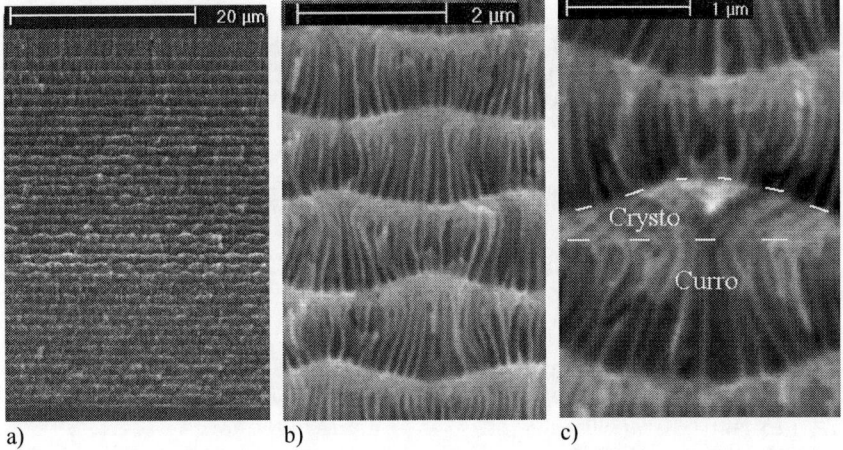

Figure 13: a) About 30 externally induced crysto-curro transitions with a self-induced pore bundle oscillation in the center layers. b), c) Details of the structure. In c) it can be seen that the interface crysto-curro is rather flat; it is only the interface curro-crysto that is wavy.

Conclusion

While InP exhibits the crysto – curro topic most prominently, it is a general feature of pore etching in semiconductors and tied to the degree of wall passivation and the kinetics of passivation. In other semiconductors some parts may be lacking at present (e.g. curro pores in GaAs), but many features or subsets of the topic have been found (and, to make a prediction, will be found). In-situ FFT impedance spectroscopy proved to be very useful for developing models and theories necessary for the understanding of this topic. A Monte Carlo model for crysto pore growth based on general current burst essentials is capable of reproducing many experimentally observed features of crysto pores and thus provides a first step for the understanding of the complete issue, in particular the growth mode transitions.

Acknowledgements

The authors gratefully acknowledge the help of Dr. Ala Cojocaru and Dipl.-Ing. E. Ossei-Wusu who provided some pictures for this paper as well as stimulating discussions. Dr. S. Langa's pioneering work is the foundation for much of what is presented here; for that we owe him. Many discussions with Prof. Tiginyanu have been helpful as well and are gratefully acknowledged.

References

1. H. Föll, S. Langa, J. Carstensen, S. Lölkes, M. Christophersen, and I.M. Tiginyanu, *Adv. Mater.* **15(3)**, 183 (2003).
2. Hamamatsu, C. Kaneshiro, H. Fujikura, and H. Hasegawa, *J. Electroanal. Chem.* **473**, 223 (1999).
3. H. Tsuchiya, M. Hueppe, T. Djenizian, P. Schmuki, and S. Fujimoto, *Sci. Technol. Adv. Mater.* **5**, 119 (2004).
4. A.M. Gonçalves, L. Santinacci, A. Eb, I. Gerard, C. Mathieu, and A. Etcheberry, *Electrochem. Solid-State Lett.* **10(4)**, D35 (2007).
5. T. Takizawa, S. Arai, and M. Nakahara, *Japan J. Appl. Phys.* **33(2, 5A)**, L643 (1994).
6. E. Kikuno, M. Amiotti, T. Takizawa, and S. Arai, *Japan J. Appl. Phys.* **34(1, 1)**, 177 (1995).
7. P. Schmuki, U. Schlierf, T. Herrmann, and G. Champion, *Electrochim. Acta* **48**, 1301 (2003).
8. H. Hasegawa and T. Sato, *Electrochim. Acta* **50**, 3015 (2005).
9. I.M. Tiginyanu, V.V. Ursaki, E. Monaico, E. Foca, and H. Föll, *Electrochem. and Sol. State Lett.* **10(11)**, D127 (2007).
10. C. O'Dwyer, D.N. Buckley, D. Sutton, and S.B. Newcomb, *J. Electrochem. Soc.* **153(12)**, G1039 (2006).
11. C. O'Dwyer, D.N. Buckley, D. Sutton, M. Serantoni, and S.B. Newcomb, *J. Electrochem. Soc.* **154(2)**, H78 (2007).
12. R.W. Tjerkstra, J. Gomesz-Rivas, D. Vanmaekelbergh, and J.J. Kelly, *Electrochem. Solid State Lett.* **5(5)**, G32 (2002).
13. J. Wloka, K. Mueller, and P. Schmuki, *Electrochem. Solid-State Lett.* **8(12)**, B72 (2005).
14. E. Monaico, I.M. Tiginyanu, V.V. Ursaki, A. Saruna, M. Kuball, D.D. Nedeoglo, and V.P. Sirkeli, *Semicond. Sci. Technol.* **22**, 1115 (2007).
15. V. Lehmann, *Electrochemistry of Silicon*, Wiley-VCH, Weinheim (2002).
16. X.G. Zhang, *Electrochemistry of silicon and its oxide*, Kluwer Academic - Plenum Publishers, New York (2001).
17. R.L. Smith and S.D. Collins, *J. Appl. Phys.* **71(8)**, R1 (1992).
18. O. Bisi, S. Ossicini, and L. Pavesi, *Surface Science Reports* **38**, 1 (2000).
19. S. Ossicini, L. Pavesi, and F. Priolo, *Light emitting silicon for microphotonics*, Springer, Berlin (2003).
20. V. Lehmann, S. Stengl, and A. Luigart, *Mat. Sci. Eng. B* **69-70**, 11 (2000).

21. J.-N. Chazalviel, R. Wehrspohn, and F. Ozanam, *Mat. Sci. Eng. B* **69-70**, 1 (2000).
22. H. Föll, M. Christophersen, J. Carstensen, and G. Hasse, *Mat. Sci. Eng. R* **39(4)**, 93 (2002).
23. C. Fang, H. Föll, and J. Carstensen, *J. Electroanal. Chem.* **589**, 259 (2006).
24. H. Föll, M. Leisner, A. Cojocaru, and J. Carstensen, *Electrochim. Acta*, in print (2009).
25. V. Kochergin and H. Föll, *Porous semiconductors: Optical properties and applications*, Springer-Verlag (2009 in print).
26. A. Cojocaru, J. Carstensen, M. Leisner, H. Föll, and I.M. Tiginyanu, *PSST 2008*, in print (2008).
27. A. Cojocaru, J. Carstensen, and H. Föll, *ECS Trans.* **16(3)**, 157 (2008).
28. S. Rönnebeck, S. Ottow, J. Carstensen, and H. Föll, *Journal of Porous Materials* **7**, 353 (2000).
29. M. Christophersen, J. Carstensen, S. Rönnebeck, C. Jäger, W. Jäger, and H. Föll, *J. Electrochem. Soc.* **148(6)**, E267 (2001).
30. S. Langa, J. Carstensen, I.M. Tiginyanu, M. Christophersen, and H. Föll, *Electrochem. and Solid State Lett.* **5**, C14 (2002).
31. S. Lölkes, M. Christophersen, S. Langa, J. Carstensen, and H. Föll, *Mat. Sci. Eng. B* **101**, 159 (2003).
32. E. Spiecker, M. Rudel, W. Jäger, M. Leisner, and H. Föll, *Phys. Stat. Sol. (a)* **202(15)**, 2950 (2005).
33. R. Lynch, C. O'Dwyer, N. Quill ans S. Nakahara, S.B. Newcomb, and D.N. Buckley, *ECS Trans.* **16(3)**, 393 (2008).
34. T. Sato and A. Mizohata, *Electrochem. Solid-State Lett.* **11(5)**, H111 (2008).
35. M. Leisner, J. Carstensen, A. Cojocaru, and H. Föll, *ECS Trans.* **16(3)**, 133 (2008).
36. M. Leisner, J. Carstensen, and H. Föll, *ECS Trans.*, this proceedings (2009).
37. J. Carstensen, R. Prange, G.S. Popkirov, and H. Föll, *Appl. Phys. A* **67(4)**, 459 (1998).
38. E. Foca, J. Carstensen, and H. Föll, *J. Electroanal. Chem.* **603**, 175 (2007).
39. S. Frey, M. Kemell, J. Carstensen, S. Langa, and H. Föll, *Phys. Stat. Sol. (a)* **202(8)**, 1369 (2005).
40. S. Langa, S. Frey, J. Carstensen, H. Föll, I.M. Tiginyanu, M. Hermann, and G. Böttger, *Electrochem. Sol. State Lett.* **8(2)**, C30 (2005).
41. K. Rottner, R. Helbig, and G. Müller, *Appl. Phys. Lett.* **62(4)**, 352 (1993).
42. S. Langa, I.M. Tiginyanu, J. Carstensen, M. Christophersen, and H. Föll, *Appl. Phys. Lett.* **82(2)**, 278 (2003).
43. S. Langa, J. Carstensen, I.M. Tiginyanu, M. Christophersen, and H. Föll, *Electrochem. Solid-State Lett.* **4(6)**, G50 (2001).
44. M. Leisner, J. Carstensen, A. Cojocaru, and H. Föll, *PSST 2008* , in print (2008).
45. J.J. Kelly and H.G.G. Philipsen, *Curr. Opin. Solid State Mater. Sci.* **9(1-2)**, 84 (2005).

ECS Transactions, 19 (3) 347-354 (2009)
10.1149/1.3120714 ©The Electrochemical Society

Production of High Aspect Ratio Single Holes in Semiconductors

M.-D. Gerngroß, H. Föll, A. Cojocaru, and J. Carstensen

Institute for Materials Science, Christian-Albrechts-University of Kiel, Kaiserstr. 2,
D-24143 Kiel, Germany

In many areas of research exists a need for single small holes with
diameters from a few nm to several μm and large aspect ratios. Existing technologies for that are complex and rather limited. It is
shown by a simple proof of principle that suitable single holes or
specific arrays of some single holes can be made by first etching a
very large number of small and deep holes or pores into semiconductors like Si or InP by established electrochemical means, followed by masking the desired holes and filling all others with, e.g.,
a metal in a galvanic process. The potential and limitations of this
technique are discussed in some detail.

Introduction

In several branches of research, single holes with small diameters in the nm - μm
range and large aspect ratios are needed. For example, detailed studies of transport and
electrochemical phenomena need holes with nanoscopic dimensions (1 - 4), while applications such as single-pore filtration of cells call for pores or holes with well-defined diameters of a few μm. In ref. (5) "patch clamping" is described, meaning the use of single
holes with diameters around 1 μm and an aspect ratio of 200 in order to avoid laborious
and difficult micromanipulations of a (micro) glass pipette under a microscope. Microfluidic applications (6) would also benefit from single holes or arrays of holes in a defined
pattern and that might also be true for some nano-optics and X-ray applications. It goes
beyond the scope of this paper (and beyond the experience of the authors) to enumerate
all possible uses of small single holes with large aspect ratios or defined arrays of such
holes.

While there are many ways of producing such structures, none of the techniques reported is easy to use and mostly very specific to certain materials and dimensions. In references (7 - 10) several methods are reviewed and described in some detail. They include
conventional lithography applied, e.g., to thin layers, ion beam and electron beam techniques as well as nuclear track etching.

In this paper we first provide a proof-of-principle for making single holes or defined
arrays of holes by first etching a multitude of suitable pores in semiconductors by established electro-chemical processes, followed by complete closing of unwanted pores (all
but one in the extreme case). Next the possibilities and limits of the technique proposed
will be discussed in some detail.

Experimental Techniques and Results

The electrochemical etching of pores or holes in standard Si substrates of n-type or p-
type was established around 1991 (11 - 13) and resulted in three major pore types: Mi-
cropores or mesopores with diameters below 2 nm or between 2 nm and 50 nm, respectively, and macropores with diameters > 50 nm. The etching of macropores with diame-

ters between about 0.5 μm (14) and up to 10 μm or larger (15) in a lithographically de-fined array and with pore depths of up to 500 μm as pioneered by Lehmann and Föll (13) is by now a well established technique and described in two books (16, 17), a number of reviews (18 - 20) and many papers (cf. for example the by now 5 proceedings of the Int. Conf. "Porous Semiconductors Science and Technology ("PSST")). **Fig. 1** gives an im-pression of what is possible; some ramification of the technique for the purpose of single hole production will be discussed later.

Given a periodic array of pores as shown in **Fig. 1a)**, a single hole can in principle be produced if all pores except the one selected are closed and the unetched part of the Si wafer is removed, so that a membrane is formed. It is also clear that closing all pores ex-cept the ones selected for a "single hole array" (SHA) can produce any SHA that is com-mensurate with the basic pore grid geometry. The task at hand is thus to i) produce a suit-able hole array, ii) mask selected holes in a way that prevents pore filling; iii) fill the un-masked holes, and iv) produce a "membrane" with just one hole or a SHA by removing the masking and the unetched part of the Si wafer.

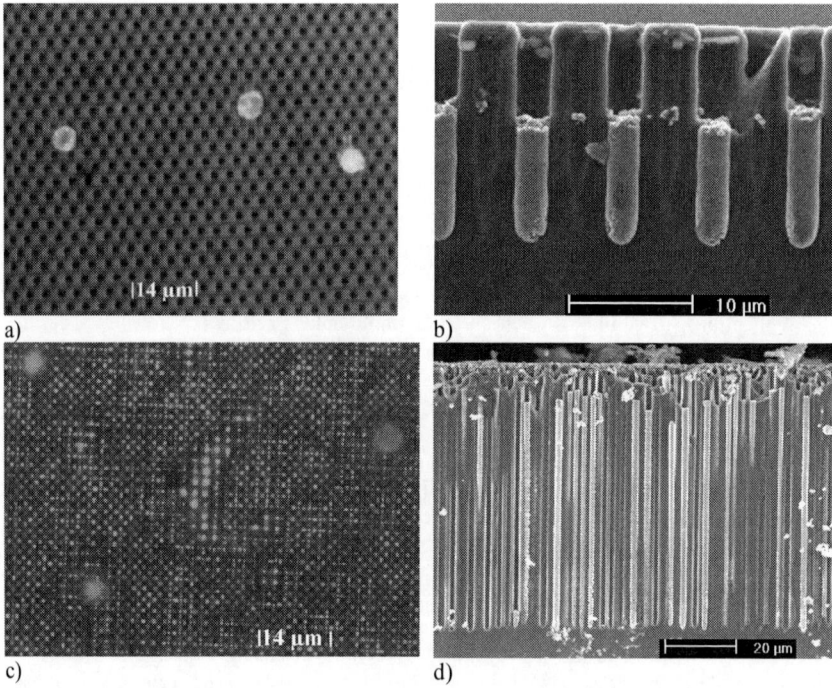

Figure 1: a) Top view, showing the defined n-Si array of pores as pre-structured by standard lithography used in the experiments. b) Cross-sectional view of a n-type Si macropore array filled with Cu. c) Top view, showing the defined p-type Si array of pores as pre-structured by standard lithography used in the experiments. d) Cross-sectional view of a p-type Si macropore array filled with Cu.

The second task of masking some pores can be done, in principle, by suitable variants of standard lithographical techniques. For the proof of principle reported here it is not important which hole exactly is kept open so a very simple masking technique was chosen. A few holes were kept open by "dusting" some toner from a Laser printer over the surface of a freshly etched pore array. Toner particles have sizes around 5 μm and some particles stick to the opening of a pore if they are "fixed", like in the printing process, by heating the specimen to (90 – 95) °C for about 10 min. **Fig. 1a)** shows three thermally fixed toner particles covering one pore completely and the neighboring pores to some extent.

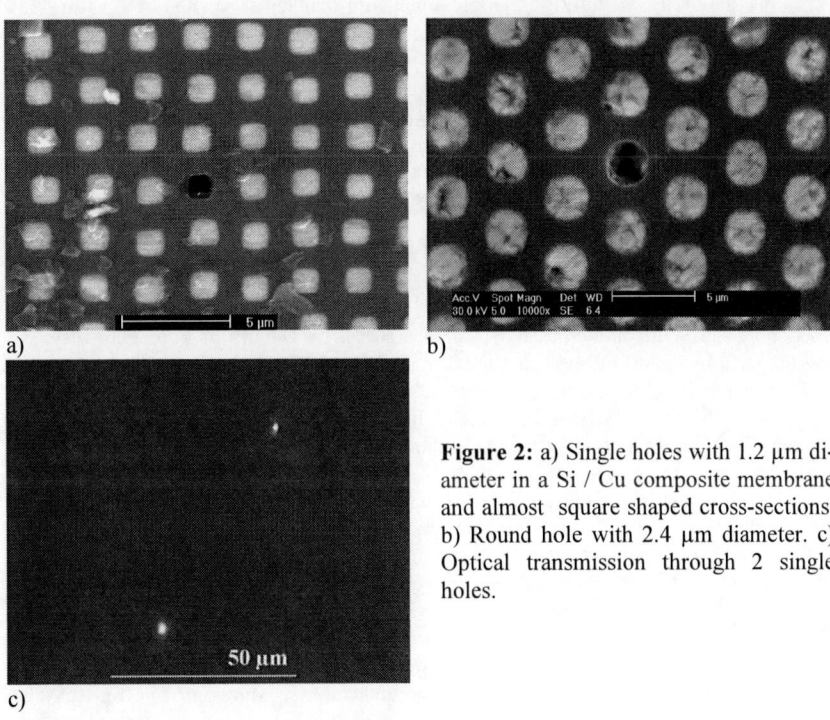

a)

b)

c)

Figure 2: a) Single holes with 1.2 μm diameter in a Si / Cu composite membrane and almost square shaped cross-sections. b) Round hole with 2.4 μm diameter. c) Optical transmission through 2 single holes.

The third task - filling the unmasked pores - was done by galvanically depositing Cu in all unmasked pores. Filling pores in Si (and in other semiconductors) by some galvanic process with metals or other materials in order to produce a nanocompound has attracted considerable interest in recent years (21 - 24), and there are many open questions with respect to filling pores completely at all and, if possible, within acceptable times. For the teak at hand Cu galvanics proved to be suitable; the experimental conditions were as follows: Electrolyte composition: 300 ml H_2O, 70 ml H_2SO_4, 5 g $CuSO_4$, 0.1 g DTAC (1-dodecyl-trimethylammoniumchloride, 97 %), 0.1 g SPS (Bis-3-sodiumsulfopropyldisulfide), 0.1 g PEG (Polyethylenglycol). The Cu deposition was done at constant 20 °C under galvanostatic conditions with a current density linearly ramped from -0.2 mA/cm^2 to -1.0 mA/cm^2 during the deposition time of about 1.000 min. The potential found changed from about -0.1 V at the beginning of the experiment to about $- 0.6$ V at the end. Complete filling of 77 μm deep pores could be achieved in this (slow) way, cf. **Fig. 1d)**; a task not yet possible in much shorter times. **Fig. 1b)** shows pores in a p-type substrate (so-called p-macro(org, litho) pores), cf. (19, 25, 26) obtained with so-called organic electrolytes, demonstrating that the technique can be used with both types of doping. The pores masked with the toner particles did not participate in the filling process as could be expected.

The fourth task of making a membrane is easy after the pores have been filled, since the sample is no longer porous and thus not very brittle. It is possible now to use standard mechanical grinding and polishing.

The single holes obtained in this way are shown in **Fig. 2**. **Figs. 2a,b)** show different pore lattice geometries resulting in single holes with diameters / depths of 1.3 μm / 150 μm and 1.7 μm / 97 μm, respectively; the aspect ratios thus were 115 and 57, respectively.

That the holes were open throughout the depth of the sample could be seen with the aid of an optical microscope in transmission. After removing all apertures and using the highest illumination intensities possible, single holes showed up as recognizable bright dots (**Fig. 2c)**).

We consider the result shown as sufficient proof of principle for making single holes or SHAs. In what follows we discuss the potential and limitation of the method described.

Potential and Limitations

The following items shall be addressed in the context of selectively filling anodically etched pores in semiconductors: i) Available pores, ii) Details of pore geometry, iii) Masking techniques, iv) Filling techniques, and v) Specialties.

While there is a huge multitude of pores available in several semiconductors (most prominent are Si, Ge (27), III-V compounds like GaAs, InP and GaP (28 - 32), SiC (33), and some II-VI compounds like CdSe (34) and in some anodically formed metal oxides (most prominent are Al_2O_3, TiO_2, and ZrO_2 (35 - 37)), only straight pores arranged in some regular pattern are of conceivable use. This excludes many of the pores produced so far but still leaves the following candidates: 1. Lithographically defined macropores in lightly doped n-type and p-type Si with minimum diameters around 0.5 μm (14) and aspect ratios > 500. 2. Self-organized macropores and possibly mesopores in Si with diameters from about 200 nm to below 50 nm and aspect ratios of > 100. The pictures in (17, 38) may serve as examples. 3. So-called "current line pores" in InP with diameters from about 1 μm to about 50 nm (28, 31, 32) and arranged in a rather well-expressed self-organized single pore crystal under certain circumstances as shown in **Fig. 3a)** (cf. also (39)). 4. Some pore types in GaAs, GaP or SiC but without offering properties that are not already covered in 1. – 3. 5. Pores in the metal oxides mentioned above with diameters typically in the 20 nm - 300 nm range. In summary, the pore diameters in suitable pore arrays span the range from around 20 nm to > 10 μm.

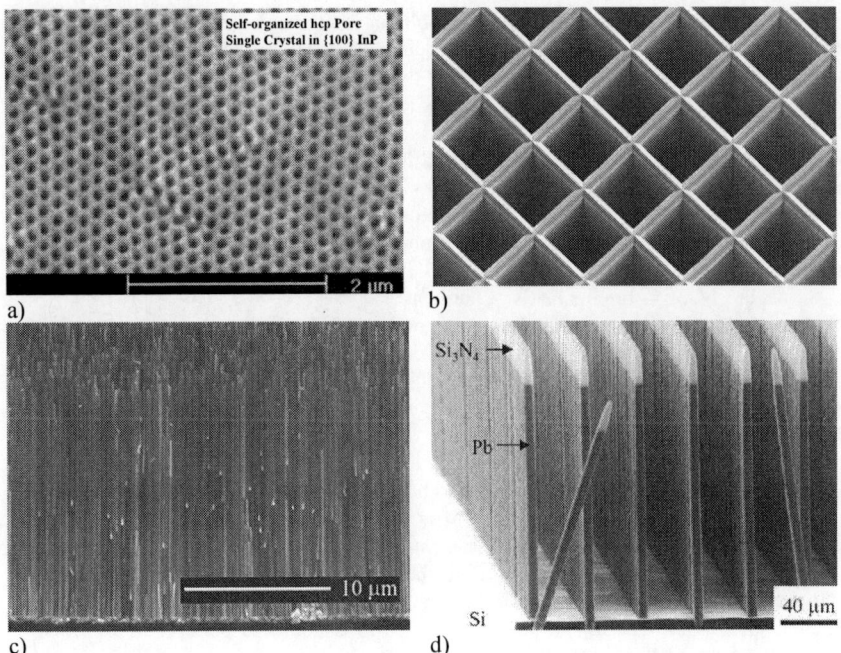

Figure 3: a) Self-organized hexagonal pore single crystal in n-type InP (40). b) Square pore array obtained from n-type Si macropores after subsequent chemical treatment in KOH (39). c) Pt tubes obtained through the filling on n-type InP pores (23). d) Pb wire array produced by filling Si macropores with liquid Pb (41). For details see the text.

With respect to the detailed single pore geometry, major parameters are straightness, cross-sectional geometry (and its variation with pore depth), and the pore wall roughness. Relative to the large number of geometrically and topologically different pore types alluded to above, only cursory information can be given here. With respect to straightness, a fundamental difference exists between pores that grow strictly in a crystallographic direction (like all n-macro(bsi, aqu) pores and many (but not necessarily all) Si-p-macro(org) pores) and thus are perfectly straight, and so-called current-line pores that grow generally in the direction of current flow and thus might deviate from a straight line. The pores in InP mentioned above belong into this category. At present it is not clear if the smaller kinds of pores can be made with deviations from a straight line much smaller than their diameter. This is not so much a principal problem but rather reflects the fact that neither measurement of this kind have been reported, nor any attempts to optimize this quantity. In our opinion, straight enough pores are possible even for very small diameters for aspect ratios not too large.

The cross-sectional geometry of pores may vary from pore to pore and within a given pore as function of depth. Generally, pore single crystals – produced by defined nucleation (all kinds of lithography) or by self-organization – show rather constant diameters from pore to pore that usually can be experimentally adjusted to a defined value within some bandwidth. More irregular arrays like small macropores or mesopores in highly doped Si tend to show larger variations of the individual diameters. That may have an inherent advantage, since it offers the possibility to select a specific pore with the right dimension for the single hole desired. As far as individual pores are concerned, there is a certain tendency with large macropores that the pore diameter decreases somewhat with depth, but this effect can be kept small, in particular for aspect ratios smaller than 100 – 200. The cross-sectional geometry in defined arrays varies from circular to almost square shaped; in more random arrays it can be multi-facetted with the symmetry of the nearest neighbor surrounding (i.e. typically fourfold - sevenfold). This may be adjusted to some extent by the etching conditions or by some post-etching treatment. Oxidation with subsequent oxide removal may make a pore "rounder" (42), while chemical treatment with highly orientation selective etchants like KOH for Si may produce rather perfect squares as shown in **Fig. 3b)** from the late V. Lehmann.

Not too much is known about the roughness of the surface of pores; for the Si – n-macro(bsi, aqu, litho) case Foca et al. performed a systematic study and showed that the roughness may vary considerably, depending on what kind of "optimizer" had been added to the electrolyte, but can be < 10 nm (43). In contrast, self-arranged arrays of small macropores and mesopores in Si are often very rough or "fuzzy". Not much is known for pores in the other semiconductors except that in typical SEM pictures the pore walls in e.g. InP current line pores appear to be not too rough.

With respect to masking techniques it is generally possible to use a somewhat adjusted standard lithography even on porous substrates as pioneered by Ottow et al. (44). While the technique has not yet been tried for masking single holes, it can be confidentially stated that this should be possible for hole dimensions of 1 μm or larger. For much smaller hole dimensions in-situ deposition of e.g. Pt should be possible in e.g. focused ion beam machines (FIB) nowadays available in many laboratories. In addition, electron-beam lithography might be used.

Filling techniques do not need to be restricted to galvanic deposition of metals. While this technique can be used even at small dimensions, cf. **Fig. 3c)**, that shows the Pt filling of pores in InP (substrate etched off) from the work of Tiginyanu et al. (23) or refer to (45) where 900 nm Al_2O_3 pores have been filled with Ni, other techniques are available, too. Besides filling pores at some elevate temperatures with liquid polymers (46), liquid metals can be used, too, as demonstrated in (41) and shown in **Fig. 3d)**.

Specialties in this context are the coating of "the" pore (or all pores before the filling) with all kinds of layers, e.g. to change the reflectance of the pore surface or intentional modulations of the pore diameter as detailed, e.g. in (47). As far as the passage of light in pore directions is concerned, ref. (48) provides a review of the many effects encountered and the possibilities of pore treatments as outlined above for optimizing optical properties (including far IR and UV). Some of these considerations are also valid for single pores.

Coating the inside of a single hole or of a SHA after it has been made with suitable layers (e.g. by using atomic layer deposition (ALD) (48)), may be used to decrease the pore diameter of more easily made larger single pores to smaller dimensions. The limits of such a technology are not clear at present but diameter reductions of a factor up to 10 (e.g. from 1 μm to 100 nm) should not be too difficult. An example is shown in **Fig. 4.**

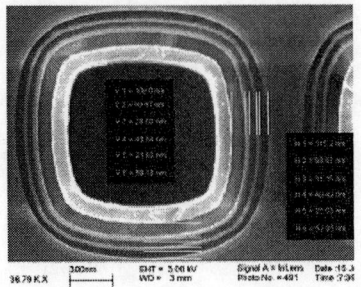

Figure 4: a) Macropore in Si coated with 5 layers of dielectric materials by atomic layer deposition (48). The diameter has been conformally reduced by 60 % and even larger reductions would be possible.

Acknowledgements

The authors are indebted to Prof. J. Bohr who suggested the topic and was always available for questions and discussions. The help of Prof. Tiginyanu, who supplied one of the pictures, is gratefully acknowledged.

References

1. S. Howorka, S. Cheley, and H. Bayley, *Nat. Biotechnol.* **19(7)**, 636 (2001).
2. O. Braha, L.-Q. Gu, L. Zhou, X. Lu, S. Cheley, and H. Bayley, *Nat. Biotechnol.* **18(9)**, 1005 (2000).
3. J. Li, D. Stein, C. McMullan, D. Branton, M.J. Aziz, and J.A. Golovchenko, *Nature* **412(6843)**, 166 (2001).
4. D.W. Deamer and D. Branton, *Nature* **35(10)**, 817 (2002).
5. N. Fertig, R.H. Blick, and J.C. Behrends, *Biophys. J.* **82(6)**, 3056 (2002).
6. J. Clayton, *Nat. Methods* **2(8)**, 621 (2005).
7. K. Healy, B. Schiedt, and A.P. Morrison, *Nanomed.* **2(6)**, 875 (2007).
8. R.E. Packard, J.P. Pekola, P.B. Price, R.N.R. Spohr, K.H. Westmacott, and Z. Yu-Qun, *Rev. Sci. Instrum.* **57(8)**, 1654 (1986).
9. C.C. Harrell, S.B. Lee, and C.R. Martin, *Anal. Chem.* **75(24)**, 6861 (2003).
10. L. Sun and R.M. Crooks, *Langmuir* **15(3)**, 738 (1999).
11. L.T. Canham, *Appl. Phys. Lett.* **57(10)**, 1046 (1990).
12. V. Lehmann and U. Gösele, *Appl. Phys. Lett.* **58(8)**, 856 (1991).
13. V. Lehmann and H. Föll, *J. Electrochem. Soc.* **137(2)**, 653 (1990).
14. J. Schilling, A. Birner, F. Mueller, R.B. Wehrspohn, R. Hillebrand, U. Goesele, K. Busch, S. John, S.W. Leonard, and H.M. van Driel, *Opt. Mater.* **17**, 7 (2001).
15. P. Kleimann, J. Linross, and S. Petersson, *Mater. Sci. Eng. B* **69-70**, 29 (2000).
16. X.G. Zhang, *Electrochemistry of silicon and its oxide*, Kluwer Academic - Plenum Publishers, New York (2001).
17. V. Lehmann, *Electrochemistry of Silicon*, Wiley-VCH, Weinheim (2002).
18. J.-N. Chazalviel, R. Wehrspohn, and F. Ozanam, *Mat. Sci. Eng. B* **69-70**, 1 (2000).
19. H. Föll, M. Christophersen, J. Carstensen, and G. Hasse, *Mat. Sci. Eng. R* **39(4)**, 93 (2002).
20. S. Ossicini, L. Pavesi, and F. Priolo, *Light emitting silicon for microphotonics*, Springer, Berlin (2003).

21. M. Jeske, J.W. Schultze, M. Thönissen, and H. Münde, *Thin Solid Films* **255**, 63 (1995).
22. K. Fukami, K. Kobayashi, T. Matsumoto, Y.L. Kawamura, T. Sakka, and Y.H. Ogata, *J. Electrochem. Soc.* **155(6)**, D443 (2008).
23. Tiginyanu, E. Monaico, and E. Monaico, *Electrochem. Comm.* **10(5)**, 731 (2008).
24. C. Fang, E. Foca, L. Sirbu, J. Carstensen, I.M. Tiginyanu, and H. Föll, *Phys. Stat. Sol. (a)* **204(5)**, 1388 (2007).
25. J. Zheng, M. Christophersen, and P.L. Bergstrom, *Phys. Stat. Sol. (a)* **202(8)**, 1402 (2005).
26. J. Zheng, M. Christophersen, and P.L. Bergstrom, *Phys. Stat. Sol. (a)* **202(8)**, 1662 (2005).
27. C. Fang, H. Föll, and J. Carstensen, *J. Electroanal. Chem.* **589**, 259 (2006).
28. H. Föll, S. Langa, J. Carstensen, S. Lölkes, M. Christophersen, and I.M. Tiginyanu, *Adv. Mater.* **15(3)**, 183 (2003).
29. J. Wloka, K. Mueller, and P. Schmuki, *Electrochem. Solid-State Lett.* **8(12)**, B72 (2005).
30. R.W. Tjerkstra, J. Gómez-Rivas, D. Vanmaekelbergh, and J.J. Kelly, *Electrochem. Solid State Lett.* **5(5)**, G32 (2002).
31. H. Hasegawa and T. Sato, *Electrochim. Acta* **50**, 3015 (2005).
32. P. Schmuki, U. Schlierf, T. Herrmann, and G. Champion, *Electrochim. Acta* **48**, 1301 (2003).
33. O. Jessensky, F. Müller, and U. Gösele, *Thin Solid Films* **297**, 224 (1997).
34. I.M. Tiginyanu, V.V. Ursaki, E. Monaico, E. Foca, and H. Föll, *Electrochem. and Sol. State Lett.* **10(11)**, D127 (2007).
35. O. Jessensky, F. Müller, and U. Gösele, *J. Electrochem. Soc.* **145(11)**, 3735 (1998).
36. J.M. Macak, H. Hildebrand, U. Marten-Jahns, and P. Schmuki, *J. Electroanal. Chem.* **621**, 254 (2008).
37. S. Berger, J. Faltenbacher, S. Bauer, and P. Schmuki, *phys. stat. sol. RRL* **2(3)**, 102 (2008).
38. P. Granitzer, K. Rumpf, P. Pölt, A. Reichmann, and H. Krenn, *Physica E* **38**, 205 (2007).
39. S. Langa, I.M. Tiginyanu, J. Carstensen, M. Christophersen, and H. Föll, *Appl. Phys. Lett.* **82(2)**, 278 (2003).
40. V. Lehmann, *Phys. Stat. Sol. (a)* **204(5)**, 1318 (2007).
41. V. Lehmann and S. Rönnebeck, *Sensors and Actuators A* **95**, 202 (2001).
42. E.V. Astrova, T.N. Borovinskaya, T.S. Perova, and M.V. Zamoryanskaya, *Semiconductors* **38(9)**, 1084 (2004).
43. E. Foca, J. Carstensen, M. Leisner, E. Ossei-Wusu, O. Riemenschneider, and H. Föll, *ECS Trans.* **6(2)**, 367 (2007).
44. S. Ottow, V. Lehmann, and H. Föll, *Appl. Phys. A* **63**, 153 (1996).
45. K. Nielsch, F. Müller, A.-P. Li, and U. Gösele, *Adv. Mater.* **12(8)**, 582 (2000).
46. K. Nielsch, F.J. Castano, C.A. Ross, and R. Krishan, *J. Appl. Phys.* **98**, 034318 (2005).
47. S. Matthias, F. Müller, J. Schilling, and U. Gösele, *Appl. Phys. A* **80(7)**, 1391 (2005).
48. V. Kochergin and H. Föll, *Mater. Sci. Eng. R* **52(4-6)**, 93 (2006).

ECS Transactions, 19 (3) 355-361 (2009)
10.1149/1.3120715 ©The Electrochemical Society

Dynamics of Macropore Growth in n-type Silicon Investigated by FFT In-Situ Analysis

J. Carstensen, A. Cojocaru, M. Leisner, and H. Föll

Institute for Materials Science, University of Kiel, Kaiserstrasse 2, 24143 Kiel, Germany

The dynamics of macropore growth in n-type silicon were investigated by in-situ FFT impedance spectroscopy and transient analysis. In particular the response to fast growing pores to current density steps in the context of so-called anti-phase diameter oscillations was investigated. These pore growth mode allows for a very fast growth of deep macropores and could for the first time be stimulated by external current steps.

Introduction

Since the discovery of macropore formation in n-type Si with backside illumination (bsi) by Lehmann and Föll (1), many investigations have been undertaken with the aim to obtain a good understanding of the underlying phenomena, cf. (2 - 4). As a result a very good control of pore geometries emerged, even for extremely deep pores; although some puzzling details remained, like the slow nucleation of macropores without lithographically pre-structured nucleation sites (5, 6), or the non-linear response of the pore diameter to modulations of the current density (7). Moreover, fast etching with high HF concentrations and correspondingly high current densities to large depths (8) produced unexpected results, in particular self-induced anti-phase diameter oscillations as shown in **Fig. 1,** often preceded and followed by several further growth mode transitions (9). This indicates that pore formation at high etching rates (and to some extend also at low etching rates / current densities) is dominated by kinetics, i.e. rate limiting time constants.

In order to investigate the underlying processes causing growth mode transitions a combination of impedance analysis and transient analysis is used in this paper. While impedance analysis, i.e. the linear response analysis of the system to a sinusodially modulated perturbation, is efficient for fast processes, it becomes very time consuming for slow processes, since an average over several oscillation periods is needed to gain a good signal to noise ratio. In contrast, transient analysis (response of the system to a step-like change of a parameter) is better suited for a reliable analysis of time constants of slow processes, since the determination of a relaxation time needs only a (large) number of measured points within a period of 3 to 4 times the relaxation time after the step and this can be done with a small sampling rate. In contrast, for fast processes a large sampling rate would be needed leading to correspondingly large errors in the measured data.

In the experiments reported here the etching current density (controlled by the backside illumination) is changed step-like with various period lengths after the macro pores have reached a certain depth.

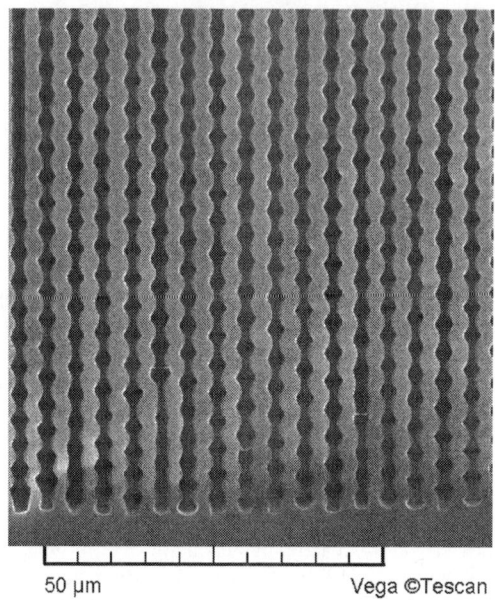

50 μm   Vega ©Tescan

Fig. 1: Self-induced "anti-phase" pore diameter oscillations in macropores in n-type Si.

This allows to investigate the slow transient processes and to check if and under which conditions anti-phase pore diameter oscillations can be triggered by external current stimuli. Simultaneously the voltage and illumination impedance (c.f. (8, 10)) are measured every 2 to 3 seconds to monitor the (fast) electrochemical processes.

Experimental Details

Low doped (5 Ωcm) (100) oriented n-type Si wafers with an n^+ layer on the back side for good ohmic contact were etched at a controlled temperature of 20 °C. The samples were pre-structured by standard photolithography; the nucleation pattern was a hexagonal lattice with a lattice constant of $a = 4.2$ μm. The electrolyte consisted of 15 wt.% HF in an aqueous-viscous solvent. Carboxymethylcellulose sodium salt (CMC) was added to the electrolyte (0.42 g/l) to increase the viscosity. The pores are etched using the basic parameters taken from the "Lehmann model" (2), i.e. the initial current density $j(t)$ and its decrease with time was chosen to yield pores with 2 μm diameter. To investigate the bsi pore formation kinetics *in-situ* FFT impedance spectroscopy (11) has been used. The current density $j(t)$ started at 27 mA/cm^2 and was reduced as a function of time to compensate for the reduction of the electrolyte concentration at longer pore tips, e.g. to 24 mA/ cm^2 after 10 min. Current steps with periods of 1 min, 2 min, and 5 min were applied in which the current was lowered by a factor of two of $j(t)$. The number of steps started from one and two up to 70.

The FFT impedance spectrometer was embedded within the etching system (ET&TE GmbH) and was used to extract in-situ information concerning the pore etching. The impedance measurements were performed at intervals of 1 second in a frequency range between 100 Hz and 20 kHz, containing 27 frequencies.

Results and Discussion

Fig. 1 shows the self-induced anti-phase diameter oscillations obtained with electrolytes of high HF concentration ($c_{HF} \gg 5$ wt%) and correspondingly high current densities. The period of the pore diameter oscillation can be estimated from the pore growth speed and the distance of two diameter maxima to be in the order of minutes. To investigate the etching conditions that allow for such anti-phase diameter oscillations, etching current steps were applied to the etching process. **Fig. 2** shows the SEM images of applying just one step by reducing the current density by a factor of two for 1 min. and 2 min., respectively, and afterwards continuing the experiment with the original current density. Neither in **Fig. 2a** (step of 1 min.) nor in **Fig. 2b** (step of 2 min.) diameter modulations in the following high current density phase are found, i.e. growth mode transition leading to self-sustained oscillations could be triggered. While the current density is reduced, only a decrease of the pore diameter is found. The diameter decreases by about $1.43 \approx \sqrt{2}$ as is expected from the Lehmann model, which predicts a change of pore tip area by a factor of two, i.e. a factor $\sqrt{2}$ for the pore diameter. In addition, for the 2 min. step the first stage of cavity formation is found soon after the current density has been increased to the full value again, but not for the 1 min. step. Such cavities are always found for the etching condition discussed here (cf. (8)), terminating the pore growth, but typically at larger pore depth. Since the cavity formation is related to strong diffusion limitation this is a first hint that the "bottle neck" induced by the current step already leads to a significantly increased diffusion limitation.

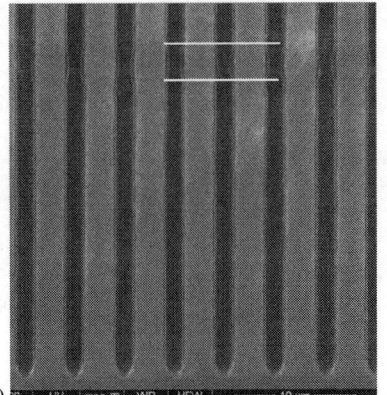 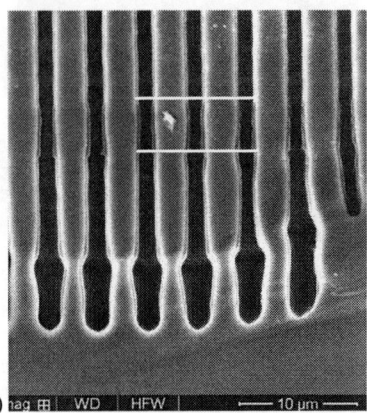

Fig. 2: SEM cross section of macropores. After 5 min. a current step with a period of a) 1 min. and b) 2 min. was initiated. The white lines indicate the region of reduced diameter.

Fig. 3 shows the pores when 2 current density steps were applied. In **Fig. 3a** (step time 1 min.) the diameter reduction as already seen in **Fig. 2a** is now found twice. In contrast to this, in **Fig. 3b** (step time 2 min.) anti-phase diameter modulations are visible. These anti-phase diameter modulations are obviously induced by the current density steps since under these etching condition *self-induced* anti-phase diameter modulation do not exist. As soon as the current density is constant again the diameter modulations stop as well (and cavity formation starts). Anti-phase diameter modulations thus are not yet stable under the applied etching condition, even if they are induced externally. This is reminiscent, to some extent, to the triggering of current oscillations in Si in a region of parameter space where these oscillations are possible but heavily damped (12).

Stable self induced anti-phase diameter oscillations occur at larger pore depths and lower current densities. The triggering of the anti-phase diameter modulation apparently needs the right period for the current steps. While it does not work for a period of 1 min. as shown in **Fig. 3a**, it does work for a period of 2 min. (**Fig. 3b**) and also for a period of 5 min. (not shown here). However, applying a step of 2 min. to pores that are significantly longer, i.e. later in the process, anti-phase diameter modulations cannot be induced. This indicates that there is only a limited region in parameter space where anti-phase diameter oscillations "resonate" with external stimuli as shown in **Fig. 2b**. Alternatively, some kind of feedback might be required to find the proper period and the modulation amplitude of the current density as a function of pore depth.

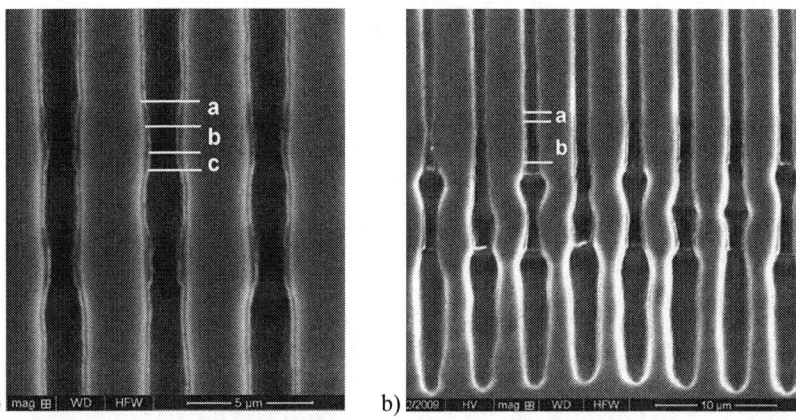

Fig. 3: SEM cross section of macropores. After 5 min. the current was modulated stepwise two times with different periods: a) 1 min. and b) 2 min. The white lines indicate the borders of regions described below and in **Fig. 5**.

To get in-situ information about the etching processes and thus also for a possible feedback for controlling the anti-phase diameter oscillations, in-situ FFT impedance analysis was performed for the whole time of all experiments. In this case only voltage impedance was used; the data obtained were analyzed using a model that allows for a very good fit of all impedance data for nearly all types of macropores formed by using backside illumination (8, 10). The most relevant parameter of the model in the context of this paper is the transfer resistance R_p that essentially describes the chemical dissolution process. The impedance results for the etching of the pores shown in **Fig. 3** are displayed in **Fig. 4**. **Fig. 4a** shows the model parameters as a function of time for the 1 min. steps and **Fig. 4b** for the 2 min. steps.

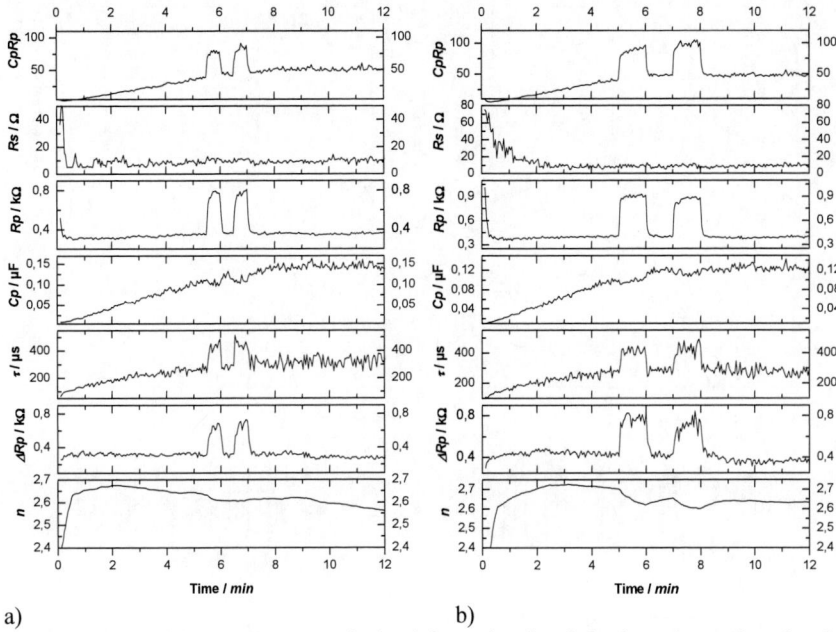

a)   b)

Fig. 4: The model parameters as obtained from the fit of the impedance data for the macropores presented in **Fig. 3**. Two current steps were used with different periods of a) 1 min. and b) 2 min.

While the series resistance R_s and the interface capacitance C_p do not change significantly when changing the current density, strong changes are found for the time constants C_pR_p and τ and the transfer resistances R_p and ΔR_p. The time constants C_pR_p and τ show a significant dependence on time, resp. pore length, while the transfer resistances R_p and ΔR_p are nearly independent of pore depth. Since the noise in R_p is much smaller then the noise in ΔR_p and the changes in the transfer resistance R_p are most easily understood; a detailed view is shown in **Fig. 5** (note the factor 2 in the scaling of the time axis).

R_p should scale inversely proportional to the integral pore tip area, i.e. the actively etched area. In the case of n-macro(bsi) pores the etched area is the sum of all macropore tips, which under perfect conditions is proportional to the current density. **Fig. 5a** and **Fig. 5b** both show an increase of R_p by roughly a factor of two after reaching steady state between the phases of high current density and of low current density. This is in good agreement to the measured decrease of the pore tip areas of a factor of two.

Steady state is reached for both experiments after roughly 0.5 min.; this is marked by dotted lines that separate the areas "A" and "B" in **Fig. 5a** and **Fig. 5b**. A very good agreement to the corresponding areas in **Fig. 3** is found: The horizontal lines mark the start point and the end point of the areas "a" to "c" in **Fig. 3**. A nearly linear transition from large pore diameter to small pore diameter is found. The areas "a" and "b" in **Fig. 3** correspond to the areas "A" and "B" in **Fig. 5**.

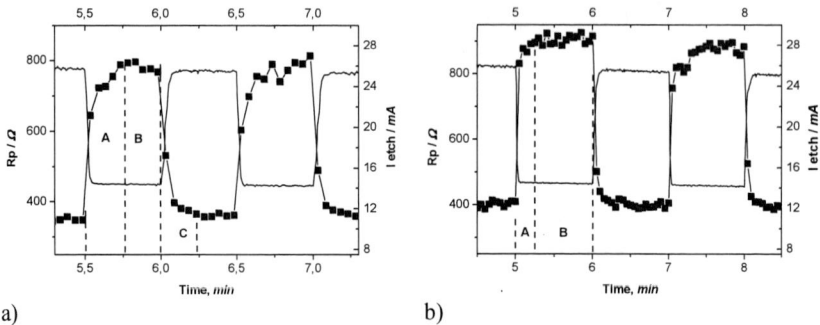

a)   b)

Fig. 5: Details of the transfer resistance R_p as presented in **Fig. 4**.

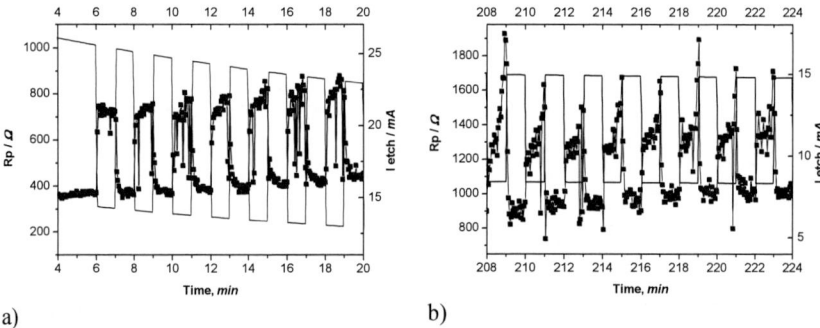

a)   b)

Fig. 6: The impedance parameter R_p for an experiment with 70 steps with a period of 2 min. a) Beginning of the experiment; b) near the end of the experiment.

Since the length for $(a+b)$ in both images in **Fig. 3** is the same, and taking into account the factor of 2 in the magnification between **Fig. 3a** and **Fig. 3b**, one can state that the pore growth speed in both cases is the same. The length of the bottle neck in **Fig. 3b** is twice as long as that in **Fig. 3a**, but the transition pore length from large to small diameter in both cases is the same. Therefore the ratio between the transition pore length and the pore length with steady state condition changes from 1 in **Fig. 3a** to 4 in **Fig. 3b**. The same ratios are found for the periods of area "A" and area "B" in **Fig. 5a** and **Fig. 5b**. Obviously this larger time for steady state in the phase of low current density is necessary to stimulate the anti-phase diameter modulation.

Already for the second current density step a significant increase in the transient time is found in **Fig. 5a** as well as in **Fig. 5b**. The transient time changes substantially as a function of pore depth; this is shown in **Fig. 6a** for R_p at the beginning of the experiment and in **Fig. 6b** for the end of the same experiment where current density steps of 2 min. period for more than 200 min. Obviously the transient time needed for the etching system to adapt to the lower or higher current densities, resp., drastically increases as a function of pore length.

Summing up the results the anti-phase diameter modulation could only be induced after 5 min. of pore etching by 2 current steps with a period of 2 min. Neither a period of 1 min. nor a period of 5 min. allowed for a stimulation. Neither staying at low current density nor staying at high current density after the two steps allowed for a continuation of the induced anti-phase growth. So in this experiments only a period of 8 times the transient time allowed for a stimulation of the anti-phase diameter modulation growth. Applying two current density steps with a period of two minutes much later in the experiment did not allow for the stimulation as well. This implies that the correct duration for applying low current densities and for applying high current densities is necessary. Staying for a too long time at high current densities cavity formation will occur. Staying for a too long time at low current densities several pores will stop growing while the remaining pores start to branch heavily.

The anti-phase diameter growth allows for an optimal current density of J_{psl} (as assumed by the Lehmann model) while the pore tip is growing and for the necessary time for recovering the electrolyte concentration while the pore tip is not growing, i.e. the neighboring pores consume the hole generated by backside illumination. If some pores completely stop to grow the distance between neighboring pores increases which is not optimal since the distance should coincide with the width of the space charge region around the pores.

To stimulate and stabilize the anti-phase diameter modulation could therefore be a promising way for growing deep macropores with the highest possible growth rates. As mentioned above, this probably needs some kind of feedback to control the modulation period. Summarizing the results of this paper a possible algorithm using the information of the FFT impedance analysis may be:

1. Start a step like modulation when a reasonable pore depth has been reached (possibly controlled by in situ impedance analysis as well).
2. Extract the relaxation τ time from the analysis of the transient e.g. of R_p.
3. Chose a step period time $t = x \, \tau$ where x is a constant (probably optimized by a series of experiments).

The near future will show, if this algorithm allows for the expected stable and fast macropore growth.

References

1. V. Lehmann and H. Föll, *J. Electrochem. Soc.* **137**, 653 (1990).
2. V. Lehmann, *Electrochemistry of Silicon*, Wiley-VCH, Weinheim (2002).
3. X.G. Zhang, *J. Electrochem. Soc.* **151**, C69-C80 (2004).
4. C. Fang, J. Carstensen, and H. Föll, *Solid State Phenomena* **121-123**, 37 (2007).
5. M. Hejjo Al Rifai, M. Christophersen, S. Ottow, J. Carstensen, and H. Föll, *J. Porous Mater.* **7**, 33 (2000).
6. M. Hejjo Al Rifai, M. Christophersen, S. Ottow, J. Carstensen, and H. Föll, *J. Electrochem. Soc.* **147**, 627 (2000).
7. S. Matthias, F. Müller, J. Schilling, and U. Gösele, *Appl. Phys. A* **80**, 1391 (2005).
8. A. Cojocaru, J. Carstensen, M. Leisner, H. Föll, and I.M. Tiginyanu, *Phys. Stat. Sol. (c)*, in print (2008).
9. J. Carstensen, A. Cojocaru, M. Leisner, and H. Föll, *ECS Trans.* **16**, 21 (2008).
10. A. Cojocaru, J. Carstensen, and H. Föll, *ECS Trans.* **16**, 157 (2008).
11. J. Carstensen, E. Foca, S. Keipert, H. Föll, M. Leisner, and A. Cojocaru, *Phys. Stat. Sol. (a)* **205**, 2485 (2008).
12. E. Foca, J. Carstensen, and H. Föll, *J. Electroanal. Chem.* **603**, 175 (2007).

CHAPTER 8

ELECTROCHEMICAL CHARACTERIZATION
AND FUNCTIONALIZATION OF SI SURFACES
AND DEVICES

364

ECS Transactions, 19 (3) 365-372 (2009)
10.1149/1.3120716 ©The Electrochemical Society

Electrochemical Passivation of (100) Silicon in Alkyl Grignard Solutions

Sri S. S. Vegunta[a], J. N. Ngungiri[a] and J. C. Flake[a]

[a] Gordon and Mary Cain Department of Chemical Engineering, Louisiana State University, Baton Rouge, Louisiana 70803, USA

Surface modification of (100) silicon with methyl groups is analyzed using electrografting and thermal hydrosilation. The surface chemistry is investigated by Fourier transform infrared (FTIR), X-ray photoelectron spectroscopy (XPS), voltammetry and atomic force microscopy (AFM). Surfaces anodically electrografted in methyl Grignard solutions show a smooth topology and improved passivation relative to passivation via hydrosilation. Functionalized surfaces are stable and hinder the formation of oxides up to 45 days after the electrografting as shown in the XPS results.

Introduction

Molecular organic layers are ideal for controlling the reactivity of silicon surfaces while maintaining the semiconductor structure (1). The applications for tuning the interfacial properties of silicon include hybrid organic/inorganic, (2, 3) biological/inorganic(4, 5) devices and fabrication of novel micro-mechanical systems (MEMS) (6, 7). Direct Si-C covalent linkages of both saturated and unsaturated organic molecules are used to modify semiconductor surfaces with various surface reactive (-COOH, -NH$_2$, -SH, -OH) or passive (hydrocarbon) groups, for hybrid organic/inorganic applications. Also the surface alkylation of silicon improves the stability against surface oxidation in ambient air (8) and aqueous redox electrolytes, (9) without increasing surface recombination rates or defect levels at silicon surfaces (10, 11).

Various methods have been adopted for covalently bonding uniform and densely packed organic monolayers on silicon surfaces. Chemical (12), thermal, photochemical(13) and electrochemical (14, 15) routes have been applied to achieve silicon derivatization. (16, 17) Precursors such as alkenes, (18) alkynes,(19) and Grignard reagents (20, 21) are activated by different techniques in each method (catalytically or chemically promoted, or *uv* irradiation) (22) in the presence of silicon surfaces to create the Si-C bonds.

Passivation of (111) silicon with methyl Grignard via hydrosilation is shown to provide >90% surface coverage; however, the reaction is relatively slow (>6 hours) and may not eliminate oxidation. In addition to new applications (e.g. bioelectronics, wet photovoltaics), an alternative to hydride termination would provide a stable surface chemistry and limit the growth of native oxides. In this work, we examine the passivation

365

performance of methyl groups electrografted onto (100) silicon. Surfaces with (100) orientation are most commonly used in manufacture of CMOS, memory and MEMS devices.

Experiment

Tetrahydrofuran (THF, 99%), methyl Grignard (CH$_3$MgCl, 3.0M in THF), hydroflouric acid (HF, 48% ACS reagent), Buffered Oxide Etch (BOE, ammonium flouride etching mixture, semiconductor grade) solution, ethanol, phosphorous pentachloride (reagent grade, 95%), benzoyl chloride (reagentplus, 99%) and benzoyl peroxide (reagent grade, 97%) were obtained from Sigma-Aldrich. The 100mm p-type (100) silicon wafers (resistivity 1-5 Ω•cm) were obtained from Montco Silicon Technologies Inc. (Springcity, PA).

The (100) Si wafers were cleaved into ~2 cm^2 pieces and immersed in the BOE solution for 30 seconds. The samples were rinsed with DI water, dried in a stream of nitrogen and transferred into a drybox. Ohmic connections were created on the backside of the silicon wafer with a small amount of Ga-In eutectic. Silicon samples were used as working electrodes against a Pt counter electrode in methyl Grignard solution and were anodically cycled from -1.4 V to +0.6 V (versus Ag/AgCl) at 10mV/s to generate the methyl coated (100) silicon.

The hydrosilation of the hydride terminated silicon samples was obtained by using a two step chlorination/alkylation process. The H-terminated samples were chlorinated by heating in a saturated solution of phosphorous pentachloride and benzoyl chloride in the presence of benzoyl peroxide, at 85°C for 45 minutes. The chlorinated silicon samples are then transferred to the methyl Grignard solution to be heated at 95°C for 8 hours to form a methyl layer on the (100) silicon surface.

Voltammetry was performed using a Princeton Applied Research 2273 potentiostat. A Nicolet 380 with ZnSe ATR multiple bounce FTIR system was used to analyze the absorption spectra of the electrografted and hydrosilated (100) silicon surface. AFM images were acquired using an Agilent 5400 AFM/SPM system. V-shaped non-conductive silicon nitride cantilevers were used in the ambient conditions for contact mode imaging of the functionalized (100) silicon surface.

Results

<u>Voltammetry</u>

As shown in the wide potential cyclic voltammograms in Figure 1, the reversible Mg couple (Mg^{++}/Mg) on Pt is observed at -1.9V (vs. Ag/AgCl) and the irreversible oxidation of the Grignard reagent is observed at -600 mV (vs. Ag/AgCl) when 1M methyl Grignard is added to THF. This oxidation behavior is also observed with the silicon electrodes; however, repeated voltammetry indicates that the silicon passivation occurs at anodic potentials along with the Grignard oxidation based on the onset potentials observed in Figures 1 and 2.

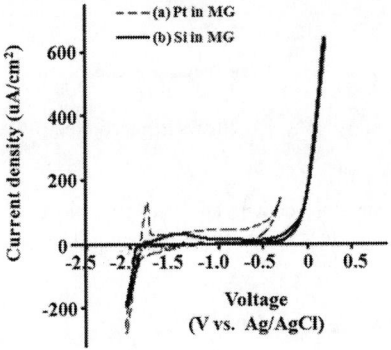

Figure 1. Cyclic voltammograms of [A] Pt working electrode in methyl Grignard reagent (1M in THF) [B] (100) Si-H working electrode in methyl Grignard reagent (3M in THF).

Figure 2 shows two voltammetric cycles of methyl Grignard electrografting on (100) silicon. At -0.6V (vs. Ag/AgCl) methyl Grignard begins to undergo oxidization as shown in Figure 2, simultaneously electrografting the silicon surface with the alkyl groups. A slight shoulder from -600mV (vs. Ag/AgCl) to -200mV (vs. Ag/AgCl) and a hysteresis between the forward and reverse scans are clearly visible in cycle 1. The hysteresis has a charge of 2960 $\mu C/cm^2$ which is approximately 14 fold higher than the theoretical charge required (217 $\mu C/cm^2$) for electrografting one methyl group per silicon via a mechanism using two electrons per each methyl group. The shoulder at -600mV (vs. Ag/AgCl) corresponds to the oxidation of methyl Grignard and a charge uptake of 577 $\mu C/cm^2$ which is approximately 2.5 fold higher than the theoretical value for electrografting methyl Grignard on the (100) silicon surface. The reduction of the hysteresis and the absence of the shoulder over the consecutive cycles, indicates the saturation of the surface sites available for electrografting and the completion of the process in cycle 1.

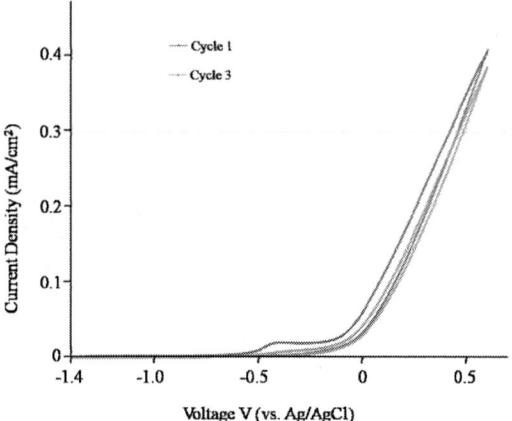

Figure 2. Cyclic voltammetry of 3M methyl Grignard electrografted onto H-terminated (100) silicon surface.

The electrografting procedure can be applied to hydride terminated silicon surfaces without using a supporting electrolyte, as the electrolyte (Grignard) is conductive in nature.

Fourier Transform Infra Red Spectroscopy (FTIR)

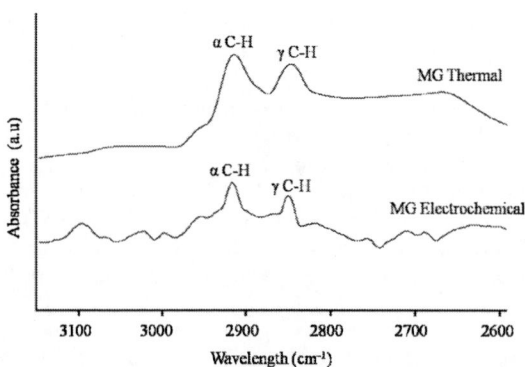

Figure 3. Absorption infrared spectra of a methyl monolayer electrografted on (100) Si.

FTIR results shown in Figure 3 confirm the presence of carbon on the surface of (100) silicon following the anodic electrografting with methyl Grignard. Two absorption peaks were observed at 2916cm^{-1} (α C-H) and 2850cm^{-1} (υ C-H) corresponding to the stretching modes of the methyl group. The electrografted surface shows relatively sharp peaks suggesting a well-organized monolayer with a higher surface density, when compared to the curved peaks of a hydrosilated surface.

Atomic Force Microscopy (AFM)

AFM surface morphology of methyl terminated (100) Si via electrografting (Figure 4a) and hydrosilation (Figure 4b) shows that the electrografting produces more uniform organic films relative to thermal hydrosilation. The homogenous contrast in Figure 4a indicates a layer with uniform and continuous thickness on the electrografted sample. For the thermally hydrosilated surface (Figure 4b), the sample shows nanoscopic oxide islands that are protruding from the surface, which are disappeared when the sample is dipped in a solution of hydrofluoric acid. The hydrosilated surface has an RMS roughness of 10.1 nm which is approximately 50 fold greater than that of an electrografted surface with an RMS roughness of 0.2 nm. This indicates a relatively smoother surface and the presence of an alkyl monolayer on the electrografted surface.

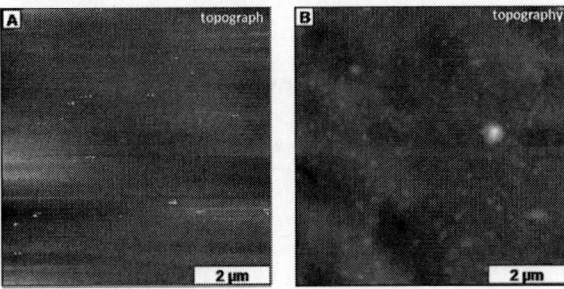

Figure 4.   Grafted methyl films on (100) Si produced via (a) electrografting (b) hydrosilation.

X-ray Photoelectron Spectroscopy (XPS)

XPS is used to investigate the passivation performance of the methyl coated silicon compared to the native oxide and hydride terminated silicon surfaces. The silicon 2p spectra are investigated for the characteristic Si-Si peak at 99.5 eV and Si-O peak at 103.5 eV after exposing the electrografted samples to ambient air for 1, 45 and 55 days. After 55 days of exposure, a slight shoulder of the silicon oxide is observed from 102-105 eV, which resembles the shape of the silicon oxide peak observed on the H-terminated surface after an exposure of 1 day to ambient air. The results indicate the stability and extraordinary passivation performance of the electrografted methyl surface against the native oxide formation for 45 days in ambient air.

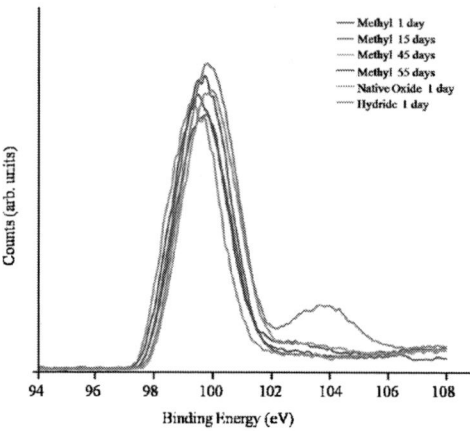

Figure 5. XPS peaks for elemental Si 2p of methyl, oxide and hydride terminated (100) silicon surfaces.

Discussion

The FTIR and AFM results indicate the presence of a relatively smooth alkyl monolayer on the electrografted surface, when compared to a hydrosilated surface. The XPS results indicate the passivation performance of the methyl monolayer electrografted on a (100) silicon surface. The cyclic voltammetry indicates the onset potentials for the oxidization of methyl Grignard and provides an insight into the potential window and a two electron mediated mechanism is proposed for the electrografting of methyl Grignard onto the (100) silicon surface.

As two molecules of methyl Grignard are simultaneously oxidized, one of the methyl radicals (CH_3^+) attack a (100) Si-H bond site, to cleave the H^+ off the surface and let the other methyl radical form the direct Si-C covalent bond with the cationic silicon surface. (13, 15, 17, 23) Each methyl group attached on a (100) silicon surface atom will sterically hinder the other methyl radicals from attacking the hydrogen bonded on the successive silicon atom, hence limiting the number of methyl groups electrografted per each (100) silicon atom to 1. This limit is lowered for other higher alkyl groups based on their chain lengths, in turn changing the onset potentials for oxidization and the passivation performance against the native oxide formation.

Conclusion

Electrografting allows the deposition of molecular organic thin films mediated by strong covalent links. This electro-initiated process only requires a charged electrode for the grafting and results in the formation of insulating and highly uniform organic films. The methyl group electrografted on the (100) silicon surface has shown passivation toward the native oxide formation for 45 days therefore suggesting a possibility of alternative microelectronics processing techniques and new technologies based upon the hybrid organic-silicon interfaces.

References

1. R. A. Wolkow, *Annual Review of Physical Chemistry*, **50**, 413 (1999).
2. E. J. Faber, L. de Smet, W. Olthuis, H. Zuilof, E. J. R. Sudholter, P. Bergveld and A. van den Berg, *Chemphyschem*, **6**, 2153 (2005).
3. Y. J. Liu and H. Z. Yu, *Chemphyschem*, **4**, 335 (2003).
4. Z. R. Scheibal, W. Xu, J. F. Audiffred, J. E. Henry and J. C. Flake, *Electrochemical and Solid State Letters*, **11**, K81 (2008).
5. K. A. Kilian, T. Bocking, K. Gaus, M. Gal and J. J. Gooding, *Biomaterials*, **28**, 3055 (2007).
6. V. K. Varadan and V. V. Varadan, *Smart Materials & Structures*, **9**, 953 (2000).
7. X. Y. Zhu and J. E. Houston, *Tribology Letters*, **7**, 87 (1999).
8. W. J. Royea, A. Juang and N. S. Lewis, *Applied Physics Letters*, **77**, 1988 (2000).
9. A. Bansal and N. S. Lewis, *Journal of Physical Chemistry B*, **102**, 4058 (1998).
10. S. R. Amy, D. J. Michalak, Y. J. Chabal, L. Wielunski, P. T. Hurley and N. S. Lewis, *Journal of Physical Chemistry C*, **111**, 13053 (2007).
11. E. J. Nemanick, P. T. Hurley, B. S. Brunschwig and N. S. Lewis, *Journal of Physical Chemistry B*, **110**, 14800 (2006).
12. J. M. Schmeltzer, L. A. Porter, M. P. Stewart and J. M. Buriak, *Langmuir*, **18**, 2971 (2002).
13. A. Faucheux, A. C. Gouget-Laemmel, C. H. de Villeneuve, R. Boukherroub, F. Ozanam, P. Allongue and J. N. Chazalviel, *Langmuir*, **22**, 153 (2006).
14. S. P. Koiry, D. K. Aswal, V. Saxena, N. Padma, A. K. Chauhan, N. Joshi, S. K. Gupta, J. V. Yakhmi, D. Guerin and D. Vuillaume, *Applied Physics Letters*, **90** (2007).
15. E. G. Robins, M. P. Stewart and J. M. Buriak, *Chemical Communications*, 2479 (1999).
16. N. Shirahata, A. Hozumi and T. Yonezawa, *Chemical Record*, **5**, 145 (2005).
17. M. P. Stewart and J. M. Buriak, *Comments on Inorganic Chemistry*, **23**, 179 (2002).
18. L. de Smet, H. Zuilhof, E. J. R. Sudholter, L. H. Lie, A. Houlton and B. R. Horrocks, *Journal of Physical Chemistry B*, **109**, 12020 (2005).
19. A. B. Sieval, R. Opitz, H. P. A. Maas, M. G. Schoeman, G. Meijer, F. J. Vergeldt, H. Zuilhof and E. J. R. Sudholter, *Langmuir*, **16**, 10359 (2000).

20. E. J. Nemanick, P. T. Hurley, L. J. Webb, D. W. Knapp, D. J. Michalak, B. S. Brunschwig and N. S. Lewis, *Journal of Physical Chemistry B*, **110**, 14770 (2006).
21. T. Yamada, K. Shirasaka, M. Noto, H. S. Kato and M. Kawai, *Journal of Physical Chemistry B*, **110**, 7357 (2006).
22. S. Takakusagi, T. Miyasaka and K. Uosaki, *Journal of Electroanalytical Chemistry*, **599**, 344 (2007).
23. S. Fellah, A. Teyssot, F. Ozanam, J. N. Chazalviel, J. Vigneron and A. Etcheberry, *Langmuir*, **18**, 5851 (2002).

ECS Transactions, 19 (3) 373-379 (2009)
10.1149/1.3120717 ©The Electrochemical Society

Electronic Properties and pH Stability of Si(111)/Alkyl Monolayers

D. Aureau,[a, b] A. Moraillon,[a] C. Henry de Villeneuve,[a] F. Ozanam,[a] P. Allongue,[a]
J.-N. Chazalviel[a] and J. Rappich[b]

[a] Physique de la Matière Condensée, Ecole Polytechnique, CNRS, 91128 Palaiseau,
France.
[b] Helmholtz-Zentrum Berlin für Materialien und Energie GmbH, Abteilung Silizium-
Photovoltaik, SE1, Kekuléstrasse 5 D-12489 Berlin, Germany.

The electronic properties of hydrogenated and alkyl-grafted silicon
surfaces have been investigated by photoluminescence, surface
photovoltage and electrochemical capacitance measurements,
which are sensitive methods for monitoring the presence of
electronic defects. On p-type silicon, the electronic quality of the
interface was found to depend critically on the surface preparation
and was studied with special care. The thermal grafting of one
monolayer of linear organic chains on atomically flat silicon
allows the formation of stable alkyl-grafted p-Si surfaces,
exhibiting a very low density of recombination centers comparable
to those measured on hydrogenated silicon surfaces.
Photoluminescence measurements show that the high electronic
quality of the grafted surfaces is preserved on a month scale in air,
indicating a much more efficient long term passivation by
comparison with hydrogenated Si surfaces. The stability of the
silicon/organic layer interface in aqueous buffer solution has been
assessed by *in situ* photoluminescence and surface photovoltage
measurements. Some degradation of the electronic properties was
evidenced during exposure in basic solution of pH > 9.

Introduction

The understanding of the electronic properties of the interfaces consisting of organic
monolayers covalently grafted onto silicon via a Si-C bond appears essential in order to
prepare different kinds of devices integrating these layers (solar cells, sensors, transistors,
etc.). Such systems were specially studied because direct Si-C bonding provides superior
chemical robustness, including in wet environment. The starting point for preparing such
modified silicon surfaces is always a hydrogenated silicon surface, obtained by
dissolution of the oxide layer (in HF or NH4F) and subsequent of silicon in the case of the
treatment in NH4F solution. The most controlled model surface in this respect is actually
the (111)Si surface, as it can be obtained atomically flat and 1×1 unreconstructed, each
surface silicon atom bearing a single Si-H bond perpendicular to the surface (1).
Nevertheless, in the case of hydrogenated p-type silicon surfaces, a strong influence of
surface preparation on the electronic properties of silicon/electrolyte interfaces has
already been observed from capacitance measurements (2). On the other hand, the
sensitivity of Si-H bonds to oxidation has motivated attempts to substituting the hydrogen
atoms by organic groups. Here, photoluminescence and capacitance measurements have

373

been used in order to determine how to prepare p-Si(111)/alkyl monolayers with very low amounts of electronic defects.

Moreover, since many of the targeted applications consist in using these assemblies in a liquid environment, it is of prime importance to determine the conditions in which the electronic properties of these Si/organic monolayer interfaces are preserved. These conditions will be worked out by using *in situ* Photoluminescence (PL) and *in situ* Surface Photovoltage (PV) measurements in contact with electrolytes (3). The latter method is used to monitor the changes in band bending at a semiconducting electrode surface (U_{PV}), while the quenching of the bandgap-related PL intensity (I_{PL}) gives direct information on the creation of interface defects, which act as non-radiative recombination centers for the photocarriers (4). This approach will lead us to redefining the pH domain where the grafted silicon surfaces are sufficiently stable to be reliably studied and used on a practical time scale.

Experimental

General information

Undecylenic acid (98%) and ammonium sulfite monohydrate (92%) were purchased from Acros Organics and Aldrich respectively. All cleaning (H_2O_2 30%, H_2SO_4 96%, acetic acid 100%) and etching (NH_4F 40%, HF 40 %) reagents were of electronic grade and supplied by Carlo Erba. Phosphoric acid 85% and potassium hydroxide 85% were purchased from SDS; boric acid 99.5%, potassium phosphate monobasic 99%, dibasic 99% from Merck and tribasic 95% from Fluka.

The compositions of the buffer solutions (ionic strength: 10^{-2} M) were as follows: pH 2-3, phosphoric acid/ potassium phosphate monobasic; pH 6-7-8, potassium phosphate monobasic/ potassium phosphate dibasic; pH 9-10, boric acid/ potassium hydroxide; pH 10-11, potassium phosphate dibasic/ potassium phosphate tribasic; pH 12, potassium hydroxide. Ultrapure water was provided by a Millipore station, which ensures a resistivity of 18.2 MΩ cm.

Preparation of the silicon surface

The silicon samples were cut from [111]-oriented p-type silicon wafers (0.2° misorientation toward [11$\overline{2}$], float zone, ρ = 30-40 Ωcm). The samples were initially cleaned in "piranha", a 3:1 mixture of concentrated H_2SO_4 (98%) and H_2O_2 (30%), in order to remove all organic contaminations. This wet oxidation was followed by copious rinsing with ultrapure water. The cleaned silicon samples were then chemically etched in NH_4F (*ca.* 0.05 mol L^{-1} ammonium sulfite was added to the etching solution) or HF solutions to remove native oxide. After etching, the surface was rinsed with ultrapure water.

Formation of alkyl monolayers on silicon by thermal hydrosilylation.

A one-step procedure was used to prepare well defined 10-carboxydecyl organic monolayers on H-Si (111) via direct thermal hydrosilylation of undecylenic acid. The neat undecylenic acid was outgassed under argon in a Schlenk tube at 90°C for 30 minutes and then cooled to room temperature under continuous argon bubbling to insert the freshly prepared H-terminated silicon sample. Grafting was performed for 20 hours at 180°C. The functionalized surface was then rinsed twice in oxygen-free hot acetic acid (60-70 °C) under argon during 15 min (to desorb hydrogen-bonded undecylenic acid molecules) (5).

Characterization methods

The electrochemical measurements were performed in a homemade three-electrode Teflon cell connected to a potentiostat (PGSTAT10, Eco Chemie BV, The Netherlands) coupled to a frequency response analyzer. The counter electrode was a Pt wire, and the reference electrode was the mercury/saturated mercurous sulphate electrode (*MSE*). The working electrode was pressed against an opening in the cell bottom using an O-ring seal (perfluorinated elastomer). Nitrogen was bubbled through the solution to remove dissolved O_2. The impedance of the interface was measured in the frequency range 1-100 kHz.

Time-integrated PL (integration time 100 μs) and time-resolved PV (time resolution 20 ns) were excited by single pulses of a dye laser (wavelength 500 nm, duration time 10 ns, intensity 200 μJ cm^{-2}) pumped by a nitrogen laser. The measured PV (U_{PV}) is sensitive to surface potential variations induced by organic surface molecules when the latter reach densities on the order of 10^{14} cm^{-2}. The *in situ* PV and PL measurement procedures were described in detail in reference (4).

Results and discussion

Capacitance measurements – Electronic quality

At negative potential vs U_{FB}, the calculated capacitance C (from measured impedance) of a p-Si electrode is reduced to the space charge capacitance of the semiconductor C_{SC}, which obeys the Mott-Schottky relation

$$C_{SC}^{-2} = \frac{2}{qN_A\varepsilon_S\varepsilon_0}\left(U_{FB} - U - \frac{kT}{q}\right)$$ [1]

where q is the electronic charge, ε_s is the dielectric constant of silicon, N_A is the net acceptor concentration, *i.e.*, the doping level of p-Si, U is the applied electrode potential, and U_{FB} is the flat-band potential. Therefore, by plotting C^{-2} as a function of U, the doping concentration and U_{FB} are obtained from the slope and the extrapolation to $C^{-2}=0$ of the linear part of the plot, respectively. Figure 1 shows such plots obtained from hydrogenated p-type silicon after different treatments.

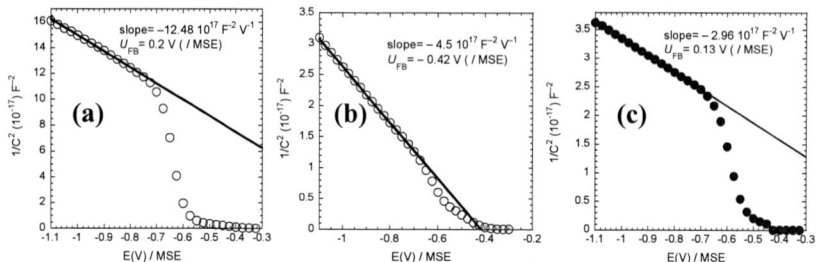

Figure 1. Mott-Schottky plots in 0.1 M H_2SO_4 electrolyte of hydrogenated p-type silicon treated by HF before (a) and after (b) an annealing at 400 °C under nitrogen or treated by NH_4F after a similar annealing (c). The uncommon forms obtained in (a) and (c) were associated with the presence of H-atoms in the sub-surface region.

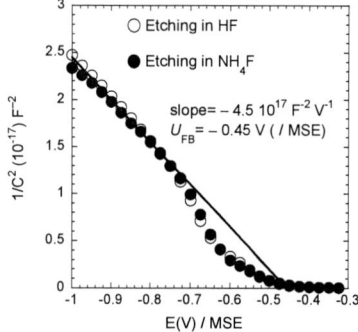

Figure 2. Mott-Schottky plots in 0.1 M H_2SO_4 electrolyte of grafted p-type silicon prepared by thermal hydrosilylation of hydrogenated surfaces prepared in HF (open symbols) and NH_4F (full symbols).

Figure 1.a shows a plot obtained for a silicon surface freshly hydrogenated by a dip in HF. Figure 1.b shows the plot obtained after the same treatment on a silicon surface previously annealed for 20 hours at 400 °C under nitrogen. The linear slope has the expected value only for the surface hydrogenated after annealing. Using equation [1], it corresponds to a doping level of $4.2 \ 10^{14}$ cm^{-3}, within the expected range for a silicon substrate of 30-40 Ωcm resistivity (6). In contrast to the case of n-type silicon, Mott-Schottky plots on p-type silicon are known to depend critically on the initial surface treatment (2). These deviations are commonly explained by the presence of hydrogen atoms in the sub-surface region. Hydrogen is known indeed for its ability to form neutral complexes with acceptors (7), which compensate or neutralize them, and lead to a decrease of the negative charge density in the space charge region for p-Si (2). The annealing removes those hydrogen atoms by diffusion, leading to reliable values for the doping level and the flat-band potential. This step then appears as essential for the preparation of p-Si surfaces.

As a matter of fact, the presence of hydrogen in the sub-surface region might result from the polishing of the Si wafer or from the wet etching surface hydrogenation treatment. As shown in Fig. 1.c, NH_4F etching appears especially critical since this treatment leads to a change in the slope of the Mott-Schottky plot. The associated penetration of hydrogen atoms into the subsurface region is likely to be associated with the dissolution of some silicon atoms during the etching (1). However, in order to obtain well-defined grafted surface, the NH_4F procedure is preferred because of its capability to yield atomically flat surfaces with regular steps and terraces (1).

Notwithstanding this problem, surfaces prepared with NH_4F have been grafted then characterized in the same way. Figure 2 shows a Mott-Schottky plot obtained for grafted surfaces prepared from hydrogenated surfaces prepared in HF and NH_4F. After grafting, no difference is discernible in the linear part of the plots, which means that thermal grafting has erased the influence of the etching process on the electrical properties of the surface. The conditions used for thermal grafting (20 hours at 180°C) appear to be sufficient to remove the impurities from the interface. Most importantly, these data show that the electrical passivation of acceptors brought along with the NH_4F treatment is not prejudicial to the surface electronic properties after thermal grafting.

ex situ Photoluminescence - Stability in air

The PL intensity (I_{PL}) of hydrogenated and grafted surfaces was investigated *ex situ* in ambient atmosphere in order to extract information on the interface passivation and stability obtained with the organic layer. I_{PL} is actually sensitive to very low concentrations of surface non-radiative recombination centers (on the order of 10^{11} cm^{-2} and below). In a wide range of surface concentration in such centers, it is simply proportional to the reciprocal of that concentration. Figure 3.a shows spectra obtained between 900 and 1500 nm for NH_4F – etched Si(111), just after hydrogenation and after one hour. The highest I_{PL} is obtained for the freshly prepared samples. However, I_{PL} decreases continuously on a time scale of a few minutes due to surface oxidation. Figure 3.b shows spectra recorded for a freshly grafted surface and the same surface after one month. I_{PL} for freshly grafted silicon is only slightly lower than for freshly hydrogenated silicon. This observation suggests that the thermal grafting does not create many electrical defects responsible for non-radiative recombination. Moreover, figure 3.b shows that this high I_{PL} is conserved along thirty days for the grafted surface. This result appears as a proof of the long-time passivation obtained with the molecular chains covalently anchored onto the surface via a Si-C bond (8).

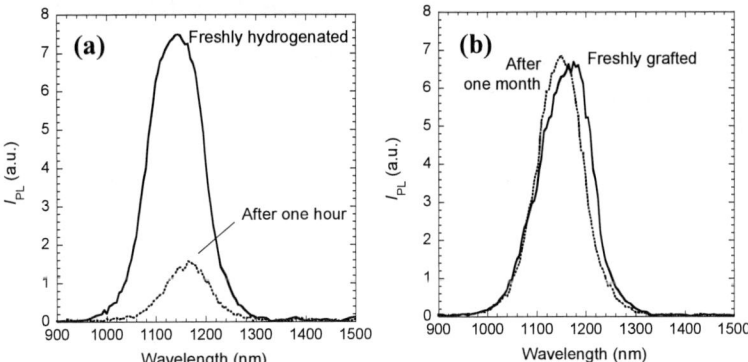

Figure 3. *Ex situ* PL spectra of hydrogenated (a) and grafted (b) surfaces (p-$SiC_{10}H_{20}COOH$) at different times in ambient air.

in situ Photoluminescence and Photovoltage - pH stability

Grafted surfaces were studied in contact with electrolytes of different pH, using different buffer solutions. Figure 4 shows the I_{PL} (measured at 1130 nm) and U_{PV} values for a surface grafted with 10-carboxydecyl groups, in contact with electrolytes of different pH, using increasing pH values. The photoluminescence intensity remains constant for pH values from 2 to 9 and decreases for more basic pH values. This decrease is attributed to the creation of interface states at the semiconductor/organic layer interface. These electronic defects are thought to result from the exposure of the surface silicon atoms to the electrolyte solution. The I_{PL} behavior appears as a fingerprint of an easier penetration of the OH⁻ ions of the alkaline solutions into the grafted layers (probably at structural defects or domain boundaries), so that etching of silicon by the OH⁻ ions sets in and new active recombination centers, associated with etch-pits or oxide precursors like Si-OH, are created.

The evolution of U_{PV} appears qualitatively similar to that of I_{PL} upon increasing pH. Nevertheless, the U_{PV} values exhibit some differences. The stability of U_{PV} up to pH 9 appears superior to that of I_{PL}, which could be explained by the lower sensitivity of U_{PV} to interface states. When stepping to a higher pH, U_{PV} decreases strongly ($\approx$-200 mV), and this jump is followed by a slow re-increase, the pH being held fixed at its new value. The abrupt change of U_{PV} between pH 9 and 10 cannot be explained by the ionization of the carboxyl groups. As it has been characterized before by infrared spectroscopy (9), the titration of such surfaces begins around pH $\approx$ 6 and occurs quite progressively. This jump can be rather attributed to the fact that up to pH 10 the silicon surface is efficiently shielded from the electrolyte (the organic layer acts as a barrier between the semiconductor and the electrolyte), but at pH 10 and 11 the electrolyte comes in contact with the surface, at least at some spots. The evolution in time at fixed pH suggests that when the electrolyte reaches the silicon, modifications occur at the interface, resulting in a change of the band bending.

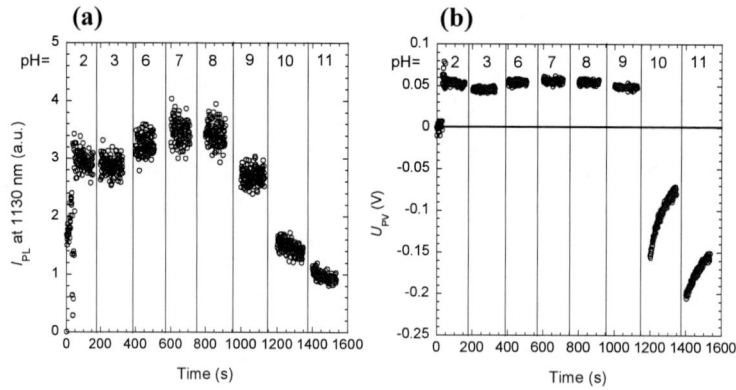

Figure 4. Evolution in time of *in situ* PL at 1130 nm (a) and *in situ* PV (b) for grafted surfaces (p-SiC$_{10}$H$_{20}$COOH).

In conclusion, the pH range where such systems can be reliably studied must be reconsidered. In a previous work, the ionization of carboxyl-terminated surfaces was studied by a direct determination of the ionization degree of the carboxyl end groups in contact with electrolytes of different pH (9). Though the organic layer appeared then to be preserved up to pH 12, the present work shows that the electrical quality of the interface is actually affected at somewhat lower pH values. For example, if the charge effect associated with carboxyl ionization has to be considered, these investigations should be limited to the "stability domain" of pH < 9, where the electronic properties of silicon are preserved (high I_{PL} and stable U_{PV}). This also represents the practical pH range where field-effect sensors based on grafted silicon have to be used. Investigations on the implementation of such interface for field-effect detection are presently in progress.

Acknowledgments

Financial support from RTRA "Triangle de la Physique" for funding this French-German collaboration is acknowledged.

References

1. P. Allongue, C. Henry de Villeneuve, S. Morin, R. Boukherroub and D.D.M. Wayner, *Electrochim. Acta*, **45**, 4591 (2000).
2. P. de Mierry, D. Ballutaud, M. Aucouturier and A. Etcheberry, *J. Electrochem. Soc.*, **137**, 2966 (1990).
3. T. Dittrich, T. Burke, F. Koch and J. Rappich, *J. Appl. Phys.*, **89**, 4636 (2001).
4. P. Hartig, J. Rappich, T. Dittrich, *Appl. Phys. Lett.*, **80**, 67-69 (2002)
5. A. Faucheux, A. C. Gouget-Laemmel, C. Henry de Villeneuve, R. Boukherroub, F. Ozanam, P. Allongue and J.-N. Chazalviel, *Langmuir*, **22**, 153 (2006).
6. S.M. Sze, *Physics of Semiconductor Devices, 2ⁿᵈ edition,* John Wiley & Sons, New York (1981).
7. M. Stavola, S.J. Pearton, J. Lopata and W.C. Dautremont-Smith, *Phys. Rev. B.*, **37**, 8313 (1988).
8. P. Gorostiza, C. Henry de Villeneuve, Q.Y. Sun, F. Sanz, X. Wallart, R. Boukherroub and P. Allongue, *J. Phys. Chem. B*, **110**, 5576 (2006).
9. D. Aureau, F. Ozanam, P. Allongue and J.-N. Chazalviel, *Langmuir*, **24**, 9940 (2008).

380

ECS Transactions, 19 (3) 381-389 (2009)
10.1149/1.3120718 ©The Electrochemical Society

Phase-relations between photocurrent and *in situ* reflectance during photoelectrochemical dissolution of silicon

M. Lublow[1a], H.J. Lewerenz[a]

[a]Division of Solar Energy, Interface Engineering Group, Helmholtz Zentrum Berlin für Materialien und Energie, Glienicker Str. 100, 14109 Berlin, Germany

The phase relation between local maxima of photocurrents and those of in-situ Brewster-angle reflectance (BAR) is investigated upon dissolution of silicon photoelectrodes in diluted NH_4F. In the region of photocurrent oscillations, charge flow and optical response can be related by a linear equation which explains the observed positive phase shift of the reflectance. In the transition regime, between porous silicon formation and electropolishing, commencing surface oxidation results in a negative phase shift of the reflectance. It is shown that this observation points to a change in the dissolution mechanism of the surface-near region. A step-function, convoluted with the charge flow, is applied to model the optical behavior. This approach is compared to optical multi-layer analysis in which simultaneously increasing sub-surface porosity and surface roughness are considered.

Introduction

The photodecomposition of semiconductor electrodes can increase the interfacial region from several nanometers up to several micrometers by modification of the surface topography and change of the chemical composition. The assessment of the respective (photo-)electrochemical reactions is considerably facilitated by application of surface and interface sensitive methods which detect the corresponding topographical changes. Ellipsometry, for instance, has already been employed for monitoring and analysis of electrochemically induced surface modifications [1]. Brewster-angle reflectometry (BAR), as a further technique, has recently been demonstrated to ensure fast and sensitive assessment of electrochemical reactions at silicon photoelectrodes [2]. In this work, the dissolution of n-type silicon in ammonium fluoride (NH_4F) containing solutions is analyzed in the principal regions of divalent and tetravalent dissolution as a model system for combined optical and electrochemical investigations. NH_4F containing electrolytes are well known to induce porous silicon (PS) formation or anodic oxidation in dependence of the applied potential [3]. While PS is produced by consumption of two charge carriers (divalent), the formation of anodic oxides requires four photo-generated holes (tetravalent). The transition regime between divalent and tetravalent dissolution is of particular interest after formation of silicon nanocrystals with varying size and density was described by several authors [4, 5]. In this work, an approach is introduced which relates charge flow and reflectance by simplified equations which allows interpretation of the phase relationship of the respective current and reflectance signals near local maxima. Details about the dissolution mechanisms are thereby accessible which cannot be assessed by interpretation of the current behavior alone. After discussing the principles of

---

[1] lublow@helmholtz-berlin.de

381

this approach in the region of photocurrent oscillations [6, 7], the interdependence of charge flow and optical response in the transition regime will be analyzed. It will be shown that nanocrystal formation in this potential region is presumably always accompanied by sub-surface porosity.

Model development

Brewster-angle reflectometry benefits from the minimum reflectance condition at the Brewster-angle, φ_B, of a substrate material: $\dfrac{dR_p(\varphi)}{d\varphi} = 0$ [8]. Data evaluation of the reflectance signal of modified electrode surfaces requires, in general, knowledge of the optical properties and film thicknesses which form during the process of (photo-)electrochemical decomposition. Provided that these quantities are known, the resulting reflectance can then be calculated by application of Fresnels formulae for multi-layer systems. To overcome these difficulties, approximate relations can be developed in the limit of thin films where the reflectance responds almost linearly to increasing film thicknesses. The resulting optical signal from a modified surface can then be considered as the bulk signal plus an additive perturbation signal. The validity of this assumption and its limitation can be assessed from Fig. 1. Here, the effects of three different types of surface films are exemplified. It can be seen that there exist, in the limit of thin films, regions where a linear approximation of the resulting reflectance can be applied.

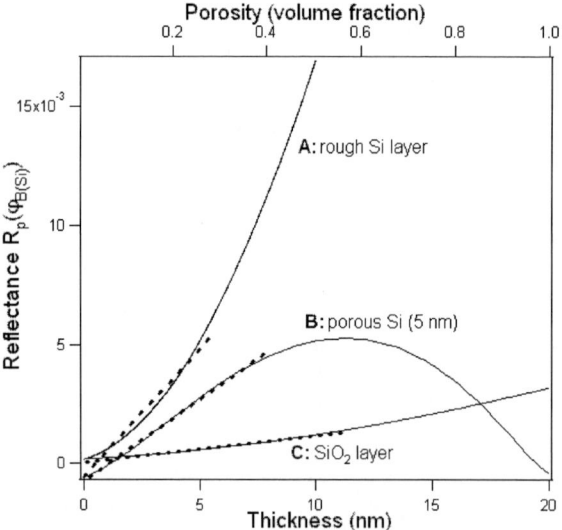

Fig. 1: Calculated response of *in situ* BAR to the presence of a rough surface layer (curve A), sub-surface porosity (curve B) and SiO₂ (curve C). The optical response R_p to the presence of surface roughness and porosity was modeled according to effective medium theory. Roughness and the SiO₂ layer were incremented in thickness (0.5 – 10nm and 0 – 20nm, respectively, see lower ordinate) while the void/substrate fraction was incremented for the 5nm thick porous layer (0-100%, see top ordinate).

Charge consumption in corrosive reactions of silicon in NH_4F containing solutions, on the other hand, directly leads to the formation of surface layers whose dielectric properties differ from those of the bulk. Therefore, an additional interface is created which gives rise to multiple reflections in corresponding optical measurements. In the following model considerations, charge transfer across the silicon/electrolyte interface without surface modification is excluded. This assumption leads to an expression that relates the charge flow to the resulting optical response:

$$R_p \approx A + BQ_{photo} = A + B \int_0^t I_{photo} dt .$$

(1)

Here, A denotes the Si bulk reflectance, $R_p(\varphi_B)$, while the charge flow is considered to be proportional to the layer thickness with proportionality factor B. Modifications of Eq. 1 will be necessary if, e.g., pure chemical reactions influence the film growth or if the optical properties of the film change with time or, during potentiodynamic scans, with the applied potential.

Experimental

Czochralski (CZ) grown n-type silicon (111) (Sico GmbH, Germany), $N_D = 10^{15}$ cm^{-3}, nominal miscut 0° was used. The samples were cleaned with ethanol and ultra-pure water (18 MΩ)) and etched in NH_4F (40%) (Merck, VLSI-grade) in a two step treatment: (i) a 100s etch step (oxide removal only); (ii) a final etch-step in a fresh solution for 10 min. This treatment leads to atomically flat (111) terraces [8]. Electrochemical preparation was carried out in 0.1 M NH_4F, pH ~4, applying a three-electrode potentiostatic arrangement (with Pt counter-, Ag/AgCl reference electrode). Photocurrents were measured by additional illumination (W/I_2 lamp). The data acquisition rate of the photocurrents was 1 data pair each 0.8 s. Reflectance data were recorded at a rate of 2 per second.

For BAR reflectance measurements polarized light ($\lambda = 500$nm, Glan-Thompson polarizer) is reflected at the sample which is adjusted on a precision goniometer table. For *in-situ* real-time monitoring, the angle of incidence was permanently kept at the Brewster-angle of the Si(111)-H(1x1) sample immersed in the electrolyte. The probe light intensity was about 4μW/cm^2.

Results and Discussion

Positive phase shift upon photocurrent oscillations at U = 6V

It is known that silicon dissolution in NH_4F containing solutions proceeds tetravalent beyond the first current maximum:

$$Si + 2H_2O + 4h_{VB}^+(hv) \rightarrow SiO_2 + 4H^+ .$$

(2)

The SiO_2 layer formed during this reaction is (partially) etched by a pure chemical reaction of the oxide with HF and HF_2^- species in the solution. For anodic potentials larger than about 3V an oscillating behavior of the photocurrent can be observed. Simultaneously, the thickness of the oxide layer varies with time. The chief feed-back mechanism during cycles of oxide formation and etching is given by diffusion limitation of reactive species towards the interface with increasing oxide thickness.

In Fig. 2, left, photocurrent oscillations and the corresponding *in situ* reflectance are shown. It can be seen that photocurrent maxima precede those of the reflectance. This behavior is interpreted as positive phase shift (of the reflectance). According to Fig.1, reflectance maxima can be understood as maximum oxide thicknesses if varying surface and interface roughness are not taken into account. The monitored silicon surface area (A) was slightly larger than $1cm^2$ which explains the relatively high photocurrents, i.e., the shown current density is multiplied by the factor A as indicated in the figure.

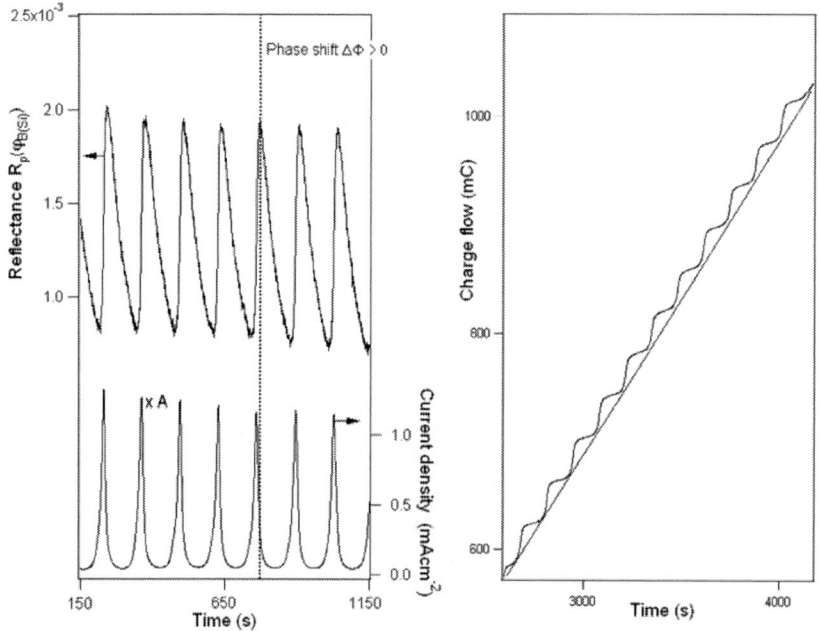

Fig. 2: Photocurrent and reflectance oscillations monitored at 6V in 0.1M NH_4F pH ~4. Left: the reflectance signal R_p is shifted by a value $\Delta\Phi > 0$ with respect to the photocurrent signal. Right: the integral charge flow shows a small modulation superimposed onto a linear function.

In Fig. 2, right, the charge flow exhibits a small modulation along a linearly increasing signal. Eq. 1 therefore does not account for the observed optical signal in its given form. In this case, an additional term, accounting for the chemical oxide etching with etch rate γ, has to be used:

$$R_p \approx A + B \int_0^t I_{photo}dt - \alpha\gamma t . \tag{3}$$

By this equation, the linear curve is subtracted from the integral and the resulting data simulates the optical data in good agreement with the experiment (see Fig.3).
The coefficients in Eq. 3 were not calculated by a least-square procedure. They were chosen such that regions are clearly visible where the validity of Eq. 3 is possibly limited.

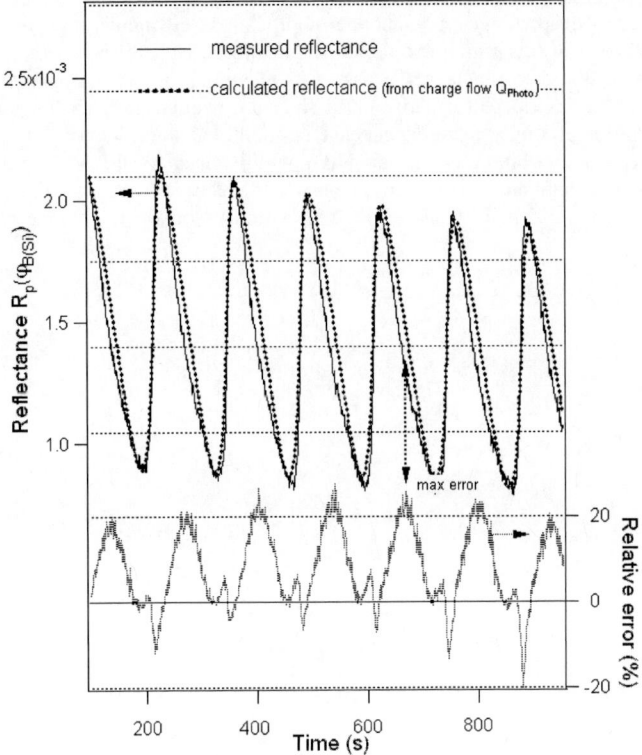

Fig. 3: Simulated optical behavior corresponding to the experimental data shown in Fig. 2. The simulated reflectance R_p (dashed curve) was modeled according to Eq. 3. The solid curve shows the experimental data. The relative error, indicated below, is highest at the negative slope of the reflectance, i.e., near photocurrent minima (see Fig. 2).

It can be seen that the most distinct deviation from experimental results is observable in the decreasing part of the reflectance cycles. These parts are associated with minimum photocurrents (see Fig. 2) and are presumably due to roughness variations at the surface and at the SiO_2/Si interfacial region.

Negative phase shift upon potential scan in the divalent dissolution region

In the divalent dissolution regime, the dissolution of a silicon surface atom proceeds without formation of SiO_2:

$$Si + 6HF + h_{VB}^+(h\nu) \rightarrow SiF_6^2 + 4H^+ + H_2 + e_{CB}^-. \tag{4}$$

This region is mainly associated with the formation of porous silicon. As mentioned before, however, results were reported about the formation of silicon nanocrystals protruding out of the surface. The potential interval between open circuit potential (OCP) and the region around the first current maximum is very small; detailed evaluation of the transition process towards tetravalent dissolution requires therefore sensitive methods which make corresponding topographical modifications distinguishable. In Fig. 4, left, the photocurrent of this region and the corresponding reflectance behavior are shown. It is remarkable that the reflectance R_p passes its maximum before the photocurrent. This behavior can be described as negative phase shift of the reflectance. The magnitude of the negative shift shows only weak dependence on the illumination intensity, i.e., density of light induced holes. Similarly, the shift is not influenced by the scan rate. It can be therefore concluded that the reflectance behavior is specific for the transition from oxide-free silicon dissolution (divalent) to SiO_2 formation (tetravalent) near the photocurrent maximum.

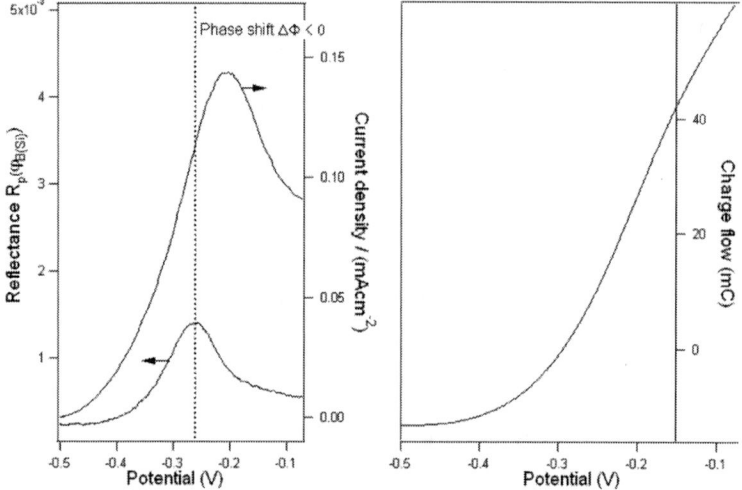

Fig. 4: Photocurrent and BAR reflectance behavior in the region of divalent Si dissolution during a potentiodynamic scan from 0.5V to 0V. Left: the reflectance signal R_p is shifted by a value $\Delta\Phi < 0$ with respect to the photocurrent signal. Right: the integral charge flow shows a monotonic increase.

The charge flow, detected in this potential region, is shown in Fig. 4, right. While the ascending part until about -0.3V resembles the corresponding initial increase of the reflectance (see Fig. 4, left), no correspondence for the reflectance maximum can be observed. It can be concluded that the interfacial topography undergoes a transformation which is, as opposed to pure chemical etching as in the case of photocurrent oscillations, induced by an electrochemical process. The local reflectance maximum could therefore be interpreted as commencing surface oxidation which would, in turn, lead to chemical oxide etching and smoothening of the surface. A smoother surface, according to Fig. 1, would then result in a lower optical reflectance signal. However, experimental observations do not confirm this conclusion. It is observed that the surface roughness continues to increase when passing the reflectance maximum. One can therefore assume that the dielectric properties of the porous film change with increasing potential such that the reflectance responds with a lower proportionality factor. As a remark, these differences in sensitivity are clearly visible in Fig. 1 where SiO_2 films induce a reflectance increase much lower than, e.g., rough films of comparable thickness.

In order to model the reflectance behavior during the transition of the dissolution valency from two to four, a step-function can be designed which accounts for the decrease in sensitivity of the in situ optical signal to the transforming porous film. This is illustrated by Fig. 5 where a so-called sigmoid function represents the transition around the first reflectance maximum.

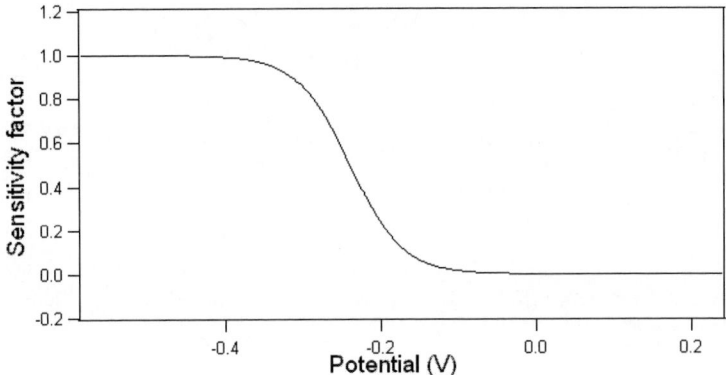

Fig. 5: Sigmoid function which models the transition from divalent dissolution (U < -0.24V to tetravalent dissolution U > -0.24V.

The sigmoid function is centered at the potential of the reflectance maximum (about $U_0 = -0.24V$) and serves as pre-factor in Eq. 5 for the appropriate modulation of the reflectance curve near the maximum:

$$R_p \approx A + B \left(\frac{1}{1+e^{b(-U+U_0)}} \right) \int_0^t I_{photo} dt . \tag{5}$$

In order to justify this approach, a multi-layer simulation was carried out which considered: (i) increasing surface roughness and (ii) increasing sub-surface porosity in a region near the reflectance maximum (-0.4V to -0.2V). The calculations according to Eq. 5 and the multi-layer simulation are compared in Fig. 6. Here, the experimental data (reflectance, dashed curve and current, thin solid curve) and the respective simulations (thick solid curve) are shown. It can be seen that both the abstract treatment by Eq. 5 and the topographic interpretation by the multi-layer model show good agreement with the experimental data in a potential range between about -0.35V and -0.2V. Outside of this potential range, other reaction mechanisms and modifications of the interfacial region have to be assumed. Although other interpretations for the reflectance behavior in the transition regime are possible, the analysis presented here is supported by experimental findings applying scanning probe microscopy. According to these results, the increasing surface roughness requires the assumption of increasing sub-surface porosity which passes a threshold value of about 60% (see Fig. 1, curve B). As a consequence, nanocrystal formation and shaping near the transition regime presumably involves a sub-surface layer which is characterized by varied porosity.

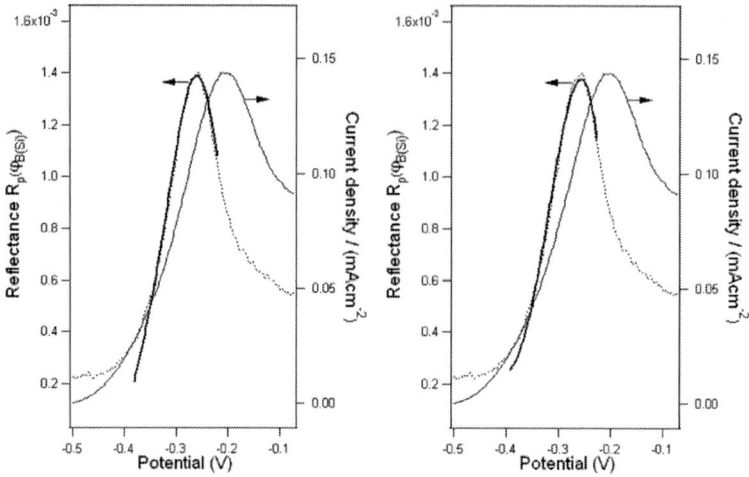

Fig. 6: Left: convolution of the sigmoid function and the charge flow Q_{Photo}. Right: simulated optical behavior corresponding to the experimental data shown in Fig. 2. The simulated reflectance R_p (solid curve) was modeled between -0.4V and the first current maximum according to a multi-layer model comprising surface roughness and sub-surface porosity. The porosity was incremented between 0 and 80%, the thickness of the porous layer from 0 to 3nm. The pointed curve shows the experimental reflectance data while the dashed curve represents the measured photocurrent.

Conclusion

Approximate formulae were presented for the reflectance behavior of Brewster-angle reflectometry in the divalent and tetravalent dissolution regime. The approach could well simulate the known processes during photocurrent oscillations and directly deduce the presence of pure chemical oxide etching. In the divalent dissolution regime, indications were found for the simultaneous increase of surface roughness, which leads to the formation of silicon nanocrystals, and sub-surface porosity.

Acknowledgments

M. Lublow acknowledges the financial support granted by the Deutsche Forschungsgemeinschaft (project no, LE 1192/4-1/2).

References

1. S. Böhm, L. M. Peter, G. Schlichthörl, R. Greef, *J. Electroanal. Chem.* **500** (2001) 178.
2. M. Lublow, H. J. Lewerenz, *Surf. Sci.* **601** (2007) 4227.
3. D. R. Turner, *J. Electrochem. Soc.*, **105**, 402 (1958).

4. M. Lublow and H. J. Lewerenz, *ECS Trans.* **6(19)**, 1 (2008).
5. T. Nychyporuk, V. Lysenko,B. Gautier, D. Barbier, *J. Appl. Phys.* **100** (2006) 104307.
6. H. Gerischer, M. Lübke, *Ber. Bunsenges. Phys. Chem.* **92** (1988) 573.
7. J. Grzanna, H. Jungblut, H.J. Lewerenz, *J. Electroanal. Chem.* **486** (2000) 181.
8. M. Lublow and H.J. Lewerenz, *Surf. Sci.* **601**, 1693 (2007).

390

CHAPTER 9

PROPERTIES AND PATTERNING OF
SEMICONDUCTORS AND RELATED
COMPOUNDS

ECS Transactions, 19 (3) 393-402 (2009)
10.1149/1.3120719 ©The Electrochemical Society

Micro-Patterning of Semiconductors by Metal-Assisted Chemical Etching through Self-Assembled Colloidal Spheres

Sachiko Ono, Fusao Arai and Hidetaka Asoh

Department of Applied Chemistry, Faculty of Engineering, Kogakuin University
2665-1 Nakano, Hachioji, Tokyo 192-0015, Japan

Macroporous semiconductors such as Si and GaAs with ordered pore intervals were fabricated by the site-selective chemical etching of a substrate using patterned Pd-Pt thin film catalyst through self-assembled colloidal spheres as a mask. The obtained macroporous silicon using noble metal catalyst was conical because the pore wall of macropores was chemically dissolved presumably due to the diffusion of positive holes injected at the metal catalyst. In contrast, relatively straight pores with uniform diameter were grown when the concentration of HF was high. With increasing concentration of HF, the etching rate at the pore bottoms in a vertical direction to the Si surface increased, while the etching rate in a lateral direction was suppressed markedly to yield high aspect ratio of pores up to 10. The change in features of macroporous silicon was thought to be involved in the difference in diffusion behavior of injected positive holes at the silicon/metal interface.

Introduction

Natural lithography, which has been proposed by Deckman and Dunsmuir (1) based on a self-organizing process has attracted increasing attention as a key method for nanofabrication owing to its relative simplicity and low cost. Recently, we reported nanopatterning processes based on a localized anodization of Si and the subsequent chemical etching of SiO_2 to fabricate an ordered dot array and a hole array on a Si surface using self-organized anodic porous alumina as a mask (2-5). Through the porous alumina mask, regularly arranged metal nanopattern on the Si surface was fabricated by the electroless deposition of Cu nanodots (5). The periodicity of such arrays could be easily controlled by changing the pore interval of the upper anodic alumina.

Several studies have been reported on the nanopatterning of solid substrates using a monolayer coating of self-assembled colloidal spheres, i.e., colloidal crystal, instead of a conventional resist as a mask (6-8). We have applied the colloidal crystal as a mask or template to fabricate hexagonally ordered hole array, convex array (9-11) and nano-sized metal patterns on a Si (12, 13) and an Al surfaces (14).

Many investigations have been carried out on the development of various techniques enabling the fabrication of nano-/micro silicon hole array structures. For applications in silicon-based optoelectronics such as solar cells, the metal-assisted chemical etching of silicon has been developed since 2000 (15-17). Such technique, which was first proposed by Li and Bohn (15), is a very simple and efficient process without using external bias. According to their report, microporous silicon layers with a thickness of ~ 3 μm were formed in HF with hydrogen peroxide (H_2O_2) using noble metal nanoparticles, i.e., Au, Pt, or Au-Pd as a catalyst for chemical etching. Tsujino and Matsumura (17) reported that straight cylindrical nanoholes up to several ten μm in Si could be obtained by immersing Si wafers loaded with Ag nanoparticles as a catalyst.

A mechanism involving a localized electrochemical process has been proposed to explain the mechanism of metal-assisted chemical etching as follows (15):
Cathode reaction (at noble metal as a local cathode):
$$H_2O_2 + 2H^+ \rightarrow 2H_2O + 2h^+,$$
$$2H^+ \rightarrow H_2\uparrow + 2h^+.$$
Anode reaction (at silicon surface):
$$Si + 4h^+ + 4HF \rightarrow SiF_4 + 4H^+,$$
$$SiF_4 + 2HF \rightarrow H_2SiF_6.$$
Overall reaction:
$$Si + H_2O_2 + 6HF \rightarrow 2H_2O + H_2SiF_6 + H_2\uparrow.$$

In other words, when the oxidants (H_2O_2) are reduced on the surfaces of noble metal catalyst, positive holes (h^+) are generated. After the removal of electrons from metal particles, the potential of the metal shifts toward a positive value to a level enabling injection of h^+ into silicon substrate. Finally, anodic oxidation and dissolution of silicon occurs in chemical etchant containing HF.

Whereas, we reported the fabrication of ordered micro convex array and hole array of silicon by a combination of colloidal crystal templating and metal-assisted chemical etching using patterned noble metal thin films such as Ag (5, 18, 19), Pt-Pd, Pt and Au (20-22) as a catalyst. By using shape-controlled circular metal thin films instead of metal nanoparticles, silicon microholes containing metal thin films at bottom of each hole could be obtained easily (21). Micro pattering of GaAs in the same manner was also performed (23). In our recent paper (22), we reported that the morphology of etched silicon microstructures and etching rate were strongly affected by catalyst species used.

In this paper, the effect of etching condition such as concentration of etching solution on the morphology and the aspect ratio of hole arrays on semiconductors, which were fabricated with patterned metal catalyst through a self-assembled colloidal crystal as a mask, was studied as

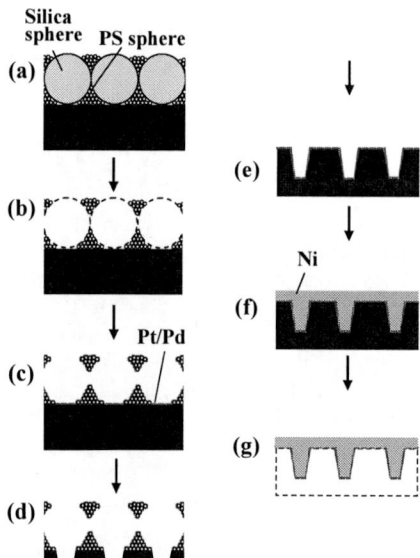

Figure 1 Schematic model of fabrication of silicon macropore arrays: (a) formation of composite colloidal crystals on silicon substrate, (b) removal of silica sphere, (c) formation of metal catalyst layer, (d) chemical etching of silicon, (e) removal of PS honeycomb mask, (f) electrodeposition of nickel layer, and (g) removal of silicon by immersion of substrate in TMAH.

well as the effect of metal species on etching behavior.

Experimental

The principle of pattern transfer, which is similar to that in our previous works (21), is schematically shown in Fig. 1. In the case of GaAs, the experimental procedure for the preparation of macropores in the substrate was almost same to that of Si. Therefore, we describe here the procedure of Si as a substrate. The silicon specimens were precleaned in 1 wt % HF to remove organic contaminants and native oxide. A mixed suspension consisting of equal volumes of a 0.2 wt % suspension of polystyrene (PS) nanospheres of 200 nm diameter (Polysciences, Inc.) and a 0.5 wt % suspension of silica microspheres of 3 μm diameter (Bangs Laboratories, Inc.) was dropped on the substrate. The suspension on the substrate was dried in air for more than one day, during which the mixture of spheres of two sizes was self-assembled into a close-packed structure by capillary forces. After the complete evaporation of the solvent, the silicon substrate with the binary colloidal crystals formed by the spheres was heated at 100 °C for 1 h to combine the adjacent PS nanospheres [Fig. 1(a)]. After heating, the silica spheres, which were used as a template, were selectively removed by immersing the specimens in 10 wt % HF for 10 min [Fig. 1(b)].

Metal thin films were deposited onto the silicon substrate through the PS honeycomb mask composed of the densely packed PS nanospheres by ion sputtering (Hitachi E-1010) at a discharge current of 15 mA in a vacuum with pressure below 10 Pa [Fig. 1(c)]. Pt-Pd (80 % Pt and 20 % Pd) was used as a sputtering target. The deposition rate of Pt-Pd was approximately 2 nm min^{-1}. After sputtering for 5 min, the specimens with locally deposited metal films were etched in a mixed solution of 5, 10 or 15 mol dm^{-3} HF and 1 mol dm^{-3} H$_2$O$_2$ at room temperature [Fig. 1(d)]. Finally, the PS honeycomb mask was removed by immersing the specimens in 97 % toluene [Fig. 1(e)].

To examine the morphology of the silicon macropores in the direction of pore depth, the formation of metal replicas were applied. After coating metal catalyst layer on the surface of silicon microstructures using ion sputtering, electrodepositon was conducted in conventional nickel plating solution [Fig. 1(f)]. Finally, silicon substrate was selectively removed in tetra methyl ammonium hydroxide (TMAH) [Fig. 1(g)]. The ordered geometric pattern formed on the silicon substrate was evaluated by scanning electron microscopy (SEM, Hitachi S-4200) and focused ion beam (FIB, Hitachi FB-2100).

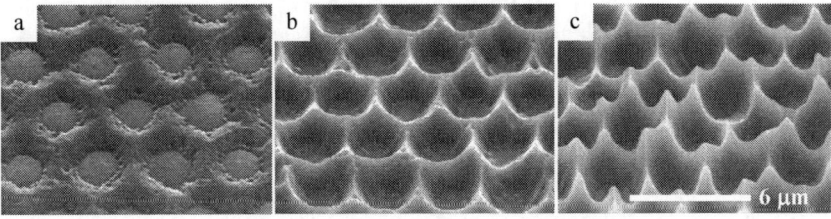

Figure 2   SEM images of patterned Pt-Pd-coated silicon after chemical etching in 5 mol dm^{-3} HF / 1 mol dm^{-3} H$_2$O$_2$ for (a) 5 s, (b) 1 min and (c) 2 min. Specimens were observed from an angle of 45° to the surface. Chemical etching was conducted before removal of polystyrene honeycomb mask.

Results and discussion

Etching process of the Si substrate by immersing in a mixed solution of 5 mol dm^{-3} HF and 1 mol dm^{-3} H$_2$O$_2$ at room temperature for (a) 5 s, (b) 1 min and (c) 2 min after sputtering of Pt-Pd catalyst through the PS honeycomb mask was shown in Fig. 2. Chemical etching was conducted before removal of polystyrene honeycomb mask in this case. When ion sputtering was conducted through the mask, isolated circular metal thin films with approximately 1.6 µm diameter and 10 nm thickness were deposited in the interspaces of the PS mask (21). The ion sputtering of Pt-Pd, which is generally used as a method of coating nonconducting materials with a metal for SEM observation, was chosen for the purpose of forming a smoothly shaped metal thin film and to avoid a granular coating (22). In fig. 2, SEM images of the tilted view of the silicon surface after chemical etching indicate that the central part of the silicon coated with Pt-Pd thin film gradually sagged downward during chemical etching yielding an ordered array of macropores with uniform diameter in the periodicity of 3 µm, which was basically determined by the diameter of the silica spheres. However, chemical etching proceeded not only at Pt-Pd thin film / substrate interface but also at the pore wall, which led to form a funnel shaped pore. Finally, the pore wall after 2 min etching was heavily solved out due to the chemical dissolution in the lateral direction as shown in Fig. 2c. Therefore, it was difficult to achieve the formation of macropores with a high aspect ratio. In this case, injected positive holes are thought to be diffused into silicon bulk, resulting in oxidization of silicon at locations away from the metal-coated silicon surface and excessive pore-widening at the outermost silicon surface as discussed in our previous paper (22). Similar phenomenon on the metal-assisted chemical etching using Ag catalyst was recently reported by Chartier et al. (24).

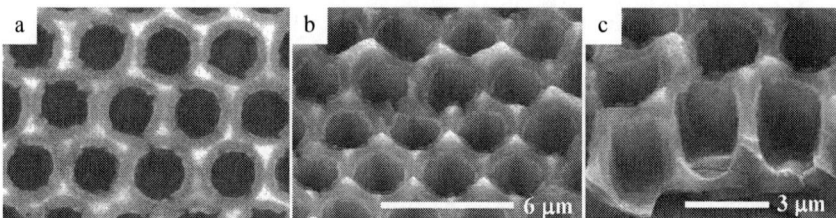

Figure 3   SEM images of patterned Pt-Pd-coated silicon after chemical etching in 5 mol dm^{-3} HF / 1 mol dm^{-3} H$_2$O$_2$ for (a) 1 min, (b) 1 min, observed from an angle of 45° to the surface and (c) 1 min, high magnification view. Chemical etching was conducted after removal of the honeycomb mask.

Next, chemical etching of the Pt-Pd coated substrate through the honeycomb mask was conducted after removal of the mask. As revealed in Fig. 3, chemical dissolution of the horizontal plane resulting excessive pore-widening was drastically suppressed. An ordered array of macropores of Si with uniform diameter having a higher aspect ratio compared to that observed in Fig. 2 was acheved by the preferential dissoution of silicon in the direction of pore depth, even in the same etching condition. The cause of excessive pore-widening due to chemical dissolution of the pore wall during the etching in HF and H$_2$O$_2$ with the honeycomb mask could be explained by the existence of Pd-Pt thin film sputtered on the wall of the honeycomb mask. It moved into pore wall during digging and acted as catalyst for chemical etching similarly to the metal / substrate interface at the pore bottom.

After Pd-Pt assisted chemical etching in 5 mol dm^{-3} HF / H$_2$O$_2$ for 2 min, however, the pore width was considerably widen by further chemical dissolution of pore wall as shown in Figs. 4a and 4c, although pore depth increased up to 7 μm suggesting the etching rate of 3.5 μm min^{-1}. On the other hand, pore-widening was drastically inhibited when 10 mol dm^{-3} HF / H$_2$O$_2$ was used as etching solution as revealed in Figs. 4b and 4d, where the pore diameter at the outer Si surface almost unchanged and was similar to that of Pd-Pt thin film located at the pore bottom. While, the pore depth increased up to 11 μm that corresponded to the etching rate of 5.5 μm min^{-1}. Notable changes in feature obtained by metal-assisted chemical etching using high concentration HF are, the suppression of chemical dissolution at the pore wall in a lateral direction resulting in a straight and sharp pore shape, finely-grooved pore wall directly reflecting the edge shape of Pd-Pt thin film and increase in etching rate. Namely, the dissolution of silicon oxide is accelerated locally at the silicon/metal interface in the direction of the pore depth, resulting in the formation of macroporous silicon with a high aspect ratio. In the case of low-concentration HF, injected positive holes are expected to diffuse into silicon bulk and oxidize silicon at locations away from the metal-coated silicon surface. In contrast, the diffusion of positive holes is thought to be suppressed in the case of high-concentration HF due to enhanced dissolution at the metal / substrate interface of pore bottom.

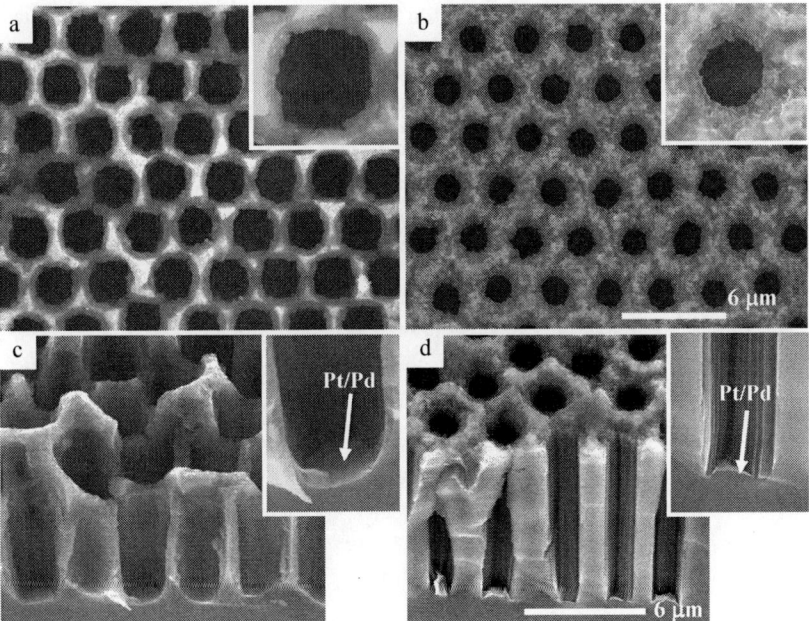

Figure 4 SEM images of patterned Pt-Pd-coated silicon after chemical etching in (a, c) 5 mol dm^{-3} HF / H$_2$O$_2$ for 2 min and (b, d) 10 mol dm^{-3} HF / 1 mol dm^{-3} H$_2$O$_2$ for 2 min. (a, b) Top view and (c, d) cross-sectional view. Chemical etching was conducted after removal of PS honeycomb mask.

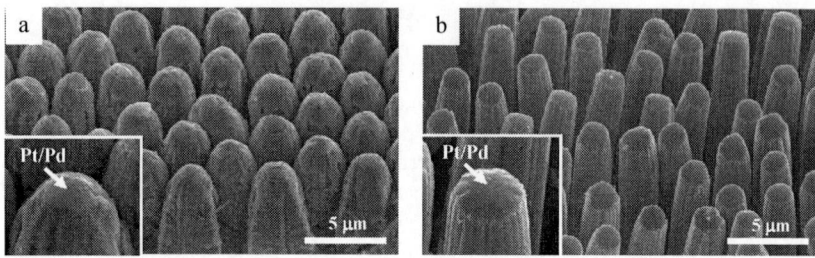

Figure 5 SEM images of nickel replica of patterned Pt-Pd-coated silicon after chemical etching in (a) 5 mol dm^{-3} HF / H$_2$O$_2$ for 2 min and (b) 10 mol dm^{-3} HF / 1 mol dm^{-3} H$_2$O$_2$ for 2 min. Specimens were observed from an angle of 45° to the surface.

To evaluate the morphology of porous structure in the direction of pore depth, nickel replica was formed by metal plating using obtained macroporous silicon as a template. From the SEM images of the obtained nickel replicas in Fig. 5, it was apparent that the chemical dissolution behavior of pores formed in low and high HF solutions was remarkably different. These images of nickel replicas were basically compatible with the SEM images of macroporous silicon; however, the difference of shape of side surface of Ni rods could be clearly confirmed in each specimen. Though the circular Pd-Pt thin film remained on the tip of rods of both replicas, the side surface of nickel rods, that is, the side wall of the silicon macropores was significantly different between both specimens. It was finely-grooved in the case of etched specimen in 10 mol dm^{-3} HF clearly reflecting the edge shape of sputtered Pd-Pt thin film because of low dissolution rate in lateral direction. When the etching time was elongated to 3 min, the rods length of the nickel replica became approximately 15 μm yealding the aspect ratio of approximately 10 as shown in Fig. 6. Inperfactions found in the rods configlation would be induced by the replication process. Thus, the higher aspect retio could be acheved by using a high concentration HF solution to avoid an excess etching of the pore walls associated with enhancement of Pd-Pt catalyst assited chemical etching at the pore bottom.

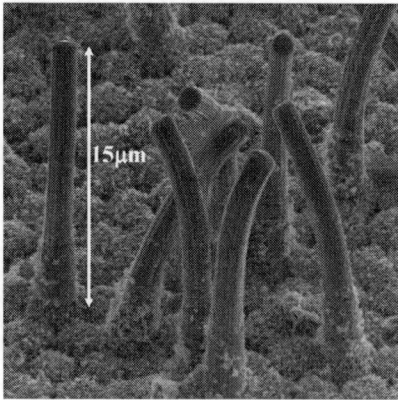

Figure 6 SEM image of nickel replica of patterned Pt-Pd-coated silicon after chemical etching in 10 mol dm^{-3} HF / 1 mol dm^{-3} H$_2$O$_2$ for 3 min. Specimens were observed from an angle of 45° to the surface.

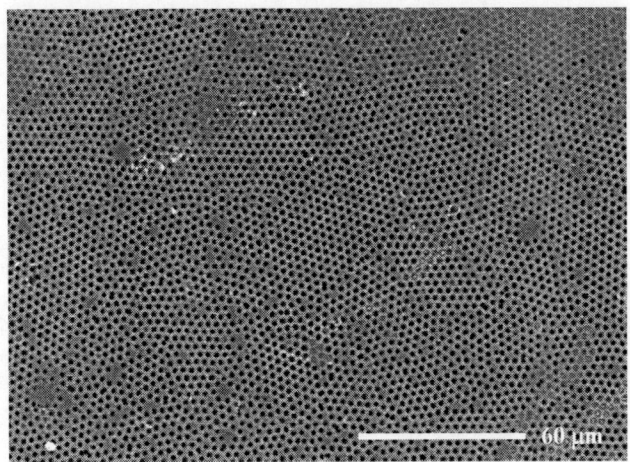

Figure 7 Top-view SEM image of macroporous silicon formed by metal-assisted chemical etching. Chemical etching was conducted in 10 mol dm^{-3} HF / 1 mol dm^{-3} H$_2$O$_2$ for 2 min using a sputtered Pt-Pd catalyst through the honeycomb mask.

Top-view SEM image of macroporous silicon formed by metal-assisted chemical etching conducted in 10 mol dm^{-3} HF / 1 mol dm^{-3} H$_2$O$_2$ for 2 min using a sputtered Pt-Pd catalyst through the honeycomb mask is shown in Fig. 7. Regularly and hexagonally arranged macrospores with the interval of 3 μm are distributed over a wide range of silicon substrate.

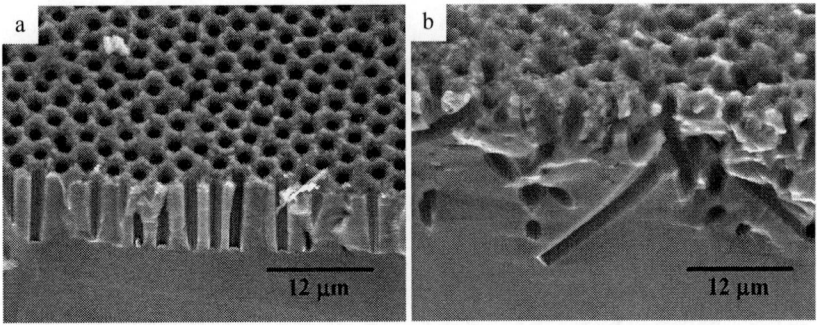

Figure 8 SEM images of cross-sectional view of patterned Pt-Pd-coated silicon after chemical etching in (a) 5 mol dm^{-3} HF / H$_2$O$_2$ for 2 min and (b) 10 mol dm^{-3} HF / 1 mol dm^{-3} H$_2$O$_2$ for 2 min. Chemical etching was conducted after removal of PS honeycomb mask.

To expand the effect of concentration of HF, metal-assisted chemical etching was conducted in 15 mol dm^{-3} HF / 1 mol dm^{-3} H$_2$O$_2$ for 2 min. The cross section is shown in fig. 8 comparing with that obtained in 10 mol dm^{-3} HF / H$_2$O$_2$. The etching speed increased up to 8.5 μm min^{-1} (Fig. 8b) from 5.5μm min^{-1} for the etching rate of 10 mol dm^{-3} HF / H$_2$O$_2$ (Fig. 8a). A notable characteristic point is the direction of pores; most pores advanced in a curved direction but not in a vertical direction to the surface of Si substrate. The reason of this phenomenon is not clear. Tsujino and Matsumura (25) found a helical nanohole borded in silicon by a 50 nm sized Pt particle when 25 mol dm^{-3} HF / H$_2$O$_2$ was used as etching solution. Therefore, 10 mol dm^{-3} HF / 1 mol dm^{-3} H$_2$O$_2$ is currently an optimum condition to realize high aspect ratio macropores in Si substrate.

In the case of GaAs, etching depths of hole arrays fabricated by Pt/Pd-assisted chemical etching for 1 min in 5 mol dm^{-3} HF / 1 mol dm^{-3} H$_2$O$_2$ evaluated by AFM measurement was less than 1 μm. Thus, etching rate of GaAs was much lower than that of Si though the etching behavior including chemical etching in the lateral direction as well as the effect of noble metal species was similar (23). Most notable feature of metal assisted etching of GaAs is anisotropic etching depending on the crystal orientation as shown in Fig. 9 for convex array obtained by electroless plated Ag assisted etching.

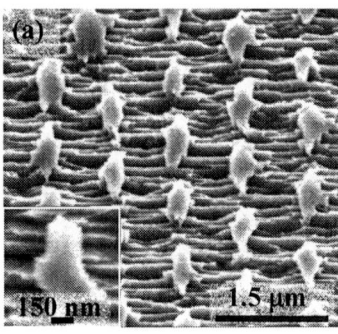 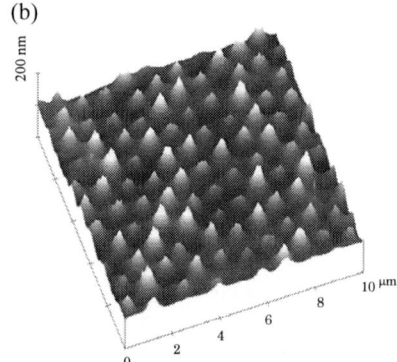

Figure 9 Surface structures of n-GaAs (100) substrate obtained by a combination of a colloidal crystal mask, electroless plating and metal-assisted chemical etching using a Ag catalyst. Typical SEM images of specimens obtained after (a) 60 s etching observed at 45° to the surface and (b) AFM image.

Further research on metal-assisted chemical etching using metal thin films as a catalyst would clarify the relationship between the mechanism for controlling the morphology of the resultant pattern and the etching conditions, such as the composition and concentration of etchant, catalytic metal species, substrate parameters, resistivity, and doping density.

Conclusions

We have investigated the effect of concentration of HF on the morphology of etched silicon microstructures. When Pt/Pd-assisted chemical etching in 5 mol dm^{-3} HF / 1 mol dm^{-3} H$_2$O$_2$ was conducted, the shape of macropores was conical because the pore wall was chemically dissolved presumably due to the diffusion of positive holes injected at the Pd-Pt catalyst. In contrast, relatively straight pores with uniform diameter were grown when the concentration of HF increased to 10 mol dm^{-3}. With increasing concentration of HF, the etching rate at the pore bottoms in a vertical direction to the Si surface increased, while the etching rate in a lateral direction was suppressed markedly. Thus, regularly ordered hexagonal array of macropores with high aspect ratio distributed over a wide range of silicon surface was obtained. The morphology of resultant porous structure is thought to be affected by the difference in etching conditions such as the composition and concentration of etchant, catalytic metal species, substrate parameters, resistivity, and doping density that influence to the diffusion behavior of injected positive holes at the silicon/metal interface.

The process presented is suitable for the large-scale production of ordered silicon macropores containing noble metal thin film that are not achievable by conventional lithographic techniques. Further research on the formation of three-dimensional silicon microstructures based on a combination of colloidal crystal templating and metal-assisted chemical etching would help us to crystallize potential applications in optical devices, chemical sensors, and silicon-based biofunctional devices.

Acknowledgments

This work was partly financially supported by a Grant-in-Aid for Scientific Research from the Japan Society for the Promotion of Science. Thanks are also due to the "High-Tech Research Center" Project for Private Universities: matching fund subsidy from the Ministry of Education, Culture, Sports, Science and Technology.

References

1. H. W. Deckman and J. H. Dunsmuir, *Appl. Phys. Lett.* **41**, 377 (1982).
2. H. Asoh, M. Matsuo, M. Yoshihama and S. Ono, *Appl. Phys. Lett.*, **83**, 4408 (2003).
3. A. Oide, H. Asoh and S. Ono, *Electrochem. Solid-State Lett.* **8**, G172 (2005).
4. H. Asoh, K. Sasaki and S. Ono, *Electrochem. Commun.*, **7**, 953 (2005).
5. S. Ono, A. Oide and H. Asoh, *Electrochim. Acta*, **52**, 2898 (2007).
6. C. Haginoya, M. Ishibashi and K. Koike, *Appl. Phys. Lett.* **71**, 2934 (1997).
7. C.-W. Kuo, J.-Y. Shin and P. Chen, *Chem. Mater.* **15**, 2917 (2003).
8. K. H. Park, S. Lee, K. H. Koh, R. Lacerda, K. B. K. Teo and W. I. Milne, *J. Appl. Phys.* **97** (2005) 024311.
9. H. Asoh, A. Uehara and S. Ono, *Jpn. J. Appl. Phys.*, **43**, 5667 (2004).
10. H. Asoh, A. Oide and S. Ono, *Appl. Surf. Sci.*, **252**, 1668 (2005).
11. H. Asoh, A. Oide and S. Ono, *Electrochem. Commun.*, **8**, 1817 (2006).
12. H. Asoh, S. Sakamoto and S. Ono, *Coll. Interf. Sci.*, **316**, 547 (2007).
13. S. Sakamoto, L. Philippe, M. Bechelany, J. Michler, H. Asoh and S. Ono, *Nanotech.*, **19**, 405304 (2008)
14. H. Asoh, K. Nakamura and S. Ono, *Electrochim. Acta*, **53**, 83 (2007)
15. X. Li and P. W. Bohn, *Appl. Phys. Lett.*, **77**, 2572 (2000).
16. S. Yae, Y. Kawamoto, H. Tanaka, N. Fukumuro and H. Matsuda, *Electrochem. Commun.*, **5**, 632 (2003).
17. K. Tsujino and M. Matsumura, *Adv. Mater.*, **17**, 1045 (2005).
18. H. Asoh, F. Arai and S. Ono, *Electrochem. Commun.*, **9**, 535 (2007).
19. H. Asoh, F. Arai and S. Ono, *Electrochem.*, **76**, 187 (2008).

20. H. Asoh, F. Arai and S. Ono, *ECS Trans.*, **6**, 431 (2007).
21. H. Asoh, F. Arai, K. Uchibori and S. Ono, *Appl. Phys. Exp.*, **1**, 067003/1-067003/3 (2008)
22. H. Asoh, F. Arai and S. Ono, *Electrochim. Acta*, (doi: 10. 1016 / j. electacta. 2009. 01. 050)
23. Y. Yasukawa, H. Asoh and S. Ono, *Electrochem. Commun.*, **10**, 757 (2008)
24. C. Chartier, S. Bastide and C. Lévy-Clément, *Electrochim. Acta*, **53,** 5509 (2008).
25. K. Tsujino and M. Matsumura, *Electrochem. Solid-State Lett.* **8,** C193 (2005).

ECS Transactions, 19 (3) 403-410 (2009)
10.1149/1.3120720 ©The Electrochemical Society

Surface Chemistry and Nanotopography of Step-Bunched Silicon Surfaces: in-system SRPES and SPM investigations

T. Stempel, A.G. Muñoz, K. Skorupska, M. Lublow, M. Kanis, H.J. Lewerenz

Institute for Solar Fuels and Energy Storage Materials, Helmholtz-Zentrum Berlin für Materialien und Energie, Glienicker Str. 100, 14109 Berlin, Germany

> The formation of step bunching was observed by atomic force microscopy on n-type Si(111) surfaces during the electrodeposition of noble metals under semiconductor depletion conditions. The surface chemical analysis performed by synchrotron radiation photoelectron spectroscopy (SRPES) indicates the formation of an ultra-thin oxide film along with the topological transformation. Step bunching is interpreted in terms of site-specific etching controlled by the reactivity of kink sites and step edges together with the surface accumulation of holes supplied by the reduction of Pt-chloride complexes via the valence band.

Introduction

The demand of more stable and efficient silicon based solar energy conversion devices requires finding novel surface electronic and structural properties. Numerous chemical and electrochemical surface treatments have been explored to modify the micro- and nanotopographies as a method to achieve new surface functionalities. For instance, electrochemical conditioning in NH$_4$F solutions was implemented for the formation of porous silicon, nanoporous SiO$_2$ layers (1) and fractal etch domains (2). Si can also be shaped by (electro)chemical treatment in alkaline solutions such as KOH, e.g. for anti-reflection patterning of solar cells or step bunching. The latter refers to the increase of step heights on (111)-orientated Si surfaces up to several atomic bilayers, as found after cathodic polarization of n-type Si electrodes. These surfaces do not only show remarkable structural properties but are also electronically modified. A stable accumulation layer is found on this type of surfaces (3), the charge of which extends predominantly along the steps. This property makes step bunched surfaces interesting candidates for applications involving electrostatic attraction, such as site specific adsorption of biomolecules or metallization (4).

Step-bunching involves selective dissolution processes, which are controlled by the reactivity and coordination of surface atoms. This type of topological evolution was observed by the electrodeposition of noble metals under semiconductor depletion conditions. This system was taken as a model to investigate the peculiarities of the chemical and electronic processes leading to distinct topographic evolutions by atomic force microscopy combined with synchrotron radiation photoelectron spectroscopy.

403

Experimental details

Electrodeposition of Pt and Ir was performed on n-type (111)-orientated Czochralski grown Si electrodes (3-5 Ω cm). Before electrochemistry, samples were cut, cleaned in acetone, ethanol and milliQ deionized water (18 MΩ cm) and etched in 40% NH$_4$F solution, which yields an atomically flat, H-terminated surface (5). GaIn eutectic was applied for ohmic back contacts. A standard three electrode potentiostatic setup was used, with a Pt-sheet and a saturated calomel electrode as counter and reference electrodes, repectively. Pt and Ir were deposited from a 1 mM H$_2$PtCl$_6$ + 0.1 M K$_2$SO$_4$ and a 2 mM IrCl$_3$ + 0.5 M KCl + 5 % iso-propanol solution, respectively.

Tapping Mode Atomic Force Microscopy (TM-AFM) was carried out in an DI multi mode II scanning probe microscope to image the surface topography.

Synchrotron Radiation Photoelectron Spectroscopy (SRPES) was performed at the Solid Liquid Interface Analysis System at the undulator beamline U49/2 at the synchrotron BessyII. To achive very low surface contamination levels and prevent contact of the sample to ambient air, electrochemistry was performed in a specially designed glas vessel under N$_2$-athmosphere, as described previously (6). The vessel was attached directly to the UHV-system. The sample was transferred into the N$_2$-filled chamber, which was then pumped down to UHV conditions immediately. SRPE spectra of the Si2p line were fitted using seven doublet peaks. The individual peaks were obtained by fits with convoluted Gaussian-Lorentzian line shapes. The width of the Lorentz-component, which originates mainly from the natural line shape of the photoionization process, was held constant for all components and is found to be about 80 meV for all spectra. The Gaussian width of the peak was used to fit the overall width of the individual component. The spin orbit splitting was taken to be 610 meV, with a 1:2 peak ratio for the 2p$_{1/2}$ and 2p$_{3/2}$ component. To achieve best fitting results, a Shirley background was fitted and subtracted simultaneously to the peak fit.

Results and discussion

Surface topological and chemical transformations by electroreduction

The electrodeposition of Pt on H-terminated n-type Si(111) at -0.35 V, i.e., under semiconductor depletion (V_{fb} = -0.575 V), is accompanied by a simultaneous etching process. It can be seen that the parallel atomic steps present on the freshly etched substrate (Fig.1a) progress towards a topography characterized by step bunches of ~ 500 nm of width and 10-15 nm of height (Fig.1b). It can be also noted, that the Pt islands nucleate preferentially on the apex of the triangular hillocks, pointing out the chemical selectivity of the process. The Pt nuclei show a rather homogeneous size distribution with a width of about 100 nm and an aspect ratio of 10:1 (Fig.1c).

The particular evolution of the surface topography involves a site-specific solvolytic attack of surface atoms occurring simultaneously with electrodeposition. Evidently, there is an interconnection between the changes of surface chemistry, the electron transfer mode in the reduction process and the etching rate. The chemical analysis carried out by *in system* SRPES after the deposition of minute amounts of Pt shows that an ultra-thin oxide layer is formed. Figure 2 shows the core level signal of Si taken with the surface sensitive excitation energy of hv=170 eV before (a) and after the deposition of Pt (b) and Ir (c). The spectral deconvolution of the signal obtained after Pt and Ir deposition shows main contributions at 99.51 eV ($2p_{3/2}$) and 103.44 eV, which are assigned to bulk Si and SiO_2 respectively (12). Intermediate oxidation states are indicated by contributions appearing at 100.33 eV (Si^+), 101.18 eV (Si^{2+}) and 101.99 eV (Si^{3+})(13,14). Two additional components appears at lower (Si_α, ΔE_b=-0.30 eV) and higher binding energies (Si_β, ΔE_b=+0.29 eV) with respect to the Si bulk line. The origin of these core level shifts was recently ascribed to strained field effects at the $Si-SiO_2$ interface (15). For comparison, spectrum (a) shows the original situation before metal deposition. Besides the main contribution at 99.51 eV, corresponding to bulk and surface Si (the signal of H-terminated Si appears at the same energy as the tensile Si_β-line), other minor contributions of Si^+ and Si^{2+} appear at higher binding energies. As shown by Lublow *et al* (5), this indicates a slightly incomplete H-termination consisting principally in small coverage fractions (up to 0.2 monolayers (ML)) of suboxides (Si_yO_x) and Si-F. This shows clearly, that the oxide is formed upon metal electrodeposition.

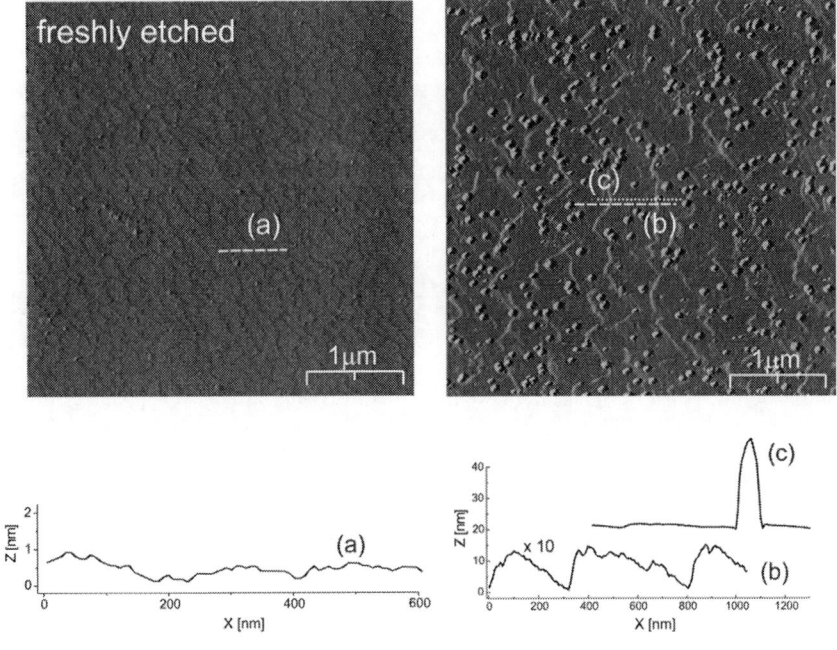

Figure 1: Tapping Mode AFM images of n-type Si(111) surfaces (a) after etching in conc. NH_4F and (b) after 3s of Pt-deposition at -0.35V (SCE).

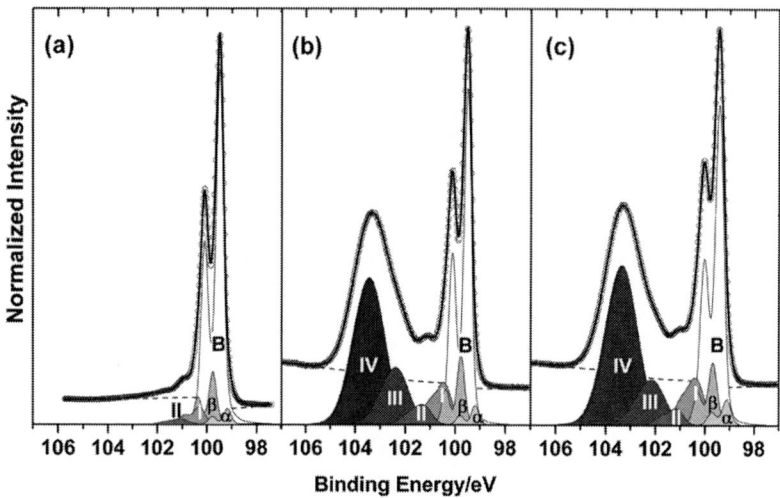

Figure 2: Si2p SRPE spectra of n-Si (111) surfaces taken at a photon energy of 170eV: (a) initial surface after NH₄F etching; (b) after initial Pt-deposition (potential scan to the rise of the first cathodic deposition peak); (c) after Ir-deposition (potential scan to the maximum of the deposition peak). IV: SiO_2, III: Si^{3+}, II: Si^{2+}, I: Si^{1+}, β: stressed Si (tensile), α: stressed Si (compressive).

The role of the electron transfer in the chemical transformation

It is necessary to find a relation between the chemical transformation of the surface and the etching process considering the electron transfer. Looking at the energy band diagram shown in figure 3, one can see that the redox levels corresponding to the reduction of $PtCl_6^{2-}$ and $IrCl_6^{3-}$ are situated below the valence band edge of the semiconductor. It means that under depletion, the electron transfer can take place (i) via filled surface states (indicated as 1) and (ii) by hole injection into the valence band (indicated as 2). The surface states are introduced by surface oxidized species such as =Si-OH-H, which are already present after the initial etching in concentrated NH₄F (5). Oxidized surface atoms, mainly at kink sites, act as nucleation sites for the growth of metal islands, as corroborated by the preferential nucleation of metal nano-islands at the apex of the triangular hillocks (7). The intermediate reduction steps of $PtCl_6^{2-}$ and $IrCl_6^{3-}$, on the other hand, are believed to be responsible for the injection of holes into the valence band (a delocalized process). Holes accumulate at the surface and lead to surface etching and oxide formation. The surface oxidation increases, in turn the density of surface states (16). This effect can be monitored by an increase of the capacitance peak appearing close to the flat band potential, as described in a previous work (7).

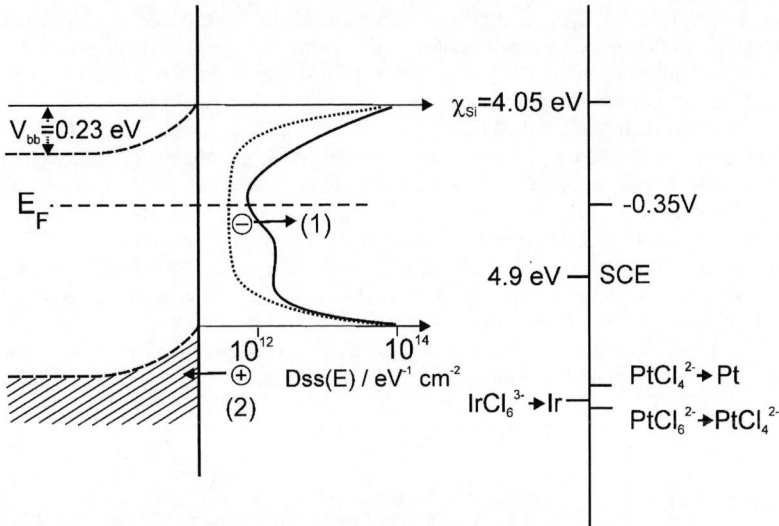

Figure 3: Energy band diagram showing the possible electron transfer processes during the electrodeposition of noble metals.

Site-specific etching of steps: the effects of Cl-ions

The etching of n-Si(111) surfaces in concentrated NH_4F solutions provided reproducible substrates characterized by parallel terraces with saw-tooth like edges and a height corresponding to an atomic bilayer (0.314 nm) (7). It is well known that the etching process is site-specific and initiates preferentially on kink sites and on the strained dihydride termination of the parallel terraces in the $< \bar{1}\bar{1}2 >$ direction (8). Therefore, the propagation of the dissolution front of atomic steps is conditioned by the chemistry of the etchant solution as well as the orientation and the electronic properties of surface. For instance, the removal of surface atoms during the photoelectrochemical dissolution of n-Si(111) in dilute NH_4F solution initiates at tri-hydride terminated kink sites and extends perpendicularly to the $< \bar{1}\bar{1}2 >$ direction (9). This dissolution sequence leads to the flattening of the initial zig-zag terrace edges.

The formation of step bunches, on the other hand, implies the stacking of the dissolving layers, which can be described in terms of the collision of steps on miscut surfaces or by kinetic instabilities caused by diffusion effects of the etchant (8,10).

The progress of the two-dimensional electronic distribution at the surface plays a decisive role in the electrochemical etching of n-Si(111) carried out by cathodic polarization in a 2 M KOH solution. In this case, large steps of up to 10 nm of height are formed, whose walls show predominantly a (100) orientation. Based on SRPES studies, Skorupska et al (11) were able to find out that the mechanism of cathodic etching initiates with the solvolytic attack of H-terminated sites and the formation of =Si-OH-H under accumulation. Additionally, Kelvin probe experiments have shown that there is a lateral modulated surface charge with an enhanced electron accumulation at the step edges (3).

In chloride solutions, a continuous increase of the capacitance peak ascribed to the response of the surface states was observed (figure 4). A similar effect, however, was not found in sulphate solutions. This implies that Cl-ions alter the electronic properties of the surface, probably by an adsorption step, inducing a solvolytic attack and subsequent oxidation of the surface atoms. The adsorption of halogenides was considered by Fujitani *et al* (17) as a probable cause for the observed flat band potential shift in concentrated HX (X: I, Br, Cl) solutions. Evidence for the formation of stable Si-X bonds was only provided for the case of iodide and the presence of chloride was not detected in our SRPES experiments. This apparent contradiction can be overcome assuming a substitution of halogenides by Si-OH after rinsing with water. The TM-AFM picture taken after 1 hour immersion in an acid chloride solution at -0.35 V shows the progress of the surface topology towards the formation of steps of ~ 100 nm of width and 1-1.5 nm of height, indicating clearly a slow etching process. It is thought that Cl⁻ ions adsorb temporarily on kink and dihydride step sites, favouring the solvolytic attack of Si back-bonds.

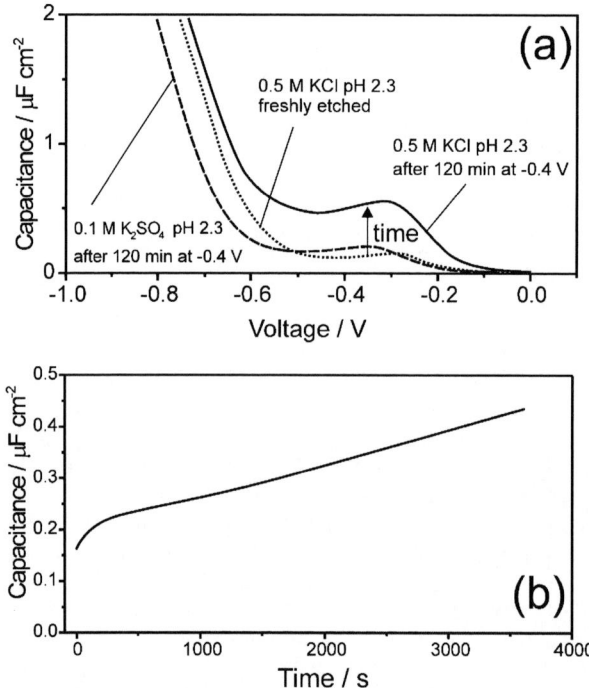

Figure 4: Potential- (a) and time-dependence (b) of the surface capacitance of n-type Si(111) in 0.5 M KCl solution. The potential dependence in K_2SO_4 solution is also compared.

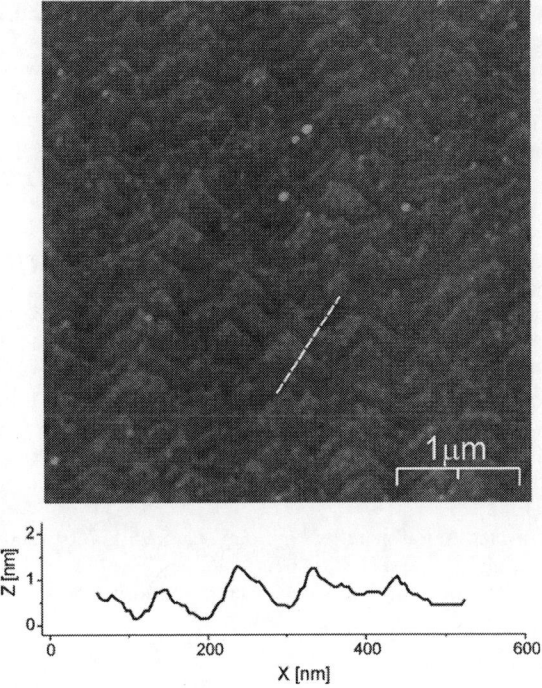

Figure 5: Tapping Mode AFM image of n-type Si(111) after immersion in a 0.5 M KCl solution (pH 2.3) at -0.35 V (SCE) for 60 min.

Conclusions

The formation of step bunches was observed during the electrodeposition of noble metals onto n-type Si(111) under depletion conditions. The particular progress of the surface topology is associated with the site-specific etching assisted by the injection of holes into the valence band. Based on the experimental evidence, the adsorption of Cl⁻ ions is proposed as an intermediate step in the removal of surface atoms during etching. Further surface hole accumulation leads to the formation of a ultra-thin oxide film.

References

1. T. Stempel. M Aggour, K. Skorupska, A. Muñoz, H. J. Lewerenz, *Electrochem. Comm.* **10** (2008) 1184.
2. M. Lublow, H. J. Lewerenz, *Electrochem. Sol. State Lett.* **10** (2007) C51.
3. K. Skorupska, *J. Solid State Electrochem.* **13**, 205 (2009).
4. K. Skorupska, F. Streicher, M. Aggour, S. Sadewasser, M. Kanis, C. Pettenkofer, H.J. Lewerenz (to be published).

5. M. Lublow, T. Stempel, K. Skorupska, A.G. Muñoz, M. Kanis and H. J. Lewerenz, *Appl. Phys. Lett.*, **93**, 062112 (2008).
6. K. Skorupska, M. Lublow, M. Kanis, J. Jungblut and H. J. Lewerenz, *Appl. Phys. Lett.*, **87**, 262101 (2005).
7. A.G. Muñoz and H.J. Lewerenz, *J. Electrochem. Soc.*, **155**, D527 (2008).
8. M.A. Hines, in *Fundamental Aspects of Silicon Oxidation*, Y.J. Chabal, Editor, p.12, Springer, Berlin (2001).
9. H.J. Lewerenz, J. Jakubowicz, H. Jungblut, *Electrochem. Commun.*, **6**, 1243 (2004).
10. S. García, H. Bao and M.A. Hines, *J. Phys. Chem. B*, **108**, 6062 (2004).
11. K. Skorupska, M. Lublow, M. Kanis, H. Jungblut, H.J. Lewerenz, *Electrochem. Commun.*, **7**, 1077 (2005).
12. C.D. Wagner, A.V. Naumkin, A. Kraut-Vass, J.W. Allison, C.J. Powell and J.R. Rumble, Jr in *NIST X-Ray Photoelectron Spectroscopy Database*.
13. H.J. Lewerenz, M. Aggour, C. Murrel, M. Kanis, H. Jungblut, J. Jakubowicz, P.A. Cox, S.A. Campbell, P. Hoffmann and D. Schmeisser, *J. Electrochem. Soc.*, **150**, E185 (2003).
14. F.J. Himpsel, F.R. McFeely, A. Taleb-Ibrahimi and J.A. Yarmoff, *Phys. Rev. B*, **38**, 6084 (1988).
15. O.V. Yazyev and A. Pasquarello, *Phys. Rev. Lett.*, **96** 157601 (2006).
16. H. Flietner, *Surf. Sci.*, **200** (1988) 463.
17. M. Fujitani, R. Hinogami, J.G. Jia, M. Ishida, K. Morisawa, S. Yae and Y. Nakato, *Chem. Lett.*, **26**, 1041 (1997).

ECS Transactions, 19 (3) 411-422 (2009)
10.1149/1.3120721 ©The Electrochemical Society

The Influence of Thermal Treatment on the Electronic Properties of a-Nb$_2$O$_5$

F. La Mantia[a], M. Santamaria[a], H. Habazaki[b], and F. Di Quarto[a]

[a] Dipartimento di Ingegneria Chimica dei Processi e dei Materiali, Università di Palermo,
Viale delle Scienze, 90128 Palermo, Italy
[b] Graduate School of Engineering, Hokkaido University, Sapporo 060-8628, Japan

The effect of thermal treatment for 1h at 250°C in air or under vacuum on the electronic structure of thick amorphous anodic niobia was characterized by electrochemical impedance, differential admittance (DA) and photocurrent spectroscopy (PCS). The analysis of anodized niobia has revealed that it behaves as a pure dielectric. The thermal treatment in air increases the value of its differential capacitance. The effect is stronger when the thermal treatment is carried out in vacuum, and can be cancelled by re-anodizing the oxide to its initial formation potential. The frequency dependence of the sample treated in vacuum exhibited behavior typical of a semiconducting amorphous material. PCS measurements were used to derive the optical band gap value and to confirm the location of the flat band potential that was derived from admittance data.

Introduction

We have recently shown (1-3) that the theory of an amorphous semiconductor Schottky barrier is able to explain, over a very large range of electrode potential and ac frequencies, both the EIS spectra and the behavior of both components (C_p and G_p) of admittance as a function of electrode potential, U_E, for both thin and thick films of semiconducting anodic oxide grown on niobium. Such investigations were designed to overcome some inconsistencies in the use of the simple Mott-Schottky analysis to interpret the impedance behavior of amorphous semiconducting (a-SC) passive films.

It has been recently proposed to use niobia anodic films instead of tantala in solid state electrolytic capacitors (4-7) or as the gate oxide in organic/inorganic field effect transistors, due to their higher dielectric constant and/or lower cost, which makes them quite attractive in large consumer electronics (8). However, compared to tantala, the electrical behavior of niobia anodic films has some disadvantages (leakage current values and thermal instabilities) that make their use rather problematic, especially at higher temperatures. Several authors have reported that the dielectric properties, particularly the capacitance values, of thick anodic oxide film on niobium are greatly changed if, after formation at high potential, they are thermally aged for some hours at a modest temperature (e.g. 250 °C) in various environments (4).

In previous work (1-3) on a-Nb$_2$O$_5$ anodic films we rationalized the complex admittance behavior of a-Nb$_2$O$_5$/electrolyte junctions as a function of frequency and electrode potential on the basis of amorphous semiconductor (a-SC) Schottky barrier theory. From such studies we have been able to derive information on the energy levels (flat band potential, U_{fb}, and conduction band, E_C, valence band, E_V, and mobility edge location) of the junction as well the density of states (DOS) distribution of such anodic films. This

411

helps provide a deeper understanding of the solid-state properties of anodic films on metals.

This paper reports a detailed characterization, using electrochemical impedance and photocurrent spectroscopy as well as differential admittance measurements, of amorphous anodic niobia before and after different treatments. An interpretation of the influence of aging conditions on solid state and electronic properties of anodic niobia will be presented, based on amorphous material theory.

Experimental Aspects

Niobium films (~ 300 nm thick) were magnetron-sputtered onto glass substrates using a 99.9% niobium target with a diameter of 100 mm. The deposited niobium had a bcc structure with a [110] preferred orientation (9). These films were anodized at a constant current density of 50 A m^{-2} in 0.1 mol dm^{-3} ammonium pentaborate electrolyte at 293 K, the potential being kept constant for 1h (called AF samples). Some of the anodized specimens were heat treated in air (TA samples) or in vacuum (TV samples) at 523 K for 1 h. Re-anodizing of the heat-treated specimens (TAR samples) was also carried out under the above conditions.

Impedance spectra and differential admittance curves were recorded in 0.5 M H$_2$SO$_4$ solution by using a Parstat 2263, connected to a computer for data acquisition. Impedance spectra were recorded after 20 minutes of polarization at the applied potential from 100 kHz down to 100 mHz, imposing an alternate potential of 10 mV. The admittance curves were recorded from 6 V down to - 0.25 V (vs. SCE), at different constant frequencies (10 Hz $\leq$ f $\leq$ 20 kHz), imposing an alternating potential of 10 mV.

For all the experiments a Pt net with a very high surface area was used as the counter electrode and a saturated calomel electrode (SCE) was employed as reference electrode.

The experimental set-up employed for the photoelectrochemical investigations has been described elsewhere (10): it consists of a 450 W UV-VIS xenon lamp coupled with a monochromator (Kratos), that allows monochromatic irradiation of the specimen surface through the electrochemical cell quartz window. A two-phase lock-in amplifier (EG&G) was used in connection with a mechanical chopper (frequency: 13 Hz) in order to separate the photocurrent from the total current circulating in the cell due to the potentiostatic control. The photocurrent spectra reported below were corrected for the relative photon efficiency of the light source at each wavelength, so that the photocurrent yield in arbitrary current units is represented on the Y axis.

Results and Discussion

Electrochemical Impedance Spectroscopy and Differential Admittance

The impedance spectra of all the samples were analyzed using the Kramer-Kronig transformation. The Kramers-Kronig (K-K) transformation relates the imaginary part of the impedance to the complete set of real values and vice versa, according to the equations:

$$Z_I(\omega) = \frac{2\omega}{\pi} \int_0^\infty \frac{Z_R(x) - Z_R(\omega)}{x^2 - \omega^2} dx$$

$$Z_R(\omega) = R_\infty + \frac{2}{\pi} \int_0^\infty \frac{x Z_I(x) - \omega Z_I(\omega)}{x^2 - \omega^2} dx$$

[1]

where Z_R is the real part of the impedance and Z_I the imaginary one. The Kramers-Kronig transformations respond to the principles of causality, linearity, and stability. In this way it is possible to eliminate possible instability and non-linearity of the system. Since it is not possible to acquire a complete experimental set of impedance values from frequency 0 to frequency ∞, a general model was used based on a wide distribution of time constants that are fixed, using different weighting factors for each time constant (11). For the analysis of each experimental spectrum, 5 time constants per decade of the investigated frequency range were used. The calculated and experimentally acquired data points showed discrepancies at very low and very high frequencies, but the relative errors were always less than 3%. In fig. 1 the real and imaginary parts of the measured impedance, and the results of the K-K transformations are shown for the TV-niobia sample at 1 V (vs. SCE) in 0.5 M H_2SO_4 aqueous solution.

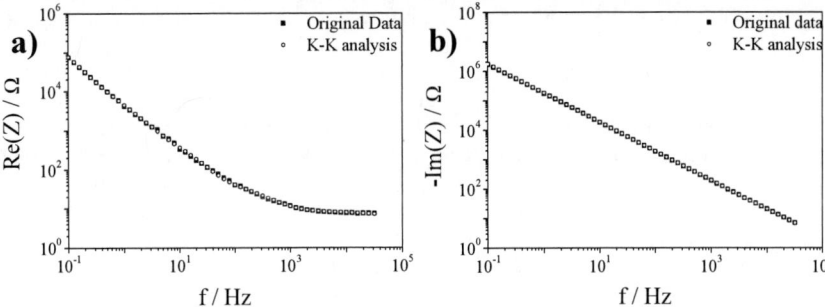

Figure 1. Real (a) and negative imaginary (b) parts of the measured and K-K analyzed impedance spectrum of TV-Nb_2O_5 at 1 V (vs. SCE) in 0.5 M H_2SO_4.

In fig. 2 the Bode plot of the specific impedance data acquired at 1 V (vs. SCE) in 0.5 M H_2SO_4 aqueous solution is represented for all the samples. The value of the modulus of the impedance data (fig. 2 a) is dependent on the thermal treatment, but the phase shift (fig. 2 b) shows that the behaviour of the Nb_2O_5 electrodes is essentially capacitive, regardless of the thermal treatment. The AF and the TAR-Nb_2O_5 electrodes behave quite similarly and the modulus of their impedance data is the highest. The modulus of the TA electrode is slightly lower (circa 30%) compared to AF-Nb_2O_5; while the modulus of the vacuum thermal treated electrode (TV-Nb_2O_5) has the lowest value (75% lower than the AF sample). According to transmission electron micrographs of ultra-microtomed sections and to the composition depth profiles provided by glow discharge optical emission spectroscopy (GDOES) analysis (4, 6), the thickness of the oxide layer does not change appreciably with the thermal treatment. Thus it seems reasonable to attribute such differences to the existence of a space-charge region inside the two samples. The width of such regions reported for amorphous semiconductors depends on the doping level (deviation from stoichiometry) of the oxide as well as on the localized DOS into the mobility gap of a-SC. Both these parameters affect the width of the space charge region in a-SC (1).

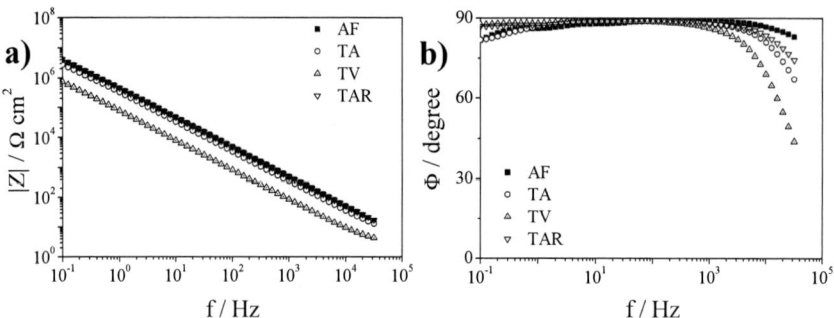

Figure 2. Bode plots of the specific impedance data acquired at 1 V (vs. SCE) in 0.5 M H_2SO_4 aqueous solution of the investigated samples.

The differences among the samples in the high frequency region (above 1 kHz) are due to the electrolyte resistance, which can change slightly with the position of the reference electrode. In the low frequency region (below 10 Hz) the major differences among the samples are due to the different polarizabilities of the electrodes.

In fig. 3 the Bode plot of the specific impedance data acquired at different potentials in 0.5 M H_2SO_4 aqueous solution is represented for TV-Nb_2O_5. The shape of the Bode plot suggests that the electrode behaviour is quite near to that of an ideally polarizable interface, and the electrical capacitance decreases appreciably by increasing the electrode polarization potential. Such behaviour indicates semiconducting behaviour with a space-charge region width much smaller than the total length of the semiconductor.

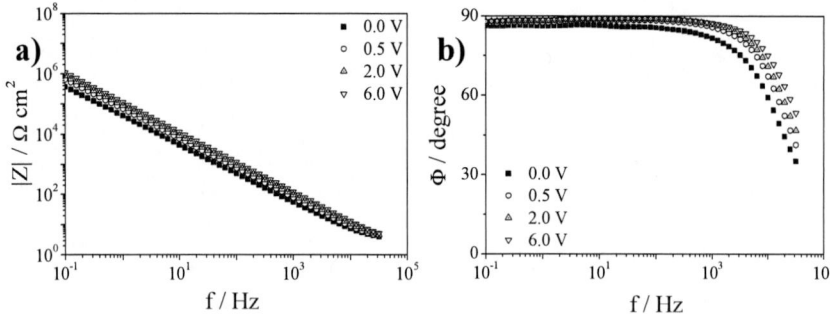

Figure 3. Bode plot of the specific impedance data acquired on TV-Nb_2O_5 at different polarization potentials in 0.5 M H_2SO_4 aqueous solution.

In fig. 4 the DA data of all the samples measured at 10 Hz are reported. The dashed line in fig. 4b is the value of the capacitance of the interface ($1/C = (L_{ox}/\varepsilon_{ox}\varepsilon_0 + 1/C_H)^{-1}$ in the case of insulator behaviour of the Nb_2O_5 sample having a thickness of 159 nm and a permittivity value of 53. The DA data stress that the AF and TAR samples are very similar and behave as insulators even at the lower frequencies. On the other hand, the TA and TV samples have semiconducting properties; the shapes of their differential capacitance curves (fig. 4b) are typical of amorphous semiconductors (1, 2, 12, 13). In fig. 4b, the difference between the dashed line and the AF and TAR samples can be due to the polarizability of the electrode and/or the influence of surface states and hopping of the electrons in the mobility gap. At higher frequencies the behaviour of these electrodes becomes typical of an insulating amorphous material, and a constant capacitance value was measured over the whole investigated electrode potential range.

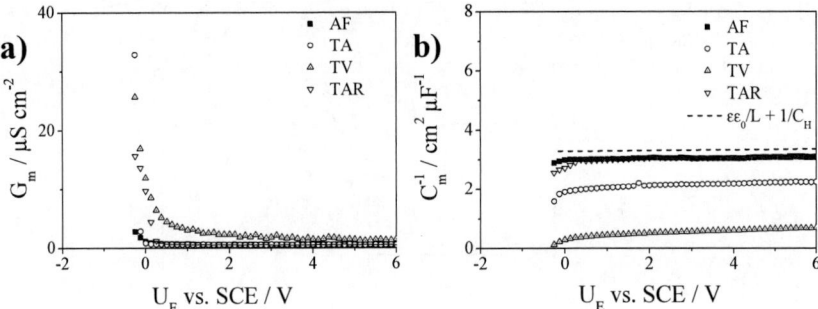

Figure 4. Equivalent conductance (a) and equivalent parallel capacitance (b) of the differential admittance data measured at 10 Hz on the different samples in 0.5 M H_2SO_4 aqueous solution.

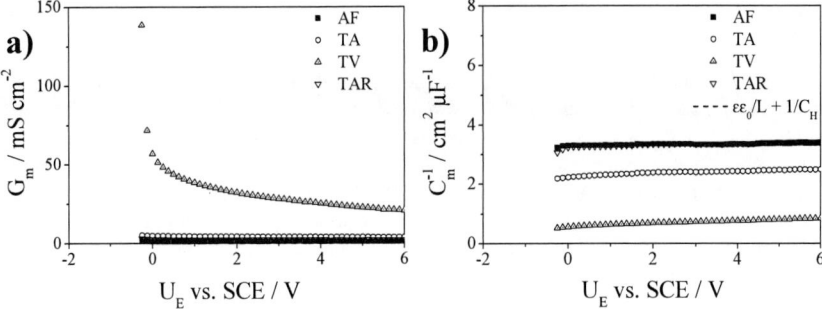

Figure 5. Equivalent conductance (a) and equivalent parallel capacitance (b) of the differential admittance data measured at 10 kHz on the different samples in 0.5 M H_2SO_4 aqueous solution.

This is shown in fig. 5, where the DA data of all the samples measured at 10 kHz are reported. As expected, in fig. 5b the capacitance data of AF and TAR samples and the dashed line coincide over a large potential range. The curves diverge only at lower polarization potentials, when the influence of a reaction can be much stronger. TA and TV samples behave as amorphous semiconductors even at 10 kHz; the Fermi level of these samples has to be not far from the bottom of the conduction band ($E_C - E_F \leq 0.40$ eV).

Fitting procedure and results. Both DA curves and EIS spectra were fitted according to an equivalent circuit, with parameters depending on the polarization potential, U_E. The EIS spectra of the a-Nb$_2$O$_5$/El junction could be fitted in most of the potential range investigated (above 0.25 V vs. SCE) by the equivalent circuit in fig. 6. The fitting of the impedance values at different frequencies was used to get a number of pieces of information. The term Z_{SC} represented the impedance of the space-charge region of the a-SC, on the electronic structure of the TV-Nb$_2$O$_5$ (U_{fb}, Fermi level, DOS in the mobility gap) and characteristic energy levels (mobility edges Ec and Ev) of the a-Sc/electrolyte junction. The equivalent circuit of fig. 6 was used to separate the influence of different processes on the total impedance measured during the experiment.

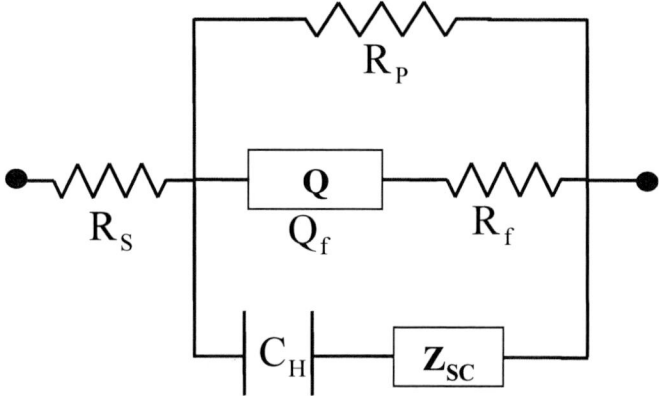

Figure 6. Equivalent circuit of the SC/El junction used to fit the impedance spectra and the differential admittance curves of TV sample.

The equivalent circuit takes into account the capacitance of the Helmholtz double layer, C_H, assumed constant and equal to 20 μF cm^{-2}; its contribution to the total measured impedance is always negligible. The parameters Q_f, R_f and R_P simulate the effect of the electrochemical reaction, surface states and transport limitations. In particular, a value of the exponent n of the constant phase element, Q_f, near to 0.80 is typical of electron transport by a hopping mechanism in localized states of the mobility gap. A value of n near to 1 can be interpreted as due to the effect of surface states. In this case, the arm Css/Rss takes into account the contribution of surface states to the measured impedance values. R_P is the polarization resistance of the a-Sc/electrolyte interface.

The relationship between the space charge impedance, Z_{SC}, and the electronic properties of the amorphous semiconducting film is obtained through the solution of the Losee's equation (14, 15), a modified version of the Laplace equation:

$$\frac{dW}{d\psi} = 1 + \frac{\rho}{\varepsilon\varepsilon_0 H}W - \frac{F}{\varepsilon\varepsilon_0 H}W^2$$

$$H(\psi) = \frac{2}{\varepsilon\varepsilon_0}\int_0^\psi \rho(\psi')d\psi'$$

$$F(\psi,\omega) = \frac{1}{1+i\omega\tau}\frac{d\rho}{d\psi}$$

$$\tau(\psi) = \tau_0 \exp\left[\frac{\Delta E_F + |e|\psi}{kT}\right]$$

[2]

where ψ is the band bending in the semiconductor, ρ the volumetric charge density, ε the relative dielectric constant of the semiconductor, ε_0 the dielectric constant of the vacuum, H the square of the electric field at band bending ψ, F the ratio between the ac component of volumetric charge density and the ac band bending ϕ, τ the time constant of capture/emission of electrons into/from the localized states in the band gap, H_S and ψ_S the value of H and ψ at the SC/El interface, ω the angular frequency, k the Boltzmann constant, T the absolute temperature, τ_0 a constant depending on the material (commonly assumes values between 10^{-10} and 10^{-14} s), and ΔE_F the distance between the mobility edge of the conduction band and the Fermi level.

Losee's equation follows directly from the Poisson equation applied to the case of a small perturbation ac signal. The function F is dependent on the nature of the semiconductor, and in this case we used the same expression for F used by Cohen et al. (15) for amorphous semiconductors. W has the dimension of an electrical potential and its relation with the impedance of the space charge region is given by:

$$W = \phi\frac{d\psi}{d\phi}$$

$$Z_{SC}(\psi_S,\omega) = \frac{W_S}{i\varepsilon\varepsilon_0\omega H_S^{0.5}}$$

[3]

with W_S the value of W at the SC/El the interface. W_S is obtained from the integration of the first order differential equation [2]. Consequently the value of impedance of the space charge region can be calculated. In this work the value of W_S was numerically evaluated by using the Runge-Kutta method. In particular, it can be observed that near zero band bending the numerical solution of equation [2] is impossible because H(0) tends to 0. Near zero band bending the solution of the ac part is trivial (14) and the value of W can be approximated by the following expression:

$$W(\psi,\omega) = [1 + i\omega\tau(0)]^{0.5}\psi$$

[4]

The value of W given by expression [4] has to be used to start the Runge-Kutta integration. The dependence of Z_{SC} from the band bending, ψ_S, is related to the shape of the DOS, but not to its absolute value. For a given DOS distribution N, Z_{SC} is inversely proportional to the square root of the absolute value of the DOS, $N^{0.5}$.

The large band bending approximation was used (1, 15, 16) for the analytical expression of the admittance in a-SC Schottly barrier (1, 2 and refs therein). This approximation considers that, at larger reverse bias $\psi_S > \psi_g$ (defined below), the quasi-Fermi level of the a-SC remains fixed near the half band gap of a-SC owing to the kinetic equilibrium between the capture/emission of minority and majority carriers from/to localized states to/from conduction/valence band edge (15, 16). In this case, for $\psi > \psi_g$, $d\rho/d\psi = 0$. Without other information, the value of ψ_g was set equal to:

$$|e|\psi_g = \frac{E_g}{2} - \Delta E_F \qquad [5]$$

where E_g is the band gap (c-SC) or the mobility gap (a-SC) of the oxide. The band gap of TV sample was measured by photocurrent spectroscopy (see next section).

The electronic structure of the TV-Nb_2O_5 electrode was obtained from the impedance spectra and the admittance curves by a trial and error method. We analyzed the DA data between -0.25 and 6 V vs. SCE at different frequencies, assuming the DOS to be similar to that obtained in previous studies (1). In this electrode potential range the R_S and C_H parameters in the equivalent circuit were considered constant. The parameters Q_f and R_f were considered exponentially dependent on the polarization potential, U_E. In this early stage of the analysis, the sample was considered homogeneous. The DOS and the electronic properties (U_{fb} and ΔE_F) were modified to obtain a good fit of the DA data without compromising the simplicity of the DOS. After obtaining the first trial shape of the DOS, the fitting of each impedance spectrum was made imposing the value of R_S and C_H and changing at each potential the value of the parameters Q_f, R_f, R_p. The DOS was eventually adjusted to take into account the spatial in-homogeneity in DOS distribution as previously reported (1, 2). Finally, the new value and shape of the DOS and the parameters of the equivalent circuit were used to fit the DA data again. The process was iterated several times to increase the accuracy of the fitting.

The fitting parameters of the impedance spectra for TV-Nb_2O_5 are reported in table I. A constant R_S value equal to 2.83 Ω cm^2 was used. We want to stress that the effects of Q_f, R_f and R_p are very small except at potentials below 0.25 V vs. SCE. In other words, in the frequency range investigated, above 0.25 V vs. SCE, the junction Nb_2O_5/El of TV niobia behaves as an ideally polarizable amorphous semiconductor/electrolyte interface.

The parameter n of the CPE element was found to be 0.8 at all the polarization potentials, except -0.25 V (vs. SCE), in which case it was equal to 0.96. As discussed above, a value of n around 0.8 is typical of the hopping mechanism of electrons in the mobility gap. When the value n is near 1, the CPE corresponds to the effect of surface states (1 and refs therein).

The electronic structure (electronic DOS distribution and Fermi Level location) of the oxide was obtained from the fitting of the impedance spectra and, in separate experiments, of the differential admittance curves as a function of electrode potential at different frequencies. From this fitting procedure a flat band potential value $U_{fb} = -0.38$ V vs. SCE could be derived, in agreement with PCS (see below) data on the same sample, and $\Delta E_F = 0.35$ eV, in agreement with previous values (1).

In fig. 7 we report the DOS distribution obtained from the fitting results. The DOS is composed by an exponentially decreasing distribution of states, which change by changing the polarization potential, and a constant distribution of states, which remains independent of the polarization potential. The change of the DOS with the polarization potential is due to the spatial in-homogeneity of the oxide. In particular, the oxide nearer to the metal interface has a higher value of the DOS, which also can be interpreted as a higher concentration of defects. In fig. 6 the Fermi level and the energies corresponding to different frequencies are reported. The latter is the energy that separates responding states (above) from non responding states (below) at the corresponding frequency. The time constant τ_0 in equation [2] was set equal to 10^{-12} s (1).

TABLE I. Fitting parameters.

U_E vs. SCE / V	R_f / Ω cm^2	Q_f / μS s^n cm^{-2}	R_P / MΩ cm^2	χ^2
-0.25	1.64	7.26	1.01	$5.05 \cdot 10^{-4}$
0	$1.72 \cdot 10^3$	$4.71 \cdot 10^{-1}$	11.4	$2.18 \cdot 10^{-4}$
0.25	$1.35 \cdot 10^6$	$8.99 \cdot 10^{-2}$	15.6	$2.85 \cdot 10^{-4}$
0.5	$6.23 \cdot 10^6$	$6.02 \cdot 10^{-2}$	34.8	$3.06 \cdot 10^{-4}$
0.75	$7.73 \cdot 10^6$	$6.09 \cdot 10^{-2}$	42.5	$3.65 \cdot 10^{-4}$
1	$8.85 \cdot 10^6$	$5.90 \cdot 10^{-2}$	48.1	$3.03 \cdot 10^{-4}$
1.5	$8.15 \cdot 10^6$	$5.01 \cdot 10^{-2}$	45.3	$2.71 \cdot 10^{-4}$
2	$7.10 \cdot 10^6$	$4.86 \cdot 10^{-2}$	49.5	$3.35 \cdot 10^{-4}$
4	$5.46 \cdot 10^6$	$4.33 \cdot 10^{-2}$	54.2	$3.71 \cdot 10^{-4}$
6	$3.74 \cdot 10^6$	$4.10 \cdot 10^{-2}$	59.5	$1.60 \cdot 10^{-4}$

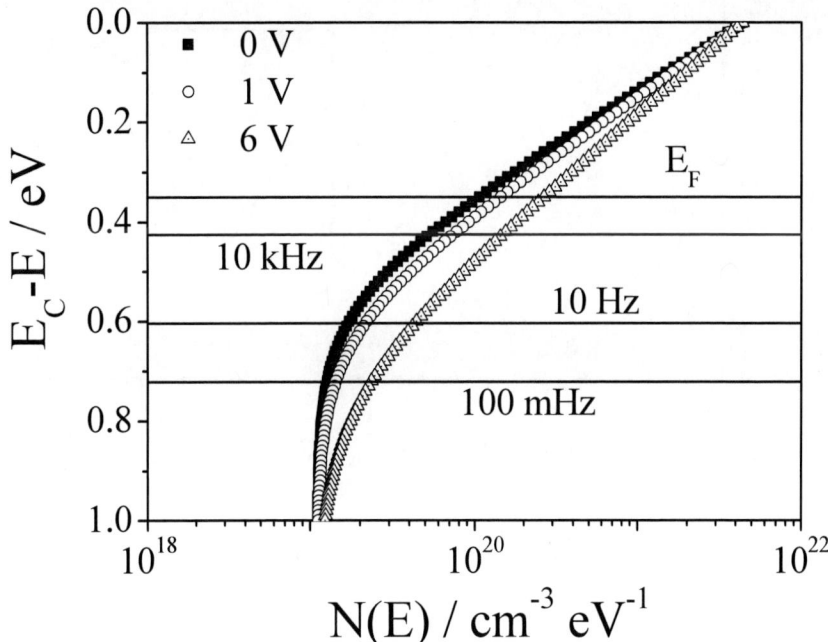

Figure 7. Electronic structure of sample 3 Nb$_2$O$_5$ electrode, as obtained from the fitting of the impedance spectra and admittance curves.

Photoelectrochemical investigation

In fig. 8 we report the photocurrent spectrum relating to the TV-Nb$_2$O$_5$, recorded by polarizing the electrode at 8 V (vs. SCE) in 0.5 M H$_2$SO$_4$. By assuming indirect (non direct for amorphous materials) optical transitions, the following equation holds (10):

$$\left(I_{ph} h\upsilon\right)^{0.5} \propto \left(h\upsilon - E_g\right) \qquad [6]$$

where I_{ph} is the photocurrent corrected for the relative photon efficiency of the light source at each wavelength and E_g is the mobility gap of the amorphous oxide. Eq. [6] allows estimation of a mobility gap of 3.45 eV for the investigated film by extrapolating the $(I_{ph}h\upsilon)^{0.5}$ vs $h\upsilon$ plot to zero. It is interesting to note the presence of photocurrent at energies lower than the estimated band gap value (see fig. 8), causing another linear region in the $(I_{ph}h\upsilon)^{0.5}$ vs $h\upsilon$ plot (see inset of fig. 8) ending at around 2.75 eV. This can be associated with optical transitions involving localized states inside the mobility gap of the oxide.

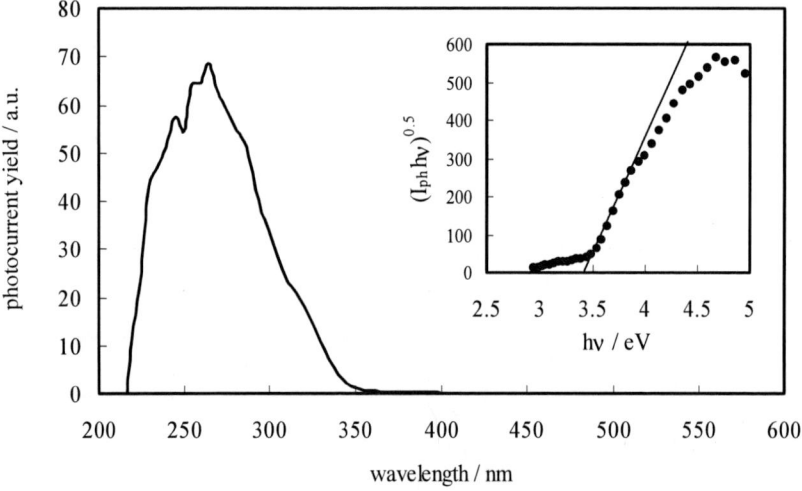

Figure 8 - Photocurrent spectrum relating to TV-Nb$_2$O$_5$, recorded by polarizing the electrode at 8 V (vs SCE) in 0.5 M H$_2$SO$_4$. Inset: $(I_{ph}h\upsilon)^{0.5}$ vs $h\upsilon$ plot.

Mobility gap values around 3.45 eV were also estimated for AF and TAR-Nb$_2$O$_5$, while a slightly lower value (3.30 eV) was estimated for TA oxide (17).
In fig 9a we report the photocurrent vs electrode potential (photocharacteristic) curve relating to TV-Nb$_2$O$_5$ recorded in 0.5 M H$_2$SO$_4$ by irradiating the oxide at 320 nm. In order to estimate the flat band potential, U_{fb}, of the film we fitted the curve according to the power law:

$$I_{ph}^n \propto \left(U_E - V^*\right) \tag{7}$$

where V*, the extrapolated zero photocurrent potential, can be assumed as a reasonable estimate of U_{fb}. According to eq. [7], for TV-Nb$_2$O$_5$ a V* = - 0.38 V (vs SCE) can be estimated, which is more cathodic than that estimated with the same procedure for anodic films on Nb (1).

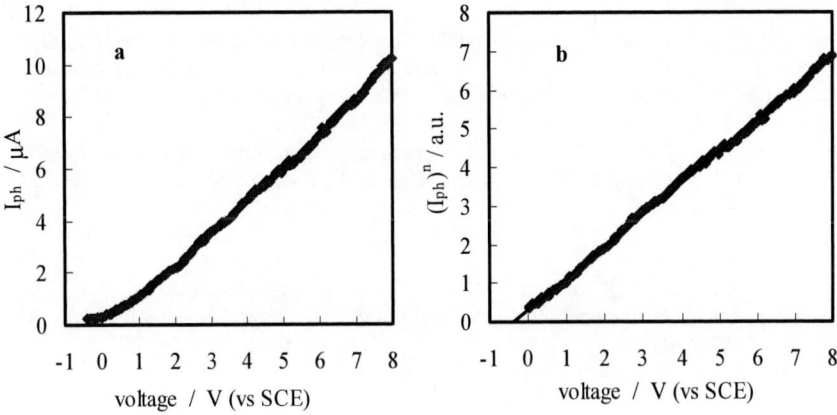

Figure 9: Photocurrent vs polarizing voltage (a) and best fitting (b) curves relating to TV Nb$_2$O$_5$. Sol: 0.5 M H$_2$SO$_4$, v_{scan} = 10 mV s^{-1}, λ = 320 nm and n = 0.83.

Summary and Conclusions

The effect of thermal treatment on the electronic properties of Nb$_2$O$_5$ thick films (160 nm) was studied through EIS, DA and PCS. Impedance spectra and admittance curves have demonstrated that the SC/El junction is almost ideally polarizable, independent of thermal treatment. The AF-Nb$_2$O$_5$ has shown insulator behavior; the capacitance of the junction is equal to $\varepsilon\varepsilon_0/L$. The effect of the thermal treatment in air (TA sample) is to shift the Fermi level nearer to the conduction band, thus showing semiconducting properties, but the capacitance is very near to $\varepsilon\varepsilon_0/L$. The thermal treatment under vacuum (TV sample) intensifies the effect of the shift of the Fermi level, thus allowing a better analysis of the sample. The effect is completely reversible, by re-anodization of the thermally treated sample (TAR sample). An detailed analysis of the impedance spectra and admittance curves was done by fitting them to an equivalent circuit. The theory of amorphous semiconductors was used to explain the change of the impedance value with the polarization potential. Finally, the electronic structure (U_{fb}, E_F, and DOS) of the TV sample was obtained. It was observed that the TV sample is not homogeneous, and the layers nearer to the metal/oxide interface have a higher density of states. This phenomenon can be interpreted as an injection of defects at the metal/oxide interface; this happens during thermal treatment under vacuum.

Photocurrent spectroscopy was used to obtain the band gap of all samples. No difference was observed between the as-formed, the re-anodized, and the vacuum-treated samples.

A lower value of the band gap was estimated for the air-treated sample. Photo-characterization curves were used to estimate the flat band potential of the oxides. The results confirmed the results obtained by fitting the differential capacity curves.

References

1. F. Di Quarto, F. La Mantia and M. Santamaria, *Electrochim. Acta*, **50**, 5090 (2005).
2. F. Di Quarto, F. La Mantia and M. Santamaria, *Corrosion Sci.*, **49**, 186 (2007).
3. F. Di Quarto, F. La Mantia and M. Santamaria, in *Passivation of Metals and Semiconductors, and Properties of Thin Oxide Layers*, P. Marcus and V. Maurice, Editors, Vol. p. 343, Elsevier, Paris (2006).
4. H. Habazaki, M. Yamasaki, T. Ogasawara, K. Fushimi, H. Konno, K. Shimizu, T. Izumi, R. Matsuoka, P. Skeldon and G. E. Thompson, *Thin Solid Films*, **516**, 991 (2008).
5. K. Kovacs, G. Kiss, M. Stenzel and H. Zillgen, *J. Electrochem. Soc.*, **150**, B361 (2003).
6. H. Habazaki, T. Matsuo, H. Konno, K. Shimizu, K. Matsumoto, K. Takayama, Y. Oda, P. Skeldon and G. E. Thompson, *Surface and Interface Analysis*, **35**, 618 (2003).
7. S. Ono, K. Kuramochi and H. Asoha, *Coor. Sci.*, doi:10.1016/j.corsci.2008.11.027
8. C. S. Kim, S. J. Jo, S. W. Lee, W. J. Kim, H. K. Baik and S. J. Lee, *J. Electrochem. Soc.*, **154**, H102 (2007).
9. H. Habazaki, T. Ogasawara, H. Konno, K. Shimizu, K. Asami, K. Saito, S. Nagata, P. Skeldon and G. E. Thompson, *Electrochim. Acta*, **50**, 5334 (2005).
10. F. Di Quarto, C. Sunseri, S. Piazza and M. Santamaria, in *Handbook of Thin Films*, H. S. Nalwa, Editors, Vol. 2, p. 373, Academic Press, San Diego (2002).
11. F. La Mantia, J. Vetter and P. Novak, *Electrochim. Acta*, **53**, 4109 (2008).
12. F. Di Quarto, S. Piazza and C. Sunseri, *Electrochim. Acta*, **35**, 99 (1990).
13. F. Di Quarto, C. Sunseri and S. Piazza, *Ber. Bunsen-Ges. Phys. Chem. Chem. Phys.*, **90**, 549 (1986).
14. D. L. Losee, *J. Appl. Phys.*, **46**, 2204 (1975).
15. J. D. Cohen and D. V. Lang, *Phys. Rev. B*, **25**, 5321 (1982).
16. C. da Fonseca, M. G. Ferreira and M. da Cunha Belo, *Electrochim. Acta*, **39**, 2197 (1994).
17. F. La Mantia, M. Santamaria, H. Habazaki and F. Di Quarto, in preparation.

Novel plasmaless photoresist removal method in gas phase at room temperature

T. Miura[a], M. Kekura[a], H. Horibe[b], M. Yamamoto[b], and H. Umemoto[c]

[a] Core Technology R&D Center, Meidensha Corporation, 515 Kaminakamizo,
Higashimakado, Numazu, Shizuoka 410-8588, Japan
[b] Department of Material Design Engineering, Graduate School of Engineering,
Kanazawa Institute of Technology, 3-1 Yatsukaho, Hakusan, Ishikawa, 924-0838, Japan
[c] Faculty of Engineering, Shizuoka University, 3-5-1 Johoku, Naka, Hamamatsu,
Shizuoka 432-8561, Japan

Novel plasmaless photoresist removal method in gas phase at room temperature has been developed using pure ozone gas which has nearly 100% of concentration. This method has enabled the removal of high dose ion implanted photoresist that had been difficult to remove so far. Conventionally, the temperature rise of 200°C or more was necessary in the photoresist removal with the ozone gas to achieve an enough effect. Because the strong reactiveness is due to the oxygen radical generated when the ozone molecules decomposed. However, Popping by the temperature rise becomes a problem in the removal of high dose ion implanted photoresist. Then, in this experiment, ethylene gas was used to control the decomposition of ozone molecule at the room temperature. As a result, it was confirmed that the ashing using ethylene gas and ozone gas was effective to remove the high dose ion implanted photoresist under unheating condition.

Introduction

Recently, ion dose quantity and frequency of the ion implantation process tend to increase along with increased device performance. Photoresist is often used as block mask in the ion implantation process, and its surface properties change by the implanted ions and its energy. Especially, the stiffened layer called crust is observed on surface of photoresist where ion dose is more than around $1e14$ atoms/cm^2. (Figure 1) This stiffened layer is flaked off and scattered by the escape gas from internal unstiffened photoresist, when it is processed at higher than 100°C which plasma ashing process requires. (so-called Popping, Figure 2) And then, a large amount of particle is generated, makes the surrounding dirty. Figure 3 shows the surface of ion implanted photoresist that was heat-treated by changing temperature under the vacuum. An ionic species: P, implanted energy: 70keV, dose: $5e15$ atoms/cm^2 and the photoresist thickness is about one μm. It is found Popping is generated in this sample at 80°C or more. In this respect, low temperature processing of less than around 100°C at least has been required.

On the other hand, the use of ozone has been extending for its strong reactiveness and low environment load because of its nature to be decomposed into oxygen easily. Here, the strong reactiveness is in many cases due to the oxygen radical generated when the ozone molecules decomposed. Heat-treatment and/or the UV light irradiation can be used for this decomposition. However, the temperature exceeds 200°C. In addition, the penetration distance of the UV light is short because the absorption with the ozone molecule is large. Therefore, it is difficult to generate the oxygen radical at the required place in case of UV light irradiation.

Then, in this experiment, the ethylene gas was used to control the decomposition of ozone at the room temperature. Ozone reacts easily with the double bond of carbon, and makes ozonide that is an unstable intermediate (1).

$$O_3 + C_2H_4 \rightarrow \begin{matrix} O-O-O \\ | \quad\quad | \\ CH_2 \text{——} CH_2 \end{matrix} \rightarrow CH_2OO + CH_2O \qquad [1]$$

In the decomposition process of CH_2OO, various activated species that react with the organism of the photoresist material such as oxygen radical, hydroxyl radical, and atomic hydrogen are generated (2). These radicals generated at room temperature would serve to remove the ion implanted photoresist. Moreover, pure ozone gas which was concentrated to about 100% by liquefaction and fractionating (3) was used to increase the number of generated active species. It was aimed to increase the probability of collision of the ozone molecule and the ethylene molecule. Here, high dose ion implanted photoresist could be removed at more than 300°C in the ashing only with pure ozone (4), but it could not be removed at all at lower than 100°C.

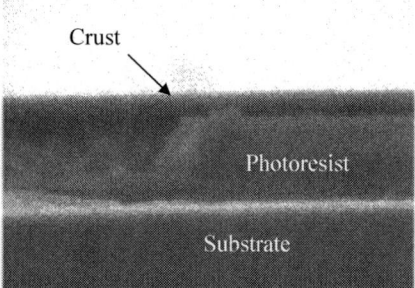

Figure 1. The stiffened layer on the photoresist by implanted ion (crust)
Ion species: P, energy: 70keV, dose: 5e15 atoms/cm^2

Figure 2. The surface of ion implanted photoresist after plasma ashing

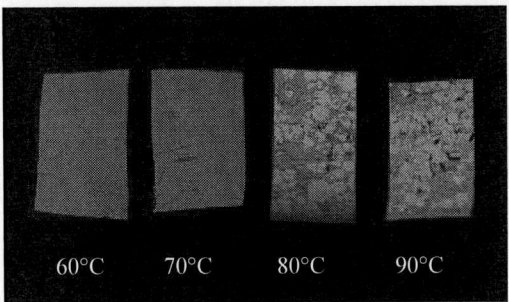

Figure 3. State of popping by annealing under the vacuum condition
(Same sample as figure 1)

Experimental

Figure 4 is the schematic diagram of the apparatus used for this experiment. To make the horizontal flow on the photoresist coated sample, the ethylene gas was introduced from the chamber sidewall and the ozone gas was introduced from the upper side of chamber so that both gases might react on the sample. It was confirmed that ozone gas was reactive with ethylene gas by QMS (Quadrupole Mass Spectrometer) set up on the exhaust pipe. The mass spectrum of ethylene is shown in Figure 5 and the mass spectrum of ozone and ethylene mixture gas is shown in Figure 6. The partial pressure ratio of both gases was adjusted to 1:1 here. The main peak 48 of ozone disappears completely, and some peak different from the ethylene is seen in mixing the gas. From the analysis of the spectrum of Figure 6, it is thought that the generated gas is composed of the elements of TABLE I.

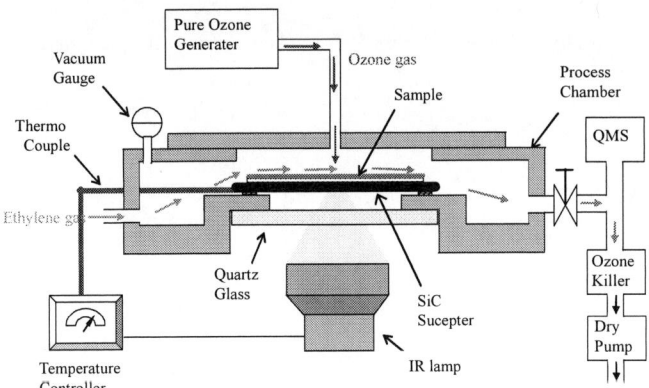

Figure 4. The schematic diagram of the apparatus.

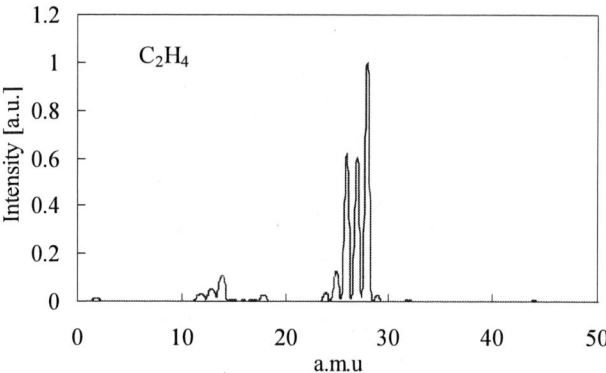

Figure 5. Mass spectrum of only ethylene gas

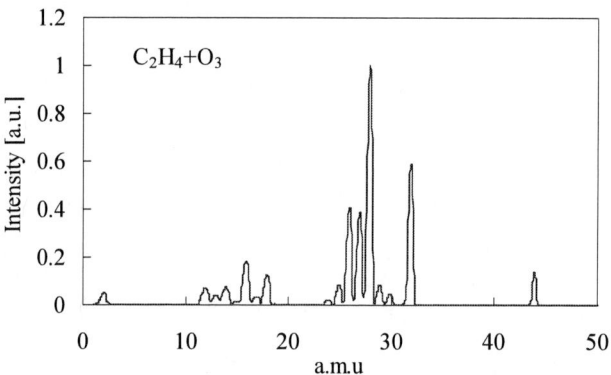

Figure 6. Mass spectrum of ozone and ethylene mixture gas

TABLE I. The elements of the generated gas

Gas name	Molecular formula	Contents [%]
Ethylene	C_2H_4	32.7
Oxygen	O_2	30.4
Carbon monoxide	CO	17.0
Carbon dioxide	CO_2	7.0
Water	H_2O	5.9
Formaldehyde	CH_2O	3.3
Hydrogen	H_2	2.0

Next, after flow-rate and the sample arrangement were optimized, the ashing of several ion implanted photoresist samples was carried out.

These samples were the novolak photoresist into which three kind of ion species were implanted on various dose, the ion species were Arsenic (As), Phosphorus (P) and Boron (B) and dose was from 5e12 to 1e16 atoms/cm^2, implant energy was 70keV. Thickness of surface stiffened layer was about 100 nm or more, when ion dose was more than 5e15 atoms/cm^2. The thickness of surface stiffened layer of each sample by spectroscopic ellipsometer is shown in Figure 7. The stiffened layer was not able to be distinguished with 1e14 atoms/cm^2 or less. These samples were processed under the condition which the stiffened layer was not etched off in a few minutes, aimed to examine the effect to the ion implanted layer.

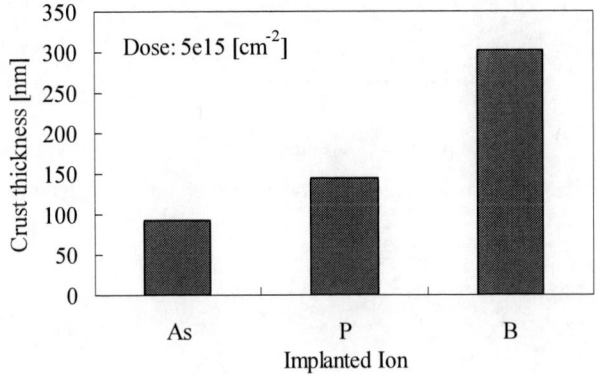

Figure 7. Thickness of surface stiffened layer

Results & discussion

The change of susceptor temperature and pressure in the Chamber during processing is shown in Figure 8. The temperature of the susceptor rose at about 25°C during the processing of three minutes by reactive heat of the gas though the sample was not heated.

Figure 9 shows processing time dependence of the photoresist thickness and the removal rate, when the sample of dose 5e15 atoms/cm^2 of As ion is processed. The film thickness decreases rapidly when processing it for four minutes or more, and the removal rate rises. Therefore, it is thought that the surface stiffened layer was lost in about four minutes.

Next, some samples that changed ion dose (5e12 to 1e16 atoms/cm^2) and species (As, P, and B) were processed. The result is shown in Figure 10. Sample was not heated while processing, and the process time was three minutes. It is found that the ashing rate becomes slow by increasing of the ion dose, it means that it becomes difficult to remove. Hence it is thought that photoresist properties are changed by the ion dose.

Figure 11 shows the surface photographs of individual samples after processing. All surfaces are not getting rough, and popping has not been generated. Meanwhile, even the unimplanted photoresist was not able to be removed at all by a similar experiment that used the ozone gas of 10vol %.

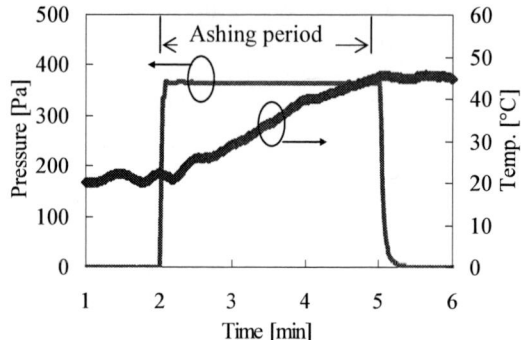

Figure 8. Pressure and temperature change with processing time.

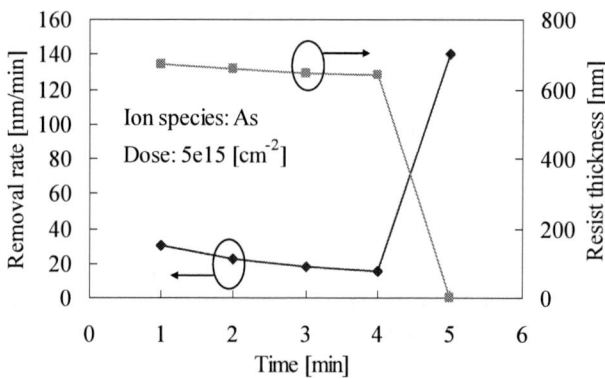

Figure 9. Processing time dependence of the photoresist thickness and the removal rate.

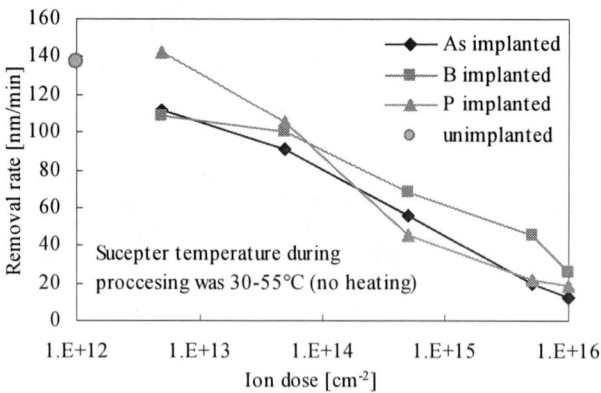

Figure 10. Relationship between removal rate and ion dose.

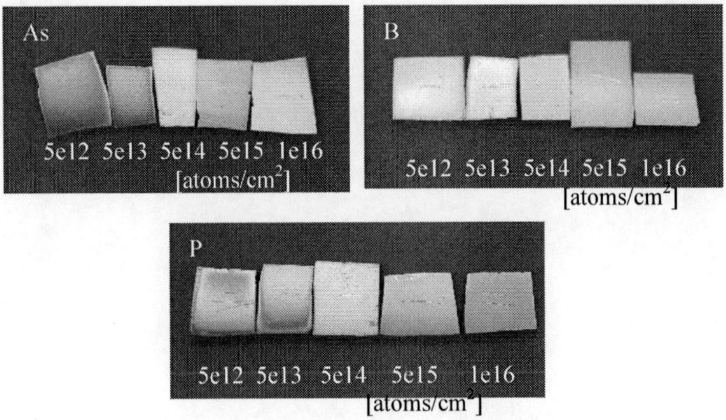

Figure 11. Surface photographs of individual samples after processing in figure 10.

Conclusions

In this experiment, ethylene gas was used to control the decomposition of ozone molecule at the room temperature for removal of high dose ion implanted photoresist.

Consequently, it was confirmed that the ashing using ethylene gas and pure ozone gas was effective to remove the high dose ion implanted photoresist under unheating condition. It is thought that this method is also effective to remove the many kinds of carbon consisting residues on the substrate.

References

1. R. Criegee, *Angew. Chem. Internat. Edit./* Vol. 14, 1975 / No. 11, p. 745-752.
2. Su, Calvert, and Shaw, *The Journal of Physical Chemistry*, Vol. 84, No. 3, 1980, p. 239-246.
3. S. Hosokawa and S. Ichimura, *Rev. Sci. Instrum.*, Vol. 62, No. 6, June 1991, p. 1614-1619.
4. T. Miura et al, *Journal of Photopolymer Science and Technology*, Volume 21, Number 2, 2008, p. 311-316.

430

Author Index

Ahmed, S.	209	Faci, S.	5
Allongue, P.	197, 283, 373	Fedderwitz, S.	5
Arai, F.	393	Flake, J. C.	365
Asoh, H.	393	Föll, H.	321, 329, 347, 355
Aureau, D.	373	Frites, M.	137
Banga, D.	245	Geaney, H.	209
Barratt, C. A.	79	Gerard, I.	313
Barrett, C. A.	209	Gerngroß, M.	347
Beckhoff, B.	227	Gila, B.	85
Benedek, M.	113	Goh, A.	85
Bouttemy, M.	221, 305, 313	Goncalves, A.	221, 235, 273, 305
Brierley, S.	113	Gong, H.	129
Buckley, D. N.	295	Gong, J.	167
		Gouget-Laemmel, A.	283
Carstensen, J.	321, 329, 347, 355	Grand, P.	189
Chang, C.	31, 57, 85, 123, 147	Gunning, R. D.	209
Chassaing, E.	189	Gupta, J.	5
Chazalviel, J.	283, 373	Gupte, A.	147
Chazelas, J.	5		
Chen, K.	57, 123	Habazaki, H.	411
Chi, G.	31	Happek, U.	245
Chiou, Y.	39, 73	Henry de Villeneuve, C.	283, 373
Chow, P.	123	Hicks, B.	31
Chu, B.	31, 85, 123, 147	Hönicke, P.	227
Chu, M.	65	Hooten, D.	123
Chung, P.	167	Hooven, A.	113
Cojocaru, A.	347, 355	Horibe, H.	423
Cortes, R.	197	Horng, R.	65
Cox, S.	245	Hsu, C.	65
		Hu, H.	65
Dabiran, A.	123	Huang, L.	73
Dale, P. J.	179	Hung, S.	31
Decorse Pascanut, C.	273		
Decoster, D.	5	Jäger, D.	5
Dennis, D.	57	Jiang, P.	147
Di Quarto, F.	411	Johnson, J. W.	31, 57, 85
Dogheche, E.	5	Joudrier, A.	273
Donley, C.	99	Jurca, H.	197
Ecin, O.	5	Kang, B.	57, 85
El Ali, O.	235	Kanis, M.	403
Etcheberry, A.	189, 221, 235, 273, 305, 313	Kekura, M.	423
		Keselowsky, B.	147

Khan, S. U.	137	Norton, D.	147
Kim, Y.	245		
Kim, Y.	161	O'Dwyer, C.	295
Kolbe, M.	227	O'Sullivan, C.	209
Kolbesen, B.	227	Ono, S.	393
Kung, C.	167	Ozanam, F.	283, 373
La Mantia, F.	411	Pearton, S. J.	31, 47, 57, 85,
Lai, H.	167		123, 147
Lampin, J.	5	Peter, L. M.	179
Le Floch, P.	235	Phalon, P.	113
Lee, C.	39, 73	Piner, E.	31, 85
Lee, C.	161	Piner, E.	57
Lee, H.	39	Poloczek, A.	5
Lee, J.	147	Prod'homme, P.	197
Lee, J.	161		
Leisner, M.	321, 329, 355	Quill, N.	295
Lele, T.	57, 123, 147		
Leu, L.	147	Rajagopa, P.	31
Lew, K.	5	Rajagopal, P.	57
Lewerenz, H.	381, 403	Ramage, J.	123
Liao, Z.	65	Rappich, J.	373
Lin, J.	85	Reinhardt, F.	227
Lin, Y.	167	Ren, F. 31, 47, 57, 85, 123, 147	
Lincot, D.	189	Roberts, J. C.	31, 57
Linthicum, K.	31, 57, 85	Ryan, K. M.	209
Lo, C.	31		
Loke, W.	5	Saadsaoud, N.	5
Loken, A.	179	Sam, S.	283
Lommel, M.	227	Santamaria, M.	411
Lou, L.	73	Santinacci, L.	273, 305, 313
Lu, D.	39	Sanyal, A.	209
Lublow, M.	381, 403	Saucedo, E.	189
Lynch, R.	295	Schauer, P.	329
		Sciullo, A.	85
Mair, L. O.	99	Scragg, J.	179
Malcoci, A.	5	Shiau, N.	73
Maroun, F.	197	Sim, Y.	5
Mathieu, C.	221, 235, 305	Simon, N.	273, 305
McAlister, S.	5	Singh, A.	209
McAnulty, R.	113	Skinner, K.	99
Mezailles, N.	235	Skorupska, K.	403
Miura, T.	423	Smith, K. V.	113
Moraillon, A.	283, 373	Stempel, T.	403
Müller, M.	227	Stickney, J. L.	245
Munoz, A.	403	Stöhr, A.	5
		Superfine, R.	99
Ng, T.	5		
Ngunjiri, J.	365	Tan, K.	5

Tilas, C.	113
Touahir, L.	283
Tournerie, N.	197
Tran-Van, P.	221
Tripon-Canseliet, C.	5
Tsai, Y.	65
Tsai, Y.	65
Tseng, Y.	123, 147
Umemoto, H.	423
Vegunta, S.	365
Vigneron, J.	221
Wang, H.	57
Wang, Y.	31, 57
Wang, Y.	123
Wei, R.	129
Weiß, M.	5
Wicaksono, S.	5
Wu, W.	85
Xu, Z.	5
Yamamoto, M.	423
Yen, K.	167
Yoon, S. F.	5
Zarkh, D.	113
Zegaoui, M.	5